AF366322

ATOMIC SPECTROSCOPY

ATOMIC SPECTROSCOPY

K. P. RAJAPPAN NAIR

Formerly of Berlin University (FU)
and ULM University

MJP Publishers

Honour Copyright
&
Exclude Piracy

This book is protected by copyright. Reproduction of any part in any form including photocopying shall not be done except with authorization from the publisher.

Cataloguing-in-Publication Data

Rajappan Nair, K. P. (1941-   ).
   Atomic Spectrocopy / by K.P. Rajappan
Nair. – Chennai : MJP Publishers, 2012
   xvi, 596 p.; 24 cm.
   Includes References and Index.
   ISBN 978-81-8094-088-0 (pb.)
   1. Spectroscopy. Atomic I. Title.
543.5 dc 22 RAJ MJP 105

ISBN 978-81-8094-088-0   **MJP PUBLISHERS**
© Publishers, 2012   New No. 5, Muthu Kalathy Street
All rights reserved   Triplicane

Publisher : J.C. Pillai
Managing Editor : C. Sajeesh Kumar
Project Editor : P. Parvath Radha
Acquisitions Editor : C. Janarthanan
Editorial Team : B. Ramalakshmi, V.R. Padma, R. Hemalatha,
M. Gnanasoundari, Lissy John, B. Annalakshmi, S. Jeevasruthi
CIP Data : Prof. K. Hariharan, Librarian
RKM Vivekananda College, Chennai.

This book has been published in good faith that the work of the author is original. All efforts have been taken to make the material error-free. However, the author and publisher disclaim responsibility for any inadvertent errors.

PREFACE

This book is the outcome of the realization that there is a dearth of books on Atomic Spectroscopy and is primarily aimed at the graduate and postgraduate students. It has been designed based on my experience in teaching atomic spectroscopy and would help students to get a fundamental knowledge on the subject, which they can use for further learning.

Latest developments in atomic spectroscopy including laser cooling, Bose–Einstein condensates and atom lasers are discussed in independent chapters. The field of atomic spectroscopy owes a great deal to the Bohr's theory and the explanation of hydrogen spectra. It still remains as the basis of any spectroscopic studies. Hence the book starts from Bohr's atom theory to the latest developments in atom cooling. The elementary atomic physics covered in the early chapters should be comprehensible to undergraduates when they are first introduced to the subject.

It should be remembered that the developments of lasers were based on the understanding of atomic and molecular energy levels. When the inversion spectrum of ammonia was studied using microwave spectroscopy it was never thought that it would lead to such a fantastic discovery that has revolutionarised the field of science and technology. Hence for the development of new lasers and for the understanding of other atomic and molecular phenomena, it is necessary to study the energy levels and the accompanying spectra. Spectroscopy plays a major role in every field of science. The book would be valuable for physicists, chemists and biologists.

Scientists from many countries helped me in many ways in my research career. I take this opportunity to thank many of my teachers, colleagues and students who had helped me in many aspects in my research in Atomic and Molecular spectroscopy. Professor N L Singh, Professor K. N. Upadhya, Professor D. K. Rai, Dr. R. N. Singh, Professor H. D. Rudolph (UlM), Professor J. E. Boggs (Texas), Professor T. Toerring, Professor J. Hoeft, Professor E. Tiemann (Berlin), Professor H. Jones (UlM), Professor Jean Demaison (Lille), Professor G. Wlodarczack (Lille, France) and Dr. J. Vogt (UlM) were in one way or the other a source of inspiration and help.

Professor P. F. Bernath (Canada), Professor S. C. Mehrotra, Professor Ann Schmiedekamp (USA), Professors S. N. Thakur, A.N. Singh, O. N. Singh, S. B. Rai, P. C. Mishra, S. Behre, S. Gunasekaharan and Dr. Renuga Devi were very helpful. Professor Jens-Uwe Grabow (Hannover), Professor Alberto Lassari (Spain) and Professor H. Maider (Kiel) have introduced me to the field of Fourier Transform Microwave Spectroscopy on Molecular Beams. They all owe special thanks.

My research students Dr. S. Shaji, Dr. Shibu, Dr. Jyotsana, Dr. Syamala, Dr. Vijayan, Dr. Sunny Kuriakose, Dr. Usha, Dr. Nandini and Dr. Thomas Zacharia were a source of inspiration to me. I would also like to thank Teachspin, Buffalo for allowing me to use some of their excellent figures on optical pumping.

My family, Dr. Radha, Dr. Rajesh, Dr. Rani, Dr. Seema, Dr. Anoop, Dr. Ranjit and the little ones Govind and Madhav deserve special thanks for their patience and understanding as I had taken much of the time which otherwise should have spent with them.

K. P. Rajappan Nair

CONTENTS

3. THE INTERACTION OF RADIATION WITH MATTER 149

4. SCHRÖDINGER EQUATION APPLIED TO ATOMS 161

5. SPECTROSCOPIC STUDIES ON THE SHELL STRUCTURE OF ATOMS 211

10   CORRELATION OF THE ATOMIC STRUCTURE WITH MORE THAN ONE ELECTRON — 379

11.   SPECTRAL TERMS FOR NEUTRAL AND IONIZED ATOMS — 403

1

EVENTS LEADING TO QUANTUM CONCEPTS

1.1 ORIGIN OF QUANTUM THEORY

The beginning of twentieth century has witnessed many developments in science. The discovery of X-rays, quantum theory of light, atomic structure, atomic theories, dual nature of matter, computers, lasers were examples of the inventions of late 19th century or of the beginning of 20th century. There is no doubt that these have revolutionarized modern technology and thereby the society.

It was realized in the beginning of the twentieth century that a number of experimental observations could not be explained by hither to known theories in classical physics. These observations were mainly related to electromagnetic radiation.

The laws of Newtonian Mechanics explained successfully the motion of celestial bodies and other earthly objects. These laws along with the laws of thermodynamics and classical electrodynamics explained successfully the physical behaviour and bulk properties of matter. This we call as Macroscopic World and the physics involved is Classical physics.

However, when the same laws were applied to the particles such as electrons, protons, neutrons, nucleus, etc., the so called Microscopic World, serious difficulties encountered. In order to explain the various phenomena arising due to the microscopic world, there arose the necessity of Quantum theory.

In the microscopic world of atoms and molecules, the energy and momentum were not having the same meaning as in the case of classical dynamics. These dynamical variables had a discrete value in particular state of an atom or molecule and did not undergo change in continuous manner from one state to another, as one would expect in classical physics. Thus to explain these phenomena classical laws were insufficient and thus the Newtonian laws of macroscopic world had to be modified though not abandoned. The incorporation of new concepts gave birth to a new mechanics called Quantum Mechanics, which made revolutionary changes in our way of thinking and new understanding on the microscopic as well as macroscopic world.

1.2 MATTER—MACROSCOPIC AND MICROSCOPIC WORLD—FUNDAMENTAL PARTICLES

"Macro" comes from the Greek word *"mackros"* which means long. Macroscopic objects are those that can be seen by humans, though it is not applicable to the big cosmic objects that are far away from us. The name *"cosm"* comes from the Greek word *kosmos* meaning universe. The word *"micro"* comes again from the Greek word *"mikros"* meaning small. In everyday usage it is used to refer to small objects, which is difficult to see. We can think of bacteria and viruses as of microscopic world. We can define a microcosm as a universe of miniature and macrocosm to the large universe or entity. When we think of the purity of water, we usually look at the dirt, leaves or grass, in the macroscopic level. If it involves measurements of bacteria, etc., it becomes microscopic measurements. Atoms are microscopic particles so also their constituents like electrons, protons and other fundamental particles. The electrons were first discovered by J.J.Thomson in 1897 through a number of experiments in electric discharges in gaseous medium.

1.3 DISCHARGE TUBE

In order to study electrical discharges, electric discharge tubes are necessary. There are many ways in which we can make electric discharge tubes. All of them need the same basic things. They are

1. Electrodes
2. Something to hold the vacuum in (a tube)
3. A vacuum pump
4. A power supply

We had the general idea that the gases are bad conductors. However, it has been found that the gases at low pressures conduct electricity at low voltages. The discharge of electricity through gases at various pressures can be studied with the help of a discharge tube.

It was by the studies of discharges of electricity through gases, the existence of negatively charged particles were established. It was found through various stages of coloured discharges that they arise because of the collision pattern between the charged particles emitted from the cathode and the atoms of the gas in the discharge tube.

The discharge tube consists of a glass tube generally of about 50 cm long and 4 cm diameter with two electrodes fused through the ends. It is connected to an exhaust pump by means of a side tube. The electrodes are then connected to the secondary of an induction coil capable of giving high potential difference. The electrode, which is connected to the negative, is called cathode while that connected to the positive is called anode. The pressure of the gas inside the discharge tube can be varied by the vacuum pump attached to the tube. As the pressure of the gas is slowly reduced the following phenomena can be observed.

1. At about 1 cm of Hg pressure, irregular flashes of light are seen passing from the cathode to anode with a cracking sound.

2. When the vacuum is reduced to 4 mm of Hg, the tube is filled with a luminous column and the colour of the glow depends on the gas inside.

3. When the pressure is still reduced to about 1 mm of mercury, a new glow at the cathode appears. This glow is known as negative glow. At the same time the positive column is separated by a dark region, which is called Faraday dark space.

4. At pressure of 0.1 mm of Hg, the negative glow gets detached and moves away from the cathode. There appears a new glow at the cathode called the cathode glow. A dark region known as Crook's dark space separates the cathode glow and the negative glow. The positive column becomes smaller in size and is broken into disclike striations.

5. At pressure of about 0.01 mm of Hg, the glow disappears and the tube becomes filled with Crook's dark space. At this stage some type of rays emerges from the cathode and produce glow on the walls facing the cathode. The colour of the glow depends on the composition and the nature of the gas inside. These invisible rays emerging from the cathode are called cathode rays.

The various stages of discharge are shown in Figure 1.1.

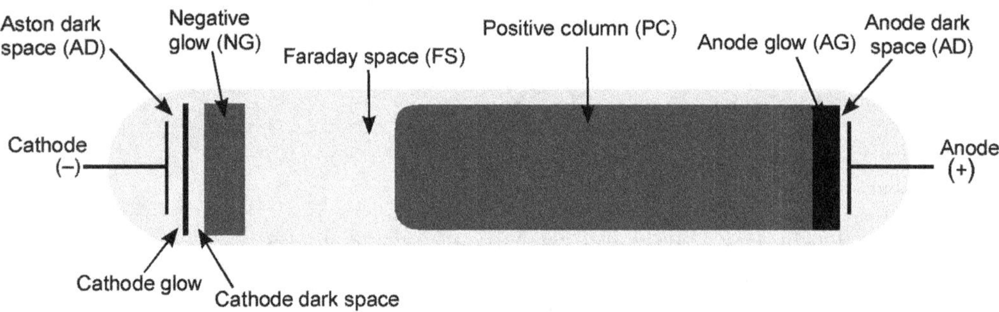

Figure 1.1   The various stages of electrical discharge

An electric glow discharge tube featuring its most important characteristics: (a) An anode and cathode at each end (b) Aston dark space (c) Cathode glow (d) Cathode dark space (also called Crookes dark space, or Hittorf dark space) (e) Negative glow (f) Faraday space (g) Positive column (h) Anode glow (i) Anode dark space.

1.3.1 Thermionic Emission

If we heat a negatively charged piece of metal, we find that some of the electrons in the conduction band get sufficient kinetic energy to escape from the surface of the wire. A metal wire can be imagined as a lattice of ions in a sea of free electrons. In effect, by heating, it will look like that we are boiling the electrons off. This effect by which electrons are released is called thermionic emission. The phenomenon of thermionic emission had been known since the middle of the nineteenth century. The experiments conducted on gases at low pressure had revealed a glow around the negative terminal. The negative terminal is called the cathode, and the rays had been named cathode rays. Some physicists were of the view that the rays were waves and others had argued that they were negatively charged particles. The Irish physicist, George Stoney, named the particles as electrons.

It was Joseph John Thomson who built a cathode ray tube (CRT) for the first time way back in 1897. This was the starting point of the CRT and the descendents of which we see every day. Essential parts of a cathode ray tube (CRT) are shown in Figure 1.2.

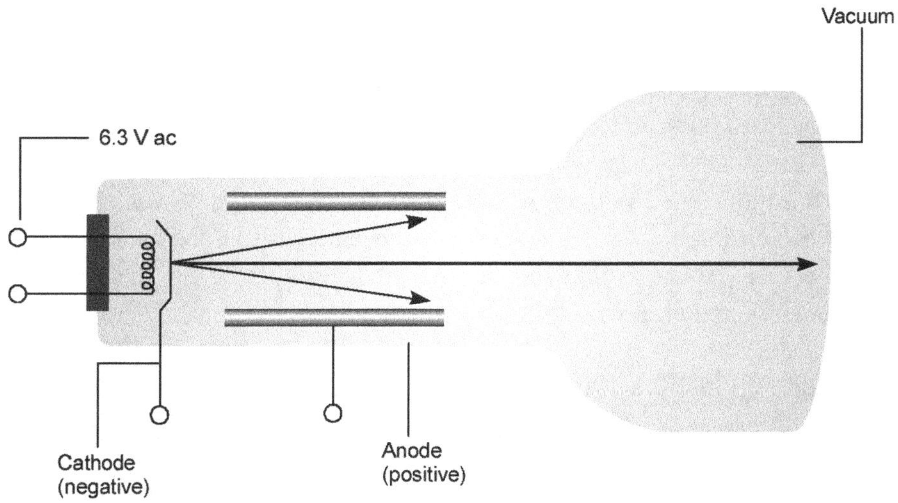

Figure 1.2   Cathode ray tube

1.3.2 Positive Rays

It was in 1897 that J.J. Thomson discovered the electron. He noticed that the electron is a negatively charged tiny particle, almost two thousand times lighter than the hydrogen atom. He referred the electrons as **corpuscles**. As the electrons started at the negative electrical terminal and moved away from it in a vacuum tube, they were previously known as "cathode rays". However, in 1907 Thomson also found that there were rays that were moving towards the cathode. Thomson noticed that these rays were positively charged. These rays were later called anode rays or positive rays. The positive rays were first observed by a German scientist Eugen Goldstein in 1886. Goldstein, in his experiments, used a gas discharge tube and cut small holes in the cathode. He produced a bunch of rays and made to stream through the holes and travel to directions opposite to the cathode rays. These positive rays were first called by Goldstein in German as *Kanalstrahlen*, meaning canal radiations as they looked like passing through a canal.

Sixteen years later, Wien deflected them in a magnetic field. Until then they remained as unknown rays. Wien found these rays much harder than cathode rays when deflected in the magnetic field. Wien also found that they were deflected in the opposite direction as that of cathode rays, suggesting they could be massive, positively charged particles.

Wien continued his experiments with electric and magnetic fields to examine the positive rays. Wien found their maximum specific charge was around two thousand times smaller than that of the electron, suggesting they could be ionized hydrogen atoms. The study of the deflection of these rays in a magnetic field revealed that the particle making up the rays were not all the same mass. The lightest, formed when there was a little hydrogen in the tube, was found to be 1837 times as massive as an electron. They were protons.

1.3.3 Cathode Rays

If the gas pressure in the discharge tube is reduced and kept at around 10^{-2} to 10^{-3} mm of Hg and a potential difference of 10,000 volt is applied between the electrodes by means of an induction coil, the whole tube is filled with darkness which is called the Crook's dark space and the wall of the tube facing the cathode is illuminated by fluorescence. The fluorescence is produced due to the falling of a particular type of invisible rays of light. As these rays emerge from the cathode, they are called cathode rays. In fact Sir William Crooks first discovered the cathode rays in the year 1878, one year before Thomson discovered electron.

1.3.4 Mechanism of Production of Cathode Rays

When a potential difference is applied between the electrodes, the positive ions created inside the tube move towards the cathode and the electrons are accelerated towards the anode. The positive ions, on colliding with cathode eject electrons from the atoms of the cathode. These electrons are also accelerated by the electric field and move towards anode. Since the electrons are light, they are highly accelerated and move towards the anode with very high velocity. These rapidly moving electrons from cathode constituted the cathode rays.

1.3.5 Properties of Cathode Rays

1. Cathode rays travel in straight lines.
2. They possess momentum and energy. When a small mica wheel is placed in the path of the cathode rays, the wheel rotates showing that the cathode rays can exert force. Hence they consist of material particles.
3. They are deflected in electric and magnetic fields.
4. The rays carry negative charge.
5. They produce heat on the material on which they fall.
6. They can produce physical and chemical changes.

7. The rays can ionize gases.

8. The rays can produce X-rays.

9. The rays produce fluorescence.

10. The rays penetrate through metal foils.

11. The e/m is a universal constant.

12. Their mass vary with velocity according to the formula

$$m = \frac{m_0}{\sqrt{1 - \dfrac{v^2}{c^2}}} \qquad (1.1)$$

where,

m_0 is the rest mass,

v is the velocity of the particle and

c is the velocity of light.

1.4 DISCOVERY OF ELECTRONS

J.J. Thomson established through a number of experiments that cathode rays consist of negatively charged particles.[1]

It was in 1897 that Sir J. J. Thomson discovered cathode rays. They were found actually consisting of negatively charged particles of very low mass and energy. He called these particles as electrons. Later it was established that they are constituent of atoms. The charge of the electron was established to be 1.6019×10^{-19} Coulomb and mass as 9.1072×10^{-31} kg.

Thus

Electron mass = 9.1072×10^{-31} kg.

Electron charge = 1.6019×10^{-19} Coulomb

If we make a hole in the anode in the cathode ray tube, while most of the electrons boiled off from the cathode will hit the anode, some will pass through the hole. Hence it is also referred to as an electron gun.

The electrons leave the cathode with negligible speed. However, the positively charged anode attracts the electrons and the attractive force of the positively charged anode accelerates them. By the time they leave the gun, the electrons have energy eV, where e is the charge on the electron (1.6×10^{-19} C) and V is the anode voltage.

Energy = Charge × Voltage = eV $\qquad (1.2)$

All the energy in the electron is kinetic, so we can say

Kinetic energy $= \frac{1}{2}\,m\mathrm{v}^2$ $\hspace{4cm}$ (1.3)

and we can combine the equations to give

$$eV = \frac{1}{2}\,m\mathrm{v}^2 \hspace{4cm} (1.4)$$

Mass of an electron is 9.11×10^{-31} kg.

Examples   We shall now see what will be the kinetic energy and speed of an electron when the anode voltage is (a) 400 V (b) 400 kV.

1. If we consider an anode voltage of 400 volt, Energy of the electron $= eV = 1.6 \times 10^{-19} \times$ 400 V $= 6.4 \times 10^{-17}$ J

    And the velocity of the electron accelerated by 400 volt can be obtained as follows:

    $\mathrm{v}^2 = 2\,eV/m = 2 \times 6.4 \times 10^{-17}$ J$/\,9.11 \times 10^{-31}$ kg $= 1.41 \times 10^{14}$ m^2 s^{-2}

    $\mathrm{v} = 1.18 \times 10^7$ m/s

2. If we consider an anode voltage of 400 kV, the energy of the electron $= eV = 1.6 \times 10^{-19}$ $\times$ 4,00 000 $V = 6.4 \times 10^{-14}$ J

    $\mathrm{v}^2 = 2\,eV/m = 2 \times 6.4 \times 10^{-14}$ J$/\,9.11 \times 10^{-31}$ kg $= 1.41 \times 10^{17}$ m^2 s^{-2}

    $\mathrm{v} = 3.74 \times 10^8$ m/s

In the second case the speed of the electron becomes larger than the speed of light. This is not possible. In this case we should have taken the relativistic mass of the electron. The electron mass increases when the speed increases by relativity theory.

1.5 THE SPECIFIC CHARGE *e/m* OF ELECTRON

The specific charge is the ratio of electron charge to its mass. An experimental set-up to determine the specific charge of the electron is shown in Figure 1.3. The method is based on the principle that an electron beam gets deflected while passing through an electric or magnetic field. The value of *e/m* can be determined by subjecting a fine beam of electrons to the combined action of crossed electric and magnetic fields. A schematic diagram is shown in Figure 1.3 and an experimental set-up is shown in Figure 1.4.

The apparatus consists of a highly evacuated tube with a filament F and anode C. The filament is heated by passing an electric current through it. The electrons emitted by the filament are accelerated towards the anode which is kept at a high positive potential with respect to the filament. The accelerated electrons emerge through the hole in the anode as a narrow beam and fall on the fluorescent screen S and produce a bright spot on it.

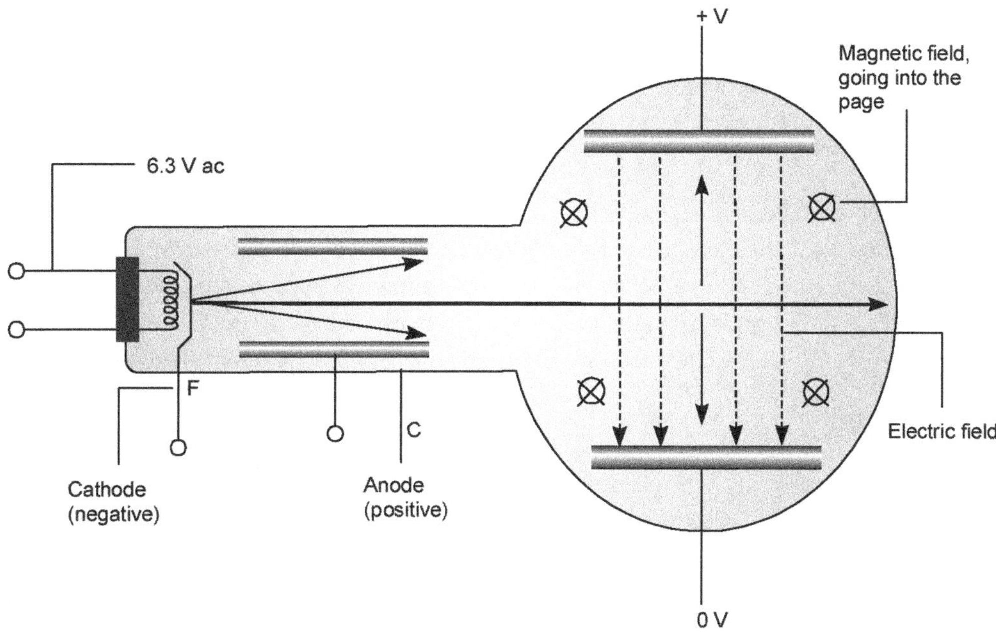

Figure 1.3 Schematic diagram of the experimental set-up in the determination of the specific charge of the electron

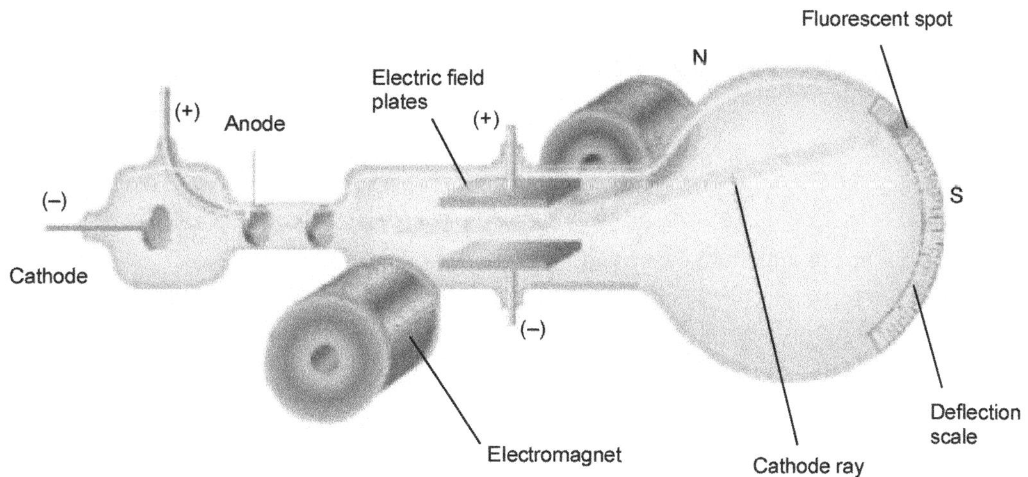

Figure 1.4 Experimental set-up to determine e/m

Let us consider the propagation of the electrons in the tube be in the X-direction. A uniform electric field is applied in the Y-direction by two metal plates arranged parallel to each other. The direction of the electric field is arranged in such a way that it is in the negative Y-direction (downwards with respect to the screen). The force on the electron is then eE where e is the charge of the electron and E is the electric field.

A magnetic field B is applied in the negative Z-direction in the same region of the electric field. The force exerted by the magnetic field on the electron is then evB where v is the velocity of the electron. This force is perpendicular to both the direction of velocity and the direction of the magnetic field. By applying Flemming's left-hand rule it can be shown that the beam will be deflected in the negative Y-direction, i.e., downward.

When the electric field alone is applied, the bright spot on the screen will be deflected upwards. When the magnetic field alone is applied the spot will be deflected downwards. When the electric and magnetic fields are applied simultaneously and if the strengths of the fields are adjusted in such a way that the spot light is undeflected, then the action of electric and magnetic fields are cancelled by each other. Thus the force exerted by the electric field is equal to the force exerted by the magnetic field.

$$\text{i.e.,} \quad eE = Bev \tag{1.5}$$

$$v = \frac{E}{B} \tag{1.6}$$

The speed of the electron can be calculated by finding the kinetic energy acquired by them in moving through the electric field maintained between the filament and the anode. If V is the potential difference between the filament and the anode, the kinetic energy acquired by the electron will be eV joules.

Hence

$$\tfrac{1}{2}\,mv^2 = eV \tag{1.7}$$

$$v^2 = \frac{2eV}{m} \quad \text{or} \quad v = \sqrt{\frac{2eV}{m}} \tag{1.8}$$

Thus by comparing equations (1.6) and (1.8),

$$\sqrt{\frac{2eV}{m}} = \frac{E}{B} \quad \text{or} \quad \frac{2eV}{m} = \frac{E^2}{B^2} \tag{1.9}$$

$$\frac{e}{m} = \frac{E^2}{2VB^2} \tag{1.10}$$

The value of e/m is 1.759×10^{11} C/kg.

The direction of deflection of the electron beam in the presence of electric and magnetic fields showed that the electrons are negatively charged.

J.J. Thomson received The Nobel Prize in Physics in 1906 *"in recognition of the great merits of his theoretical and experimental investigations on the conduction of electricity by gases"*.

1.6 MILLIKAN'S OIL DROP METHOD

It was in 1909 Robert A. Millikan[2] (1868–1953) carried out the most precise determination of the electronic charge. Millikan's experiment is repeated with varying degrees of success by generations of physics students now. He published the results in the year 1913 and is known as Millikan's oil drop experiment. The experiment is prescribed in the laboratory course for physics students ever since. He did his experiment by balancing the gravitational and electric forces on tiny charged droplets of oil supported between two metal electrodes.

This is thus a unique method in which the rate of motion of a small oil drop is investigated under free fall due to gravity and under the influence of an electric field.

An experimental arrangement is shown in Figure 1.5. The main part of the experiment consists of two horizontal metal discs A and B arranged parallel to each other. The upper plate contains a tiny pinhole H. The plates are made optically plain and are held by quartz pieces. The parallel plates are at a separation of nearly 1.5 cm and are placed in a chamber C. The chamber is filled with purified dry air and is placed in a constant temperature oil bath. Potential differences as high as 10 kV can be applied to the parallel plates. An atomizer is placed at a top corner of the chamber and fine droplets of oil are sprayed near the tiny pinhole of the upper plate. A few droplets are passed through the hole. The space between the plates is illuminated by means of an arc lamp passing the light through a window W_1. The light from the arc lamp is passed through cells containing water and cupric chloride to remove the heat waves. A telescope is mounted on the other side of the chamber to observe the movement of the oil droplets. The eyepiece of the telescope is equipped with three cross wires at known separations and the time taken by the droplets to pass between them is measured. The droplets from the atomizer get charged due to friction. Additional ions can also be produced if needed by ionizing the air between the plates by means of X-rays.

A high-tension battery is connected between the parallel plates. The upper plate is connected to the positive terminal in order to make the upper plate positive. The negatively charged drops make an upward force and the magnitude of the field is adjusted so that a suitable drop moves upwards. The drop then acquires a terminal velocity v_e due to the viscous drag of the air. This terminal velocity of the droplet is measured by measuring the time taken by the droplets to cover a distance of two cross wires of the eyepiece of the telescope. The electric field is then switched off and the same droplet is allowed to fall freely under gravity and its terminal velocity v_g is then determined by measuring the time it takes to travel two pieces of the cross wires in the eyepiece.

Initially the oil drops are allowed to fall between the plates under gravity with the electric field turned off. If we consider the motion of the droplet under gravity, we know that the viscous force opposes the motion under gravity. As the velocity of the drop increases, the viscous force also increases and a stage will be reached where these two forces are equal. That means the force due to viscosity becomes equal to the weight of the droplet. The drop then attains the terminal velocity. Thus the oil drops reach a terminal velocity because of the friction with the air in the chamber. By applying Stoke's formula the viscous force is $6\pi r \, \eta \, v_g$ where r is the radius of the droplet and η is the viscosity of air.

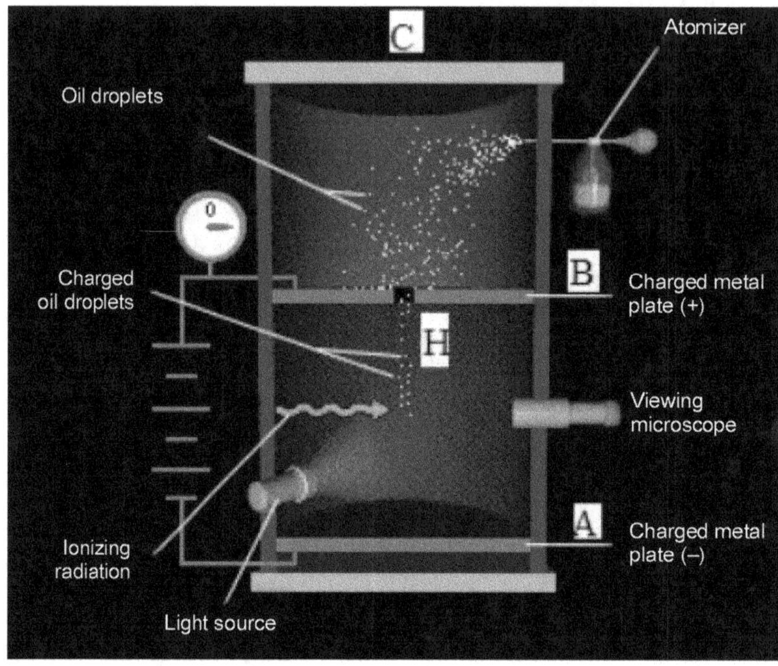

Figure 1.5   Millikan's oil drop experimental arrangement

Thus

$$mg = 6\pi r\, \eta\, v_g \tag{1.11}$$

where m is the mass of the droplet and g is the gravitational constant and mg is the weight. η is the fluid viscosity (dyne sec/cm^2), v_g is the particle velocity (velocity in the absence of the electric field or in other words the terminal velocity), and r is the radius of the drop.

The weight W is the volume V multiplied by the density ρ and the acceleration due to gravity g. The apparent weight in air is the true weight minus the upthrust (which equals the weight of air displaced by the oil pump.)

If ρ is the density of the oil and σ the density of the air, then for a perfectly spherical droplet the apparent weight can be written as:

$$W = mg = \frac{4}{3}\pi r^3\,(\rho - \sigma)g \tag{1.12}$$

where r is the radius of the drop.

Hence

$$\frac{4}{3}\pi r^3(\rho - \sigma)g = 6\pi r\, \eta\, v_g \tag{1.13}$$

$$r^2 = \frac{9\eta v_g}{2g(\rho - \sigma)}$$

$$r = \sqrt{\frac{9\eta v_g}{2g(\rho - \sigma)}} \qquad (1.14)$$

Once r is determined W can be very easily calculated. Thus W is

$$W = \frac{4}{3}\pi \left(\frac{9\eta v_g}{2g(\rho - \sigma)}\right)^{\frac{3}{2}}(\rho - \sigma)g \qquad (1.15)$$

Now the field is turned on. If E is the intensity of the electric field applied between the parallel plates and q is the negative charge carried by the droplets, then the upward force exerted by the electric field on the droplet is

$$F_E = qE \qquad (1.16)$$

where q is the charge on the oil drop and E is the electric field between the plates. For parallel plates

$$E = \frac{V}{d} \qquad (1.17)$$

V is the potential difference and d is the distance between the plates.

The net upward force is then $qE - mg$ where m is the mass of the droplet and g is the gravitational constant. As soon as the drop attains the terminal velocity the upward force becomes equal to the force due to the viscous drag. Thus

$$qE - mg = 6\pi r \, \eta \, v_e \qquad (1.18)$$

where $6\pi r \, \eta \, V_e$ is the force due to viscosity.

Dividing equation 1.18 by equation 1.11 we get

$$\frac{qE}{mg} - 1 = \frac{v_e}{v_g} \qquad (1.19)$$

$$\frac{qE}{mg} = \frac{v_e}{v_g} + 1 = \frac{(v_e + v_g)}{v_g} \qquad (1.20)$$

$$q = \frac{mg}{Ev_g}(v_e + v_g) \qquad (1.21)$$

Let t_g and t_e be the time interval taken by the drop to travel the distance S between the two cross wires, then $v_g = S/t_g$ and $v_e = S/t_e$.

The experiment is repeated on a number of drops and also at different intervals and can also be repeated for different distances of the parallel plates. In all the cases it had been seen that the change on the drop is an integral multiple of a smallest fundamental unit. This fundamental unit is called the charge of the electron. Here $q = ne$. This also established the quantum nature of the electric charges.

The presently accepted value for the elementary charge is $1.602176487(40) \times 10^{-19}$ Coulombs, where 40 indicates the uncertainty of the last two decimal places. [NIST Reference on Constants]. In his Nobel lecture,[3] Millikan gave his measurement as $4.774(5) \times 10^{-10}$ statcoulombs which is equal to $1.5924(17) \times 10^{-19}$ coulombs. The mass of the electron is 9.1×10^{-31} kg and e/m is 1.759×10^{11} C kg^{-1}.

Millikan obtained Nobel Prize for Physics in 1923. Millikan is mostly remembered for his measurements of the elementary charge, but his Nobel prize was awarded for two achievements: "for his work on the elementary charge of electricity and on the photoelectric effect". His Nobel lecture came almost 20 years after Einstein's paper. In his lecture he gave a summary of his work, but also he included a summary of the role of the experiment and the theory in physics.

"The fact that science walks in two feet, viz., theory and experiment ... sometimes it is one foot which is put forward first, sometimes the other, but continuous progress is only made by the use of both—by theorizing and then testing, or by finding new relations in the process of experimenting and then bringing the theoretical foot up and pushing it on beyond, and so on in unending alternations."

1.7 ELECTROMAGNETIC SPECTRUM

The electromagnetic radiation is a distribution of energy which are classified in terms of wavelength and frequency

$$\nu = \frac{c}{\lambda} \tag{1.22}$$

The electromagnetic spectrum is a continuum of all electromagnetic radiation (waves) arranged according to frequency and wavelength. The sun, earth, and other bodies emit electromagnetic energy of varying wavelengths. The electromagnetic energy passes through space at the speed of light in the form of sinusoidal waves. Figure 1.6 shows the wavelength, which is the distance from the wave crest to the next wave crest, i.e., wavelength is the distance between one wave crest to the next.

Radio waves, television waves, and microwaves, visible rays are all part of electromagnetic waves. The only difference is their wavelengths.

Visible light is a particular type of electromagnetic radiation that can be seen and sensed by the human eye; but electromagnetic radiation exists at a wide range of wavelengths other than visible.

The **micron** is the basic unit for measuring the wavelength of electromagnetic waves. The spectrum of waves is divided into sections based on wavelength. The shortest waves are gamma rays, which have wavelengths of 10^{-6} microns or less. The longest waves are radio waves which have wavelengths of many kilometres. The visible region of the spectrum covers from 0.4 micron (blue) to 0.7 microns (red).

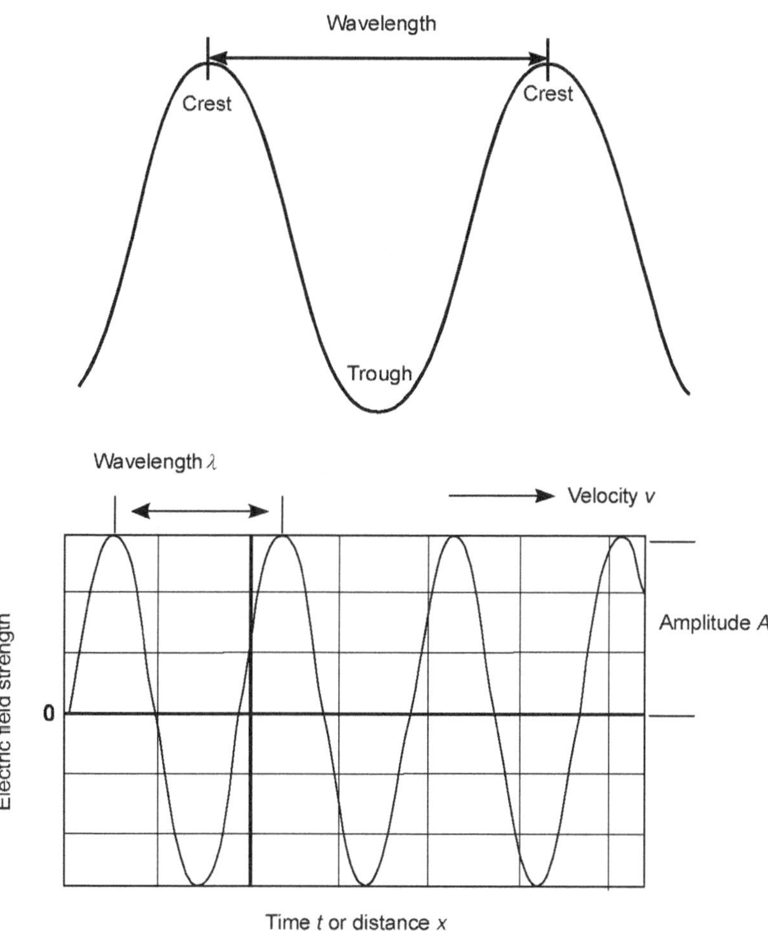

Figure 1.6   Propagation of wave in a medium

It was Newton who demonstrated in the late 1600s that the sunlight consists of seven colours by dispersing it with a prism. He called it as a spectrum. He concluded that light consists of tiny particles. Later a Dutch astronomer Christian Huygens suggested that light travels in the form of waves rather than as particles. In 1801 the wave nature of light was demonstrated by Thomas Young. The wave motion is shown in Figure 1.6. In the 1860s the Scottish mathematician James Clerk Maxwell attributed light as having the properties of electricity and magnetism. He demonstrated that electric and magnetic effects are the two aspects of the same phenomena and called it as electromagnetism. Because of the electric and magnetic characteristics, light is called as electromagnetic radiation. In a set of four equations, Maxwell provided a complete description of electromagnetic radiation and also predicted other forms of electromagnetic radiation other than visible light which were later discovered as infrared, radio waves, ultraviolet and X-rays. Thus the electromagnetic spectrum ranges from the radio waves on one side and X-rays and gamma rays on the other side. It should be mentioned that gravity and electromagnetism were the only forms of forces that were known then and combining the theories of Newton and Maxwell, one could describe the every day physical world. Then came many problems connected with the atomic structure and the stability of the nucleus. The question of how light travels as waves without a medium was also questioned. The obvious analogy of light travelling as a wave is that of ripples of water. There it is the water that moves to make the ripples. For sometimes the physicists really thought of the existence of a media called **aether** which filled the entire universe to provide a medium for light waves to travel. If the aether concept were true it would have been very difficult for the earth and the other bodies to pass through it unimpeded. The experiments conducted in 1880s by Michelson and Morley conclusively disapproved the existence of the aether theory and the aether hypothesis was totally abandoned. But it remained as a puzzle how light as waves was produced. It was then obvious that just as one could make ripples in water by wiggling one's finger about in water, the electromagnetic waves must be produced by some kind of vibrations or oscillations of electrically charged entities in matter, which consist of atoms. Then came the stumbling block that, according to the classical mechanics of Newton, charged particles radiate very suddenly. This was then circumvented by the German physicist Max Planck in 1900s who suggested that light could be emitted in lumps of certain size which he called as photon. He had derived this idea of photons during his attempt to derive a mathematical formula for explaining the radiation from a black body, the so called the black body curve. Earlier it was in 1879, that the Austrian physicist Josef Stefan described the law governing the emission of energy and the temperature of the object. He derived an expression connecting the energy flux that is the energy emitted from each square centimetre of an object's surface in each second and its temperature as

$$E = \sigma T^4 \tag{1.23}$$

where σ is a constant and T is the temperature in Kelvin. Five years later another Austrian physicist Ludwig Boltzmann derived the law mathematically from basic assumptions about atoms and molecules. This law is known as Stefan–Boltzmann law. This law is obeyed by a body, which is perfect absorber of energy. Such an object is called a black body. A black body does not reflect light, only absorbs and its energy flux depends on temperature. Ordinary objects are not perfect black bodies. We see them as they reflect light. The energy flux they emit slightly deviates from the amount calculated from Stefan–Boltzmann law. It was in 1900 that Max Planck suggested that he could derive a mathematical formula for a black body curve by assuming that the electromagnetic radiations are emitted in light quanta. Albert Einstein verified this in 1905. He used the idea of the particle like nature of light to explain the photoelectric effect.

Light and sound are the two principal media through which we gather information about our surroundings even that of the external world. The earlier notion that light is something we can see and sound as something we can hear. Later this notion was shattered and physicists learned that there are light which we can not see and sound which we can not hear. This breakthrough came when in 1800 a British astronomer William Herschel discovered infrared radiation.

When sunlight passes through a prism, it breaks into rays of different colours like that of a rainbow (*See* Figure 1.7). Each colour represents a wavelength (or frequency region). The solar spectrum consists of seven colours. These are violet, indigo, blue, green, yellow, orange and red (VIBGYOR). Studying the heat energy, William Herschel discovered radiation just beyond the red end of the visible spectrum and called it as infrared. In 1801 John Ritter, a German physicist discovered rays beyond violet which is now known as ultraviolet. It thus turned out that sunlight consists of not only visible light, but also invisible as well. In 1895 yet another German physicist Heinrich Hertz succeeded in producing light with wavelengths much longer than infrared which are called radio waves or microwaves. In 1895 William Roentgen another German physicist invented a machine that produced radiation with wavelengths shorter than 100 Å, now known as X-rays or *Roentgen rays*. Over the years radiation was discovered in many other wavelength regions, visible region is only a fraction of the electromagnetic radiation.

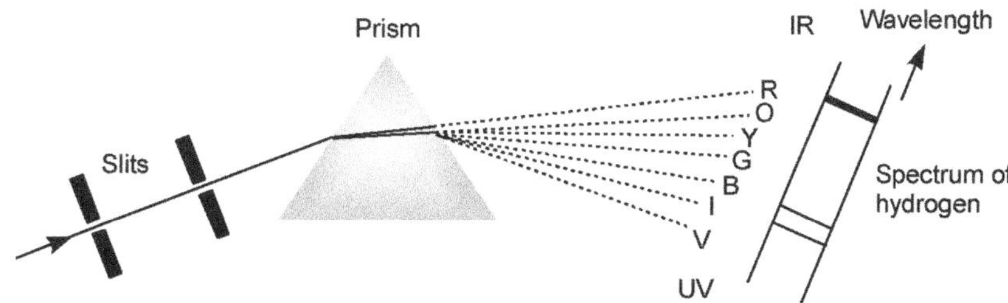

Figure 1.7   Dispersion of light through a prism

The waves in the electromagnetic spectrum thus vary in size. They range from very long radio waves having the size of buildings, to very short gamma-rays smaller than the size of the nucleus of an atom. The complete electromagnetic spectrum is shown in Figure 1.8.

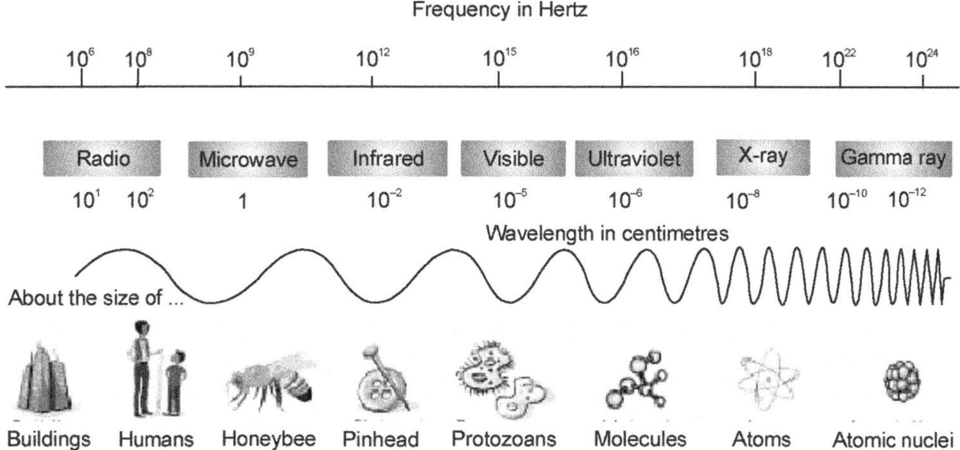

Figure 1.8   Waves in the electromagnetic spectrum

The electromagnetic waves can not only be described by their wavelength, but also by their energy and frequency. All three of these are related to each other mathematically by the relation $v = \dfrac{c}{\lambda}$. And energy $= hv = h\dfrac{c}{\lambda}$. This means we talk about the energy of an X-ray in eV or the wavelength of a microwave or the frequency of a radio wave in MHz or kHz or Hz and the visible or ultraviolet light in Angstrom or in nanometre. Thus we describe the electromagnetic radiation by one of these terminologies, viz., energy, wavelength or frequency. Scientists use the units in a way that are most convenient for them in describing whatever energy of light they are looking at. Note that there is a large difference in energy between radio waves and gamma-rays. Radio waves have energy of the order of 4×10^{-10} eV and the gamma-rays are of energy 4×10^{9} eV. That is they have an energy difference of the order of 10^{19} eV.

We can calculate these energies in the following way:

Radio waves

$$E = hv = 6.624 \times 10^{-34} \times 10^{6} \text{ Joule} = \frac{6.624 \times 10^{-34} \times 10^{6}}{1.6 \times 10^{-19}} \text{eV} = 4.14 \times 10^{-10} \text{ eV} \qquad (1.24)$$

Gamma-rays

$$E = hv = 6.624 \times 10^{-34} \times 10^{24} \text{ Joule} = \frac{6.624 \times 10^{-34} \times 10^{24}}{1.6 \times 10^{-19}} \text{eV} = 4.14 \times 10^{9} \text{ eV} \qquad (1.25)$$

Thus the electromagnetic spectrum includes, from longest wavelength to shortest: radio waves, microwaves, infrared, optical, ultraviolet, X-rays, and gamma-rays, which differ in energy, wavelength or frequency. We have already seen that the wavelength is the distance between two peaks of a wave and it is usually measured in metres whereas the frequency is the number of cycles of a wave to pass some point in a second. Thus the unit of frequency is cycles per second and is named as Hertz (Hz).

1.8 FREQUENCY, WAVE NUMBER, WAVELENGTH

Most of us have seen waves on water. The motions of sound and light are also like waves. Sound is a form energy, which produces sensation of hearing. Light produces sensation of vision. Wave motions and vibrations are closely related. The sound is produced by the vibrations of solids, liquids or gases. When a violin or guitar is played, the sound is heard when the wires are made to vibrate. We can stimulate these vibrations into a musical symphony, which gives pleasant sensation to ears. A horn is heard when the air vibrates, or a thunder is heard when the air expands or vibrates. The vibrations are propagated through the air. Thus a vibrating body is a source of sound. The number of vibrations made by a body in one second is called frequency and is usually denoted by the Greek letter v. If a body vibrates 50 times per second, the frequency is 50 cycles per second or 50 Hz. The modern unit of frequency is Hertz or in short Hz given in honour of the German Scientist Heinrich Rudolf Hertz who invented radio waves, i.e., one cycle per second is one Hertz. The different nomenclature in the units of frequency or wavelength is listed in Table 1.1.

Table 1.1   Units of frequency and wavelength

1 kilohertz	$=$ 1 kHz $=$ 1000 Hz
1 megahertz	$=$ 1MHz $=$ 1000 000 or 10^6 Hz (one million)
1 Gigahertz	$=$ 1 GHz $=$ 1000 000 000 or 10^9 Hz
1 Terahertz	$=$ 1 THz $=$ 1000 000 000 000 or 10^{12} Hz

Wavelength is the distance between the wave crests as shown in Figure 1.6.

1 Micrometre	$=$ 1 μm $=$ 10^{-6} m (one millionth) $=$ micron
1 Nanometre	$=$ 1 nm $=$ 10^{-9} m (one billionth) $=$ 10 Å
1 Angstrom	$=$ 1 Å $=$ 10^{-10} m $=$ 10^{-8}cm $=$ 10^{-1} nm $=$ 0.1 nm
10 Angstrom	$=$ 1 nm

It means that we talk of the visible region as between 4000 Å – 7000 Å, i.e., between 400–700 nm. Micrometre is generally used in infrared, sometimes Terahertz is also used in IR, nanometre and Angstrom in visible, UV and X-ray regions; microns (abbreviated as μ) used in the case of infrared is $1\mu = 10^{-6}$ m $= 10^{-4}$ cm $= 10^4$ Å. Those who work in microwaves and radio frequency regions often refer the spectrum in terms of Hz, kHz, MHz or in GHz. As said earlier these are units only for the convenience of the user.

1.9 NANOPARTICLES

Nanoparticles consist of combination of atoms which has the size from 1–100 nm. The diameter of hydrogen atom is approximately 1 Å which is 0.1 nm and hence nanoparticle of one nanometre size is the size of ten hydrogen atoms put together. It is equivalent to 1000 bacteria. 1 nm is also the size of glucose molecule. DNA has a size of approximately 2.5 nm. If we give Table1.1 in another form, it will be as shown in Table 1.2.

We have seen that the frequency is the number of vibrations per second. The time taken for one complete oscillation is called the period. It is $T = 1/\nu$ or $\nu = 1/T$. The distance travelled by a wave during one vibration is called wavelength denoted by the Greek letter λ. It's unit is Angstrom which is 10^{-8} cm. The distance travelled in one second is called the velocity. Velocity is the product of frequency and wavelength, i.e., $c = \nu\lambda$. The frequency of light is thus related by $\lambda = c/\nu$. It can be readily seen that when the frequency increases, the wavelength decreases. The frequency is inversely proportional to wavelength.

During wave propagation if the particles are moving perpendicular to the propagation, it is called transverse waves. When the motions of particles are in the direction of propagation, it is called longitudinal motion. The sound waves are longitudinal waves whereas the light waves are transverse in nature. The amplitude of wave is the maximum displacement of the particle from their equilibrium position. Thus the characteristics of a wave are its wavelength λ , its frequency ν, its velocity and the amplitude. When a wave propagates as in the case of water waves, there are elevations and depressions on the surface during their propagation. The elevations are called crests and the depressions as troughs. As described earlier, wavelength is the distance between two consecutive crests or troughs and the frequency is the number of such crests or troughs in a second. The velocity of sound in air is 340 metres per second. Objects moving with a speed greater than sound are called supersonic. The unit of speed of sound is **1 mach**. Noise is a sound produced by irregular succession of disturbances. They are always unpleasant to ears. The sensation of sound of course depends mainly on two factors, the intensity of sound or noise and the sensitiveness of the ears. Depending upon the intensity we can characterize the sound to various loudness. The unit of loudness is **decibel** (dB). It is named after Alexander Graham Bell, the inventor of telephones. One decibel is 1/10th of bell. Normal speaking voice is around 60 dB.

Table 1.2   Units in metre in terms of nanometre

1 metre	=	1000,000,000 nanometre
1 decimetre	=	100,000,000 nanometre
1 centimetre	=	10,000,000 nanometre
1 millimetre	=	1,000,000 nanometre
1 micrometre	=	1,000 nanometre
1 nanometre	=	$0.000\ 000\ 001$ metre $= 10^{-9}$ metre
1 Å $= 0.1$ nm	=	$0.0\ 000\ 000\ 001$ metre $= 10^{-10}$ metre
1 picometre	=	10^{-12} metre
1 femtometre	=	10^{-15} metre
1 atometre	=	10^{-18} metre

1.10   DUAL NATURE OF ELECTROMAGNETIC RADIATION

Scientists in earlier days found it difficult to ascertain whether electromagnetic radiation is wavelike or particlelike. The phenomena like diffraction and interference could be explained only by considering electromagnetic radiation as wavelike whereas the phenomena like photoelectric effect and black body radiation could be explained only by considering the electromagnetic radiation as particlelike. It was then suggested that light has a dual nature, i.e., it can behave like wave as well as particle.

In order to understand a light wave and the electromagnetic spectrum we shall take an analogy of a rope tied on a frictionless table, which is shaken by our hand. If we shake our hand from side to side we will see that a wave travels along the rope. Light waves are similar to the waves formed by a rope. They form a periodic disturbance. There are two varying quantities in a light wave, one is the electric field and the other is the magnetic field. Light wave is thus vibrating in electric and magnetic fields and is hence called electromagnetic wave. This covers not only the visible light, but also radio waves, infrared, ultraviolet, X-rays, cosmic rays and so on. The differences of regions are only in wavelength, i.e., in frequency. Radio waves are of several metres in wavelength whereas X-rays are of one hundred millionth of a centimetre and cosmic rays are still shorter.

The electric field is generally denoted as E and the magnetic field as B. An ordinary light consists of a wave having such perpendicular oscillations in all directions. In such a case the electric and magnetic fields are in mutually perpendicular planes. The frequency is v, which is c/λ. If we consider the red radiation at a wavelength, for example 600 nm = 6000 Å, the corresponding frequency is

$$v = \frac{c}{\lambda} = \frac{2.99792458 \times 10^8\,\mathrm{ms}^{-1}}{6000 \times 10^{-10}\,\mathrm{m}} = 4.542 \times 10^{14}\ \mathrm{Hz} = 454200\ \mathrm{GHz}$$

The hydrogen gas emits radio waves with a wavelength of 21.12 cm. This is the famous 21 cm-line of atomic hydrogen. The frequency of radiation is then,

$$v = \frac{c}{\lambda} = \frac{2.99792458 \times 10^8\,\mathrm{ms}^{-1}}{0.2112\,\mathrm{m}} = 1.420 \times 10^9\ \mathrm{Hz} = 1420\ \mathrm{MHz}$$

The frequency which we generally use in microwave oven is 2450 MHz which is equivalent to

$$\lambda = \frac{c}{v} = \frac{2.99792458 \times 10^8\,\mathrm{ms}^{-1}}{2450\,\mathrm{MHz}} = 12.236\ \mathrm{cm}$$

The wave number is another unit which is commonly used especially in visible or in ultraviolet regions. It is $1/\lambda$ and is in cm^{-1}. It is the number of waves that fit in a centimetre. For red light of 6000 Å, the wave number is $\dfrac{1}{6.0 \times 10^{-5}} = 16666.6$ cm^{-1}.

1.11   INADEQUACY OF CLASSICAL MECHANICS AND ORIGIN OF QUANTUM THEORY

Prior to 1900, it was believed that all experimental results could be explained in terms of what is now known as the classical physics. The laws proposed by the Newton's equations could successfully explain the motions of mechanical objects of terrestrial or celestial scale. The dynamical properties like position, momentum, energy, etc. of a particle could be determined by the laws of mechanics. The classical mechanics does not put any restrictions on the dynamical properties of a particle. However, the experimental measurements on the atomic and molecular systems showed that only a discrete set of values are possible. For example, it was shown that only a discrete set of energy values are possible for electrons moving round the nucleus. These observations have paved the way to the knowledge of quantization.

There were difficulties in explaining the phenomena like spectral distribution of thermal radiations from a black body, the specific heat at low temperatures, photoelectric effect and so on by the classical physics. The Newtonian mechanics which could explain the macroscopic phenomena had to be modified. Max Planck explained the black body spectrum in terms of emission and absorption of electromagnetic radiation in discrete quanta each of which contains an amount of energy E proportional to the frequency v given by the expression $E = hv$. The proportionality factor h is named as the Planck's constant. The quantum idea was later used by Einstein to explain photoelectric effect. The dual character of electromagnetic radiation was also established. These ideas put forward in the first quarter of the 20th century to explain the phenomena of black body radiation, photoelectric effect, emission and absorption of radiation, dual character of electromagnetic radiation, Compton Effect, Combination Principle—the Ritz Rydberg formula, specific heats, Franck–Hertz experiments, Stern and Gerlach experiments, Bohr atomic structure and so on, paved the way to quantum physics.

1.12 BLACK BODY RADIATION

Max Planck first introduced the Quantum theory in the year 1900 in his attempt to explain and derive laws governing black body radiation. Till then the classical theory could not explain the black body radiation. The Wien's law and the Rayleigh–Jean's law did not precisely agree with the experimental results.

A perfect black body is defined as one, which absorbs all the heat radiations (corresponding to all wavelengths) incident on it. When such a body is heated, it emits radiations of all types of wavelengths. In practice, a perfect black body is not available. A body showing close approximation to a perfect black body can be constructed. The origin of radiation from a heated body is the rapidly vibrating particles (known as oscillators) composing the body. According to Maxwells electromagnetic theory, these oscillations emit radiant energy in the form of electromagnetic waves.

The variation of the intensity of radiations with wavelength at different temperatures is as shown in Figure 1.9.

1. At a given temperature, the energy is not uniformly distributed.
2. At given temperature, the intensity of radiations increases with increase in wavelength and at a particular λ it's value is maximum. With further increase in wavelength, the intensity of heat radiation decreases.
3. With increase in temperature, λ_{max} decreases.
4. For all λ, an increase in temperature causes an increase in energy emission.
5. The area under which each curve represents the total energy emitted for the complete spectrum at a particular temperature.

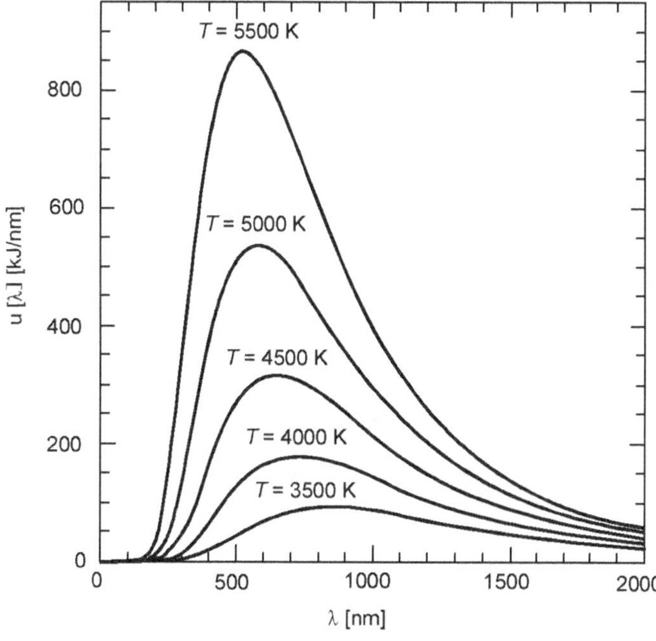

Figure 1.9   Variation of the intensity of radiations with wavelength

1.13   STEFAN–BOLTZMANN LAW

Stefan experimentally established this law. It states that the intensity of total radiation (area under cover)

$$E \propto T^4 \quad \text{or} \quad E = \sigma T^4 \tag{1.26}$$

where σ is called Stefan's constant and its value is 5.672×10^{-8} JK^{-4}. Subsequently this law was derived by Boltzmann using thermodynamic principles.

Wien's displacement law   It says in short form that $\lambda_{max} \propto 1/T$ or

$$\lambda_{max} \times T = \text{Constant} = 0.2892 \text{ cm K} \tag{1.27}$$

Wien also showed that $E_{max} \propto T^5$

$$E_{max} = \text{Constant} \times T^5$$

$$E = \frac{a}{\lambda^5} \exp\left(\frac{-b}{\lambda T}\right) \tag{1.28}$$

This gave good fitting in low-wavelength regions.

Rayleigh–Jean's law $E = kT$ for average vibrational mode.

Energy density in the range $\lambda + \Delta\lambda$ is $dn \cdot \varepsilon$ where dn is the number of oscillations in the range λ and $\lambda + \Delta\lambda$.

$$dn = 8\pi \frac{d\lambda}{\lambda^4}$$

$$E_\lambda \, d\lambda = dn. \, \varepsilon = \left(8\pi \frac{d\lambda}{\lambda^4} \right) kT$$

$$\mathrm{E}_\lambda = 8\pi \frac{kT}{\lambda^4} \tag{1.29}$$

It is found that the above expression is valid for long wavelength region but fails in low-wavelength regions.

Planck' contributions   Planck deduced an intermediate expression based on the following assumptions.

i.   A black body radiator contains simple harmonic oscillators of molecular dimensions, which can vibrate with all possible frequencies.

ii. The classical principle of equipartition of energy is not applicable to black body oscillators because this assumption would lead to Rayleigh–Jean's expression. Instead it is assumed that the oscillator of the black body cannot have any amount of energy but has a discrete energy equal to the integral multiple of some minimum energy, i.e. ,

$$E = n \, h\nu \tag{1.30}$$

where ν is the frequency of an oscillator and n is an integer.

iii. The oscillators do not radiate or absorb energy continuously. However, they exchange energy (emission or absorption) with its surroundings in discrete values, viz., $0, h\nu, 2\,h\nu, 3\,h\nu \ldots n\,h\nu$, i.e., multiples of some small unit called the quantum.

Let N be the total number of Planck's oscillators and E their total energy, then the energy per oscillator is

$$\overline{E} = \frac{E}{N} \tag{1.31}$$

If $N_0, N_1, N_2, \ldots N_r$ is the number of oscillators having energies $0, \, \varepsilon, 2\varepsilon, 3\varepsilon, \cdots r \, \varepsilon$ respectively, then

$$N = N_0 + N_1 + N_2 + \cdots N_r + \cdots \tag{1.32}$$

$$E = 0 + \varepsilon N_1 + 2\,\varepsilon N_2 + \cdots r\,\varepsilon\,N_r + \cdots \tag{1.33}$$

From Boltzmann's law, the number of oscillators having energy $r\varepsilon$ will be

$$N_r = N_0\,e^{-r\varepsilon/kT}$$
$$\text{i.e.,}\quad N_1 = N_0\,e^{-\varepsilon/kT}, \quad N_2 = N_0\,e^{-2\varepsilon/kT}, \text{ etc.} \tag{1.34}$$

Thus we have

$$N = N_0 + N_0 e^{-\varepsilon/kT} + N_0 e^{-2\varepsilon/kT} + \ldots + N_0 e^{-r\varepsilon/kT} + \ldots$$
$$= N_0 + [1 + e^{-\varepsilon/kT} + e^{-2\varepsilon/kT} + \ldots + e^{-r\varepsilon/kT} + \ldots]$$

$$N = \frac{N_0}{1 - e^{-\frac{\varepsilon}{kT}}} \tag{1.35}$$

$$E = 0 + \varepsilon N_0 e^{-\varepsilon/kT} + 2\varepsilon N_0 e^{-2\varepsilon/kT} + \ldots + r\varepsilon N_0 e^{-r\varepsilon/kT} + \ldots$$
$$= N_0\varepsilon + [e^{-\varepsilon/kT} + 2e^{-2\varepsilon/kT} + \ldots + r\,e^{-r\varepsilon/kT} + \ldots]$$
$$= N_0\varepsilon\,e^{-\varepsilon/kT}[1 + 2e^{-2\varepsilon/kT} + \ldots + r\,e^{-(r-1)\varepsilon/kT} + \ldots]$$

and
$$= \frac{N_0\varepsilon e^{-\frac{\varepsilon}{kT}}}{(1 - e^{-\frac{\varepsilon}{kT}})^2} \tag{1.35a}$$

Here we have made use of the formula $\dfrac{1}{(1-x)^2} = 1 + 2x + 3x^2 + \ldots$

The average energy of oscillator is then

$$\overline{E} = \frac{E}{N} = \frac{N_0\varepsilon e^{-\frac{\varepsilon}{kT}}}{N\left(1 - e^{-\frac{\varepsilon}{kT}}\right)^2}$$

However $N = \dfrac{N_0}{1 - e^{-\frac{\varepsilon}{kT}}}$ and hence

$$\overline{E} = \frac{E}{N} = \frac{N_0\varepsilon e^{-\frac{\varepsilon}{kT}}}{(1 - e^{-\frac{\varepsilon}{kT}})^2} \times \frac{(1 - e^{-\frac{\varepsilon}{kT}})}{N_0}$$

$$= \frac{\varepsilon e^{-\frac{\varepsilon}{kT}}}{(1 - e^{-\frac{\varepsilon}{kT}})} = \frac{\varepsilon}{(e^{\frac{\varepsilon}{kT}} - 1)} = \frac{h\nu}{(e^{\frac{h\nu}{kT}} - 1)} \tag{1.36}$$

The number of oscillator per unit volume in frequency range ν and $\nu + d\nu$ is given by

$$n = \frac{8\pi\nu^2}{c^3} d\nu$$

The above expression was first calculated by Rayleigh and can be obtained from Saha and Srivastava.[4]

The average energy per unit volume or the energy density $E_\nu d\nu$ is then

$$E_\nu d\nu = \frac{8\pi\nu^2}{c^3} d\nu \times \frac{h\nu}{\left(e^{\frac{h\nu}{kT}} - 1\right)} = \frac{8\pi h\nu^3}{c^3} \frac{1}{\left(e^{\frac{h\nu}{kT}} - 1\right)}$$

This is known as the Planck's radiation formula.

In terms of wavelength,

$$\nu = c / \lambda \text{ and } d\nu = -\frac{c}{\lambda^2} d\lambda, \left|d\nu\right| = \frac{c}{\lambda^2} d\lambda$$

$$E_\lambda \cdot d\lambda = \left(\frac{8\pi h\nu^3}{c^3}\right) \frac{1}{e^{\frac{h\nu}{kT}} - 1} \frac{c}{\lambda^2} d\lambda$$

$$= \frac{8\pi hc}{\lambda^5} \frac{1}{e^{\frac{h\nu}{kT}} - 1} d\lambda = \frac{8\pi hc}{\lambda^5} \frac{1}{e^{hc/\lambda kT} - 1} d\lambda \tag{1.37}$$

By integrating over frequency, the expression yields the total energy density. The radiation field of a black body may be thought of as a photon gas, in which case this energy density would be one of the thermodynamic parameters of that gas.

$$u(\nu, T) = \frac{8\pi h\nu^3}{c^3} \frac{1}{\left(e^{\frac{h\nu}{kT}} - 1\right)} d\nu$$

and in terms of wavelength:

$$u(\lambda, T) = \frac{8\pi hc}{\lambda^5} \frac{1}{e^{\frac{hc}{\lambda kT}} - 1} \tag{1.38}$$

Wien's law and Rayleigh–Jean's law can be derived from Planck's law.

i.   For small temperatures, λT is small and therefore $1/e^{hc/\lambda kT} \geq 1$ and hence

$$E_\lambda \, d\lambda = \frac{8\pi hc}{\lambda^5}\frac{1}{e^{\frac{hv}{kT}}}d\lambda = \frac{8\pi hc}{\lambda^5}\frac{1}{e^{hc/\lambda kT}}d\lambda = \frac{8\pi hc}{\lambda^5}e^{-\frac{hc}{\lambda kT}}d\lambda. \text{ This is Wien's law} \qquad (1.39)$$

ii.   For large temperatures, λT is large and

$$e^{hc/\lambda kT} = 1 + \frac{hc}{\lambda kT} + \frac{hc}{2!\lambda kT} + \dots = 1 + \frac{hc}{\lambda kT}$$

$$E_\lambda d\lambda = \frac{8\pi hc}{\lambda^5}\cdot\left(\frac{1}{1+\dfrac{hc}{\lambda kT}-1}\right)d\lambda$$

$$= \frac{8\pi hc}{\lambda^5}\cdot\frac{\lambda kT}{hc}d\lambda = \frac{8\pi kT}{\lambda^4}d\lambda. \text{ This is Rayleigh-Jean's law} \qquad (1.40)$$

iii.   If we integrate the above expression for all wavelengths we obtain

$$\int_0^\infty E_\lambda d\lambda = \frac{8\pi h}{c^3}\frac{k^4 T^4}{h^4}\left(\frac{1}{1^4}+\frac{1}{2^4}+\frac{1}{3^4}+\dots\right) = \frac{8\pi h}{c^3}\frac{k^4 T^4}{h^4}1.082 = \sigma T^4 \qquad (1.41)$$

The above expression leads to Stefan's fourth power law.

The success of Planck lies in identifying the elementary energy packets of radiation in the form of hv. This is the quantum of energy.

In the expression $E = hv$, h is called the Planck's constant.

Planck's hypothesis is that all electromagnetic radiations are quantized and occurs infinite "bundles" of energy. These "bundles" of energy are called photons. The quantum of energy contained in a photon is hv, i.e., the product of Planck's constant and the frequency of the electromagnetic radiation. The quantization have the meaning that the blue light of a particular frequency (or wavelength) has always the same quantum of energy. Thus a photon of a blue light having a wavelength $\lambda = 5000$ Å will have always the energy 2.48 eV.

$$E = hv = h\frac{c}{\lambda} = \frac{6.624\times10^{-34}\times3\times10^8}{5000\times10^{-10}}\text{Joule} = \frac{6.624\times10^{-34}\times3\times10^8}{5000\times10^{-10}\times1.6\times10^{-19}} = 2.48 \text{ eV}$$

That means the wavelength 500 nm has always the same energy "bundle" of value 2.48 eV.

The dimension of h is the same as that of angular momentum. We know from the principle of dimensions that $E = \text{Force} \times \text{Acceleration} = F.s = ma.s$ and the dimension is ML^2T^{-2}. And the dimension of h is obtained from the expression

$$h = \frac{\text{Energy}}{\text{Frequency}} = ML^2T^{-2} / T^{-1} = ML^2T^{-1} = MLT^{-2} \cdot L = ML^2T^{-1}$$

which is the dimension of angular momentum.

Thus h is defined as the smallest quantum angular momentum of a particle.

The ideas proposed by Planck were immediately accepted and were used to solve many physical problems which could not be solved by classical theory. Einstein applied the concept of Planck to explain the photoelectric effect in 1905 and in 1907, he could explain the specific heat of solids applying Planck's concepts. Bohr in 1913 used the concept of Planck to explain the origin of spectral lines and Compton in 1922 used it to explain the phenomenon of X-ray scattering known as Compton effect. Thus the photon emission still exhibiting a wave nature having frequency v its energy contents are given to the atoms in the form of quanta, which is hv.

When photon is regarded as a particle of radiation, then we have to make an attempt to find out the characteristics of a particle such as mass, momentum, energy, statistics, etc.

Energy of a photon is in multiples of hv. If a photon undergoes interaction with matter, either it can be completely absorbed, transferring all or part of its energy, and its frequency is then adjusted to a lower level.

$$\text{Mass of a photon is Mass} = \frac{hv}{c^2}$$

$$\text{Momentum} = \text{Mass} \times \text{Velocity} = \left(\frac{hv}{c^2}\right) \times c = \frac{hv}{c} \tag{1.42}$$

$$\text{Energy} = hv = mc^2 = \frac{m_0 c^2}{\sqrt{1 - \dfrac{v^2}{c^2}}} \tag{1.43}$$

where,

$$m_0 = \sqrt{1 - \frac{v^2}{c^2}} \times \frac{hv}{c^2} \tag{1.44}$$

Photon always travels with the velocity of light. However, its energy content is always finite, viz., hv. It means that the rest mass m_0 of the photon approaches zero when its velocity approaches that of light. However, its energy always approaches to hv. Hence photon is always having a wave structure, sometimes behaving in compact form like a wave.

Photons are electrically neutral. Photons travel with the same velocity, though with different energies. Photons follow Bose–Einstein statistics.

1.14 PHOTONS

The name *photon* derives from the Greek word for light *phos*. The word was first coined in 1926 by the physical chemist Gilbert N Lewis, who published a speculative theory in which photons were considered to be "uncreatable and indestructible". Although Lewis theory was never accepted, being contradicted to many experiments, the new name *photon* was readily adopted.

Photon is usually denoted by the symbol γ (the Greek letter gamma). Remember that it is also the symbol for gamma-rays. In chemistry and optical engineering, photons are usually symbolized by hν, the energy of the photon. h is Planck's constant and ν is photon's frequency. In some books it is also denoted by *hf* where its frequency is denoted by *f*.

In modern physics, the photon is considered as elementary particle responsible for electromagnetic phenomena. It is the particle, which mediates electromagnetic interactions and makes up all forms of light. The photon has zero invariant mass and travels at a constant speed, the speed of light in empty space. In the presence of matter, a photon can be absorbed or emitted, transferring energy and momentum proportional to its frequency. Photon has both wave and particle properties thereby exhibiting wave–particle duality.

The modern concept of the photon was developed by Albert Einstein. He used this concept to explain experimental observations that did not fit with the classical wave model of light. The photon model could account for the frequency dependence of light's energy. The photon was originally called as "light quantum" (das Licht Quantum, plural—Licht Quanta) by Albert Einstein. It could also explain the ability of matter and radiation to be in thermal equilibrium. Other scientists tried to explain the anomalous observation on semiclassical models in which light was still described by Maxwell's equations but the objects, which emit and absorb light, were quantized. The semiclassical models contributed to the development of quantum mechanics and further experiments proved Einstein's hypothesis that light is quantized and the light quanta are called photons.

The photon concept has led to many advances in experimental and theoretical physics. Atomic and molecular theories and transitions from quantized energy levels, Bose–Einstein condensation, quantum field theory and the probability concept in quantum mechanics, lasers and laser cooling of atoms are some examples. The intrinsic properties of photons such as charge, momentum and spin are determined by the properties of gauge symmetry in particle physics. Photons have many applications, in spectroscopy, photochemistry, photo dissociation of atoms and molecules, optoelectronics and so on. A new field, viz., photonics has already been originated. Recently photons have been studied as elements of quantum computers and applications in optical communications.

The photon is massless, has no electric charge and does not decay spontaneously in empty space. As the photon is massless, it moves at the speed of light in empty space and its energy and momentum are related by $E = c\,p$ where p is the magnitude of momentum and c is the velocity of light. For comparison, the corresponding equation for particles with an invariant mass m would be $E^2 = c^2 p^2 + m^2 c^4$, as shown in special theory of relativity. A photon has two possible polarization states. They are described by the components of its wave vector, which determine its wavelength and its direction of propagation.

Photons are emitted in many natural processes, e.g., when a charge is accelerated, when an atom, molecule or a nucleus jumps from a higher to lower energy level, or when a particle and its antiparticle are annihilated.

The energy and momentum of a photon depend only on its frequency ν or, equivalently, its wavelength λ. The energy and momentum of photon can also be written in the following way.

$$E = h\nu = h\frac{\omega}{2\pi} = \frac{h}{2\pi}\omega = \hbar\omega = \frac{h}{2\pi}\omega = \frac{hc}{\lambda} \tag{1.45}$$

and consequently the magnitude of the momentum is

$$p = \hbar k = \frac{h}{\lambda} = \frac{h\nu}{c} \tag{1.46}$$

where $\hbar = \dfrac{h}{2\pi}$ (known as Dirac's constant of Planck's reduced constant); k is the wave vector (with the wave number $k = \dfrac{2\pi}{\lambda}$ as its magnitude) and $\omega = 2\pi\nu$ is the angular frequency. The wave vector k points in the direction of the photon's propagation. The photon also carries spin angular momentum that does not depend on its frequency. The magnitude of its spin is $\sqrt{2}\hbar$ and the component measured along its direction of motion, its helicity, must be $\pm\hbar$. These two possible helicities correspond to the two possible circular polarization states of the photon (right-handed and left-handed).

Though Maxwell put forward wave theory of electromagnetic radiation, his theory, however, does not account for all properties of light. The Maxwell theory implies that the energy of light depends only on its intensity and not on frequency. However several experiments have unequivocally proved that the energy of light depends only on its frequency and not on intensity contrary to Maxwell's theory.

For example, some chemical reactions can be provoked only by light of frequency higher than a certain threshold; light of lower frequency, no matter how intense, is incapable of exciting the reaction. Similarly, electrons can be ejected from a metal plate by shining light of sufficiently high frequency on it (the photoelectric effect); the energy of the ejected electron is related only to the light's frequency, not to its intensity.

At the same time, investigations of black body radiation carried out over four decades (1860–1900) by various researchers culminated in Max Planck's hypothesis that the energy of any system that absorbs or emits electromagnetic radiation of frequency ν is an integer multiple of an energy quantum $E = h\nu$. As shown by Albert Eienstein some form of energy quantization must be assumed to account for the thermal equilibrium observed between matter and electromagnetic radiation.

1.14.1 Basic Properties of Photons

According to the photon theory of light, photons

♦   move at a constant velocity, $c = 2.99792458 \times 10^8$ m/s (i.e., "the speed of light"), in free space.

♦   have zero mass and rest energy.

✧   carry energy and momentum, which are also related to the frequency ν and wavelength *lamda* of the electromagnetic wave by $E = h\nu$ and $p = h/\lambda$.

✧   can be destroyed/created when radiation is absorbed/emitted.

✧   can have particle-like interactions (i.e., collisions) with electrons and other particles, such as in the Compton effect .

1.15   PHOTONS AND ELECTRONS

The essential comparison and differences of photons and electrons are given in the following Table 1.3.

Table 1.3   The essential comparison and differences of photons and electrons

Electrons	Photons
Electronics is the science dealing with electrons.	Photonics is the science dealing with photons.
Charged	No charge
Follow Fermi–Dirac statistics.	Follow Bose–Einstein statistics.
The charged nature of electrons makes them easier to contain and guide.	As photons do not have charge they can crowd together and hence generate enormous energy density.
	This does not happen in the case of photons.
Restricts concentration because of mutual repulsion.	
As electrons are charged particles they can not cross each other.	Photons are not charged particles. Crossing of beams possible.
Electrons have properties of interacting at a distance leading to inductive and capacity effects which cause delays in transmission. Can be used as switches.	Do not interact, but can be used for switches by the quality of non-linear effect— bistability.

1.15.1   Photon Facts

We know that the photon is an elementary particle, despite the fact that it has no mass. It cannot decay on its own, although the energy of the photon can transfer (or be created) upon interaction with other particles. Photons are electrically neutral and are one of the rare particles that are identical to their antiparticle, the antiphoton.

Photons are spin-1 particles (making them bosons), with a spin axis that is parallel to the direction of travel (either forward or backward, depending on whether it's a "left-hand" or "right-hand" photon). This feature is what allows for polarization of light.

At the beginning of the twentieth century itself the quantum concept of radiation was applied in explaining

- ✧ Max Planck's quantum concepts on black body radiation in 1900;

- ✧ Photoelectric effect by Einstein in 1905;

- ✧ Specific heat by Einstein in 1907;

- ✧ Atomic structure by Niels Bohr in 1913;

- ✧ X-ray scattering by Compton in 1922;

- ✧ Louis deBroglie theory of matter waves in 1924.

All these paved the way to Quantum Physics or Quantum Mechanics.

1.16 PHOTOELECTRIC EFFECT

The phenomenon of emission of electrons from a substance when exposed to electromagnetic radiations of suitable wavelengths is called photoelectric effect.

A quartz window is sealed to a glass tube which permits light to pass through and irradiate the plate P. The plate P is kept at zero potential while the plate C can be maintained at a desired + or –ve potential. The potential difference between the plates is measured using a voltmeter and the current flowing through the tube is measured using a micrometer. The experimental set-up is shown in Figure 1.10.

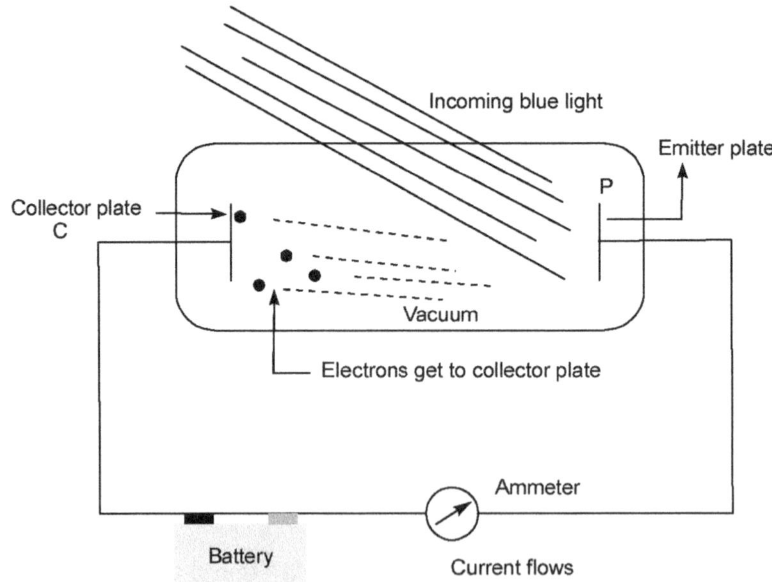

Figure 1.10   Photoelectric effect

Ultraviolet light of a particular frequency is allowed to fall upon the plate. The electrode C is made positive. The electrons are ejected from P and attracted towards C. Keeping the frequency of the incident light and the accelerating potential constant, the intensity of the incident light is varied and photoelectric current is measured in each time. It is found that the photoelectric current is directly proportional to the incident light.

Keeping the intensity and frequency of incident light constant, C is given gradually increasing positive potential and the photoelectric current is measured each time. As the accelerating potential increases, the photoelectric current increases and reaches a maximum value. This maximum value of current is called saturation current.

The potential is now reduced to zero and then given an increasing –ve potential. The current decreases and at a particular potential the current decreases to zero. This potential is called stopping potential or cut off potential. If the light of same frequency is used at a higher intensity, the saturation current increases but the stopping potential remains the same. Thus the stopping potential is independent of the intensity of the incident light.

If the experiment is repeated with a monochromatic light of another frequency, it is observed that stopping potential will change. As the frequency of the incident light is increased the stopping potential is increased.

Thus when the experiment is conducted using different frequencies, it is observed that the stopping potential uniformly increases as the frequency of the incident light is increased.

As the frequency of the incident light is decreased continuously, we find that the photoelectric current stops abruptly if the frequency is below a particular value v_0.

Most commonly observed phenomena with light can be explained by waves. But the photoelectric effect suggested a particle nature for light. A schematic representation of photoelectric phenomena is shown in Figure 1.11(a). A graph showing the frequency vs. photoelectron energy in the case of sodium metal is shown in Figure 1.11(b).

Light below a frequency of 4.39×10^{14} Hz or wavelength longer than 683 nm would not eject electrons in this case. The plot above was not dependent upon the intensity of the incident light which implied that the interaction was like a particle which gave all its energy to the electron and ejected it with that energy minus that which it took to escape the surface.

Thus for getting a photoelectric current it is seen that the frequency of the incident light should be more than a particular minimum value v_0. It is called the threshold frequency. Thus the threshold frequency v_0 may be defined as the minimum frequency of the incident light required to obtain a photoelectric current. $\lambda = c / v_0$ is called the critical wavelength of cut off wavelength.

The fact that the above plot was not dependent on the intensity of the incident light implies that the interaction is like a particle which gave all its energy to the electron and ejected it with that energy minus that which it took to escape the surface (work function).

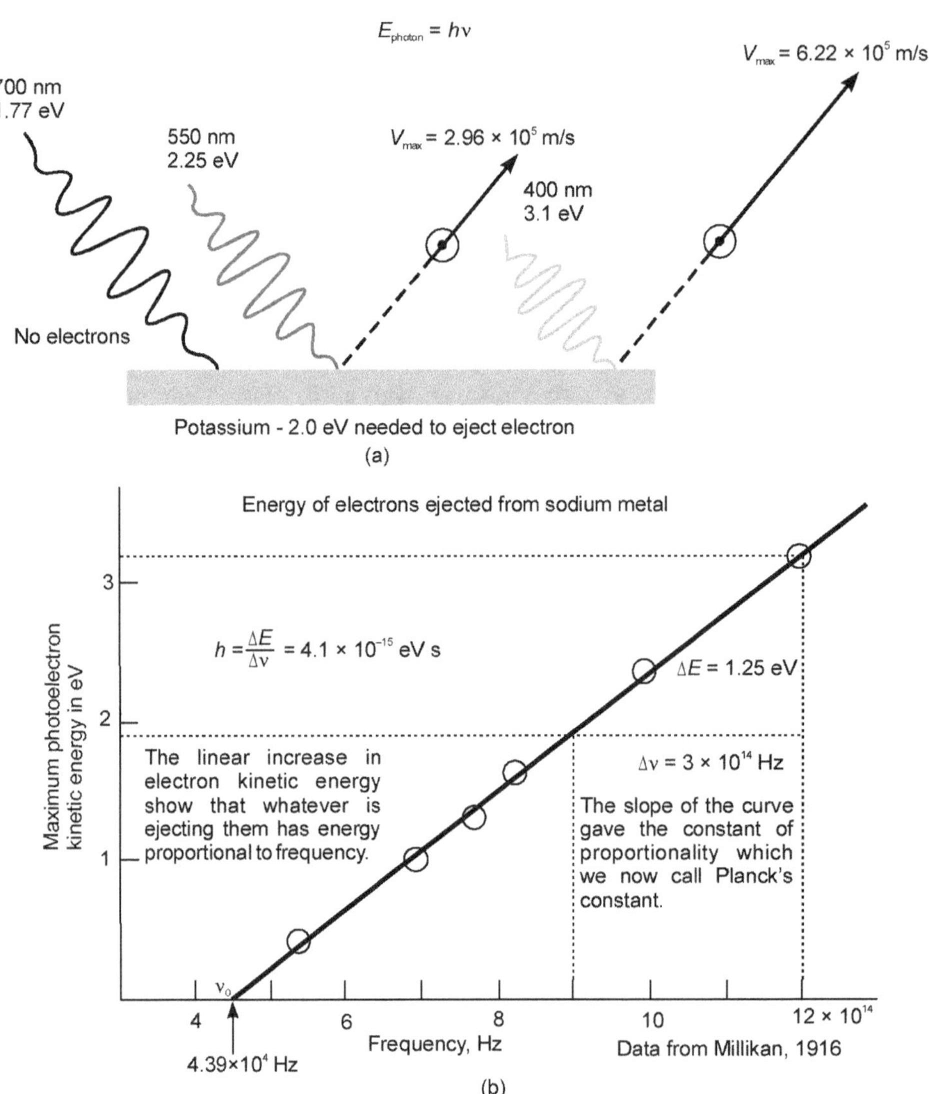

Figure 1.11   (a) Schematic representation of photoelectric effect (b) Frequency vs. photoelectron energy

The value of the threshold frequency of the critical wavelength will not be the same for all metals because they are found to be the properties of the metal surfaces used. Stopping potential depends on the kinetic energy of photoelectrons.

$$E_k = \frac{1}{2}mv^2 = V_0 e \tag{1.47}$$

As it is seen that the stopping potential increases uniformly as the frequency of the incident radiation increases, there should be a direct relation between the kinetic energy of the electrons emitted and the frequency of the incident radiation. This was first explained by Einstein in 1905.

1.17   LAWS OF PHOTOELECTRIC EMISSION

Law I   The number of photoelectrons ejected is directly proportional to the intensity of the incident light. Thus if the intensity of the incident light is increased, the number of electrons emitted is increased, but their velocity remains constant.

Law II   The velocity of the emitted electrons increases with the increase of frequency of the incident light. This law implies that there is a limiting frequency for a metal below which no electrons are emitted. This is known as threshold frequency.

The laws of photoelectric effect were explained by quantum theory.

When a photon of energy $h\nu$ is incident on a metal surface, the energy is used in two ways.

i.   a part is used up in ejecting the electron just out of the surface. The energy depends on the nature of the metal and is called work function.

ii.   The rest part of energy of the striking photon is used up in imparting kinetic energy $\left(\dfrac{1}{2}mv^2\right)$ of the ejected electrons.

Thus
$$h\nu = \varphi_0 + \frac{1}{2}mv^2 \tag{1.48}$$

$$\varphi_0 = h\nu_0 \tag{1.49}$$

ν_0 is the threshold frequency and $\varphi_0 = h\nu_0$ is the threshold energy to eject the electrons from the metal which is called the work function.

$$h\nu = h\nu_0 + \frac{1}{2}mv^2 \tag{1.50}$$

$$\frac{1}{2}mv^2 = h(\nu - \nu_0) \tag{1.51}$$

This equation is called Einstein's photoelectric equation. The equation was tested by Millikan in 1915. The photoelectric effect established the particle-like behaviour of light.

1.18   MILLIKAN'S EXPERIMENT

Millikan was the first to verify Einstein's photoelectric equation by accurate measurement of cut off voltages of sodium metal by using monochromatic light of known frequencies. For a given frequency ν the stopping potential V measures the maximum kinetic energy of the photoelectrons which reach the anode.

Millikan applied a negative potential to the cylinder (*See* Figure 1.12) and a positive potential to the plate.

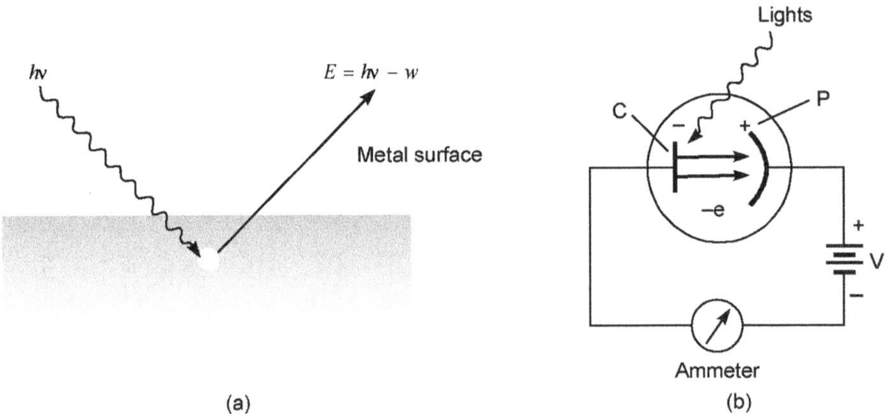

Figure 1.12   Millikan's experiment

He studied the effect of light for a range of frequencies of sodium, potassium and lithium. The positive potential applied to the plate P were such that no electron reached the cylinder C. This was detected with the help of a galvanometer. When the cylinder C is positive with respect to P, the galvanometer shows a deflection and as V is increased there is no further change in deflection. When the cylinder C is negative with respect to P the deflection in the galvanometer decreases and for a particular value of V_0, the deflection is zero. It means that no electron reaches the cylinder C. This potential V_0 is known as the stopping potential. When the intensity of the light is increased for the same frequency, the deflection of the galvanometer is more when C is positive with respect to P. This shows that the number of photoelectrons ejected depends upon the intensity of the incident light. When C is made negative with respect to P, it is found that the deflection in the galvanometer decreases and for a particular value of V_0, the deflection is zero. It is found in both these cases, the stopping potential V_0 is the same and is independent of the intensity of the incident light. The fact that the stopping potential V_0 is independent of the intensity of the incident light can be interpreted only by assuming that the kinetic energy of the electrons emitted by the surface does not exceed a certain maximum value.

$$eV_0 = \frac{1}{2}mv^2 \tag{1.52}$$

All other electrons which leave the surface with kinetic energy less than the maximum are stopped by smaller values of potential difference between C and P. This explains the decrease in deflection in the galvanometer when potential difference between C and P is made negative. From equations (1.51) and (1.52),

$$h(v - v_0) = eV_t \tag{1.53}$$

The value of the stopping potential V_0 depends upon the frequency of the incident light and the threshold frequency for the particular metal. If a graph is plotted between v and V_0, a straight line graph is obtained. The slope of the line gives h/e. As the value of e is known from Millikan's experiment, the value of h can be determined. Einstein's equation has been found valid for electrons ejected under the action of X-rays, γ-rays and also visible light.

$$\frac{1}{2}mv^2_{\max} = K \cdot E_{\max} = eV = h(v - v_0) \tag{1.54}$$

$$eV = h(v - v_0)$$

$$V = \frac{hv}{e} - \frac{hv_0}{e} \tag{1.55}$$

A graph between the stopping potential V and the frequency v gives a straight line, the slope of which is h/e.

The experimental results are as follows:

1. If all the electrons emitted by the photoelectric effect is collected at the anode by setting a high voltage, the electric current is proportional to the intensity of light illumination on the cathode.
2. There is a threshold frequency to release electrons from the cathode. If the frequency is less, no photoelectric electrons are released no matter how intense the illuminated light is.
3. The maximum kinetic energy of the photoelectron is independent of the intensity of the light falling on the cathode.
4. The maximum kinetic energy of the emitted electron is a linear function of frequency which is exactly in accordance with Einstein's hypothesis

$$E = hv - w$$

Millikan obtained a value $h = 6.58 \times 10^{-34}$ J.s which is in conformity with the modern values.

Specific heat of solids and Compton effect were explained by Planck's quantum theory.

1.19 COMPTON EFFECT

In 1923 A H Compton[5] observed that when a beam of monochromatic X-rays was scattered by a light element such as carbon, the scattered radiations consisted of two components, one of the longer wavelength than the incident radiation and the other of the same wavelength as that of incident radiation.

It has also been observed that the difference in wavelength of the two scattered radiation increases with the angle of scattering. For the scattered radiation at right angles to the incident beam, the difference in wavelength was found to be 0.0236×10^{-10} m. It was also found that in this case, the difference in wavelength is independent of the incident beam and also of the nature of the scattering material.

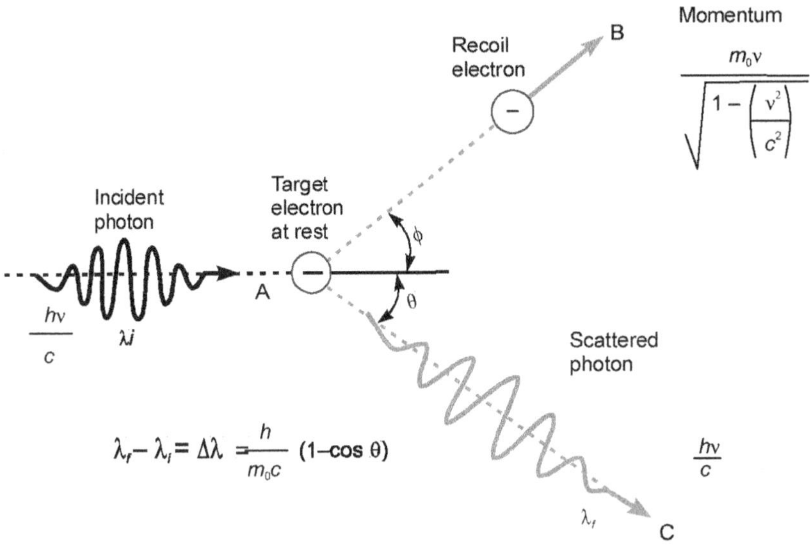

Figure 1.13   Compton scattering

In order to explain this scattering phenomenon, Compton assumed that the scattering is occurred by a collision process between the incident photon and the electrons on the target. He could explain the phenomena on the basis of quantum theory proposed by Max Planck and Einstein. The Compton scattering demonstrated the particle concept of electromagnetic radiation and earned Compton Nobel Prize in Physics in 1927.

1.19.1 Theory of Compton Effect

Let us consider an incident photon hν falling on a target at the point A. Each photon has an energy $h\nu$ and momentum $h\nu/c$ where c is the velocity with which the photon moves. Consider the incident photon strikes a free electron at rest. In this process, the energy of the photon is shared between the photon and the electron at rest. The scattered photon has the energy $h\nu$ and makes an angle θ with the original direction and the electron at rest is scattered with a velocity v at an angle ϕ with the original direction of the photon beam (*See* Figure 1.13). We can apply the conservation of momentum and energy in the above process. The energy for the electron is obtained from the incident photon and hence the scattered photon has less energy than the incident photon. Hence the scattered photon has a lower frequency than the incident one. That means the scattered photon has a longer wavelength than that of the incident photon.

Let us consider a photon colliding with an electron at rest and after collision let them move in the directions as shown in Figure 1.13.

The energy of the incident photon $= h\nu$ and hence the momentum is $\dfrac{h\nu}{c}$.

And the energy of scattered photon $= h\nu'$ and its momentum is $\dfrac{h\nu'}{c}$.

The electron is assumed to be at rest before collision and its energy according to the theory of relativity is mc^2 and the momentum is zero. After collision the velocity of the recoil electron is v and the corresponding mass

$$m = \frac{m_0}{\sqrt{1 - \left(\dfrac{v^2}{c^2}\right)}} \tag{1.56}$$

The energy of the recoil electron is $mc^2 = \dfrac{m_0 c^2}{\sqrt{1 - \left(\dfrac{v^2}{c^2}\right)}}$ and its momentum is $\dfrac{m_0 v}{\sqrt{1 - \left(\dfrac{v^2}{c^2}\right)}}$

Here m_0 is the rest mass of the electron. If we apply now the principle of conservation of energy, we get

$$h\nu + m_0 c^2 = h\nu' + mc^2$$

and momentum $\quad \dfrac{h\nu}{c} + 0 = \dfrac{h\nu'}{c}\cos\theta + mv\cos\varphi \tag{1.57}$

We can see later that the energy of the scattered electron $m_0 c^2 \left[\dfrac{1}{\sqrt{1 - \dfrac{v^2}{c^2}}} - 1\right] \tag{1.58}$

Momentum of the incident photon $= h\nu/c$ along AX

Momentum of the scattered photon $= h\nu'/c$ along AC

Momentum of the electron $= \dfrac{m_0 v}{\sqrt{1 - \left(\dfrac{v^2}{c^2}\right)}}$ along AB $\tag{1.59}$

Component of the momentum of the scattered photon along X-direction is

$$= \frac{h\nu'}{c}\cos\theta \tag{1.60}$$

Component of the momentum of the scattered electron in the X-direction is

$$= \frac{m_0 v}{\sqrt{1 - \left(\dfrac{v^2}{c^2}\right)}} \cos \varphi \qquad (1.61)$$

Sum of the components of momentum of the scattered photon and the electron in the X-axis

$$= \frac{hv'}{c} \cos \theta + \frac{m_0 v}{\sqrt{1 - \left(\dfrac{v^2}{c^2}\right)}} \cos \varphi \qquad (1.62)$$

But the momentum of the incident photon along X-axis = hv/c

Hence
$$hv/c = \frac{hv'}{c} \cos \theta + \frac{m_0 v}{\sqrt{1 - \left(\dfrac{v^2}{c^2}\right)}} \cos \varphi \qquad (1.63)$$

Sum of the components of momentum of the scattered photon and the scattered electron in the Y-axis

$$= \frac{hv'}{c} \sin \theta - \frac{m_0 v}{\sqrt{1 - \left(\dfrac{v^2}{c^2}\right)}} \sin \varphi \qquad (1.64)$$

As the initial momentum of the incident photon in the Y-axis is zero, we get

$$\frac{hv'}{c} \sin \theta - \frac{m_0 v}{\sqrt{1 - \left(\dfrac{v^2}{c^2}\right)}} \sin \varphi = 0 \qquad (1.65)$$

Substituting $\lambda = c/v$, $\lambda' = c/v'$ and $v/c = \beta$ in the principle of conservation of energy,

$$hv + m_0 c^2 = hv' + mc^2$$

we have $\dfrac{hc}{\lambda} = \dfrac{hc}{\lambda'} + mc^2 - m_0 c^2,$ $\dfrac{hc}{\lambda} = \dfrac{hc}{\lambda'} + \dfrac{m_0 c^2}{\sqrt{1 - \dfrac{v^2}{c^2}}} - m_0 c^2$

we have substitued above, the energy of the recoil electron $mc^2 = \dfrac{m_0 c^2}{\sqrt{1 - \left(\dfrac{v^2}{c^2}\right)}}$

Therefore
$$\frac{h}{\lambda} = \frac{h}{\lambda'} + m_0 c \left(\frac{1}{\sqrt{1-\beta^2}} - 1 \right) \tag{1.66}$$

where $\beta = \dfrac{v}{c}$

Thus $\dfrac{h}{\lambda} - \dfrac{h}{\lambda'} = m_0 c \left(\dfrac{1}{\sqrt{1-\beta^2}} - 1 \right) = \dfrac{m_0 c}{\sqrt{1-\beta^2}} - m_0 c \tag{1.67}$

$$\frac{h}{\lambda} - \frac{h}{\lambda'} + m_0 c = \frac{m_0 c}{\sqrt{1-\beta^2}} \tag{1.68}$$

Squaring $\left(\dfrac{h}{\lambda} - \dfrac{h}{\lambda'} \right)^2 + m_0^2 c^2 + 2 m_0 c \left(\dfrac{h}{\lambda} - \dfrac{h}{\lambda'} \right) = \dfrac{m_0^2 c^2}{1-\beta^2} \tag{1.69}$

$$\left(\frac{h^2}{\lambda^2} + \frac{h^2}{\lambda'^2} - \frac{2h^2}{\lambda\lambda'} \right) + 2 m_0 c \left(\frac{h}{\lambda} - \frac{h}{\lambda'} \right) = \frac{m_0^2 c^2}{1-\beta^2} \qquad m_0^2 c^2 = \frac{m_0^2 c^2 \beta}{1-\beta^2} \tag{1.70}$$

We have from equation (1.63)

$$\frac{h\nu}{c} = \left(\frac{h\nu'}{c} \right) \cos\theta + \frac{m_0 v}{\sqrt{1 - \left(\dfrac{v^2}{c^2} \right)}} \cos\varphi \tag{1.71}$$

$$\frac{h}{\lambda} = \left(\frac{h}{\lambda'} \right) \cos\theta + \frac{m_0 v}{\sqrt{1-\beta^2}} \cos\varphi \tag{1.72}$$

$$\frac{h}{\lambda} - \left(\frac{h}{\lambda'} \right) \cos\theta = \frac{m_0 v}{\sqrt{1-\beta^2}} \cos\varphi \tag{1.73}$$

From equation (1.65)

$$\frac{h\nu'}{c} \sin\theta - \frac{m_0 v}{\sqrt{1 - \left(\dfrac{v^2}{c^2} \right)}} \sin\varphi = 0 \tag{1.74}$$

$$\left(\frac{h}{\lambda'}\right)\sin\theta - \frac{m_0 v}{\sqrt{1-\beta^2}}\sin\varphi = 0 \tag{1.75}$$

$$\left(\frac{h}{\lambda'}\right)\sin\theta = \frac{m_0 v}{\sqrt{1-\beta^2}}\sin\varphi \tag{1.76}$$

Squaring (1.73) and (1.76) and adding

$$\frac{h^2}{\lambda^2} + \frac{h^2}{\lambda'^2}\cos^2\theta - \frac{2h^2}{\lambda\lambda'}\cos\theta + \frac{h^2}{\lambda'^2}\sin^2\theta = \frac{m_0^2 v^2 \cos^2\varphi}{1-\beta^2} + \frac{m_0^2 v^2 \sin^2\varphi}{1-\beta^2} \tag{1.77}$$

Therefore $\dfrac{h^2}{\lambda^2} + \dfrac{h^2}{\lambda'^2}(\sin^2\theta + \cos^2\theta) - \dfrac{2h^2}{\lambda\lambda'}\cos\theta$

$$= \frac{m_0^2 v^2 (\sin^2\phi + \cos^2\phi)}{1-\beta^2} \tag{1.78}$$

$$\frac{h^2}{\lambda^2} + \frac{h^2}{\lambda'^2} - \frac{2h^2\cos}{\lambda\lambda'}\cos\theta = \frac{m_0^2 v^2}{1-\beta^2} \tag{1.79}$$

here $v/c = \beta$, $v = \beta c$ and hence,

$$\frac{h^2}{\lambda^2} + \frac{h^2}{\lambda'^2} - \frac{2h^2}{\lambda\lambda'}\cos\theta = \frac{m_0^2 \beta^2 c^2}{1-\beta^2} \tag{1.80}$$

Subtracting (1.80) from (1.70)

$$\left(\frac{2h^2}{\lambda\lambda'}\right)(\cos\theta - 1) + 2\,m_0 c\left(\frac{h}{\lambda} - \frac{h}{\lambda'}\right) = 0 \tag{1.81}$$

$$\left(\frac{2h^2}{\lambda\lambda'}\right)(\cos\theta - 1) = -2\,m_0 c\left(\frac{h}{\lambda} - \frac{h}{\lambda'}\right) \tag{1.82}$$

$$\left(\frac{2h^2}{\lambda\lambda'}\right)(\cos\theta - 1) = -2\,m_0 ch\left(\frac{1}{\lambda} - \frac{1}{\lambda'}\right) \tag{1.83}$$

$$\left(\frac{2h^2}{\lambda\lambda'}\right)(\cos\theta - 1) = 2\,m_0 ch\left(\frac{\lambda' - \lambda}{\lambda\lambda'}\right) \tag{1.84}$$

$$h(\cos\theta - 1) = -m_0 c\,\Delta\lambda \quad \text{where} \quad \Delta\lambda = (\lambda' - \lambda) \tag{1.85}$$

$$\Delta\lambda = \frac{h(1 - \cos\theta)}{m_0 c} = \frac{2h}{m_0 c}\sin^2\frac{\theta}{2} \tag{1.86}$$

$$\lambda' = \lambda + \frac{2h}{m_0 c}\sin^2\frac{\theta}{2} \tag{1.87}$$

The Compton relationship is thus

$$\lambda' - \lambda = \frac{h}{m_0 c}(1 - \cos\theta) = \frac{2h}{m_0 c}\sin^2\frac{\theta}{2}$$

where

λ is the initial wavelength,

λ' is wavelength after scattering

m_0 is the rest mass of the electron,

c is the speed of light and

θ is the scattering angle.

It shows that the wavelength λ' of the scattered photon is greater than the wavelength λ of the incident quantum by an amount $\dfrac{2h}{m_0 c}\sin^2\dfrac{\theta}{2}$.

To obtain in terms of frequency, we use equation 1.86 and write

$$\lambda' - \lambda = \frac{h}{m_0 c}(1 - \cos\theta), \quad \text{i.e.,} \quad \frac{c}{v'} - \frac{c}{v} = \frac{h}{m_0 c}(1 - \cos\theta) \tag{1.88}$$

$$\frac{1}{v'} - \frac{1}{v} = \frac{h}{m_0 c^2}(1 - \cos\theta) \tag{1.89}$$

$$\frac{1}{v'} = \frac{1}{v} + \frac{h}{m_0 c^2}(1 - \cos\theta) = \frac{m_0 c^2 + hv(1 - \cos\theta)}{m_0 c^2 v} \tag{1.90}$$

$$v' = \frac{m_0 c^2 v}{m_0 c^2 + hv(1 - \cos\theta)} = \frac{v}{1 + \left(\dfrac{hv}{m_0 c^2}\right)(1 - \cos\theta)} = \frac{v}{1 + \left(\dfrac{2hv}{m_0 c^2}\right)\sin^2\dfrac{\theta}{2}}$$

$$\text{Thus} \quad v' = \frac{v}{1 + 2\alpha \sin^2 \frac{\theta}{2}} \quad \text{where} \quad \alpha = \frac{h v}{m_0 c^2} \tag{1.91}$$

When $\theta = 90°$, $\cos \theta = 0$ and hence $\Delta\lambda = h/m_0 c$

Substituting the value of

$$h = 6.624 \times 10^{-34} \text{ J.s}$$

$$m_0 = 9.1 \times 10^{-31} \text{ kg}$$

$$c = 3 \times 10^8 \text{ m/s}$$

$$\Delta\lambda = 0.0242 \times 10^{-10} \text{ m}$$

The experimentally observed value of $\Delta\lambda = 0.0236 \times 10^{-10}$ m.

The equation (1.86) shows $\Delta\lambda$ is independent of the wavelength of the incident photon.

The above theory does not explain the behaviour of those photons without change in wavelength. This is because that an atom contains free electrons and bound electrons. When the incident photon strikes a free electron, wavelength of the scattered photon is more than the incident photon. When the incident photon strikes a bound electron in the atom, the electron is not detached from the atom, the photon experiences negligible loss of energy and the scattered photon has the same wavelength or frequency as that of the incident photon.

The recoil electron produced in the scattering process acquires kinetic energy equal to the amount of energy lost by the photon, viz., $(hv - hv')$.

Using (1.63) and (1.65)

$$\tan \varphi = \frac{hv' \sin \theta}{hv - hv' \cos \theta} = \frac{v' \sin \theta}{v - v' \cos \theta} \tag{1.92}$$

But

$$v' = \frac{v \sin \theta}{1 + 2\alpha \sin^2 \frac{\theta}{2}}$$

Therefore

$$\tan \varphi = \frac{\dfrac{v \sin \theta}{1 + 2\alpha \sin^2 \frac{\theta}{2}}}{v - \dfrac{v \cos \theta}{1 + 2\alpha \sin^2 \frac{\theta}{2}}} = \frac{\sin \theta}{1 + 2\alpha \sin^2 \frac{\theta}{2} - \cos \theta}$$

$$= \frac{2\sin\dfrac{\theta}{2}\cos\dfrac{\theta}{2}}{1+2\alpha\sin^2\dfrac{\theta}{2}-\left(1-2\sin^2\dfrac{\theta}{2}\right)} = \frac{2\sin\dfrac{\theta}{2}\cos\dfrac{\theta}{2}}{2(1+\alpha)\sin^2\dfrac{\theta}{2}}$$

$$\therefore \tan\varphi = \frac{1}{(1+\alpha)\tan\dfrac{\theta}{2}} \tag{1.93}$$

This shows that as θ varies from 0 to 180, φ varies from 180 to zero

The energy of the recoil electron $W = h\nu - h\nu' = h\nu\left(1-\dfrac{1}{1+2\alpha\sin^2\dfrac{\theta}{2}}\right)$

$$W = h\nu \times \frac{2\alpha\sin^2\dfrac{\theta}{2}}{1+2\alpha\sin^2\dfrac{\theta}{2}}$$

From equation (1.93) $\tan\dfrac{\theta}{2} = \dfrac{\sin\dfrac{\theta}{2}}{\cos\dfrac{\theta}{2}} = -\dfrac{1}{(1+\alpha)\tan\varphi}$

$$\frac{\cos\dfrac{\theta}{2}}{\sin\dfrac{\theta}{2}} = -(1+\alpha)\tan\varphi, \qquad \frac{\cos^2\dfrac{\theta}{2}}{\sin^2\dfrac{\theta}{2}} = (1+\alpha)^2\tan^2\varphi,$$

$$\frac{\cos^2\dfrac{\theta}{2}}{\sin^2\dfrac{\theta}{2}} + 1 = (1+\alpha)^2\tan^2\varphi + 1$$

$$\frac{1}{\sin^2\dfrac{\theta}{2}} = (1+\alpha)^2\tan^2\varphi + 1 \tag{1.94}$$

$$\sin^2\dfrac{\theta}{2} = \frac{1}{(1+\alpha)^2\tan^2\varphi + 1}$$

$$W = hv \times \frac{2\alpha \sin^2 \frac{\theta}{2}}{1 + 2\alpha \sin^2 \frac{\theta}{2}} = hv \times \frac{\dfrac{2\alpha}{(1+\alpha)^2 \tan^2 \phi + 1}}{1 + \dfrac{2\alpha}{(1+\alpha)^2 \tan^2 \phi + 1}} = hv \times \frac{2\alpha}{(1+\alpha)^2 \tan^2 \phi + 2\alpha + 1} \qquad (1.95)$$

Example 1   An X-ray photon of wavelength 0.6 Å scattered from a block of carbon is viewed at an angle of 90° to the incident beam. Find the (1) Compton shift (2) Kinetic energy of the recoil electron.

Solution

1. $\Delta\lambda = \dfrac{h(1 - \cos\theta)}{m_0 c}$

    Here $\lambda = 0.6$ Å $= 6 \times 10^{-11}$ m and $\theta = 90°$ and $m_0 = 9.1 \times 10^{-31}$ kg

$$\Delta\lambda = h / m_0 c = \frac{6.624 \times 10^{-34}}{9.1 \times 10^{-31} \times 3 \times 10^8}$$

$$= 0.0242 \times 10^{-10} \text{m} = 0.0242 \text{ Å}$$

2. Energy of the recoil electron

$$U = hv - hv' = \frac{hc}{\lambda} - \frac{hc}{\lambda'} = \frac{hc[\lambda' - \lambda]}{\lambda\lambda'}$$

$$= hc \frac{\Delta\lambda}{\lambda^2} = \frac{6.624 \times 10^{-34} \times 3 \times 10^8 \times 0.0242 \times 10^{-10}}{(6 \times 10^{-11})^2}$$

$$= 1.34 \times 10^{-16} \text{J} = \frac{1.34 \times 10^{-16}}{1.6 \times 10^{-19}} = 837.5 \text{ eV}$$

1.20 WAVE NATURE OF MATTER

Nature manifests itself in two fundamental forms, viz., matter and energy. We know that optical phenomena like reflection, refraction, interference, diffraction and polarization are explained on the basis of Maxwell's electromagnetic theory. But the photoelectric effect could be explained only on the basis of photon theory. The Planck's quantum theory and Einstein's photoelectric effect made it clear that light possess particle character as well. Thus radiant energy has dual character, particle and wave. In view of the dual nature of light scientists were looking for a theory which could be developed in which electrons behave as a particle as well as wave. In the process the wave mechanics or quantum mechanics came into existence.

In 1924 the French physicist Louis de Broglie put forward suggestions that matter which is ordinarily considered as made up of discrete particles—molecules, atoms, protons, electrons, etc. might exhibit wavelike properties under appropriate conditions. This means that matter, like radiation has a dual nature. Thus he put forward the concept of matter waves.

1.20.1 de Broglie Wavelength

de Broglie theory can be summed up in the very simple but fundamental equation

$$\lambda = \frac{h}{mv} \text{ or } \frac{h}{p} \tag{1.96}$$

Thus whenever a particle of mass m is moving with a velocity v, it can be associated with a matter wave of wavelength given by the above equation. Just as the light waves are known to have corpuscular properties, the particles are having wave properties as well, as proposed by the above equation. An electron having the mass $m = 9.1 \times 10^{-31}$ kg having a velocity 10^6 m/s would have a wavelength

$$\lambda = \frac{h}{mv} = \frac{6.62 \times 10^{-34} \, \text{Js}}{9.1 \times 10^{-31} \, \text{kg} \times 10^6 \, \text{ms}^{-1}} = 7 \times 10^{-10} \text{ m} = 70 \text{ nm} \tag{1.97}$$

Thus the order of magnitude of wavelength shows that the wave effect can be felt in electron. For heavier particles, however, the wavelength will be too small to have observable wave effects. A particle of 1 g moving with a velocity 10 cm/s have wavelength

$$\lambda = \frac{h}{mv} = \frac{6.62 \times 10^{-34} \, \text{Js}}{1.0 \times 10^{-3} \, \text{kg} \times 10^{-1} \, \text{ms}^{-1}} = 6.62 \times 10^{-30} \text{ m} \tag{1.98}$$

Thus electrons, protons, atoms, molecules and thus matter under suitable conditions can be diffracted in much the same manner as light and sound. The wavelength associated with any moving particle is inversely proportional to the momentum mv. de Broglie further argued that the electron in a stationary orbit of Bohr may be regarded as a stationary wave; otherwise it may destroy itself by destructive interference. And for a stationary wave to exist along an orbit, its circumference must contain an integral number of wavelengths. An illustration of matching and mismatching of de Broglie wave travelling in Bohr orbit is shown in Figure 1.14.

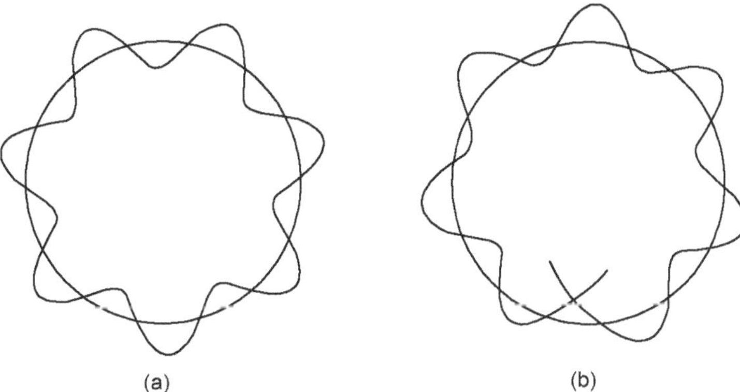

Figure 1.14   Matching (a) and mismatching (b) of the de Broglie wave in Bohr orbit

If the wavelengths of the de Broglie waves are such that an integral number of them fit around the circle , then they match after a complete revolution (Figure 1.14a). Figure 1.14b shows the case when the wave does not match after a complete revolution.

This condition requires that

$$2\pi r = n\lambda \quad n, \quad 1,2,3, \cdots \tag{1.99}$$

From de Broglie's theory, $\lambda = h/mv$ and hence $2\pi r = n\dfrac{h}{mv}$

$$mv\, r = n\frac{h}{2\pi} \tag{1.100}$$

which was the first postulate in Bohr's theory. It should be noted that Bohr postulated it even before the de Broglie's theory.

Considering wave aspect, $E = hv$

Considering particle aspect, $E = mc^2$

$$E = hv = mc^2$$

$$m = \frac{hv}{c^2} \tag{1.101}$$

Momentum of the photon $p = mc = h\dfrac{v}{c} = \dfrac{h}{\lambda}$ $\qquad\qquad$ (1.102)

Extending wave–particle duality of matter we say that if we have a particle of mass m moving with a velocity v then the momentum of the particle $p = mv = \dfrac{h}{\lambda}$

$$\text{or} \quad \lambda = \frac{h}{p} = \frac{h}{mv} \tag{1.103}$$

This equation is called de Broglie wave equation.

In fact Bohr's theory of quantization of energy in hydrogen atom was considered to be quite arbitrary. Bohr has given no reason as why only discrete values of energy were allowed for electrons in the atom. Quantization is of course an inherent phenomenon in wave motion. A string fixed at the two ends can vibrate at frequencies $v, 2v, 3v$ and so on. We can consider that the frequencies here are quantized. Vibration is synonymous to a wave motion.

When a beam of X-rays is directed at a crystalline substance, the beam is scattered in a definite manner characteristic of the atomic structure of the crystalline substance. This phenomenon is called X-ray diffraction and occurs because the interatomic spacings in the crystal are about the same as the wavelength of the X-rays. The X-rays scatter and make rings of different diameters. The distance between the rings are determined by the interatomic spacings in the metal. An electron when diffracted for example by an aluminium foil also shows rings similar to the X-ray diffraction. The electrons do behave analogously in this experiment. The wavelike property of electrons is used in electron microscopes.

An interesting historical antecedent in the concept of wave–particle equality of matter is that the first person to show that the electron was a subatomic particle was the English physicist Sir Joseph J. Thomson (J.J. Thomson) in 1895 and then his son Sir George P. Thomson (G.P. Thomson) was among the first to show experimentally in 1926 that the electron could act as a wave. The father won Nobel prize for showing the electron as a particle in 1906 and the son won the Nobel prize in 1937 for showing that it is a wave.

Example 2   Calculate the wavelength of electrons that have been accelerated through a potential difference of 100 kV.

We have to first calculate the linear momentum p of the electron as we know that $\lambda = h / p$. When an electron is accelerated through a potential difference V, the energy acquired by the electron is eV. At the end of the period of acceleration the acquired energy is converted into the form of kinetic energy. The kinetic energy is $p^2/2m$ where m is the mass of the electron. Thus $p^2/2m = eV$

$$p = (2\ meV)^{1/2} \text{ and hence } \lambda = h / (2\ meV)^{1/2}$$

$$\lambda = \frac{6.626 \times 10^{-34}\,\text{Js}}{(2 \times 9.109 \times 10^{-31}\,\text{kg} \times 1.609 \times 10^{-19}\,\text{C} \times 100.0 \times 10^3\,\text{V})^{1/2}} = 0.038\ \text{Å}$$

For 40 kV it is $6.1 \times 10^{-12}\,\text{m} = 0.061\text{Å}$

1.21 ELECTRON WAVE

Let us consider an electron of mass m and charge e being accelerated through an electric potential V volt. If v is the velocity acquired by the electron then

$$eV = \frac{1}{2}mv^2 \text{ and hence } v = \sqrt{\frac{2eV}{m}} \tag{1.104}$$

Wavelength associated with the electron $\lambda = \dfrac{h}{mv} = \dfrac{h}{m\sqrt{\dfrac{2eV}{m}}} = \dfrac{h}{\sqrt{2meV}}$ $\tag{1.105}$

If we substitute the values of *h, e* and *m*

$$h = 6.625 \times 10^{-34} \text{ J.s}$$

$$e = 1.6 \times 10^{-19} \text{ Coulombs}$$

$$m_0 = 9.1 \times 10^{-31} \text{ Kg}$$

then $\lambda = \sqrt{\dfrac{150}{V}} \, \text{Å} = \dfrac{12.3}{\sqrt{V}} \, \text{Å}$

1.22 DAVISSON AND GERMER EXPERIMENT

In 1925, Davisson and Germer conducted experiments in low-voltage electron diffraction and established the wave nature of the electron and confirmed the theory put forward by de Broglie upon which wave mechanics is based. At almost the same time (in 1927) G.P. Thomson, son of J.J. Thomson discovered electron diffraction using much higher-voltage electrons[6]. G.P. Thomson shared the Nobel Prize with Davisson in 1937 for these discoveries. As mentioned earlier that it is interesting to note that J.J. Thomson got the Nobel Prize for discovering that electrons are particles, and G.P. Thomson got it for discovering that they aren't.

Davisson and Germer Experiment was the first experiment to verify the de Broglie's relation. They later studied the electron scattering from various materials. The basic experimental set-up is shown in Figure 1.15.

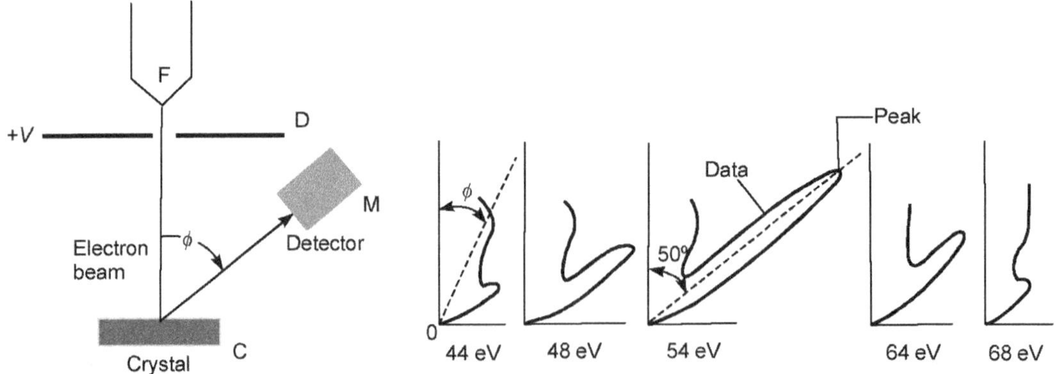

Figure 1.15   Electron diffraction using low voltage

Electrons from a filament F are collimated into a fine beam by applying a positive potential to the cylinder D. The kinetic energy of electrons was controlled by the voltage V applied to the plate D. A fine narrow beam of electrons is incident on the nickel crystal C. The electrons are reflected and the intensity of the reflected electrons is measured at various angles with the help of a current meter M. Different positive potentials are applied to the cylinder D and the velocity of the electrons were measured. The curve represents a polar graph showing the intensity of the reflected electrons at various angles. It was found that the intensity was maximum at 50° for a critical energy of 54 eV. By altering the potential applied to the cylinder D, the energy of the electrons can be changed. It was found that the nature of the intensity graph is similar and the maximum intensity (spur) occurred always at 50°. This experiment thus showed that the electrons behave like matter waves as suggested by de Broglie.

Their important discovery was made when nickel was used as the target (Figure 1.16).

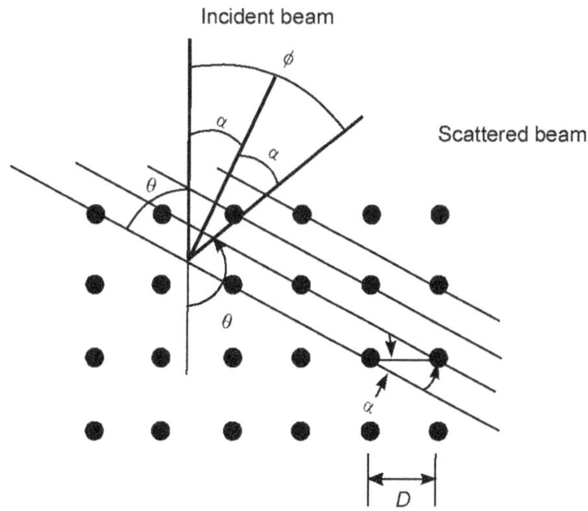

Figure 1.16   Electron diffraction by a metal

When we apply Bragg's condition, the maximum constructive interference occurs when

$$2d\sin\theta = m\lambda, \ m = 1,2,3,\cdots$$

From the Figure1.16, we can see that $\phi = 2\alpha \ = \ \pi - 2\theta$

Hence $\sin\theta = \cos\alpha$ and the de Broglie condition can also be expressed in terms of α and is

$$2d\cos\alpha = m\lambda, \ m = 1,2,3, \cdots$$

As $d = D\sin\alpha$, $2d\cos\alpha = 2D\sin\alpha\cos\alpha = D\sin 2\alpha = D\sin\varphi = m\lambda$

Thus $D\sin\varphi = m\lambda$ or $\lambda = \dfrac{D\sin\phi}{m}$

Davisson and Germer found that the peak intensity in nickel is at $\varphi = 50°$ and its $D = 0.215$ nm. With $m = 1$, the electron wavelength $\lambda = 0.167$ nm . It was found that the maximum intensity corresponds to $V = 54$ and hence the kinetic energy $K = 54$ eV and hence $mc^2 = \sqrt{2mc^2 K}$.

$$mc^2 = pc = \sqrt{2 \times 0.511 \times 10^6 \, \text{eV} \times 54 \text{eV}} = 7.429 \text{ keV}$$

$$\lambda = \frac{h}{p} = \frac{hc}{pc} = 0.167 \text{ nm}$$

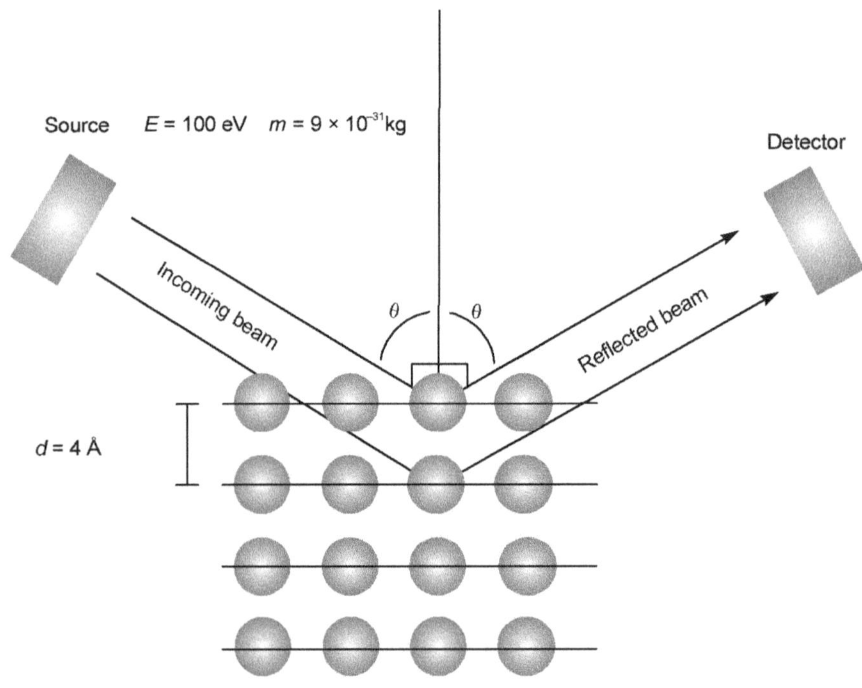

Figure 1.17   Electron scattering

1.23 GP THOMSON EXPERIMENT

The experimental arrangement is as shown in Figure 1.18. A beam of cathode rays is produced in a discharge tube by means of an induction coil. The electron beam is passed through a diaphragm tube to obtain a narrow beam of electrons which is then allowed to fall on a very thin metallic film F of gold, aluminium, etc. The film is normally of thickness of 10^{-6} cm. A photographic plate is placed and is received after the electrons pass through the metallic film. A fluorescent screen can also be used for visual examination of the passage of electrons through the metal foil. The camera part is kept under high vacuum and air is allowed in the discharge tube section. A symmetrical pattern of concentric rings about a central spot is obtained which is due to the diffraction electrons by the metallic foil.

The apparatus is evacuated with the help of a vacuum pump. The filament F acts as a source of electrons. The electrons from the filament are accelerated through a potential difference of 50 kV by applying a positive potential to the cylinder D. The energy of each electron is 50 keV. The beam of electrons is incident on a thin gold foil G and after passing through the foil, it is incident on a photographic plate P. The pattern obtained on the photographic plate consists of concentric circular rings of varying intensity (See Figure 1.19). If the electrons behaved as particles, these particles should have been scattered through a wide angle while passing through the foil. The pattern obtained here is similar to the Laue pattern obtained with X-rays, which is possible only if the electrons behave as waves.

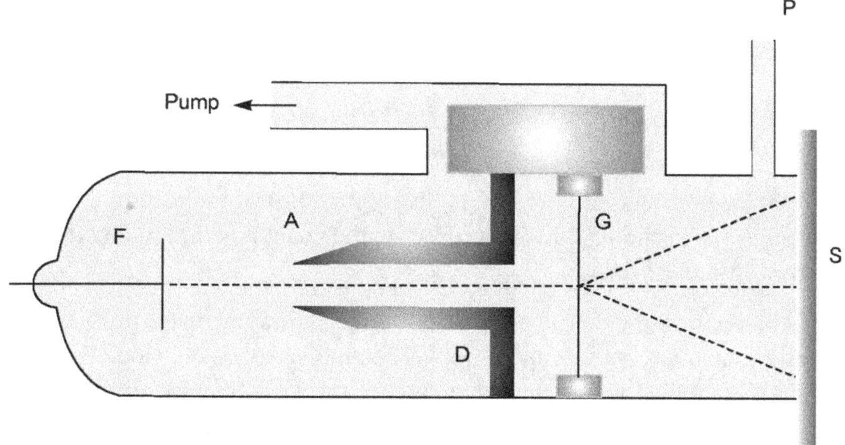

Figure 1.18   G P Thomson experiment

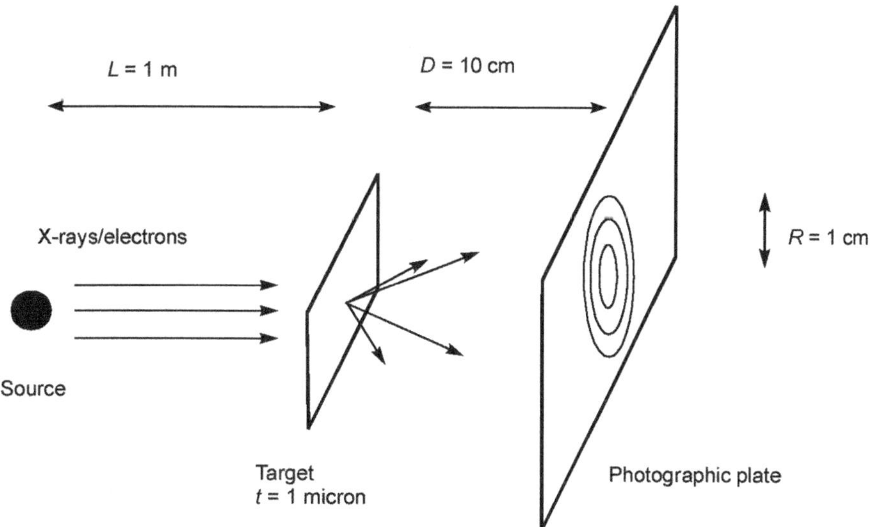

Figure 1.19   Experiment showing that the electrons behave as waves

To make sure that the pattern is produced by diffracted electrons and not by secondary X-rays generated by the electrons in their passage through the foil, a magnetic field can be applied which would clearly show a total shift of the rings which cannot happen if the diffraction patterns were by secondary X-rays. Hence we may conclude from the above experiment that the electrons also behave as waves.

1.24 HEISENBERG UNCERTAINTY PRINCIPLE

In the experiments for the determination of the specific charge of an electron, it was seen that the electrons are deflected by the application of electric and magnetic fields. It was hence shown that the electrons behave as a particle and it has definite mass, momentum and energy, which can be measured with desired accuracy. But the experiment of Davisson and Germer and of G.P. Thomson showed clearly the wave nature of electrons and the wavelength of the electron is

$$\lambda = \frac{h}{mv} = \frac{h}{p}$$

where h is the Planck's constant and $p = mv$ is the momentum of the electron. As we know that a wave extends through space and it is difficult to locate the exact position of an electron behaving as a wave at any given instant of time.

Thus the dual behaviour of electron as a particle and as a wave creates problems in locating the exact position and momentum at the same time. When an electron is considered as a wave, it is not possible to know its exact position as the wave extends throughout a certain region in space. Thus the following question arises, "If the electron is exhibiting dual nature (wave and particle) is it possible to know the exact position of the electron in space at some given instant?" In order to overcome the difficulty, Heisenberg proposed his uncertainty principle, which is known as Heisenberg's uncertainty principle.

The principle states that the exact position and momentum of a particle (say electron) cannot be determined simultaneously with desired accuracy. Thus "it is impossible to determine simultaneously the position and momentum of electron with any desired accuracy".

1.24.1 Mathematical Relations

If Δx is the error in determining the position and Δp the error in determining the momentum at the same instant, then these quantities are related by the following relation:

$$\Delta x\, \Delta p \approx \frac{h}{2\pi}$$

The product of uncertainty in the simultaneous determination of the position and momentum of a particle is equal to or greater than $\hbar$. The above relation is of fundamental nature because it sets a limit in the accurate and simultaneous measurement of position and momentum. From the above expression we can see that when Δx is small Δp is large and vice versa as $\Delta x \cdot \Delta p \sim \dfrac{h}{2\pi}$. It means that if one quantity is measured accurately, the other quantity is less accurate. The above equation implies that as we try to locate more closely a particle, the greater is the uncertainty in its momentum and vice versa.

Due to the small value of h, this kind of uncertainty relation has not been found relevant in macroscopic world. We can also put the relation as

$$\Delta x \cdot \Delta p \ \approx \frac{h}{2\pi} \quad \text{or} \quad \Delta x \cdot \Delta p \geq \frac{1}{2}\frac{h}{2\pi} \tag{1.106}$$

Thus the product of the uncertainties is of the order of the Planck's constant.

We can also put the principle in terms of the uncertainties in time and energy.

The uncertainty principle states that the position and velocity cannot both be measured, exactly, at the same time (actually pairs of position, energy and time).

Dividing and by multiplying by v (velocity) the above expression,

$$\frac{\Delta x}{v} \cdot \Delta pv = \frac{\Delta x}{v} \cdot \Delta mv \cdot v \ = \ \Delta t \cdot \Delta E \approx \frac{h}{2\pi} \quad \text{or} \quad \Delta t \cdot \Delta E \ \geq \frac{1}{2}\frac{h}{2\pi} \tag{1.107}$$

ΔE represents the uncertainty in the measurement of energy of a particle and Δt, the uncertainty in the measurement of time. A similar argument holds well for in the measurement of energy and time as in the case of position and momentum. As we try to determine the energy more exactly, the greater is the uncertainty in the time and vice versa. This is more important when we consider the life time of an electron (simply called the life time of the energy state) in an energy state and its energy. If the life time is large, the energy state is narrow and the transition observed will be a narrow spectral line.

A simple mathematical formulation of the uncertainty principle can be obtained by considering the diffraction of a beam of electrons through a single slit—the so called single slit experiment. Before doing so let us consider an imaginary double-slit diffraction experiment in which a beam of electrons is passed through a screen having two slits for the electron to pass. The other part of the screen is opaque for the electron. When we close the first slit and observe the passage of the electron through the other slit, we obtain an intensity pattern on a screen kept behind the first screen containing the slits. In the same manner we close the second slit and observe the intensity pattern on the screen by the passage of the electron through that slit. On opening both slits at the same time we expect an intensity pattern, which is the superposition of the two intensity patterns observed earlier when one slit is opened according to classical mechanics. This is so because the electrons moving through one slit do not have any influence on the electrons passing through the other slit. However, the phenomenon of electron diffraction gives a different pattern, which does not, at all, correspond to a simple superposition of patterns of the single slit. This observation is against the usual knowledge in the classical picture of the electrons moving in well-defined paths. In quantum mechanics there is no concept of a well-defined path for electrons. This forms the basis of uncertainty principle which states that "there exists no predictive law which gives information about the simultaneous position and momentum of a particle".

A mathematical formulation for the uncertainty principle can be obtained from the diffraction pattern of electron beams through a single slit experiment.

In Figure 1.20 consider a beam of electrons travelling in the direction as shown.

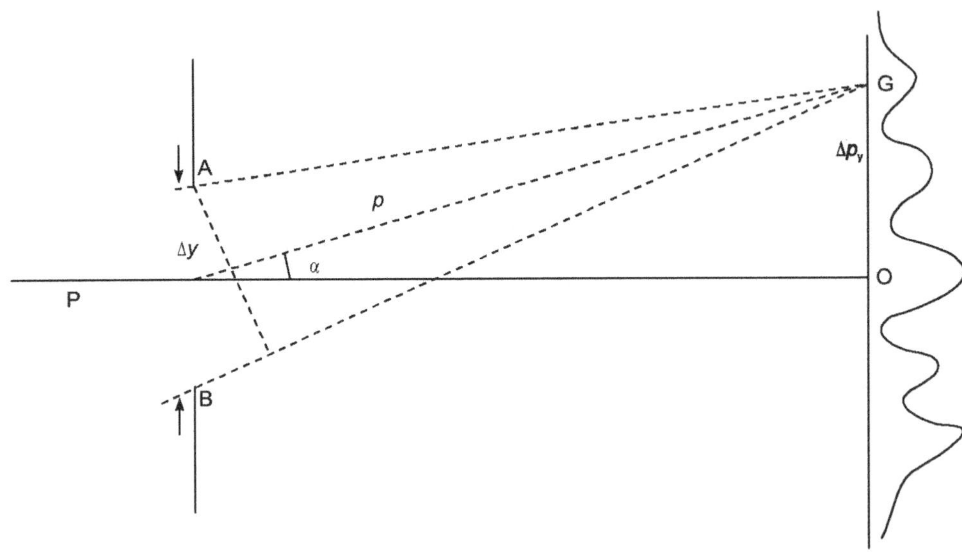

Figure 1.20   Diffraction of an electron at a slit

The electron is allowed to pass through a slit AB of width Δy kept perpendicular in its path. Before entering the slit, the electron has a definite momentum $p = mv$. While passing through the slit, the electron gets diffracted and acquires a momentum along OG . The angular deflection θ depends upon the component of the momentum parallel to the slit, i.e., $\Delta p = p\ \sin\theta = p\theta$ as we consider small angle deflection. For the first minimum of the diffraction pattern the angle θ_0 is

$$\sin\theta = \frac{\lambda}{\Delta y}$$

$$\theta = \frac{\lambda}{\Delta y} \text{ (for small values of } \theta_0)$$

Here Δy is the width of the slit. Hence the error in the measurement of the position of the particle is Δy.

We shall now consider the particle incident on the screen at its first minimum position of the diffraction pattern. The particle can fall anywhere between this minimum and the centre of the screen. If p is the momentum of the particle in the extreme y-direction which makes an angle a as shown in the Figure 1.20. Then $\sin\alpha = \dfrac{\lambda}{\Delta y}$.

Component of the momentum in y direction $\Delta p_y = p \sin\alpha = p\dfrac{\lambda}{\Delta y}$.

If Δp_y is the error in momentum, the product of two errors $\Delta y \cdot \Delta p_y = p \cdot \lambda \approx p \cdot \dfrac{h}{p} \approx h$

The Planck's constant h is larger than $\hbar$ and hence $\Delta y \cdot \Delta p \geq \dfrac{h}{2\pi} \geq \dfrac{1}{2}\dfrac{h}{2\pi}$

$$\Delta y = \frac{\lambda}{\theta_0}$$

$$\Delta p \; \Delta y \; \sim \; p\lambda \cdot \frac{\theta}{\theta_0}$$

But $p\lambda = h$

Hence $\Delta p \; \Delta y = h \cdot \dfrac{\theta}{\theta_0}$

Considering $\theta = \theta_0$

$$\Delta p \; \Delta y \sim h$$

It is obvious that the probable deflection θ of the electron is less than θ_0 and according to Heisenberg, the uncertainty relationship

$$\Delta p \; \Delta y \geq \frac{1}{2}\frac{h}{2\pi} \tag{1.108}$$

Thus any experimental measurement determining the position or momentum introduces an uncertainty in the other. The instrument cannot measure the quantities more accurately than predicted by Heisenberg's uncertainty principle.

Thus the uncertainty principle envisages that the product of uncertainties in determining the position and momentum of the particle can never be smaller than the number of the order $\dfrac{1}{2}\dfrac{h}{2\pi}$.

These are the important expressions

$$\Delta x \; \Delta p \geq \frac{1}{2}\frac{h}{2\pi}$$

$$\Delta t . \; \Delta E \geq \frac{1}{2}\frac{h}{2\pi}$$

$$\Delta J \cdot \Delta \varphi \geq \frac{1}{2}\frac{h}{2\pi}$$

ΔJ is the error in the measurement of the angular momentum and $\Delta\varphi$ is the uncertainty in angular displacement. Here, y is changed as x to suit the usual expressions.

1.25 WAVE–PARTICLE DUALITY

Though the behaviour of electrons may be regarded as a wave in diffraction experiments, the electrons are best described as particles in many experiments like observation of electron tracks in cloud chambers, impact of electrons on fluorescent screens, photoelectric effect and so on. These may look like puzzles. Thus the electrons themselves are not waves, but their motion can be represented by wave equations which are to be regarded as methods of calculating probability of finding the electrons on a photographic plate or a fluorescent screen. Dual nature of photons and electrons may be believed to come from the variation in their behaviour caused by the act of observation rather than they changing their properties according to the method used by the observer. By an observation in quantum mechanics we measure a process of interaction between classical objects and quantum objects. Objects are large in dimension classically and quantum objects are small in dimensions. For example electrons or photons are quantum objects. In carrying out a measurement on the quantum object, the very process of measurements affect the quantum object and it is impossible to make the effect arbitrarily small for a given accuracy of measurements. As we proceed for more accuracy in measurements, the stronger is the effect of measurement. For a low accuracy the effect of measuring process on the quantum object is small.

Light has properties of both a wave and a particle. Thus photons act both as a wave and a particle all the time (even though it looks like a paradox, to say that it's "sometimes a wave and sometimes a particle". It depends upon which features are more obvious at a given time).

We have seen wave–particle duality (or particle–wave duality). Photons, though treated as particles, can be calculated to have frequency, wavelength, amplitude, and other properties inherent in wave mechanics.

SOLVED PROBLEMS

1. The limits of the microwave region are approximately 1 GHz and 100 GHz and that of millimetre region is approximately 100 GHz to 300 GHz. Convert these frequencies into wavelengths.

Solution

$$1 \ \text{GHz} \qquad \lambda = \frac{c}{v} = \frac{3\times10^8\,\text{ms}^{-1}}{1\times10^9\,\text{s}^{-1}} = 0.3 \ \text{m}$$

$$100 \ \text{GHz} \qquad = \frac{3\times10^8\,\text{ms}^{-1}}{100\times10^9\,\text{s}^{-1}} = 3 \ \text{mm}$$

$$300 \ \text{GHz} \qquad = \frac{3\times10^8\,\text{ms}^{-1}}{300\times10^9\,\text{s}^{-1}} = 1 \ \text{mm}$$

2. What is the energy of 1 photon of radiation of the following frequencies? (a) 4.6 GHz (b) 2450 MHz (c) Wave number of 37000 cm^{-1}. What is the energy of 1 mole of these photons?

Solution

(a) $E = hv = 6.6 \times 10^{-34}$ J.s $\times 4.6 \times 10^9$ s$^{-1} = 3.04 \times 10^{-24}$ J

Avagadro number $N_A = 6.02214199 \times 10^{23}$ mol^{-1}

For 1 mol, energy $E = 3.04 \times 10^{-24}$ J $\times 6.02214199 \times 10^{23}$ mol^{-1}

$$= 1.8 \text{ J}$$

(b) $E = hv = 6.6 \times 10^{-34}$ J.s $\times 2.45 \times 10^9$ s$^{-1} = 1.62 \times 10^{-24}$ J

For 1 mol, energy $E = 1.62 \times 10^{-24}$ J $\times 6.02214199 \times 10^{23}$ mol^{-1}

$$= 0.97 \text{ J}$$

(c) $E = hc\,v = 6.6 \times 10^{-34}$ J.s $\times 3 \times 10^{10}$ cm s$^{-1} \times 37000 \times$ cm^{-1}

$$= 7.3 \times 10^{-19} \text{ J}$$

For 1 mol, energy $E = 7.3 \times 10^{-19}$ J $\times 6.02214199 \times 10^{23}$ mol^{-1}

$$= 4.4 \times 10^5 \text{ J} = 440 \text{ kJ mol}^{-1}$$

3. Calculate the number of photons emitted by a 200 watt yellow lamp in 1 second. (Wavelength of the yellow lamp is 560 nm).

Solution

Energy of one photon is hv

Number of photons needed to produce an energy of $E = E/hv$

$E =$ Power $\times$ Time

Thus the number of photons $N = E/hv = E/h(c/\lambda) = \lambda Pt/hc$

$$= \frac{(5.60 \times 10^{-7} \text{ m}) \times (200 \text{ Js}^{-1}) \times (1.0 \text{ s})}{(6.626 \times 10^{-34} \text{ Js}) \times (2.998 \times 10^8 \text{ ms}^{-1})} = 5.6 \times 10^{20}$$

4. An X-ray tube produces a continuous spectrum of radiation with its short wavelength end at 0.45 Å. What is the maximum energy of a photon in the radiation in electron volt.

Solution

Wavelength of photon $\lambda = 0.45$ Å $= 0.45 \times 10^{-10}$ m

Frequency $v = c/\lambda = \dfrac{3 \times 10^8}{0.45 \times 10^{-10}} = 6.66 \times 10^{18}$ Hz

$$E = hv = 6.625 \times 10^{-34} \times 6.66 \times 10^{18} = 44.16 \times 10^{-16} \, \text{J} = \frac{44.16 \times 10^{-16}}{1.6 \times 10^{-19}} \, eV$$

$$= 27.6 \text{ keV} \quad [1 \text{ eV} = 1.6 \times 10^{-19} \text{ J}]$$

5. In an accelerator experiment on high-energy collisions of electrons with positron, a certain event is interpreted as annihilation of an electron–positron pair of total energy 10.2 BeV into two γ-rays of equal energy. What is the wavelength associated with each ray?

Solution

Energy of each γ-ray photon $= 10.2/2 = 5.1 \text{ BeV} = 5.1 \times 10^9 \text{ eV}$

$$= 5.1 \times 10^9 \times 1.6 \times 10^{-19} \text{ J}$$

$$= 8.16 \times 10^{-10} \text{ J}$$

Energy of each γ-ray $= hv = \dfrac{hc}{\lambda} = 8.16 \times 10^{-10} \, \text{J}$

$$\lambda = \frac{hc}{8.16 \times 10^{-10}} = \frac{6.625 \times 10^{-34} \times 3 \times 10^8}{8.16 \times 10^{-10}}$$

$$= 2.435 \times 10^{-16} \text{ m}$$

6. Find out the number of photons emitted per second by a microwave transmitter of 10 kW power emitting radio waves of wavelength 500 m.

Solution

Total energy emitted per second $=$ Power $\times$ Time $= 10 \text{ kW} \times 1 \text{ sec.} = 1 \times 10^4$

Energy of each photon $= hv = hc / \lambda = \dfrac{6.625 \times 10^{-34} \times 3 \times 10^8}{500}$

$$= 39.75 \times 10^{-29} \text{ J}$$

Number of photons emitted per second $= \dfrac{10^4}{39.75 \times 10^{-29}} = 2.51 \times 10^{31}$

7. Find the number of photons entering the pupil of our eye per second corresponding to minimum intensity of white light that human being can perceive (10^{-10} W m^{-2}). If the area of the pupil is about 0.4 cm^2 and the average frequency of the white light is about 6×10^{14} Hz.

Solution

$$v = 6 \times 10^{14} \text{Hz}$$

Energy of the photon $= hv = 6.625 \times 10^{-34} \times 6 \times 10^{14} = 3.975 \times 10^{-19}$

Number of photons crossing 1 m^2 area per second (photon flux)

$$= \frac{\text{Intensity of photon}}{\text{Energy of photon}}$$

$$= \frac{10^{-10}\,\text{Wm}^{-2}}{3.975 \times 10^{-19}\,\text{J}} = 2.5 \times 10^{8}\,\text{m}^{-2}\text{s}^{-1}$$

Number of photons entering the pupil per second

$$= 2.5 \times 10^{8} \times 0.4 \times 10^{-4} = 10^{4} \text{ per second}$$

8. Photoelectric threshold wavelength of sodium is 680 nm. Find the work function of sodium in electron volts.

Solution

$$\lambda_0 = 680 \text{ nm} = 680 \times 10^{-9}\,\text{m}$$

$$\nu_0 = c/\lambda_0 = \frac{3 \times 10^{8}}{680 \times 10^{-9}} = \frac{3 \times 10^{16}}{68}\,\text{Hz}$$

Work function $W = h\,\nu_0 = 6.625 \times 10^{-34} \times \dfrac{3 \times 10^{16}}{68}\,\text{J}$

$$= 6.625 \times 10^{-34} \times \frac{3 \times 10^{16}}{68 \times 1.6 \times 10^{-19}} = 1.827 \text{ eV}$$

9. Radiations of wavelength 3000 Å falls on a surface of work function 2 eV. Calculate the maximum velocity of the ejected electron.

Solution

$$W = 2 \text{ eV} = 2 \times 1.6 \times 10^{-19}\,\text{J}$$

$$\lambda = 3000 \times 10^{-10}\,\text{m}$$

$$\nu = \frac{c}{\lambda} = \frac{3 \times 10^{8}}{3000 \times 10^{-10}} = 10^{15}\,\text{Hz}$$

$$h\nu = W + \frac{1}{2}mv^{2}$$

$$\frac{mv^{2}}{2} = h\nu - W = 6.625 \times 10^{-34} \times 10^{15} - 2 \times 1.6 \times 10^{-19} = 3.425 \times 10^{-19}$$

$$v = \frac{\sqrt{2 \times 3.425 \times 10^{-19}}}{m(kg)}\,m/s = \frac{\sqrt{2 \times 3.425 \times 10^{-19}}}{9.10 \times 10^{-31}} = 8.67 \times 10^{5}\,ms^{-1}$$

10. The surface of a certain metal is illuminated by light of wavelength 300 nm. The photoelectrons emitted are stopped by a retarding potential of 1 volt. Calculate the threshold frequency of the metal.

Solution

$$\lambda = 300 \times 10^{-9}\,\text{m}, \quad v = c/\lambda = \frac{3 \times 10^{8}}{3000 \times 10^{-10}} = 10^{15}\ \text{Hz}$$

Since the photoelectrons are stopped by a retarding potential of 1 volt, the kinetic energy of the electron is 1 eV.

$$mv^2/2 \ = \ 1\,\text{eV} = 1.6 \times 10^{-19}\ \text{J}$$

but

$$hv \ = \ hv_0 + mv^2/2$$

$$v_0 = \frac{\left(hv - \dfrac{1}{2}mv^2\right)}{h} = \frac{6.625 \times 10^{-34} \times 10^{15} - (1.6 \times 10^{-19})}{6.625 \times 10^{-34}}$$

$$= \ 7.58 \times 10^{14}\ \text{Hz}$$

11. The threshold wavelength of certain metal surface is 600 nm. Find the maximum kinetic energy and velocity of the ejected electron when light of wavelength 300 nm falls on it.

Solution

Threshold frequency $v = \dfrac{c}{\lambda} = \dfrac{3 \times 10^{8}}{600 \times 10^{-9}} = 5 \times 10^{14}\ \text{Hz}$

Energy $\qquad hv = 6.625 \times 10^{-34} \times 5 \times 10^{14}\ \text{J} = 3.3125 \times 10^{-19}\,\text{J}$

$$\lambda_0 = 300 \times 10^{-9}\,\text{m}, \quad v_0 = c/\lambda = \frac{3 \times 10^{8}}{3000 \times 10^{-10}} = 10^{15}\,\text{Hz}$$

$$hv = hv_0 + \frac{1}{2}mv^2$$

$$\frac{1}{2}mv^2 = 6.625 \times 10^{-34} \times 10^{15} - 3.3125 \times 10^{-19}$$

$$= \ 3.3125 \times 10^{-19}$$

$$v = \sqrt{\frac{2 \times 3.3125 \times 10^{-19}}{9.1 \times 10^{-31}}} = \ 8.532 \times 10^{5}\,\text{ms}^{-1}$$

12. Electrons are emitted with a maximum speed of 6×10^{5} ms^{-1} when a light of frequency 9×10^{14} Hz is incident on a metal surface. Calculate the threshold frequency of the metal.

Solution

$$hv = hv_0 + \frac{1}{2}mv^2 \qquad hv_0 = hv - \frac{1}{2}mv^2$$

$$hv = 6.625 \times 10^{-34} \times 9 \times 10^{14} = 59.625 \times 10^{-20}$$

$$hv_0 = 59.625 \times 10^{-20} - \frac{9.1 \times 10^{-31}}{2} \times (6 \times 10^{5})^{2} = 59.625 \times 10^{-20} - 16.38 \times 10^{-20}$$

$$v_0 = \frac{59.625 \times 10^{-20} - 16.38 \times 10^{-20}}{6.625 \times 10^{-34}} \text{Hz}$$

$$= \frac{43.245 \times 10^{-20}}{6.625 \times 10^{-34}} = 6.527 \times 10^{14} \text{Hz}$$

13. Light of wavelength 400 nm falls on a metal having a work function of 1.5 eV. Find the maximum kinetic energy of the ejected electron.

Solution

$$hv = hv_0 + \frac{1}{2}mv^2$$

$$\lambda = 400 \times 10^{-9} \text{m}, \quad v = c / \lambda = \frac{3 \times 10^{8}}{400 \times 10^{-9}} = 0.75 \times 10^{15} \text{Hz}$$

$$hv = 6.625 \times 10^{-34} \times 0.75 \times 10^{15} = 6.625 \times 0.75 \times 10^{-19} \text{J}$$

$$= \frac{6.625 \times 0.75 \times 10^{-19}}{1.6 \times 10^{-19}} \text{eV} = 3.1055 \text{ eV}$$

$$\frac{1}{2}mv^2 = hv - hv_0 = 3.1055 - 1.5 = 2.605 \text{ eV}$$

14. Ultraviolet light of wavelength 150 nm is incident on the surface of molybdenum. Find the maximum velocity of the ejected electron if the work function of the molybdenum is 4.15 eV.

Solution

$$\lambda = 150 \times 10^{-9} \text{m}, v = c / \lambda = \frac{3 \times 10^{8}}{150 \times 10^{-9}} = 2 \times 10^{15} \text{Hz}$$

Incident energy $= hv = 6.625 \times 10^{-34} \times 2 \times 10^{15} = 13.25 \times 10^{-19} \text{J}$

$$= \frac{13.25 \times 10^{-19}}{1.6 \times 10^{-19}} \text{eV} = 8.28 \text{ eV}$$

$$\frac{1}{2}mv^2 = hv - hv_0 = 8.28 - 4.15 = 4.13 \text{ eV}$$

$$v = \sqrt{\frac{13.25 \times 10^{-19}}{9.1 \times 10^{-31}}} = 1.2 \times 10^{6} \text{ms}^{-1}$$

15. The work function of aluminium is 4.2 eV. Calculate the kinetic energy of the fastest and the slowest photoelectrons, the stopping potential and cut off wavelength when light of wavelength 2000 Å falls on a clean aluminium surface.

Solution

$$h\nu = h\nu_0 + \frac{1}{2}m\mathrm{v}^2$$

$$h\nu = W + \frac{1}{2}m\mathrm{v}^2$$

$$\text{K.E.} = h\nu - W = h\nu - h\nu_0$$

$$\lambda = 2000 \text{ Å}$$

$$\nu = \frac{3\times10^8}{2000\times10^{-10}} = 1.5\times10^{15}\,\text{Hz}$$

$$h\nu = 6.625 \times 10^{-34} \times 1.5\times10^{15} = 6.625\times1.5\times10^{-19}\ \text{J}$$

$$W = 4.2\times1.6\times10^{-19}\,\text{J} = 6.72\times10^{-19}\,\text{J}$$

$$\textit{\textbf{K.E.}}_{\text{max}} = 9.9375\times10^{-19} - 6.72\times10^{-19} = 3.2175\times10^{-19}\,\text{J} = \frac{3.2175\times10^{-19}}{1.6\times10^{-19}}\text{eV}$$

$$= 2.011 \text{ eV}$$

Stopping potential = $\text{K.E.}_{\text{max}} = eV_0 = 2.011$ eV or $V_0 = 2.011$

$$W = h\nu_0 = hc\,/\,\lambda = 6.625\times10^{-34}\times3\times10^8\,/\,\lambda = \frac{6.625\times10^{-34}\times3\times10^8}{4.2\times1.6\times10^{-19}}$$

$$= 2.958\times10^{-7}\ \text{m} = 2958\ \text{Å}$$

16. UV light of wavelength 2271 Å from a 100 W mercury source irradiates a photocell made of molybdenum metal. If the stopping potential is 1.3 V, estimate the work function of the metal. How would the photocell respond to high intensity ($\sim 10^5$ Wm^{-2}) red light of wavelength 6328 Å of He-Ne laser.

Solution

$$h\nu = h\nu_0 + \frac{m\mathrm{v}^2}{2}$$

$$\lambda = 2271 \text{ Å}, \nu = c\,/\,\lambda = \frac{3\times10^8}{2271\times10^{-10}} = 1.32\times10^{15}\,\text{Hz}$$

Kinetic energy of electron $\frac{1}{2}m\mathrm{v}^2 = eV_0$ where V_0 is the stopping potential

$$= 1.3 \times 1.6 \times 10^{-19} \text{ J} = 2.08 \times 10^{-19} \text{ J}$$

$$h\nu_0 = \text{Work function} = h\nu - \frac{1}{2}m\text{v}^2 = 6.625 \times 10^{-34} \times 1.32 \times 10^{15} - 2.08 \times 10^{-19} \text{ J}$$

$$= 6.665 \times 10^{-19} \text{ J} = \frac{6.665 \times 10^{-19}}{1.6 \times 10^{-19}} \text{eV} = 4.165 \text{ e}$$

$$\text{Threshold frequency} = \nu_0 = \frac{6.665 \times 10^{-19}}{6.625 \times 10^{-34}} = 1.006 \times 10^{15} \text{ Hz}$$

For red light $\lambda = 6328$ Å, $\quad \nu = c/\lambda = \dfrac{3 \times 10^8}{6328 \times 10^{-10}} = 4.74 \times 10^{14}$ Hz

Since the frequency of red light is less than the threshold frequency, photoelectric emission will not take place.

17. An electron has a speed of 600 m/s with an accuracy of 0.005%. Calculate the certainty with which we can locate the position of the electron.

Solution

$$h = 6.6 \times 10^{-34} \text{ Joules second}$$

$$m = 9.1 \times 10^{-31} \text{ kg}$$

$$\text{v} = 600 \text{ m/s}$$

Momentum of the electron $= m\text{v} = 9.1 \times 10^{-31} \text{ kg} \times 600 \text{ m/s}$

$$\Delta p = 9.1 \times 10^{-31} \text{kg} \times 0.6 \text{ km/s} \times \left(\frac{0.005}{100} \right)$$

$$\Delta p \, \Delta x \geq \frac{h}{4\pi} \text{ and hence } \Delta x \geq \frac{h}{4\pi \cdot \Delta p}$$

$$\Delta x \geq \frac{6.6 \times 10^{-34}}{4\pi \times 5 \times 10^{-3} \times 9.1 \times 10^{-31} \times 600} \geq 0.001923 \text{ m} \geq 1.923 \text{ mm}$$

18. Obtain the de Broglie wavelength of a neutron of kinetic energy 150 eV. Would a neutron beam of same energy is suitable for crystal diffraction. ($m_n = 1.675 \times 10^{-27}$ kg.). Obtain the de Broglie wavelength associated with thermal neutron at room temperature 27°C.

Solution

$$\text{K.E.} = \frac{1}{2}m\text{v}^2 = \frac{p^2}{2m}$$

$$p = \sqrt{2m\,K.E.} \qquad\qquad \lambda = \frac{h}{p} = \frac{h}{\sqrt{2m\,K.E.}}$$

$$\lambda = \frac{6.625 \times 10^{-34}}{\sqrt{2 \times 1.675 \times 10^{-27} \times 150 \times 1.6 \times 10^{-19}}}$$

$$\lambda = 2.33 \times 10^{-12}\,\text{m}$$

The interatomic spacing is of the order of 1 Å = 10^{-10} m and is about 100 times greater than the above λ. Hence a neutron beam of 150 eV is not suitable for diffraction experiments.

Average kinetic energy of a neutron at absolute temperature is given by the kinetic energy

$K.E. = 3kT/2$

$$\lambda = \frac{h}{p} = \frac{h}{\sqrt{2m\,\text{K.E.}}} = \frac{6.625 \times 10^{-34}}{\sqrt{2 \times 1.675 \times 10^{-27} \times 1.5 \times 1.38 \times 10^{-23} \times 300}} = 1.45\ \text{Å}$$

Hence thermal neutrons can be used for diffraction experiments.

19. Crystal diffraction experiments can be performed using X-rays or electrons accelerated through appropriate voltage. Which probe has greater energy? An X-ray photon or electron? (X-ray = 1 Å, $m_e = 9.1 \times 10^{-31}$ kg.).

Solution

For electron 1 Å = 10^{-10} m, $m_e = 9.1 \times 10^{-31}$ kg

and hence $\quad p = h/\lambda = \dfrac{6.625 \times 10^{-34}}{10^{-10}} = 6.625 \times 10^{-24}$

$$\text{Energy} = \frac{p^2}{2m} = \frac{(6.625 \times 10^{-24})^2}{2 \times 9.1 \times 10^{-31}}\,\text{J}$$

For electron, $\quad E = \dfrac{(6.625 \times 10^{-24})^2}{2 \times 9.1 \times 10^{-31} \times 1.6 \times 10^{-19}}\,\text{eV} = 150\ \text{eV}$

For photon $\quad E = h\nu = \dfrac{hc}{\lambda} = \dfrac{6.625 \times 10^{-34} \times 3 \times 10^{8}}{10^{-10}}\,\text{J} = 12.42\ \text{eV}.$ So, X-ray electron has greater energy.

20. Calculate the de Broglie wavelength of an electron whose kinetic energy is 50 eV. ($h = 6.625 \times 10^{-34}$ J.s, $m = 9.1 \times 10^{-31}$ Kg).

Solution

$$\left[eV = \frac{1}{2}mv^2,\ v^2 = 2\,eV/m \quad v = \sqrt{\frac{2eV}{m}} \right.$$

$$\lambda = \frac{h}{mv} = \frac{h}{m\sqrt{\dfrac{2eV}{m}}} = \frac{h}{\sqrt{2meV}} = \sqrt{\frac{150}{V}} = \frac{12.3}{\sqrt{V}}$$

$$= \frac{12.3}{\sqrt{50}} = 1.73\text{Å}$$

21. An electron microscope uses 40 keV electrons. Find it's resolving limit on the assumption that it is equal to the wavelength of electrons.

Solution

$$\lambda = \sqrt{\frac{150}{V}}\ \text{Å} = \sqrt{\frac{150}{40\times10^3}} = 0.061\ \text{Å}$$

22. Calculate the uncertainity in momentum of photon confined in a box of length 1 Å.

Solution

$$\Delta p_x \cdot \Delta x = \frac{h}{2\pi} \quad \Delta p_x = \frac{6.625\times10^{-34}}{2\times3.14\times10^{-10}} = 1.05\times10^{-24}\ \text{kg m/s}$$

23. Determine the velocity and kinetic energy of a neutron having de Broglie wavelength 1 Å. (Mass of neutron 1.67×10^{-27} g, $h = 6.625 \times 10^{-34}$ Js $= 6.625 \times 10^{-27}$ erg).

Solution

$$\lambda = \frac{h}{mv} \quad v = \frac{h}{m\lambda} = \frac{6.625\times10^{-34}}{1.67\times10^{-27}\times10^{-10}} = 3.97\times10^3\ \text{m/s}$$

$$\text{K.E.} = \frac{1}{2}mv^2 = \frac{1}{2}\times1.67\times10^{-27}\times(3.97\times10^3)^2$$

$$= 13.16\times10^{-21} = 8.22\times10^{-2}\ \text{eV}$$

24. An excited atom has an average life time of 10^{-8} sec. Calculate the uncertainty in energy of the emitted photon in the minimum uncertainty in the frequency of this photon.

Solution

$$\Delta E.\Delta t \ge \frac{h}{2\pi}$$

Minimum $\Delta E = \dfrac{h}{2\pi\Delta t} = \dfrac{6.625\times10^{-34}}{2\times3.14\times10^{-8}} = 1.05\times10^{-26}\ \text{J}$

$$\Delta v = \frac{\Delta E}{h} = \frac{1.05 \times 10^{-26}}{6.625 \times 10^{-34}} = 1.6 \times 10^{7}\,\text{Hz}$$

25. If $\Delta\lambda$ is the width of the spectral line of wavelength λ then what is the time for which the atom is in excited state?

Solution

$$\Delta E \cdot \Delta t \approx \frac{h}{2\pi} \qquad\qquad \Delta t \ \geq \frac{h}{2\pi\Delta E}$$

$$E \ = \ h v = \ hc\,/\,\lambda \qquad \Delta E \ = \ hc\left(-\frac{\Delta\lambda}{\lambda^2}\right)$$

$$\Delta t \ = \ \frac{h}{2\pi\Delta E} = \frac{h}{2\pi\Delta}\frac{\lambda^2}{\lambda hc} = \frac{\lambda^2}{2\pi c\Delta\lambda}$$

26. Duration of a laser pulse is 10^{-8} s. What is uncertainty in its energy?

Solution

$$\Delta E \cdot \Delta t \approx \frac{h}{2\pi} \qquad \Delta E \approx \frac{h}{2\pi\Delta t} = \frac{6.625 \times 10^{-34}}{2 \times 3.14 \times 10^{-8}}$$

$$= \frac{6.625 \times 10^{-26}}{6.28}$$

$$= 1.05 \times 10^{-26}$$

ENDNOTES

1. Carriers of negative electricity, Nobel Lecture, December 11, 1906; In: *Nobel Lectures: Physics, 1901–1921* (Amsterdam: Elsevier, 1967), pp. 145–153.

2. Robert Andrews Millikan, 22 March 1868–1953. December 1953, Nobel Prize in 1923.

3. The Electron and Licht-quanta from the Experimental Point of View, Robert Millikan, Nobel Lecture May 23, 1924.

4. Saha, M.N. and Srivastava, B.N. (1969). *Treatise on Heat*, 5th edn. Indian Press, Allahabad. pp. 534–537.

5. Arthur Compton. (1923). A quantum theory of scattering of X-rays by light elements. *Phys.Rev.* 21(5) 483–502.

6. George Paget Thomson (Nobel Prize 1937), Nobel Lecture, June 7, 1938, Electronic Waves.

REFERENCES

Ehrenhaft, F. (1910). Über die Kleinsten Messbaren Elektrizitätsmengen, *Phys. Zeit.* 10. p. 308.

Millikan, R.A. (1910). A new modification of the cloud method of determining the elementary electrical charge and the most probable value of that charge. *Phys. Mag.* XIX, 6. p. 209.

Millikan, R.A. (1913). On the elementary electric charge and the Avogadro constant. *Phys. Rev.* II, 2. p. 109.

2

ATOMIC STRUCTURE AND SPECTRA

2.1 DALTON'S ATOMIC THEORY

It was John Dalton[1] (1766–1844) who placed the atom on a solid foothold as a fundamental chemical object although Democritus first suggested the existence of the atom almost two millennia before. Even after two centuries now, Dalton's atomic theory remains still valid in modern chemical thought. John Dalton, reasoned that if atoms really exist, they must have certain properties to account for the two laws of chemical combination. These properties are now called Dalton's Atomic Theory.

The main points of Dalton's Atomic Theory:

✧ Matter consists of definite particles called atoms or in other words the elements are made of tiny particles called atoms.

✧ All atoms of a given element are identical.

✧ The atoms of a given element are different from those of any other element; the atoms of different elements can be distinguished from one another by their respective relative weights.

✧ Atoms of one element can combine with atoms of other elements to form chemical compounds; a given compound always has the same relative numbers of types of atoms.

✧ Atoms are indestructible. In chemical reactions, the atoms rearrange but they do not themselves break apart. Atoms cannot be created, divided into smaller particles, nor destroyed in the chemical process; a chemical reaction simply changes the way atoms are grouped together.

Till 19th century it was believed that atom is the smallest, indivisible, indestructible particle of any element as proposed by Dalton in early 1800s.

The picture that the atom as a solid, indivisible and impenetrable sphere started collapsing towards the end of the 19th century. It was observed then that the atom could be penetrated in by suitable particles of appropriate energy. The discovery of electron by J.J. Thomson in 1897 and the realization that all atoms contained electrons gave first significant insight into the probable structure of atoms. Atom was found to be electrically neutral. And as the atom is electrically neutral it must then contain an equal number of positively charged particles as that of the negatively charged electrons in order to balance the charges to make it neutral. As the mass of the electron was found to be very small, it was also suggested that the entire mass of the atom must comprise of positively charged particles. Various atom models were then proposed and all were aimed at describing the arrangement of positive and negative charges. Today it is known that the atoms consist of electrons, protons and neutrons and are electrically neutral. Dalton's theory is thus found to be not perfectly correct.

It has been found later that atoms are made up of three subatomic particles—protons, neutrons, and electrons. The atom has a nucleus made up of protons and neutrons with electrons "orbiting" around it. The internal structure of the atom allows us to understand its properties. Protons and neutrons are in the nucleus and they are collectively called nucleons. Protons have a mass of approximately 1 atomic mass unit (amu—1.67×10^{-24} g) and have a positive charge. Neutrons have a mass of approximately 1 amu but have no charge (neutral). Electrons have a mass of approximately 1/1836 amu and have a negative charge. Protons repel each other so also electrons repel each other because of identical charges, keeping them spread out throughout the volume of the atom. Protons and electrons have opposite charges, and therefore attract each other. This attraction holds the electrons around the nucleus. The atomic number or number of protons determines what the element is. For example, an element with 10 protons is neon, and every element with 79 protons is gold. Isotopes are variations on atoms.

Atoms of different elements differ from one another according to the number of protons, they have and the value is called atomic number and characterized by the letter Z. If we know the number of protons we can identify that element. For example an element having six protons can only be carbon. Atoms are electrically neutral and hence have no net charge. It means that the number of positively charged protons and the number of negatively charged electrons must be the same. Thus the atomic number is also the same as the number of electrons. The atomic number is the number of protons (or the number of electrons) in an atom. There are also equal or more number of neutrons in an atom (except in normal hydrogen). The sum of the protons and neutrons in an atom is called atom's mass number denoted by A. Thus atoms of a given element have the same number of protons. However, some atoms of the same element can have different numbers of neutrons and therefore have different mass numbers. Atoms with identical atomic numbers but having different mass numbers are called isotopes. The difference is in the number of neutrons.

The properties are also slightly different and may be sometimes radioactive. The atomic masses of isotopes are different because the number of neutrons in the atom is different. You can not have a different number of protons because then by definition it is a different atom. As mentioned earlier the mass number is the sum of the protons and the neutrons. In a neutral atom (an atom with no electrical charge), the number of electrons always equals the number of protons. Isotopes are represented with a mass number as superscript and atomic number as subscript. For example, an atom with 82 protons and 126 neutrons is written as $^{208}_{82}Pb$. Pb is the symbol for lead. An atom with 82 protons is always lead. Oxygen has 8 protons and 8 neutrons and is written as $^{16}_{8}O$ so is nitrogen $^{14}_{7}N$. The two chlorine isotopes are written as $^{35}_{17}Cl$ and $^{37}_{17}Cl$.. The natural abundances of isotopes are also different. It has also been found that the abundance ratios of isotopes in interstellar media are sometimes different from the terrestrial ratio. Copper isotopes are written as $^{63}_{29}Cu$ and $^{65}_{29}Cu$.

It has already been mentioned that in an element, number of protons and electrons are the same. Difference in numbers can be in the number of neutrons. For example in $^{14}_{7}N$, there are seven electrons and 7 protons, but in $^{15}_{7}N$ there are fifteen neutrons. In $^{27}_{13}Al$, there are 13 protons, 14 neutrons and 13 electrons. In $^{115}_{47}Ag$ there are 47 protons, 47 electrons and 68 neutrons. The most abundant hydrogen isotope is also called protium and has no neutrons and thus has a mass number 1. The hydrogen isotope deuterium has one proton and one neutron and has mass number 2. The third isotope of hydrogen called tritium has one proton and two neutrons and hence has a mass number 3. Tritium does not exist naturally and it is unstable, but they are made in nuclear reactors.

To find the number of neutrons in any atom, substract the atomic number from the mass number. For example in phosphorus $^{31}_{15}P$,

Mass number (sum of protons and neutrons) – Atomic number (number of protons)

$$31 - 15 = 16 \text{ neutrons.}$$

In Uranium $^{235}_{92}U$, the number of neutrons is

$$235 - 92 = 143$$

and in $^{238}_{92}U$ the number of neutrons is

$$238 - 92 = 146$$

We have seen that the atoms are composed of tiny subatomic particles called protons, neutrons and electrons. The mass of the proton is $1.672\ 622 \times 10^{-24}$ g and has a positive charge. The neutron has also almost the same mass as proton, more exactly $1.674\ 927 \times 10^{-24}$g and is electrically neutral. The electron has a mass $9.109\ 328 \times 10^{-28}$g that is only 1/1836 that of proton and carried a negative charge. Electrons are much lighter than protons and neutrons and hence in some calculations its mass is usually ignored.

The masses of atoms and the subatomic particles are very small numbers when given in grams. It is hence expressed in yet another relative unit called atomic mass unit (amu). In this case a particular atom is assigned to have a mass and all other atoms are measured relative to that particular atom. The basis for the relative atomic scale is taken as that of carbon having 6 protons and 6 neutrons and thus the atomic mass is exactly 12. Masses of all other atoms are assigned relative to carbon. This is equivalent of considering the relative mass of a foot ball in terms of a golf ball. If we decide that the golf ball of 46 g is assigned a number one, then a volley ball of 270 g can be assigned with a number 270/46 = 5.87. Similarly assuming a number 12 to carbon, the magnesium atom can be assigned a number 24 and a hydrogen which is about one twelfth of carbon can be assigned a mass close to 1 amu. 1 amu is the mass of N number of atoms and hence it is the mass of one atom in gram.

$$1/6.022 \times 10^{23} \text{ g} = 1.660\ 539 \times 10^{-24} \text{ g}$$

Name	Symbol	Grams	amu	Charge
Proton	p	$1.672\ 622 \times 10^{-24}$	$1.007\ 276$	$+$
Neutron	n	$1.674\ 927 \times 10^{-24}$	1.008665	0
Electron	e^-	$9.109\ 328 \times 10^{-28}$	$5.485\ 799 \times 10^{-4}$	$-$

We now know that atoms can be broken into smaller pieces (Nuclear Reactions-Nuclear Physics) and most elements occur as mixtures of two or more isotopes (atoms of an element with slightly different masses). Atoms join together to form molecules or compounds (molecular physics and chemistry).

2.1.1 Thomson's Model

J. J. Thomson (1856–1940)[2] suggested the first atom model in 1897. According to this model the atom is supposed to be a sphere filled with positively charged matter of uniform density in which just sufficient number of electrons were embedded to balance the positive charge. It was also proposed that the emission of light occurs due to the vibrationary motion of electrons and the electrons were believed to be at rest when no light was emitted. A few years later experiments by Rutherford and his associates Geiger and Marsden on the scattering of alpha particles by thin metal foils produced evidence which gave a severe set back to Thomson's picture of the atom. Alpha particles are helium atoms, which had lost both their electrons.

When alpha particles were shot at thin metal foils, it was found that most of the alpha particles were scattered through small angles. But some of the alpha particles were scattered through large angles and a few were even deflected back. According to Thomson's atom model, alpha particles could be scattered through only small angles. The positive charge in Thomson's model is spread too thin to exert strong forces on the alpha particles to deflect them back in large angles. To account for the large angle scattering Rutherford suggested the nuclear atom model.

2.1.2 Rutherford's Nuclear Atom Model

Experiments by Ernest Rutherford (1871–1937)[3] and others in the early part of the 20th century has revealed that the atoms consisted of a diffuse cloud of negatively charged electrons surrounding a small, dense, positively charged nucleus. Thus a planetary model for the atom had been considered with electrons orbiting the nucleus like the planets around the sun. However, this model of the atom experienced serious difficulties. The most serious was that an accelerating electron emits energy (synchrotron radiation); thus with each orbit around the nucleus the electrons radiate away its energy thereby gradually spiralling inwards to the nucleus until the atom is no more. It almost happens instantly; thus atom cannot survive. Also the 19th century experiments with electric discharges had shown that the atoms emit electromagnetic radiations, but only at certain discrete frequencies. The proposed model could not explain the then already known atomic spectra.

2.1.3 Drawbacks of Rutherford's Atom Model

In Rutherford's model the electron is continuously moving around the nucleus in circular orbits. But according to classical electrodynamics, any charged particle, which is accelerating, should radiate energy. Hence the electron loses energy during every circular motion. Consequently the radius of the circular path decreases. Hence the electron takes a spiral path and falls into the nucleus within 10^{-8} seconds. Hence the atom is not stable.

According to classical theory, as the electron revolves around the nucleus in smaller and smaller orbits it should emit electromagnetic waves of gradually increasing frequency in a continuum range. But experiments show that atoms emit radiations of discrete frequencies. Thus the consequences of Rutherford's model were in contradiction to experiment, which led to a new model proposed by Bohr.

2.2 THE BOHR MODEL OF THE ATOM

To overcome these difficulties, Niels Bohr (1885–1962)[4] proposed, in 1913, what is now known as the **Bohr atom model**. The model's key success was in explaining the Rydberg formula (*See* Section 2.8.1) for the spectral emission lines of atomic hydrogen. Though Rydberg formula was tested experimentally it did not gain any theoretical basis until Bohr model of atom was introduced.

Bohr accepted Rutherford's concept of nuclear atom model, i.e., all the positive particles are concentrated at a point known as the nucleus. The electrons are revolving around the nucleus in circular orbits, the electrostatic force of attraction between the nucleus and the electron is compensated by the necessary centripetal force for the electron revolving around the nucleus. Thus the Bohr model depicts the atom as a small, positively charged nucleus surrounded by waves of electrons in orbit—similar to the structure to the solar system, but with electrostatic forces providing attraction in atoms rather than gravity and with waves spread over entire orbit instead of localized planets.

According to this model, all positive particles are concentrated at a point known as the nucleus. The electrons are moving around the nucleus in circular orbits, the electrostatic force between the electrons and the nucleus is compensated by the necessary centripetal force.

We can thus assume that the single electron in hydrogen atom revolves as a pointlike particle in a circular orbit of radius r around the proton; the stability of the orbit requires equilibrium between the attractive Coulomb force on the electron and the centrifugal force.

If q_1 and q_2 are the two point charges separated by a distance r, the magnitude of the force $F \alpha \dfrac{q_1 q_2}{r}$ or $F = A\dfrac{q_1 q_2}{r}$ where A is a constant which depends on the nature of the medium and the unit in which charges are measured. If the charges are placed in free space $A = \dfrac{1}{4\pi\varepsilon_0}$ where ε_0 is called the permittivity of the space and its value is

$$\varepsilon_0 = 8.854\ 187\ 817 \times 10^{-12}\ \mathrm{C^2/Nm^2} = \mathrm{Farad\ m^{-1}}$$

$$\frac{1}{4\pi\varepsilon_0} = 8.987\ 551\ 787 \times 10^{9}\ \mathrm{Newton\ m^2/C^2}\ (= 1.439\ 96 \times 10^{-9}\ \mathrm{eV\ m})$$

Historically, the physical constant ε_0 has been known by many different names. The terms "vacuum permittivity" or its variants, such as "permittivity in/of vacuum", "permittivity of empty space", or "permittivity of free space" are widespread. Worldwide, "electric constant" is the uniform term used for this quantity.

Niels Bohr proposed his atomic model using Planck's quantum theory. He put forward the concept of special orbits (which are now known as stationary orbits) and quantization of energy of atoms was brought in. The Bohr model is based on some postulates. As usual the postulates generally are incapable of a direct proof. However, we shall see that it appears reasonable in character. The ultimate justification for a postulate is generally sought in the correspondence between the predictions made with its aid and the observed behaviour of actual systems. The Bohr concept is depicted in Figure 2.1.

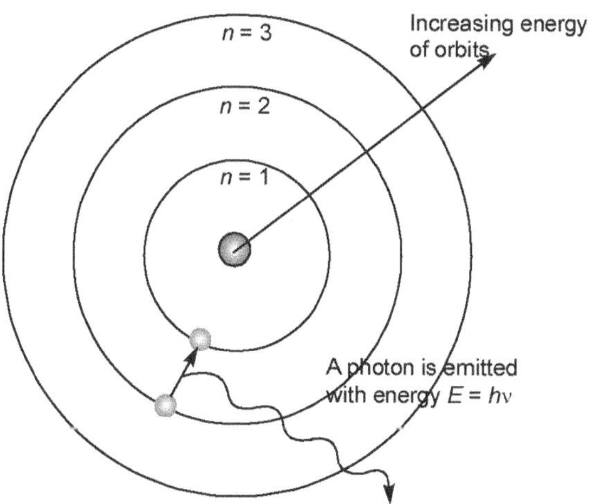

Figure 2.1   The Bohr's concepts of special orbit

Thus the electrons cannot revolve in any arbitrary orbit. They can move only in certain orbits, which satisfy certain quantum conditions. According to Bohr model, only those orbits are allowed in which the angular momentum of electron is an integral multiple of $\dfrac{h}{2\pi}$. While moving along these orbits around the nucleus, an electron does not radiate energy. These non-radiating orbits are called stationary orbits.

Thus the important postulates of this model are:

Postulate I   The electrons revolve around the nucleus only in circular orbits, the necessary centripetal force for circular motion is provided by the electrostatic attraction between the positively charged nucleus and negatively charged electrons. Thus the electron is held in orbit by the Coulomb force, which is balancing with the centripetal force. If Z is the atomic number, then

$$\frac{1}{4\pi\varepsilon_0}\frac{Ze.e}{r^2}=\frac{mv^2}{r} \tag{2.1}$$

A pictorial representation of balancing of these forces is shown in Figure 2.2.

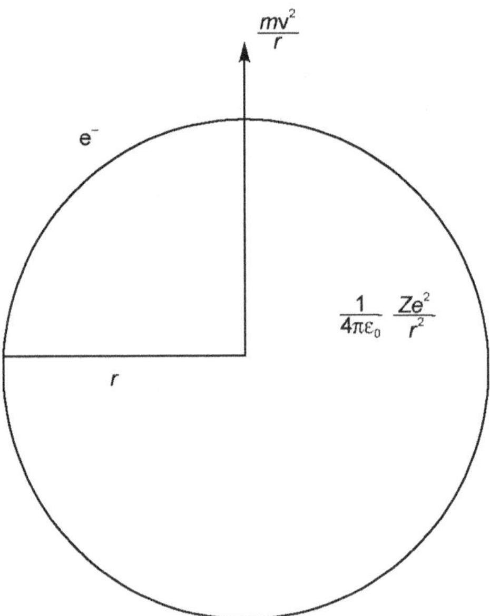

Figure 2.2   Revolution of electron around the nucleus

Postulate II   The orbiting electrons existed in orbits that had discrete quantized energies. That means, not every orbit is possible but only certain specific ones. Or in other words one can say that out of several number of orbits an electron can revolve only in those orbits in which it's angular momentum is multiple of $\dfrac{h}{2\pi}$

We know the angular momentum of an electron $L = I\omega = mr^2\omega = mr^2\dfrac{v}{r} = m\,r\,v.$

Thus as per Postulate II

$$mvr = n\dfrac{h}{2\pi}, \quad n = 1, 2, 3\ldots \tag{2.2}$$

The key ideas of the Bohr atom model:

1. The electrons orbit around the nucleus in discrete quantized orbits. That is every orbit is not possible but only certain specific ones.

2. The laws of classical mechanics do not apply when electrons make a jump from one quantized orbit to the other.

3. When an electron makes a jump from one quantized orbit to the other, it is accomplished by a single light quantum called photon which has an energy which is equal to the energy difference between the two orbits.

4. The allowed orbits has orbital angular momentum L which is characterized by the relation

$$L = n \cdot \hbar = n\dfrac{h}{2\pi}$$

where n is called principal quantum number and can take integer values $n = 1, 2, 3, \ldots.$

This implies that the circumference of electron's orbit must be an integer multiple of its wavelength.

$$\text{i.e., } 2\pi r = n\lambda. \left[\text{Note that } 2\pi r = n\dfrac{h}{mv} = n\lambda \right]$$

Though Bohr did not make the assumption of de Broglie's hypothesis in his original derivation because the de Broglie hypothesis had not been proposed at that time, the equation (2.2) is only the consequence of de Broglie's hypothesis. The equation (2.2) states that

$$mvr = n\dfrac{h}{2\pi}, \quad \text{i.e., } 2\pi r = n\dfrac{h}{mv} \text{ and hence } 2\pi r = n\lambda \tag{2.3}$$

From equation (2.1)

$$mv^2 r = \dfrac{Ze^2}{4\pi\varepsilon_0} \tag{2.4}$$

From equation (2.2)

$$mvr = n\frac{h}{2\pi}$$

Thus by dividing equation (2.4) by equation (2.2)

$$v = \frac{Ze^2}{4\pi\varepsilon_0} \cdot \frac{2\pi}{nh} \tag{2.5}$$

From (2.3)

$$r = \frac{nh}{2\pi mv} \tag{2.6}$$

Substituting v from (2.5)

$$r = \frac{nh}{2\pi m} \cdot \frac{nh.4\pi\varepsilon_0}{Ze^2.2\pi}$$

$$= \frac{n^2}{m} \cdot \frac{4\pi\varepsilon_0 \cdot \left(\dfrac{h}{2\pi}\right)^2}{Ze^2} = \frac{n^2 h^2 \varepsilon_0}{\pi m Ze^2} \tag{2.7}$$

For hydrogen $Z = 1$ and $\qquad r_n = \dfrac{n^2 h^2 \varepsilon_0}{\pi me^2} = n^2 \dfrac{h^2 \varepsilon_0}{\pi me^2} \tag{2.8}$

where n is the principal quantum number

For $n = 1$ $\qquad\qquad\qquad r_1 = \dfrac{h^2 \varepsilon_0}{\pi me^2} \tag{2.9}$

$$h = 6.625 \times 10^{-34} \text{ Js}$$
$$\varepsilon_0 = 8.85 \times 10^{-12} \text{ C}^2 / \text{Nm}^2$$
$$m_e = 9.11 \times 10^{-31} \text{ kg}$$
$$e = 1.6 \times 10^{-19} \text{ C}$$

$$r_1 = \frac{(6.625 \times 10^{-34})^2 \times 8.85 \times 10^{-12}}{3.14 \times 9.11 \times 10^{-31} \times (1.69 \times 10^{-19})^2} = 0.529 \times 10^{-8} \text{cm} = 0.529 \text{ Å}$$

This is called first Bohr radius and is generally denoted as a_0. r_n is also known as a_n.

The orbital radius for any orbit n from equation (2.8), then gives $r_n = n^2 a_0$.

where $a_0 = \dfrac{h^2 \varepsilon_0}{\pi m e^2}$

From equation (2.3) the electron wavelength in the nth orbit is $\lambda = \dfrac{2\pi r_n}{n} = 2\pi n\, a_0$ (See Figure 2.2).

2.2.1 Energy in Terms of Other Constants

Total energy of the electron = Kinetic energy + Potential energy

$$\text{T.E.} = \frac{1}{2}m v^2 + \frac{Ze.(-e)}{4\pi\varepsilon_0 r} \tag{2.10}$$

From (2.1) we have
$$\frac{1}{2}m v^2 = \frac{Ze^2}{8\pi\varepsilon_0 r} \tag{2.11}$$

Thus
$$E = \frac{Ze^2}{8\pi\varepsilon_0 r} - \frac{Ze^2}{4\pi\varepsilon_0 r} = -\frac{Ze^2}{8\pi\varepsilon_0 r} \tag{2.12}$$

$$\text{i.e., } E_n = -\frac{Ze^2}{8\pi\varepsilon_0 r_n} = -\frac{Ze^2}{8\pi\varepsilon_0} \times \frac{\pi m e^2 Z}{\varepsilon_0 n^2 h^2} = -\frac{Z^2 e^4 m}{8\varepsilon_0^2 n^2 h^2}$$

$$= -Z^2 \frac{m e^4}{8 h^2 \varepsilon_0^2} \cdot \frac{1}{n^2} = -\frac{Z^2 m e^4}{8\varepsilon_0^2} \cdot \frac{1}{n^2 h^2} \tag{2.13}$$

The minus sign indicates that the electron has the least energy in the first orbit. In other words the electron is bound in the field of nucleus in this orbit.

From equation 2.7, radius of Bohr orbit is $r = \dfrac{n^2 h^2 \varepsilon_0}{\pi m Z e^2} = a_n$ and the first radius $a_0 = \dfrac{h^2 \varepsilon_0}{\pi m Z e^2}$ and is called Bohr radius.

In terms of Bohr radius a_0,
$$E_n = -\frac{Z^2 e^2}{8\varepsilon_0} \frac{m e^2}{h^2 \varepsilon_0} \frac{1}{n^2} = -\frac{Ze^2}{8\pi\varepsilon_0 a_0} \frac{1}{n^2} \tag{2.14}$$

We know that $E = h\nu$ hence $\dfrac{E}{hc} = \dfrac{h\nu}{hc} = \dfrac{\nu}{c} = \dfrac{1}{\lambda} = $ wave number unit (cm^{-1}).

Thus the energy of the atom in wave number unit (cm^{-1}) which is called the term value

$$\frac{E_n}{hc} = -\frac{Z^2 e^4 m}{8\varepsilon_0^2 n^2 h^2 ch}\,\text{cm}^{-1} = -Z^2 \frac{m e^4}{8\varepsilon_0^2 ch^3} \cdot \frac{1}{n^2} = -\frac{m e^4}{8\varepsilon_0^2 ch^3} \frac{Z^2}{n^2} = -\frac{RZ^2}{n^2}\,\text{cm}^{-1} \tag{2.15}$$

$$R = \frac{me^4}{8\varepsilon_0^2 ch^3} = \text{Rydberg constant} = \frac{9.11 \times 10^{-31} \times (1.6 \times 10^{-19})^4}{8 \times (8.85 \times 10^{-12})^2 \times 3 \times 10^8 \times (6.625 \times 10^{-34})^3}$$

$$= 1.0973731 \times 10^7 \, \text{m}^{-1} \tag{2.16}$$

Rydberg constant R can also be written as $\quad R = \left(\frac{1}{4\pi\varepsilon_0}\right)^2 \frac{2\pi^2 me^4}{ch^3} \tag{2.17}$

You will find R as $\dfrac{2\pi^2 me^4}{ch^3}$ in some books omitting the constant term.

We can write the energy term in joules as $\quad E_n = -\dfrac{Rhc}{n^2} Z^2 \, \text{Joule} \tag{2.18}$

For hydrogen $Z = 1$ and the energy for hydrogen atom can be written as

$$\textbf{Energy} \qquad E_n = -\frac{Rhc}{n^2}\text{Joule} \quad R = 1.09 \times 10^7 \, \text{m}^{-1}$$

$$h = 6.625 \times 10^{-34} \, \text{Js}$$

$$c = 3 \times 10^8 \, \text{m/s}$$

$$\textbf{Hence} \qquad E_n = -\frac{1.09 \times 10^7 \times 6.625 \times 10^{-34} \times 3 \times 10^8}{n^2} \text{J}$$

$$= -\frac{21.635 \times 10^{-19}}{n^2}\text{J} = -\frac{21.635 \times 10^{-19}}{n^2 \times 1.6 \times 10^{-19}}\text{eV} \tag{2.19}$$

$$= -\frac{13.6}{n^2} Z^2 \, \text{eV}$$

Thus the energy of hydrogen is written in simple expression as

$$E_n = -\frac{13.6}{n^2} \, \text{eV} \tag{2.20}$$

(Note that 1 eV = 1.6×10^{-19} Joule = 8065.5 cm^{-1})

The energy levels are as shown in Figure 2.3.

The energy of the atom in terms of Bohr radius can be derived as follows. From equation (2.13)

$$E_n = -Z^2 \frac{me^4}{8h^2 \varepsilon_0^2} \cdot \frac{1}{n^2} \tag{2.21}$$

$$\text{i.e.,} \quad E_n = \frac{Ze^2}{8\pi\varepsilon_0} \div n^2 \frac{h^2\varepsilon_0}{\pi me^2} = -\frac{Ze^2}{8\pi\varepsilon_0} \frac{\pi me^2}{h^2\varepsilon_0} \frac{1}{n^2} = -\frac{Ze^2}{8\pi\varepsilon_0} \frac{1}{a_0} \frac{1}{n^2} = -\frac{Ze^2}{8\pi\varepsilon_0 a_0} \cdot \frac{1}{n^2}$$

The energy of the hydrogen atom in the stationary states is negative. This means that the electron is bound. In the lowest state (i.e., the ground state) in the hydrogen atom is bound to the nucleus by 13.6 eV. It requires 13.6 eV to take away the electron from the nucleus, i.e., to remove it to $n = \infty$. The minimum energy required to remove an electron from the atom is called ionization potential. The ionization potential of the hydrogen atom is 13.6 eV. The $n = 1$ state is called the ground state. The ionized atom is denoted by a plus sign as superscript. Thus the ionized hydrogen atom is denoted as H^+. It has only one proton and has no electron.

The unit eV means an electron volt and it is an easy way to represent the kinetic energy gained by a charged particle when it is accelerated through an electrical potential difference of 1 volt. The charge of an electron is 1.6×10^{-19} coulombs and if it is accelerated through a potential difference of 1 volt (1 J/C) then the kinetic energy gained is 1 eV.

Thus
$$1 \text{ eV} = \frac{1.6 \times 10^{-19} C}{1 \text{ J/C}} = 1.6 \times 10^{-19} \text{ J} \tag{2.22}$$

The energy is thus given in different units like Joules, eV, cm^{-1}, and also in calorie. When one uses different formulae concerning energies, note that they must be converted into Joules in order to use it further. But it is for the convenience of the user to use any units he wishes.

$$n = 1 \qquad E_1 = -\frac{13.6}{1^2} = -13.6 \text{ eV}$$

$$n = 2 \qquad E_2 = -\frac{13.6}{2^2} = -3.4 \text{ eV}$$

$$n = 3 \qquad E_3 = -\frac{13.6}{3^2} = -1.51 \text{ eV}$$

$$n = 4 \qquad E_4 = -\frac{13.6}{4^2} = -0.85 \text{ eV}$$

$$n = 5 \qquad E_5 = -\frac{13.6}{5^2} = -0.54 \text{ eV}$$

$$n = 6 \qquad E_6 = -\frac{13.6}{6^2} = -0.38 \text{ eV}$$

$$\cdots \qquad \cdots$$

$$n = \infty \qquad E_\infty = -\frac{13.6}{\infty^2} = 0 \text{ eV}$$

Figure 2.3   The energy levels of hydrogen atom

Ionization potential   We have seen from expression (2.20) that $E_n = -\dfrac{13.6}{n^2}$ eV. Thus the lowest state of the hydrogen atom (ground state, $n = 1$) has an energy -13.6 eV. It means that the electron is bound to this state by -13.6 eV. It needs 13.6 eV to take away the electron from the nucleus (i.e., to raise it to the orbit vis-à-vis level $n = \infty$). The wave number at the series limit is dependent of Z^2 as is also the ionization potential. The ionization potential is the minimum amount of energy which must be supplied to an electron in the $n = 1$ level to remove it completely from the influence of the nucleus. When $n_2 = 1$ and $n_1 = \alpha$

$$hc\ v_\alpha = hc\ RZ^2 \tag{2.23}$$

Thus the ionization potential in He$^+$ is four times the value in hydrogen, that in Be$^+$ is sixteen times greater. The lowest wave number v_α where the continuum begins, gives a direct measurement of energy required for the removal of the electron and hc v_α is termed as ionization potential. Alkali atoms have minimum and noble gases have maximum ionization potential.

2.2.2 Excited State

Normally hydrogen atom will be in its ground state. When a hydrogen atom receives energy, the electron can make a transition to a state with higher energy. These states of higher energy are called excited state or excited electronic states.

2.3 QUANTUM PROCESSES

Quantum processes are inevitable properties of atomic and molecular systems. The radiations are quantized. The energy levels of atoms and molecules can have only certain quantized values. Transitions between these quantized states occur by absorption, spontaneous emission, and stimulated emission of photons. (There can be also radiationless transition which are not discussed here). All of these processes require that the photon energy given by the Planck relationship is equal to the energy separation of the participating pair of quantum energy states $\Delta E = h\nu$. For a given frequency of radiation, there can be only one value of quantum energy for the photons of that radiation. Way back in 1917, Albert Einstein derived relationship between absorption, spontaneous emission, and stimulated emission, which have later, became the foundation stone in explaining laser processes (*See* Chapter 3).

2.3.1 Quantized Energy States

The quantized Bohr orbits and their corresponding energy levels are shown in Figure 2.4.

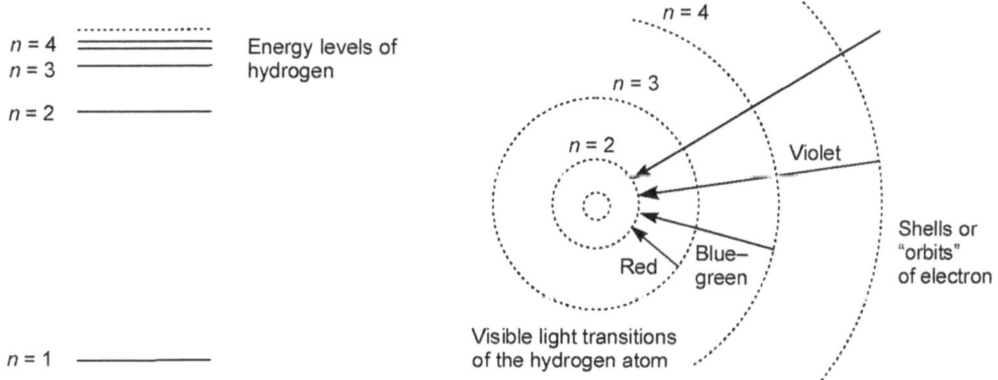

Figure 2.4   Quantized Bohr orbits and corresponding energy levels of hydrogen

2.3.2 Energy Level Diagrams

The stationary states of an atom are generally represented by horizontal lines plotted to energy as it is always difficult to represent them by circular orbits as proposed by Bohr's theory. If circular orbits represent the stationary energy states, it becomes difficult to represent orbits corresponding to large n values. Thus using energy level diagrams as horizontal lines solve these difficulties (Figure 2.3, 2.4 and 2.5). The above description is only for convenience. The term values E/hc are generally plotted and hence the difference between two levels gives directly the frequency of the absorbed or emitted radiation in wave numbers.

2.3.3 Hydrogen Energy Levels

The basic hydrogen energy level structure is in agreement with the Bohr model. Common pictures are those of a shell structure with each main shell associated with a value of the principal quantum number n.

This Bohr model picture of the orbits has some usefulness for visualization so long as it is realized that the "orbits" and the "orbit radius" just represent the most probable values of a considerable range of values. If the radial probabilities for the states are used, then the Bohr picture can be superimposed on that as a kind of conceptual skeleton.

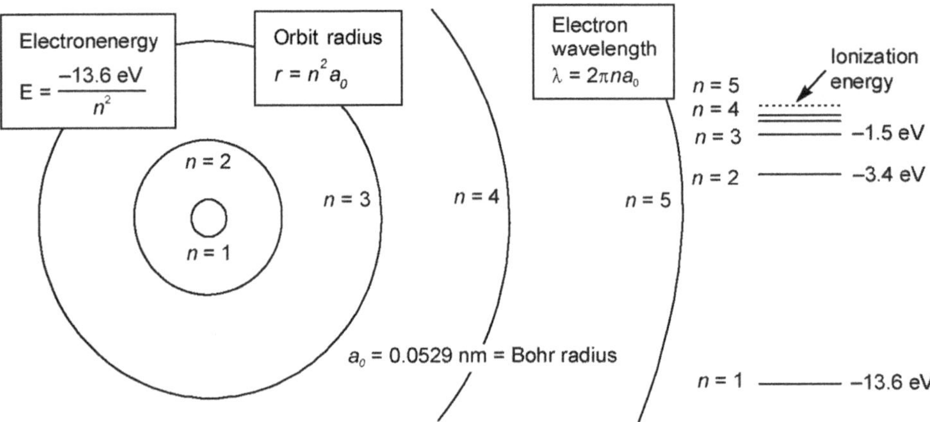

Figure 2.5   Bohr model for an electron transition in hydrogen

2.3.4 Electron Transitions

1. The electrons in free atoms will be found in only certain discrete energy states. These energy states are related with the orbits or shells of electrons in the atom, e.g., a hydrogen atom. The allowed orbits depend on quantized (discrete) values of orbital angular momentum, L according to the equation

$$L = n \cdot \hbar = n \cdot \frac{h}{2\pi}$$

where $n = 1,2,3,\ldots$ and is called the principal quantum number, and h is Planck's constant.

2. A spectral line is obtained when an electron makes a "jump" (transition) from one orbit (energy level) to another. A transition from an upper state to a lower state is termed emission and a transition from a lower state to an upper state is called absorption. The absorption is possible only in the presence of an external field.

3.   The laws of classical mechanics do not apply when electrons make the jump from one allowed orbit to another. It implies that only certain photon energies are allowed when electrons jump down from higher levels to lower levels or vice versa.

4.   When an electron makes a jump from one orbit to another, the energy difference in the orbits is carried off (or supplied) by a single quantum of light (called a photon) which has energy equal to the energy difference between the two orbitals.

As mentioned earlier, the assumption (1) above clearly implies that the lowest value of n is 1. This corresponds to the smallest possible radius of 0.0529 nm in hydrogen. This is known as the Bohr radius. Once an electron is in this lowest orbit, it can no longer get closer than this to the proton. As mentioned earlier, the lowest energy state is called the ground state of the atom.

The Bohr model for an electron transition in hydrogen between quantized energy levels with different quantum numbers n yields photon by emission with quantum energy. Similarly absorption is also possible, when the electron moves from a lower state to an upper state. Suppose the energy of the electron in orbit of quantum number n_1 is E_1 and in the orbit of quantum number n_2 is E_2 and the electron makes a jump from n_2 to n_1, then the difference in energy give the frequency of transition by the expression (*See* Figure 2.6).

$$E_2 - E_1 = h\nu$$

We have seen from the expression (2.18) that for hydrogen atom $E_n = -\dfrac{Rhc}{n^2}$ Joule

We get,

using Rydberg constant

$$R = \left(\frac{1}{4\pi\varepsilon_0}\right)^2 \frac{2\pi m e^4}{ch^3}$$

$$E_n = -\left(\frac{1}{4\pi\varepsilon_0}\right)^2 \frac{2\pi^2 m e^4}{h^2} \frac{1}{n^2} \tag{2.24}$$

Thus $R = 1.0973731 \times 10^7 \text{ m}^{-1}$ and $Rhc = 21.635 \times 10^{-9}$ J

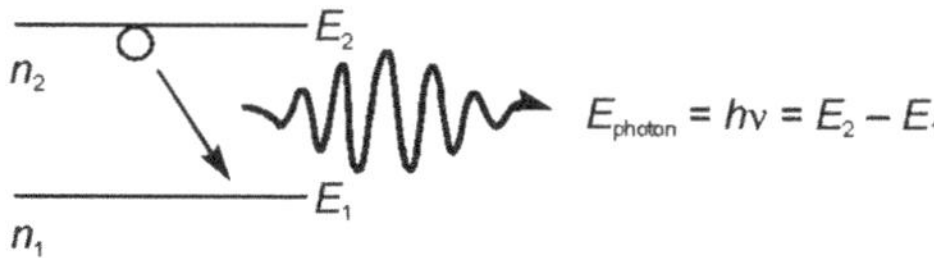

A downward transition involves emission of a photon of energy

An upward transition involves an absorption of a photon of energy

Figure 2.6   Electron transition between quantized energy levels

Hence $E_1 = -\dfrac{21.63 \times 10^{-9}}{1.6 \times 10^{-19}} \text{eV} = 13.6 \text{ eV}$

$$h\nu = \left(\frac{1}{4\pi\varepsilon_0}\right)^2 \frac{2\pi^2 m Z^2 e^4}{h^2} \left[\frac{1}{n_1^2} - \frac{1}{n_2^2}\right] \text{Joule} = \left(\frac{1}{4\pi\varepsilon_0}\right)^2 \frac{2\pi^2 m Z^2 e^4}{ch^3} \left[\frac{1}{n_1^2} - \frac{1}{n_2^2}\right] \text{cm}^{-1}$$

$$= \left(\frac{1}{4\pi\varepsilon_0}\right)^2 \frac{2\pi^2 m Z^2 e^4}{h^2} \frac{1}{1.6\times10^{-19}} \left[\frac{1}{n_1^2} - \frac{1}{n_2^2}\right] \text{eV} = 13.6 \, Z^2 \left[\frac{1}{n_1^2} - \frac{1}{n_2^2}\right] \text{eV} \qquad (2.25)$$

in Joules it is $\quad h\nu = -\dfrac{RhcZ^2}{n_2^2} - \left(-\dfrac{RhcZ^2}{n_1^2}\right) = Rhc \, Z^2 \left[\dfrac{1}{n_1^2} - \dfrac{1}{n_2^2}\right] \text{Joule} \qquad (2.26)$

This is often expressed in terms of the inverse wavelength or "wave number" as follows:

$$h\nu = \left(\frac{1}{4\pi\varepsilon_0}\right)^2 \frac{2\pi^2 m Z^2 e^4}{h^2} \left[\frac{1}{n_1^2} - \frac{1}{n_2^2}\right] \text{Joule}$$

$$\left(\frac{h\nu}{hc}\right) = \left(\frac{1}{4\pi\varepsilon_0}\right)^2 \frac{2\pi^2 m Z^2 e^4}{ch^3} \left[\frac{1}{n_1^2} - \frac{1}{n_2^2}\right] \text{cm}^{-1}$$

$$\left(\frac{\nu}{c}\right) = \left(\frac{1}{\lambda}\right) = \left(\frac{1}{4\pi\varepsilon_0}\right)^2 \frac{2\pi^2 m Z^2 e^4}{ch^3} \left[\frac{1}{n_1^2} - \frac{1}{n_2^2}\right] \text{cm}^{-1} \qquad (2.27)$$

Thus in wave numbers

$$\frac{1}{\lambda} = R_H \left[\frac{1}{n_1^2} - \frac{1}{n_2^2}\right] \text{ where } R_H = \left(\frac{1}{4\pi\varepsilon}\right)^2 \frac{2\pi^2 m e^4}{ch^3} \text{ is called the Rydberg constant.}$$

$$R_H = 1.0973731 \times 10^7 \, \text{m}^{-1} \qquad (2.28)$$

The above formula (2.25, 2.26) is known as the Rydberg formula. It was first developed by Balmer in 1884 and was known to scientists even before there was no theoretical perception until Bohr derived it.

Thus R_H in wave numbers $= \dfrac{me^4}{8\varepsilon_0^2 ch^3} = \dfrac{9.1\times10^{-31}\,\text{kg} \times (1.6\times10^{-19}\,\text{C})^4}{8 \times (8.85\times10^{-12})^2 \times 3\times10^8 \times (6.625\times10^{-34})^3}$

$$R_H = 1.0973731 \times 10^7 \, \text{m}^{-1} \qquad (2.29)$$

2.4 ABSORPTION AND EMISSION

Using the Bohr model, the energy levels of an atom with atomic number Z are evaluated with the simple formula $E_n = -\dfrac{13.6}{n^2}Z^2$ eV.

E_n correspond to the energy levels. The ground state is represented by $n = 1$, the first excited state is the one with $n = 2$, the second excited state with $n = 3$ and so on.

For H atom $Z = 1$ and hence the energy of the nth level is given by

$$E_n = -\frac{13.6}{n^2} \text{ eV}$$

Thus $E_1 = -13.6$ eV

$$E_2 = -3.4 \text{ eV}; \ E_3 = -1.51 \text{ eV}; \ E_4 = -0.85 \text{ eV}; \ E_5 = -0.54 \text{ eV}; \ E_6 = -0.38 \text{ eV and so on.}$$

Taking these energies on a linear scale, horizontal lines can be drawn, which represent the energies of H atom. The different transitions are given in the Figure 2.7. If ΔE is the energy difference in electron volt and λ is the wavelength of photon emitted or absorbed in Angstrom then

$$\lambda = \frac{12375}{\Delta E} \tag{2.30}$$

To remember it we can approximate it to $\dfrac{12345}{\Delta E}$ $\hspace{2em}$ (2.31)

2.4.1 Absorption

Normally all atoms are in their lowest energy level known as the ground state. When we pass white light through a tube filled with hydrogen, then the hydrogen atoms absorb photons of appropriate energies and rise to various energy levels following certain quantum conditions and selection rules. These transitions are called absorption transitions or simply absorption.

2.4.2 Emission

In emission the reverse process occurs. The atoms, which are being excited to the excited states drop back to lower states. The atoms excited means the valence electrons are brought to the excited levels by some methods and then they are emitted back.

If the electrons in absorption start from $n = 2,3,4\dots$ levels (*See* Figure 2.8) then members of Balmer, Paschen, Brackett..., etc. series are absorbed. The group of transitions from $n - 1$ to different higher levels $n = 2,3,4$, etc. are known as Lyman series.

In ordinary temperatures almost all atoms remain in their lowest level $n = 1$. Hence only Lyman series is absorbed.

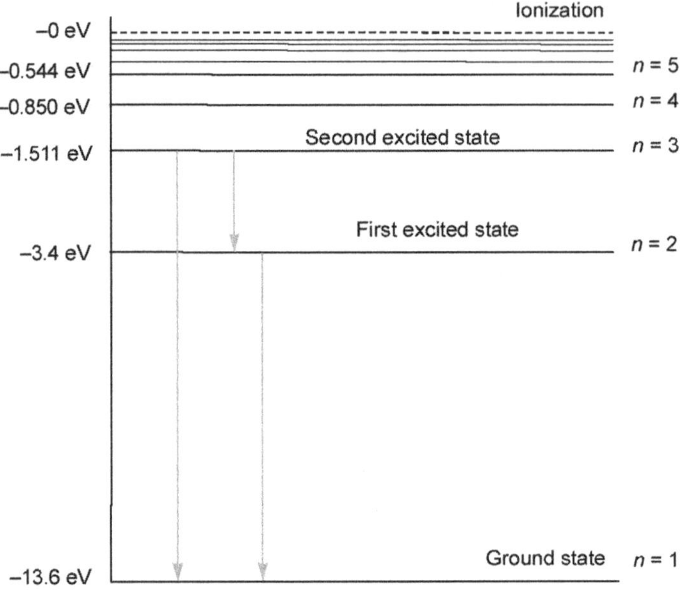

Figure 2.7   Hydrogen energy levels and transitions

$$E = -\frac{13.6 \text{ eV}}{n^2}$$

Hydrogen atoms are also present in the sun where the temperature is of the order of 2×10^7 Kelvin. Therefore Balmer series are also obtained from the spectrum of sun. They are also observed in other stars and a typical example is shown in Figure 2.16. It should be noted that the lines are shifted towards red when we observe stellar spectra due to Doppler Effect. The stars are receding away from us, an experimental evidence for expanding universe.

2.5 FREQUENCY OF EMITTED OR ABSORBED RADIATION

Suppose the energy of the electron in the state with orbital quantum number n_1 is E_1 and in the state with orbital quantum number n_2 is E_2. A transition is obtained when the electron jumps from an orbit to another. If the electron moves from the upper state n_2 to a lower state n_1 (emission) or from n_1 to n_2 (absorption) then the transition frequency in the case of any atom with atomic number Z is

$$E_2 - E_1 = h\nu$$

$$-\frac{RZ^2hc}{n_2^2} + \frac{RZ^2hc}{n_1^2} = h\nu$$

or $v = RcZ^2 \left[\dfrac{1}{n_1^{\,2}} - \dfrac{1}{n_2^{\,2}} \right]$ and for hydrogen $v = Rc \left[\dfrac{1}{n_1^{\,2}} - \dfrac{1}{n_2^{\,2}} \right]$

$$\frac{1}{\lambda} = \frac{v}{c} = R\,Z^2 \left[\frac{1}{n_1^{\,2}} - \frac{1}{n_2^{\,2}} \right] \tag{2.32}$$

Thus in wave number $v = \dfrac{1}{\lambda} = \dfrac{v}{c} = R\,Z^2 \left[\dfrac{1}{n_1^{\,2}} - \dfrac{1}{n_2^{\,2}} \right]$

For H atom $\qquad v = R \left[\dfrac{1}{n_1^{\,2}} - \dfrac{1}{n_2^{\,2}} \right] \text{cm}^{-1}$ \hfill (2.33)

2.6 SPECTRUM OF H ATOM

Selection rule is the change in principal quantum number for allowed transitions. The selection rule is that n can change to any value from the original state. Thus we have a number of series spectra in hydrogen atom.

2.6.1 Lyman Series

Group of transitions involving $n = 1$ to or from different higher energy levels, $n = 2,3,4$, etc. is known as Lyman series. The series is named after its discoverer Theodore Lyman (1874–1954). Consider that the electron jumps from any outer orbit to the first orbit. The general formula for wavelength of emitted radiation is

$$\frac{1}{\lambda} = \bar{v} = R \left[\frac{1}{n_1^{\,2}} - \frac{1}{n_2^{\,2}} \right], \; n_2 = 2,3,4,5... \infty \;\; \text{and} \;\; n_1 = 1 \tag{2.34}$$

This series lies in the ultraviolet region. Longest wavelength of Lyman series can be obtained as follows

$$\frac{1}{\lambda} = R \left[\frac{1}{1^2} - \frac{1}{2^2} \right] = R \left[1 - \frac{1}{4} \right] - \frac{3R}{4} = \frac{3}{4} \times 1.09 \times 10^7 \text{ m}^{-1} \tag{2.35}$$

$$\lambda = \frac{4}{3R} = \frac{4}{3} \times \frac{1}{1.09 \times 10^7} \approx 1216 \text{ Å} \tag{2.36}$$

Shortest wavelength $\dfrac{1}{\lambda} = \left[\dfrac{1}{1^2} - \dfrac{1}{\infty}\right] = R,\ \lambda = \dfrac{1}{R} \approx 911\ \text{Å}$ (2.37)

The transitions are named sequentially by Greek letters: from $n = 2$ to $n = 1$ is called Lyman-alpha (1215.67 Å), 3 to 1 is Lyman-beta (1025.18 Å), 4 to 1 is Lyman-gamma (972.02 Å), etc. with the limit 911.267 Å. They are shown in Figure 2.8.

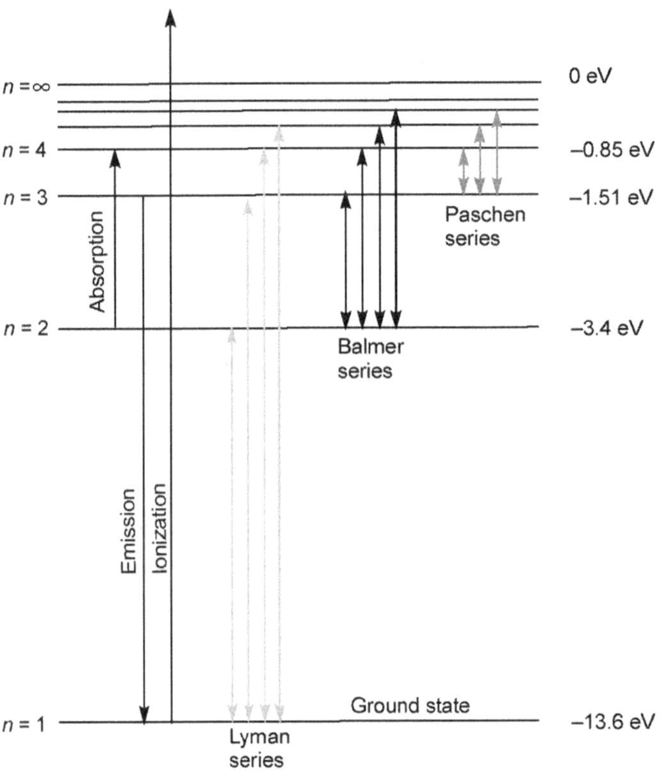

Figure 2.8 Series spectra in hydrogen atom

Among all elements the hydrogen atom is the simplest with only one proton and a single electron and hence can be treated theoretically with the utmost precision. Quantum electrodynamics (QED) allows determining energy levels with 12 significant digits of accuracy.

The $1S$-$2S$ transition frequency is known now with almost 14 digits of accuracy while all other transition frequencies have been measured with more or less similar accuracies.

The first member of the Lyman series L^α (i.e., the $1S$–$2S$ transition frequency) is very accurately measured by Hänsch[5] and co-workers in Muenchen as[6]

$$\nu(1S-2S) = 2466\ 061\ 102\ 474\ 851\ (25)\ \text{Hz}$$ (2.38)

Some of the relevant references on hydrogen atom by Hänsch's group in Muenchen are given at the end of the chapter.

Thus the spectroscopic measurements of the ultraviolet 1S–2S two-photon resonances in hydrogen and deuterium in their laboratory have led to new tests of quantum electrodynamics theory, and they have yielded accurate values of the Rydberg constant.

The most precise value of the Rydberg constant to date are derived from optical frequency measurements in atomic hydrogen by Hänch and co-workers.

The Rydberg constant thus determined is

$$R_\infty = \frac{\alpha^2 m_e c}{2h} = 10973731.568527\,(73)\ \text{m}^{-1} \tag{2.39}$$

It is the most accurate one hitherto measured.

2.6.2 Balmer Series

This is the most important series in hydrogen atom as almost all lines of this series lie in the visible region.

A series of emission or absorption lines of hydrogen atom in the visible region due to transition between $n = 2$ and higher states constitute Balmer series. This series was first discovered by the Swiss physicist Johan Balmer (1825–1898)[7]. The transition from the third level $n = 3$ to the second level $n = 2$ yields the red H_α emission line at 6563 Å; The next transition from $n = 4$ to $n = 2$ which lies in the green part of the spectrum at 4861 Å gives the H_β line, H-gamma is the transition from the $n = 5$ to $n = 2$ and lies in the violet at 4342 Å, and H-delta is the transition from $n = 6$ to $n = 2$ and is at 4101 Å and the series extends further. It ends relatively abrupt at the continuous spectrum at about 3650 Å and the absorption lines in the Balmer series crowding at the series limit. Balmer series formula was developed by Balmer in 1884 relating the frequencies of those lines to one another. In terms of wave number (inverse of wavelength), the formula is

$$\frac{1}{\lambda} = R\left(\frac{1}{2^2} - \frac{1}{n^2}\right) \tag{2.40}$$

where R is Rydberg's constant, whose value is given above and n is 3, 4, 5, 6, or 7 for the four lines studied by Balmer. The spectrum and transitions are given in Figure 2.9.

In this series, the electron jumps from any other orbit to the second orbit. Thus

$$\frac{1}{\lambda} = R\left[\frac{1}{2^2} - \frac{1}{n_2^2}\right], \; n_2 = 3,4,5,\dots \infty \tag{2.41}$$

Longest wavelength of Balmer series $\dfrac{1}{\lambda} = R\left[\dfrac{1}{2^2} - \dfrac{1}{3^2}\right] = R\left[\dfrac{1}{4} - \dfrac{1}{9}\right] = \dfrac{5R}{36}$

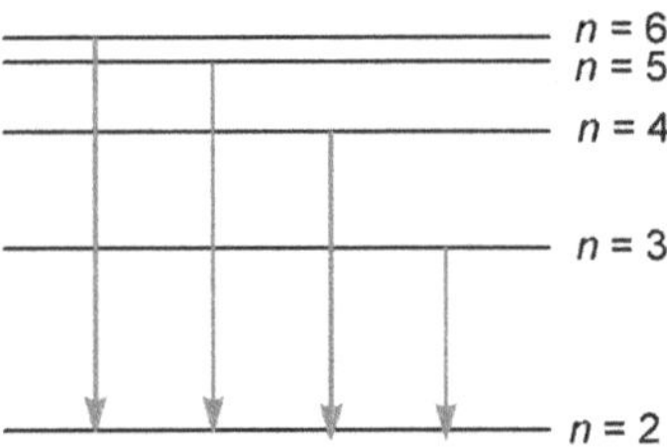

Figure 2.9   Balmer series

$$\lambda = \frac{36}{5R} = 6563 \text{ Å}$$

Series limit of Balmer series $\dfrac{1}{\lambda} = R\left[\dfrac{1}{2^2} - \dfrac{1}{\infty}\right] = \dfrac{R}{4},$

$$\lambda = \frac{4}{R} = 3648 \text{ Å} \tag{2.42}$$

The first, second, third members of the Balmer series are called H_α, H_β, H_γ, etc.

The Balmer series is important while almost all members in this series lie in the visible region.

Measured lines in Balmer series of hydrogen spectrum   The measured lines of the Balmer series of hydrogen in the nominal visible region are given in Table 2.1.

Table 2.1 Measured lines in Balmer series of hydrogen spectrum

Wavelength (nm)	Relative intensity	Transition	Colour
656.2852	180	$3 \rightarrow 2$	Red
656.272	120	$3 \rightarrow 2$	Red
486.133	80	$4 \rightarrow 2$	Blue-green (cyan)
434.047	30	$5 \rightarrow 2$	Violet
410.174	15	$6 \rightarrow 2$	Violet
397.0072	8	$7 \rightarrow 2$	Violet
388.9049	6	$8 \rightarrow 2$	Violet
383.5384	5	$9 \rightarrow 2$	Violet

The red line of deuterium is measurably different at 656.1065 (0.1787 nm difference).

2.6.3 Paschen Series

This series occurs when the electron makes a transition from any other outer orbit to third orbit, i.e., $n_1 = 3$. The series is known after its discoverer, Louis Karl Heinrich Friedrich Paschen[8] (1865–1947), German physicist, who is also equally known for his work on electrical discharges. He observed this series of hydrogen lines in 1908 and is known as Paschen series. The series lie mostly in the infrared region.

$$\frac{1}{\lambda} = R\left[\frac{1}{3^2} - \frac{1}{n_2^2}\right], \; n_2 = 4,5,6\dots \infty \tag{2.43}$$

Longest wavelength in Paschen series

$$\frac{1}{\lambda} = R\left[\frac{1}{3^2} - \frac{1}{4^2}\right] = R\left[\frac{1}{9} - \frac{1}{16}\right] = \frac{7R}{144} \tag{2.44}$$

$$\lambda = \frac{144}{7R} = 18761.1 \text{ Å}$$

Series limit of Paschen series

$$\frac{1}{\lambda} = R\left[\frac{1}{3^2} - \frac{1}{\infty}\right] = \frac{R}{9}, \; \lambda = \frac{9}{R} = 8208 \text{ Å} \tag{2.45}$$

2.6.4 Brackett Series

The Brackett series is obtained when the electron jumps from any other orbit to the fourth orbit, i.e., $n_1 = 4$ and $n_2 = 5,6,7$ ... The series was first observed by Frederick Sumner Brackett (1896–1988)[9] in 1922.

$$\frac{1}{\lambda} = R\left[\frac{1}{4^2} - \frac{1}{n_2^{\,2}}\right], \quad n_2 = 5,6,7,... \ \infty \tag{2.46}$$

Longest wavelength in Brackett series

$$\frac{1}{\lambda} = R\left[\frac{1}{4^2} - \frac{1}{5^2}\right] = R\left[\frac{1}{16} - \frac{1}{25}\right] = \frac{9R}{400} \tag{2.47}$$

$$\lambda = \frac{400}{9R} = 40533.3 \ \text{Å}$$

The shortest wavelength or the series limit of the Brackett series is

$$\frac{1}{\lambda} = [\frac{1}{4^2} - \frac{1}{\infty}] = R[\frac{1}{16} - 0] = \frac{R}{16} = 14592 \ \text{Å} \tag{2.48}$$

2.6.5 Pfund Series

This series is emitted when the electron jumps from any outer orbit to the fifth orbit.

$$\frac{1}{\lambda} = R\left[\frac{1}{5^2} - \frac{1}{n_2^{\,2}}\right], \quad n_2 = 6,7,8... \ \infty \tag{2.49}$$

Longest wavelength of Pfund series

$$\frac{1}{\lambda} = R\left[\frac{1}{5^2} - \frac{1}{6^2}\right] = R\left[\frac{1}{25} - \frac{1}{36}\right] = \frac{11R}{900}$$
$$\lambda = \frac{900}{11R} = 74618.18 \ \text{Å} \tag{2.50}$$

Series limit of Pfund series (shortest wavelength)

$$\frac{1}{\lambda} = R\left[\frac{1}{5^2} - \frac{1}{\infty}\right] = \frac{R}{25} \ \text{and hence } \lambda = \frac{25}{R} = 22800 \ \text{Å} \tag{2.51}$$

August Herman Pfund (1879–1949)[10] was an American spectroscopist.

Figure 2.10 shows the different spectral series in hydrogen atom.

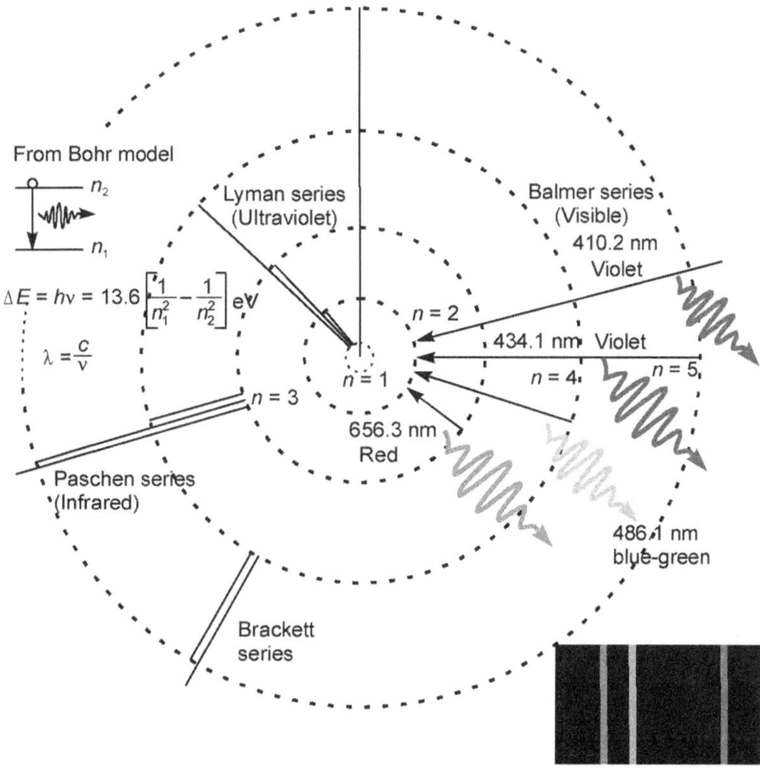

Figure 2.10   Spectrum of hydrogen atom

Based on the Bohr atomic model, we can now construct the energy level diagram of Li^{++} with $Z = 3$.

We know that $E_n = -\dfrac{13.6}{n^2}Z^2$ eV

For Li^{++}, $Z = 3$ and hence $E_n = -\dfrac{13.6}{n^2}3^2$ eV $= -\dfrac{122.4}{n^2}$ eV

Energy in the ground state $E_1 = -122.4$ eV

Ionization energy $= 122.4$ eV

Ionization potential $= 122.4$ Volt

It will give the energy levels as shown in Figure 2.11.

Now we can see that the lowest energy level is −122.4 eV whereas it was −13.6 eV in hydrogen atom. The energy level closer and closer to the nucleus is more and more negative. If we have to remove this electron from the trapped "energy well" we will have to irradiate the atom with a photon of energy 122.4 eV or greater. Thus we can ionize the atom by supplying energy of atleast 122.4 eV. This energy is used to lift the electron from energy of –122.4 eV to 0 eV. This is the only incident where energy need not have the exact match of the difference in energy of two states. In other cases of transitions (emission or absorption), it should exactly match the energy differences. Any excess energy (in this example greater than 122.4 eV) will remain as kinetic energy of the ionized electron.

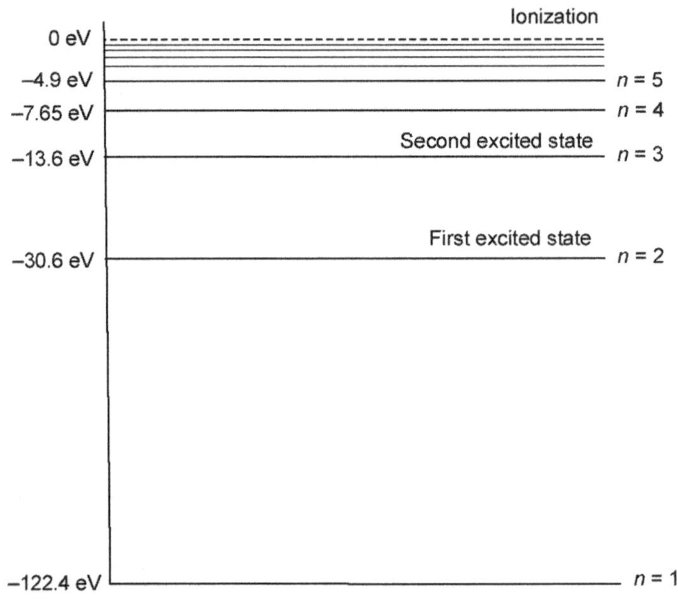

Figure 2.11   Energy level diagram of Li^{2+}

We consider an electron of the above atom (with $Z = 3$) has been excited to its second excited state ($n = 4$). When it falls back to its ground state it can do so through a total of five possible transitions as shown in Figure 2.12.

1. Third excited state ($n = 4$) directly to 4 its ground state ($n = 1$) releasing: ($n = 4$ to $n = 1$)
    In this case $\Delta E = -7.65 - (-122.4) = 114.75$ eV

2. Third excited state ($n = 4$) to first second excited state ($n = 3$), i.e., ($n = 4$ to $n = 3$) releasing: $\Delta E = -7.65 - (-13.6) = 5.95$ eV and then

3. Second excited state ($n = 3$) directly to its ground state ($n = 1$) releasing: ($n = 3$ to $n = 1$)
    In this case $\Delta E = -13.6 - (-122.4) = 108.8$ eV

4. Second excited state ($n = 3$) to first first excited state ($n = 2$) ($n = 3$ to $n = 2$) releasing: $\Delta E = -13.6 - (-30.6) = 17$ eV and then

5. First excited state ($n = 2$) to its ground state ($n = 1$) ($n = 2$ to $n = 1$) releasing: $DE = -30.6 - (-122.4) = 91.8$ eV

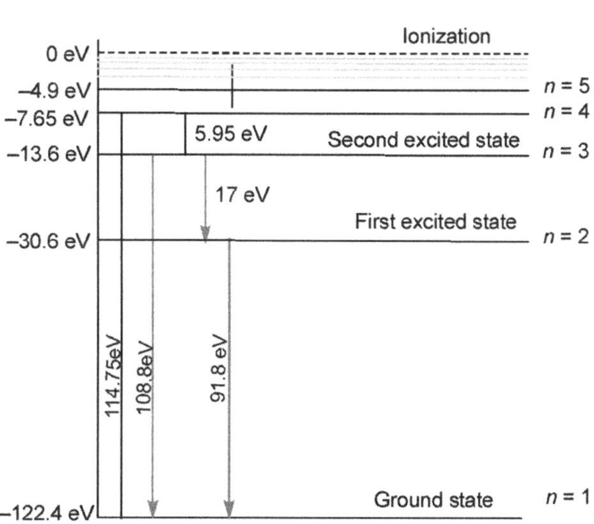

Figure 2.12   Lithium energy levels and transitions

Electron	114.75 eV	$=$	108.8 eV+5.95 eV
	$E_{4\to\text{ground}}$	$=$	$E_{4\to3}+E_{3\to\text{ground}}$
Photon	114.75eV	$=$	108.8 eV + 5.95 eV
	$h\nu_{4\to\text{ground}}$	$=$	$h\nu_{4\to3}+h\nu_{3\to\text{ground}}$

Converting our energy values from eV to Joules and using the formula $\nu = E/h$, our frequencies would be:

$$\to \text{ground } \nu_4 = \frac{114.75 \text{ eV} \times 1.6 \times 10^{-19}}{6.63 \times 10^{-34}} \text{Hz} = 276.92 \times 10^{14} \text{ Hz}$$

$$\nu_{4\to3} = \frac{5.95 \text{ eV} \times 1.6 \times 10^{-19}}{6.63 \times 10^{-34}} \text{Hz} = 14.36 \times 10^{14} \text{Hz}$$

$$\nu_{3\to\text{ground}} = 108.8 \, (1.6 \times 10^{-19}) \text{ Joule}/(6.63 \times 10^{-34}) = 262.56 \times 10^{14} \text{ Hz}$$

$$\nu_{4\to\text{ground}} = \nu_{4\to3} + \nu_{3\to\text{ground}}$$

$$276.92 \times 10^{14} \text{ Hz} = 14.36 \text{ Hz} + 262.56 \text{ Hz}$$

$$v_{3\to2} = 17(1.6\times10^{-19})\ \text{Joule}\ /\ (6.63\times10^{-34}) = 41.0\times10^{14}\ \text{Hz}$$

$$v_{2\to\text{ground}} = 91.8\ (1.6\times10^{-19})\ /\ (6.63\times10^{-34}) = 221.5\times10^{14}\ \text{MHz} \tag{2.52}$$

$$v_{3\to\text{ground}} = v_{3\to2} + v_{2\to\text{ground}}$$

$$262.56\times10^{14}\ \text{Hz} = 41.0\times10^{14}\ \text{Hz} + 221.5\times10^{14}\ \text{Hz} \tag{2.53}$$

It may be noted that in all formulae, when energy is given in eV, it should be converted into Joules. $1\ \text{eV} = 1.6\times10^{-19}\ \text{J}$

Many times we talk about the emitted wavelengths instead of frequencies. We use the formula $\lambda = \dfrac{c}{v}$ to convert the frequencies into wavelengths.

$$\lambda_{4\to\text{ground}} = \frac{3\times10^{8}}{276.92\times10^{14}} = 1.08\times10^{-8}\ \text{m}$$

$$\lambda_{3\to\text{ground}} = \frac{3\times10^{8}}{262.56\times10^{14}} = 1.14\times10^{-8}\ \text{m}$$

$$\lambda_{3\to2} = \frac{3\times10^{8}}{41.0\times10^{14}} = 7.32\times10^{-8}\ \text{m}$$

$$\lambda_{2\to\text{ground}} = \frac{3\times10^{8}}{221.5\times10^{14}} = 1.35\times10^{-8}\ \text{m} \tag{2.54}$$

We could see that all of these emissions are in the ultraviolet region. We can also note that though the frequencies add together correctly, the wavelengths do not.

2.7 RITZ COMBINATION PRINCIPLE

Walter Ritz discovered a pattern in the frequencies of spectral lines even before Bohr's theory was announced. The theory was proposed in 1908 to explain relationship of the spectral lines for all atoms. The principle states that the spectral lines of the elements have frequencies that are either the sums or the differences of the frequencies of two other lines.

The Bohr's atomic model supports this addition pattern of the frequency of emission lines. The energies lost, or released, by the electron are the energies present in each emitted photon. This principle is also known as Rydberg–Ritz Combination Principle.

We have seen from equation that the energies of stationary states are negative and $T_n \to 0$ as $n \to \infty$. Hence the term $n = \infty$ is plotted at the top and the other terms come below. The transitions are often drawn as vertical or oblique lines and the arrows represent whether the transitions are in emission or in absorption.

2.8 HYDROGEN SPECTRA AND RYDBERG CONSTANT

2.8.1 Rydberg Formula

While trying to explain the wavelengths of the spectral lines in alkali metals, Janne Rydberg, in the year 1888, noticed that the lines could be very well be explained by a formula if one could represent the lines in wave numbers (the number of waves in unit length which is $1/\lambda$ the inverse of wavelength) rather than the wavelength. This was even before the atom models were proposed and the Balmer formula was derived. He obtained a formula

$$n = n_0 - \frac{C_0}{(n+m')^2} \tag{2.55}$$

where n is the line in wave number unit, n_0 is the series limit, m' is a constant. m' differs for different series and C_0 is a universal constant. It has been found that for hydrogen $m' = 0$ and $C_0 = 4n_0$ and thus the formula for hydrogen is

$$n = n_0 - \frac{4n_0}{(n)^2} \tag{2.56}$$

The Rydberg formula describes the transitions or quantum jumps between energy levels. When an electron "jumps" from one energy level to another lower level, a photon is emitted. We have already seen that using the derived formula for the different energy levels of hydrogen, we can determine the wave number and then the wavelength of light the hydrogen atom emits. We have also seen the energy of photon is given by the difference of the two energy levels involved and is given by

$$E - E_{final} - E_{initial} = \frac{m_e e^4}{8h^2 \varepsilon_0^2}\left(\frac{1}{n_f^2} - \frac{1}{n_i^2}\right) \tag{2.57}$$

where n_f corresponds to the final level and n_i initial energy level. The energy of the photon $E = \dfrac{hc}{\lambda}$ and hence the wavelength of the photon is given by

$$\frac{1}{\lambda} = \frac{m_e e^4}{8ch^3 \varepsilon_0^2}\left(\frac{1}{n_f^2} - \frac{1}{n_i^2}\right) \tag{2.58}$$

$$\text{i.e.,}\quad \frac{1}{\lambda} = R\left(\frac{1}{n_f^2} - \frac{1}{n_i^2}\right) \tag{2.59}$$

R is called the Rydberg constant. As described earlier, this formula was known from 1888 to scientists studying spectroscopy, however there was no theoretical justification for the formula till Bohr derived it as described in the earlier sections.

2.8.2 Rydberg Constant

The Rydberg constant is one of the most well determined physical constants with relatively high accuracy. Each chemical element has its own Rydberg constant, which can be derived from the "infinity" Rydberg constant.

The "infinity" Rydberg constant according to 2002 CODATA is

$$R_\infty = \frac{m_e e^4}{(4\pi\varepsilon_0)^2 \hbar^3 4\pi c} = 1.097\ 373\ 156\ 8527\ (73) \times 10^7\ \text{m}^{-1} \tag{2.60}$$

where,

$\hbar$ is the reduced Planck's constant.

m_e is the rest mass of the electron.

e is the elementary charge.

c is the speed of light in vacuum.

ε_0 is permittivity of free space.

The Rydberg constant is often used in the form of energy and can be written as

$$hcR_\infty = 13.6056923\ (12)\ \text{eV} = 1\ R_y \tag{2.61}$$

The Rydberg constant for a particular atom with one electron is given by

$$R_M = \frac{R_\infty}{1 + \dfrac{m_e}{M}} \tag{2.62}$$

where M is the mass of the particular atomic nucleus.

It was later pointed out by Bohr that the electron mass in the theory should be better replaced by $\dfrac{mM}{m+M}$ which is called the reduced mass μ. It can be proved that when two masses of m and M rotate, it can be considered as the rotation of a single mass μ which is $mM/m+M$.

Consider the mass of the nucleus is M and that of the electron is m (*See* Figure 2.13).

We can consider the electron e^- and the nucleus e^+ at the positions in the Figure 2.14. r is the distance between the electron and the nucleus, r_1 is the distance from the electron to the centre of mass and r_2 the distance from the centre of mass to the nucleus. The moment equation is then

$$mr_1 = Mr_2 \tag{2.63}$$

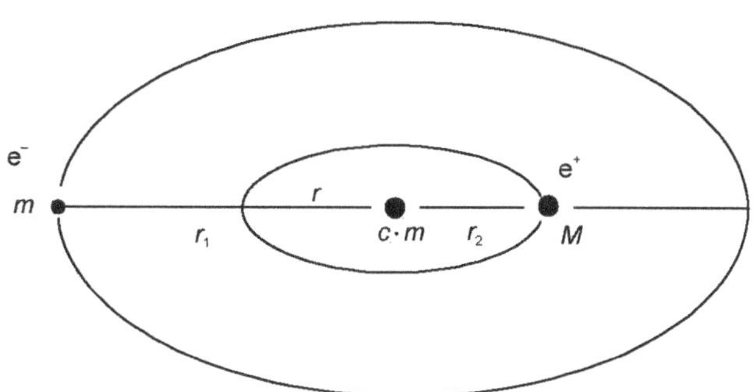

Figure 2.13   Diagrammatic description of an atom

and the moment of inertia of rotation $I = mr_1^2 + Mr_2^2$

$$r_1 + r_2 = r$$

$$m(r - r_2) = Mr_2 \text{ and}$$

hence $$mr - mr_2 = Mr_2, \quad (m + M)r_2 = mr$$

$$r_2 = \left(\frac{m}{m + M}\right)r \tag{2.64}$$

Similarly

$$r_1 = \left(\frac{M}{m + M}\right)r \tag{2.65}$$

Thus $$I = m\,r_1^2 + Mr_2^2 = m\left(\frac{M^2}{(m + M)^2}\right)r^2 + M\left(\frac{m^2}{(m + M)^2}\right)r^2$$

$$= \left(\frac{1}{(m + M)^2}\right)r^2\left\{mM^2 + Mm^2\right\}$$

$$= \left(\frac{mM}{(m + M)^2}\right)r^2\left\{M + m\right\} = \left(\frac{mM}{(m + M)}\right)r^2 = \mu r^2 \tag{2.66}$$

where $\mu = \dfrac{mM}{(m + M)}$ is called the reduced mass.

The moment of inertia and the angular momentum of this single mass point revolving around the nucleus would be the same as the electron of mass m and nuclear mass M rotate around their common centre of mass.

It can be seen from the equation that when M is infinitely large μ becomes the same as m. Thus we can consider that the Rydberg constant in equation corresponds to $M = \infty$.

We have seen in equation that

$$R = \frac{m_e e^4}{8ch^3 \varepsilon_0^2} \tag{2.67}$$

It can be shown that $\dfrac{m_e e^4}{8ch^3 \varepsilon_0^2} = \left(\dfrac{1}{4\pi\varepsilon_0}\right)^2 2\pi^2 me^4 / ch^3$ $\tag{2.68}$

$$R = \left(\frac{1}{4\pi\varepsilon_0}\right)^2 2\pi^2 me^4 / ch^3 = \left(\frac{1}{4\pi\varepsilon_0}\right)^2 \left(\frac{2\pi^2 e^4}{ch^3}\right) m = R_\infty \tag{2.69}$$

When the nuclear mass is considered as μ then

$$R = \left(\frac{1}{4\pi\varepsilon_0}\right)^2 \left(\frac{2\pi^2 e^4}{ch^3}\right) \frac{mM}{M+m} \tag{2.70}$$

$$R = \frac{e^4}{8ch^3 \varepsilon_0^2} \frac{mM}{M+m} = \left(\frac{1}{4\pi\varepsilon_0}\right)^2 \left(\frac{2\pi^2 e^4}{ch^3}\right) \frac{mM}{M+m}$$

$$= \left(\frac{1}{4\pi\varepsilon_0}\right)^2 \left(\frac{2\pi^2 e^4}{ch^3}\right) m \frac{M}{M+m} = R_\infty \frac{M}{M+m} \tag{2.71}$$

$$R_m = R_\infty \left(\frac{M}{M+m}\right) \tag{2.72}$$

In the case of hydrogen and helium the Rydberg constants derived by Houston[11] way back in 1927 were $R_H = 109677.759$ cm^{-1} and $R^+_{He} = 109722.403 \pm 0.004$ cm^{-1}.

The atomic mass of helium is four times that of hydrogen and we can infer that $R_\infty = R_H\left(1+\dfrac{m}{M}\right)$

which is equal to $R_{He+}\left(1+\dfrac{m}{4M}\right)$.

For a hydrogen atom, in principle, the effective mass is the reduced mass obtained from the proton and the electron. The mass of a proton is 1.672 621 71(29) × 10^{-27} kg and the mass of an electron is 9.1093897 × 10^{-31} kg. Thus proton is about 1836 times the mass of an electron.

In MKS, this gives the Rydberg constant as follows:

Reduced mass of the hydrogen atom

$$\mu = \frac{m_e m_p}{m_e + m_p} = \frac{9.10938 \times 10^{-31}\,\text{kg} \times 1.67262 \times 10^{-27}\,\text{kg}}{9.10938 \times 10^{-31} + 1.67262 \times 10^{-27}} = 9.10442 \times 10^{-31}\ \text{kg} \tag{2.73}$$

$$R_{\text{H}} = \frac{m_e m_p}{m_e + m_p} \frac{e^4}{8\varepsilon_0^2 c h^3} = 10967758\ \text{cm}^{-1} = 1.0967758 \times 10^5\ \text{cm}^{-1} \tag{2.74}$$

The best value for $R_{\text{H}} = 10967758 \pm 1$ cm^{-1}

where m_e is the electron mass, m_p is the proton mass, e is the electron charge, c is the speed of light, ε_0 is the permittivity of free space, and h is Planck's constant.

For heavy hydrogenic ions $m \gg m_e$, so the reduced mass can be taken as that of the electron m_e, as

$$\frac{m_e m_p}{m_e + m_p} = \frac{m_e m_p}{m_p} = m_e \tag{2.75}$$

Thus $R_\alpha = \dfrac{m_e e^4}{8\varepsilon_0^2 c h^3} = 1.0973731568527\,(73) \times 10^5\ \text{cm}^{-1}$

$$= 10973731.568527\,(84)\ \text{m}^{-1}$$

$$= 1.0973731568527\,(84) \times 10^7\ \text{m}^{-1} \tag{2.76}$$

CODATA Value of $R_\alpha = 10\ 973\ 731.\ 568\ 527\ (73)$ m^{-1}.

2.8.3   Energy in Terms of Other Constants

We have seen that from expression (2.13)

$$E_n = -Z^2 \frac{m_e e^4}{8h^2 \varepsilon_0^2} \cdot \frac{1}{n^2}$$

Or putting in valucs of

$$\varepsilon_0 = 8.85 \times 10^{-12}\ \text{C}^2\,/\,\text{Nm}^2$$

$$m_e = 9.11 \times 10^{-31}\ \text{Kg}$$

$$e = 1.6 \times 10^{-19}\,\text{C}$$

$$h = 6.625 \times 10^{-34}\,\text{Js}$$

$$E_n = \frac{-13.6}{n^2} Z^2 \text{ eV}$$

And for hydrogen

$$E_n = \frac{-13.6}{n^2} \text{eV}$$

Thus the lowest energy level of hydrogen ($n = 1$) is about –13.6 eV. The next energy level ($n = 2$) is $-\dfrac{13.6}{2^2} = -3.4\text{ eV}$. The third ($n = 3$) is –1.51 eV, and so on. Note that these energies are less than zero which means that the electron is in a bound state with proton. Positive energy states correspond to ionized atom where the electron is no longer bound, but is in a scattering state. These are shown in Figure 2.14.

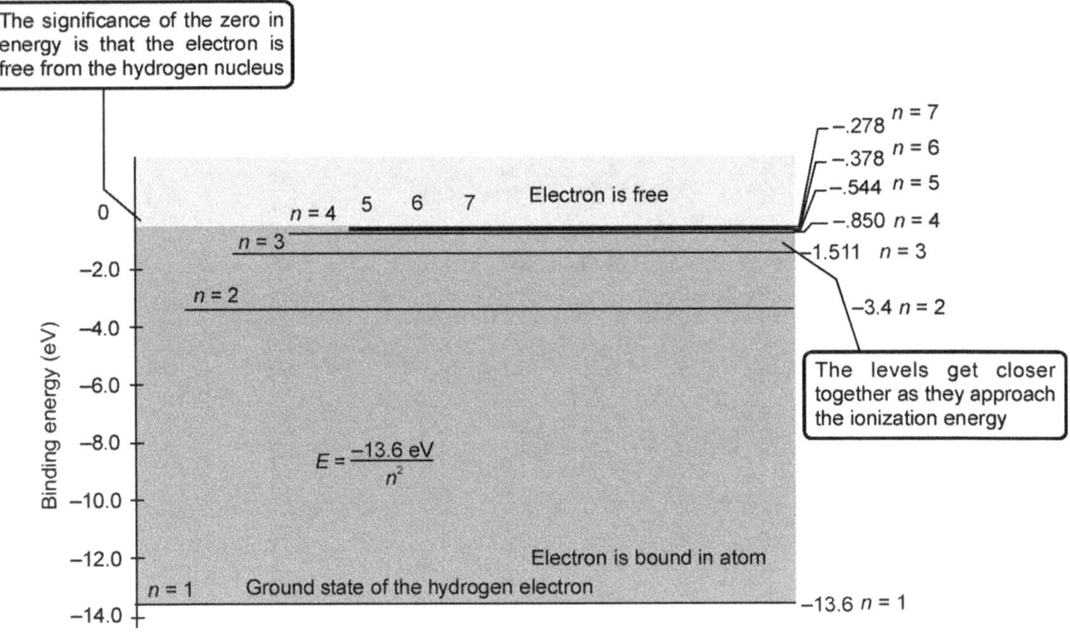

Figure 2.14   Scaled plot of hydrogen atom energy levels in electron volts

We have found that $E_n = -\dfrac{Z^2 m_e e^4}{8 h^2 \varepsilon_0^2} \cdot \dfrac{1}{n^2}\,.$

We multiply the expression $E_n = -\dfrac{Z^2 m_e e^4}{8h^2 \varepsilon_0^2} \cdot \dfrac{1}{n^2}$ by c^2 and divide by c^2, then we get

$$E_n = -\frac{Z^2 m_e c^2 e^4}{8h^2 c^2 \varepsilon_0^2} \cdot \frac{1}{n^2} \tag{2.77}$$

By re-grouping

$$E_n = -\frac{1}{2} m_e c^2 \left(\frac{Z^2 e^4}{4h^2 c^2 \varepsilon_0^2} \right) \frac{1}{n^2}$$

$$= \frac{-E_r Z^2 \alpha^2}{2n^2} \tag{2.78}$$

where,

$\alpha = \dfrac{e^2}{2\varepsilon_0 hc}$ is the fine-structure constant, also known as **Sommerfeld fine-structure constant**.

E_r is the rest energy of the electron $= m_e c^2$.

n is the principal quantum number.

E_n is the energy level.

Thus,

$$\alpha = \frac{e^2}{2\varepsilon_0 hc} \tag{2.79}$$

and we can represent R_∞ in terms of fine structure constant and Compton wavelength as

$$R_\infty = \frac{\alpha^2 m_e c}{2h} = \frac{\alpha^2}{2\lambda_e} \tag{2.80}$$

where,

α is fine-structure constant and

λ_e is the Compton wavelength of the electron.

The fine-structure constant or Sommerfeld fine-structure constant, usually denoted as α is considered to be a fundamental physical constant characterizing the strength of the electromagnetic interaction.

$$\alpha = \frac{e^2}{2\varepsilon_0 hc}$$

$$\varepsilon_0 = 8.85 \times 10^{-12} \ C^2 / Nm^2$$

$$e = 1.6 \times 10^{-19} \ C$$

$$h = 6.625 \times 10^{-34} \ Js$$

$$c = 3 \times 10^8 \ m/s$$

The fine-structure constant is a unitless numerical constant, i.e., it is dimensionless. The value of α is approximately 1/137.

$$\alpha \sim \frac{1}{137} \tag{2.81}$$

The defining expression and the value recommended by 2006 CODATA as reported by NIST reference on constants, units and uncertainty[12] is:

$$\alpha = \frac{e^2}{2\varepsilon_0 hc} = \frac{(1.6 \times 10^{-19})^2}{2 \times 8.85 \times 10^{-12} \times 6.625 \times 10^{-34} \times 3 \times 10^8} = 7.2973525376 \ (50) \times 10^{-3} \tag{2.82}$$

$$= \frac{1}{137.035999679(94)}$$

(numbers within parentheses are uncertainties.)

The physical origin of fine-structure constant is related to the so called fine structure-closely spaced groups of optical spectrum lines of elements like: hydrogen and helium. With each optical spectrum line one can associate three numbers:

✧ n, which is the principal quantum number

✧ l, which is the azimuthal quantum number

✧ j, which is the angular momentum quantum number and

✧ m, which is the magnetic quantum number and is an additional quantum number.

Since the energy levels are quantized, the optical spectrum appears as a series of lines instead of continuous spectrum. Each group of optical lines has same n number, but different values of l and j numbers. The quantum number $j = l \pm \frac{1}{2}$ (*See* Section on spin–orbit interaction). According to the relativistic quantum mechanics proposed by P.A.M. Dirac [13] ,the energy levels of a one-electron atom (example hydrogen atom) which have the same n and j numbers will coincide exactly—but the energy value is different from what is predicted by the Bohr's theory by an amount that is proportional to the square of the fine-structure constant (alpha).

There have been deviations from Dirac's theory discovered in 1947. It has been thought that level having same n, different l but with same j were degenerate, i.e., have the same energy. This was the case with $n = 2$, $l = 0$ and 1 and $j = 1/2$ state in hydrogen atom. The $n = 2$, $l = 0$, $j = 1/2$ state is called $2 \ ^2S_{1/2}$ and $n = 2$, $l = 1$, $j = 1/2$ state is called $2 \ ^2P_{1/2}$. That means the levels $2P_{1/2}$ and $2S_{1/2}$ were thought to be having the same energy. However in 1947 Lamb and Rutherford employed the technique of microwave spectroscopy and found that these levels infact are split by an amount of 1058 MHz. This discrepancy was later named as Lamb shift (after its discoverer Willis Eugene Lamb, Jr.[14] [1913–2008]) and is due to the interaction of an electron with the zero-point fluctuations of the electromagnetic field.

As mentioned above in a multi-electron atom system, the spin–orbit interaction becomes important too and the number of optical spectrum lines in the groups becomes even more numerous.

Historically, the fine-structure constant $\left(\alpha = \dfrac{v}{c} \right)$ was interpreted as the ratio of the velocity of the electron in the first circular orbit of the relativistic Bohr atom to the speed of light in vacuum. It was Arnold Sommerfeld who first introduced the fine-structure constant in physics in 1916, as a measure of the relativistic deviations in atomic spectral lines from the predictions of the Bohr atom model.

In Sommerfeld's analysis it determines the amount of the splitting or fine-structure of the hydrogenic spectral lines. Though the name of the fine-structure constant has originally used in the theory for the fine structure of atomic spectra, its modern use is far from being as specialized. It is also the quotient between the maximum angular momentum allowed by relativity for a closed orbit and the minimum angular momentum allowed for it by quantum mechanics.

The fine-structure constant is now considered as a universal constant. Scientists have been wondering for many years whether the fine-structure constant is really a constant, i.e., whether or not its value is different at different times or in different places. It has been perceived it many times as a varying constant in order to solve many of the cosmological problems like the String theory and Standard Model in particle physics. However, the radioactive decay observed in Oklo natural nuclear fission reactor and observation of spectral lines of distant astonomical objects have revealed that the fine structure constant is a real constant and does not vary with place or time[15–20].

As mentioned earlier the fine-structure constant is dimensionless and it has long fascinated physicists. As a dimensionless constant it does not seem to be directly related to any mathematical constant. Richard Feynmann[21], one of the founders of quantum electrodynamics, referred to it as "one of the greatest damn mysteries of physics: a magic number that comes to us with no understanding by man."

The fine-structure constant can also be defined as

$$\alpha = \frac{k_e e^2}{\hbar c} = \frac{e^2}{2\varepsilon_0 hc} \tag{2.83}$$

where k_e is the electrostatic constant which is

$$1/4\pi\varepsilon_0 = \mu_0 c_0^2 / 4\pi = 8.987\ 551\ 787 \times 10^9 \ \text{C}^{-2}\text{Nm}^2 (= 1.439\ 96 \times 10^{-9}\,\text{eV m}),$$

where e is the elementary charge, $\hbar$ is the reduced Planck constant, c is the speed of light in a vacuum, and ε_0 is vacuum permittivity or electric constant.

$$\varepsilon_0 = 1/\mu_0 c^2 \tag{2.84}$$

where μ_0 is magnetic constant.

Its value is $8.854\ 187\ 817 \times 10^{-12}$ A^2 s^4 kg^{-1} m^{-3} or $5.526\ 35 \times 10^7$ eV m^{-1}.

In electrostatic cgs units, the unit of electric charge (the Statcoulomb or esu of charge) is defined in such a way that the permittivity factor, $4\pi\varepsilon_0$, is the dimensionless constant 1. Then the fine-structure constant becomes

$$\alpha = \frac{e^2}{\hbar c} \tag{2.85}$$

The value of Rydberg constant in some of the atoms are given in Table 2.2. and is shown graphically in Figure 2.15.

Table 2.2   Value of Rydberg constant in some of the atoms

Atom	Atomic number (Z)	Atomic weight (A)	R_A (in cm^{-1})	$R_\infty - R_A$ (in cm^{-1})
H	1	1	109677.759	59.665
D	1	2	109707.387	30.037
He	2	4	109722.403	15.021
Li	3	7	109728.84	8.58
Be	4	9	109730.74	6.68
B	5	11	109731.96	5.46
O	8	16	109733.66	3.76
Ne	10	20	109734.42	3.00
Se	21	45	109736.08	1.34
As	33	75	109736.62	0.80
∞	–	∞	109737.424	0.00

Today the hydrogen atom plays an important role in determining fundamental constants and atomic parameters. The hydrogen spectroscopy has been revolutionized by the work of Hänsch and co-workers in Muenchen.

In some books electronic charge e is described as q_e and thus E_n is described as

$$E_n = \frac{-m_e q_e^4}{8h^2 \varepsilon_0^2} \cdot \frac{1}{n^2} \quad \text{for hydrogen where } Z = 1 \tag{2.86}$$

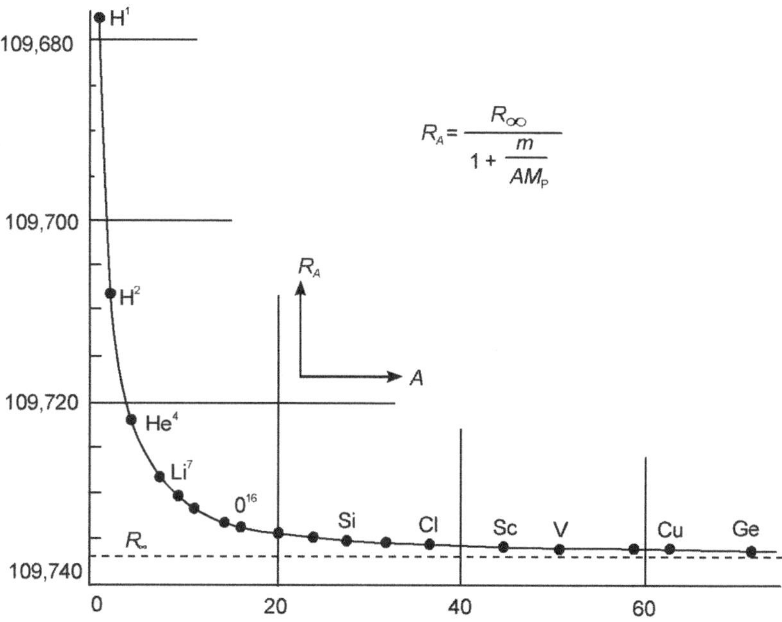

$$R_A = \frac{R_\infty}{1 + \dfrac{m}{AM_P}}$$

Figure 2.15   Variation of the Rydberg constant R with the atomic weight A

And also E_n is described as $E_n = h\nu = \dfrac{2\pi^2 me^4}{h^2} Z^2 \dfrac{1}{n^2}\,\mathrm{eV}$ (2.87)

And immediately written as $E_n = -\dfrac{13.6}{n^2}\,\mathrm{eV}$

In expression 2.87, $\left(\dfrac{1}{4\pi\varepsilon_0}\right)^2$ is omitted, as it is a constant. Generally it should be taken into account.

$$\frac{2\pi^2 me^4}{h^2} = \frac{2\times(3.14)^2 \times 9.11\times10^{-31}\times(1.6\times10^{-19})^4}{(6.625\times10^{-34})^2}\mathrm{J} = 26.823\times10^{-39}\,\mathrm{J}$$

$$(\frac{1}{4\pi\varepsilon_0})^2 = \frac{1}{4\times3.14\times(8.85\times10^{-12})^2} = \frac{1}{12355.655\times10^{-24}}$$

$$\left(\frac{1}{4\pi\varepsilon_0}\right)^2 \times \frac{2\pi^2 me^4}{h^2} = \frac{26.823\times10^{-39}}{12355.655\times10^{-24}} = 2.170918\times10^{-18}\ \mathrm{J}$$

And in eV it is $\dfrac{2.170918\times10^{-18}}{1.6\times10^{-19}} = 13.6\ \mathrm{eV}$ (2.88)

2.8.4 Hydrogen like Atoms

Bohr's theory of hydrogen atom can be applied to other hydrogen like atoms, e.g. atoms in which one electron is bound to the nucleus. Examples are He^+, Li^{++}, Be^{+++}. It was Pickering who observed a series of seven lines in the spectrum of ς- Puppis at approximate wavelengths 384.4, 385.7, 392.3, 402.8, 420.3, 450.5 and 514.4 nm. These lines were thought of originating from hydrogen as they converged to the same limit as the Balmer series.

Figure 2.16 shows the spectra of hydrogen atom (Balmer series) observed in a radio source 3C 273 called a quasi stellar radio source (quasars).[22] Stellar spectra exhibit comparatively small Doppler shifts because a star in our galaxy can not move extremely fast in relative to the sun. However, in quasars which belong to another solar system, the shifts are large enough to be identified. According to Hubble law, the huge red shift implies that the distance to 3C 273 is roughly 3 billion light years.

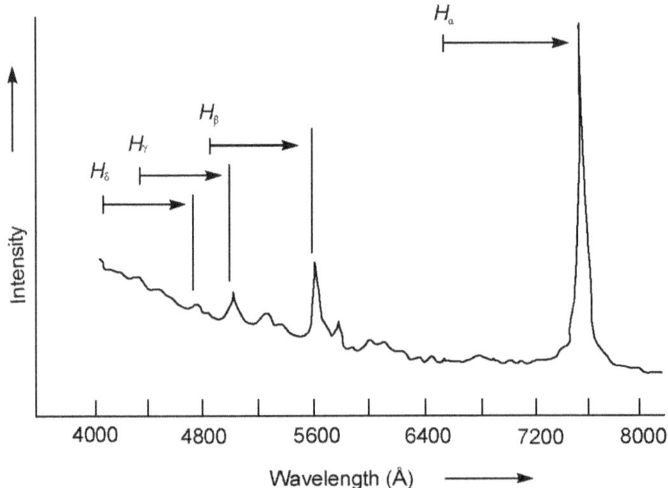

Figure 2.16   Spectrum of 3C 273-Four bright emission lines of hydrogen atom dominate the spectrum. The arrows indicate how far these spectral lines are red shifted from their visual laboratory wavelength.

We have seen from the Bohr's theory of hydrogen atom, Rydberg constant is

$$R = \frac{Z^2 e^2}{2a_0 hc} = \left(\frac{1}{4\pi\varepsilon_0}\right)^2 \frac{2\pi^2 me^4 Z^2}{ch^3} \tag{2.89}$$

In the case of helium $Z = 2$, the Rydberg constant R in the case of hydrogen atom can be replaced by $4R$ and the spectrum of ionized helium would be given by

$$v = 4R\left[\frac{1}{n_1^2} - \frac{1}{n_2^2}\right]$$

The various series observed in He$^+$ is given in Table 2.3.

Table 2.3   Spectrum of ionized helium

n_1	n_2	Spectral series	Spectral region
1	2,3...	Lyman series	Extreme ultraviolet
2	3,4...	Lyman series	Extreme ultraviolet
3	4,5...	Fowler series	Ultraviolet and visible
4	5,6...	Pickering series	Near ualtraviolet and visible

The successful interpretation of the spectral features of the ionized helium was the first conformation of the success of Bohr theory.

In the case of Li^{++} and Be^{+++}, the main features of the spectra can be obtained substituting $Z = 3$ and $Z = 4$ in equation placing $n_1 = 1$ and $n_2 = 2,3,4,...$ The spectral lines in the extreme ultraviolet region, the first members of which have been found by Ericsson and Edler at 135.02 and 75.94 nm in the case of Li^{++} and Be^{+++} respectively.

We can see from the expression of electronic transitions (*See* for example, equation 2.28) that if Rydberg constant is the same for both hydrogen and ionized helium, the alternate members of the Pickering series in He$^+$ should exactly coincide with the Balmer series of hydrogen atom. However experimental observation had revealed that they are not exactly coincident which shows that the Rydberg constant in both the cases is different. A careful examination had shown that the series limit of these two series differ by about 11 cm^{-1} which implies that the Rydberg constant in ionized helium is 44 cm^{-1} more than that of hydrogen. The Table 2.4 shows the differences in the altering members of the Pickering series of He$^+$ and the Balmer series in hydrogen.

The photon energy due to an electron transition between an upper atomic level of energy E_2 and a lower level E_1 is

$$\Delta E = E_2 - E_1 = h\nu$$

We have seen that $\nu = \dfrac{c}{\lambda}$ and hence $\Delta E = E_2 - E_1 = h\nu = \dfrac{c}{\lambda} = hc\bar{\nu}$ where ν is the frequency,

λ is the wavelength in vacuum, c is the velocity of light and h is the Planck's constant. In most spectroscopic work especially in radio frequency, microwave or laser spectroscopy, the transition frequencies are measured in Hertz (1 Hz = 1 cycle/second) or in multiples example: in kHz , MHz or GHz. However, measurements in any units, frequency, wavelength (in vacuum) or in wave numbers are equally employed and they are related to each other through the speed of light which is accurately determined. The most common wavelength units are Angstrom or nanometre (1 nm = 10 Å) and micrometre (μm). The SI unit of wave number is the inverse metre. But in practice wave numbers are expressed in inverse centimetre: 1 cm^{-1} = 10^2 m^{-1}. 1 cm^{-1} is equivalent to 2.997 924 58 $\times$ 10^4 MHz.

For a rough approximation 1 cm^{-1} = 30 GHz. More exactly it is 29.9792458 GHz. In addition to frequency and wave number units, atomic energies are also often expressed in electron volts (eV).

Table 2.4 Wavelengths of the Balmer and Pickering series

Pickering series of He$^+$		Balmer series in H		Difference in wavelength (A)
n_2	$\lambda(A)$	n_2	$\lambda(A)$	
6	6560.19	3	6562.852	2.66
7	5411.60			
8	4859.40	4	4861.33	1.93
9	4541.66			
10	4338.74	5	4340.47	1.73
11	4199.90			
12	4100.10	6	4101.74	1.64

One eV is equivalent to[23]

2.417 988 4(7) × 10^{14} Hz

8 065.54445(69) cm^{-1}

1 239.84191(11) nm

11 604.505(20) K (Kelvin)

1.602 1765 3(14) × 10^{-19} J (joule)

The basic unit of temperature is Kelvin, which is equivalent to 0.695 039(6) cm^{-1}, i.e., the value of the Boltzmann constant k expressed in wave number units per Kelvin is 0.7 cm^{-1}/K. The earlier idea of using the symbol K (1 kayser) for 1 cm^{-1} is nowadays discontinued.

Again

1 Hatree = 2625.5 kJ = 627.52 kcal/mol = 27.2116 eV = 219 474.6 cm^{-1}

a_0 = 0.529 Å = 0.0529 nm = 52.9 pm

1 cal = 4.184 J

1 eV = 96.485 kJ/mol = 8065.541 cm^{-1} = 1.60217653 (14) × 10^{-19} Joule

1 Joule = 6.241506363 × 10^{18} eV

1 cm^{-1} = 11.962 J/mol = 29.9792458 GHz ≈ 30 GHz

kT/hc = 207.226 cm^{-1} at 298.2 K

The unit of energy in the system of atomic units (a.u.) often used for theoretical calculations is the Hatree. One hatree is equal to 2 rydbergs. We have already seen that the rydberg for an atom having nuclear mass M is

$$1R_y \ = \ R_M = M(M+m_e)^{-1} \ R_\infty,$$

with
$$R_\infty \ = \ \frac{\alpha^2 m_e c}{2h} = 10973731.568\ 62(9)\ \mathrm{m}^{-1}$$

which is equivalent to 13.605 698(4) eV. The Rydberg constant R_∞ is thus the limit value for infinite nuclear mass.

2.9 BOHR'S MODEL IN ABSTRACT

2.9.1 Electron Energy Levels in Hydrogen

The derivation starts with three simple assumptions:

i.   All particles are wavelike, and an electron's wavelength λ, is related to its velocity v by:

$$\lambda = \frac{h}{m_e \mathrm{v}} \tag{a}$$

where h is Planck's constant , and m_e is the mass of the electron. This is de Broglie's hypothesis of matter waves, but Bohr did not make use of this assumption in his original derivation, because it hadn't been proposed at that time. However it allows the following intuitive statement.

ii. The circumference of the electron's orbit must be an integer multiple of its wavelength:

$$2\pi r = n\lambda \tag{b}$$

where r is the radius of the electron's orbit, and n is a positive integer.

iii. The electron is held in orbit by the coulomb force and the stability of the atom is by a balance between the coulomb and the centripetal force.

$$\left(\frac{1}{4\pi\varepsilon_0}\right)\frac{Ze^2}{r^2} = \frac{m_e \mathrm{v}^2}{r} \tag{c}$$

where m_e is the mass of the electron and e is its charge.

(a), (b) and (c) are three equations with three unknowns: $\lambda, r,$ v. After solving this system of equations to find an expression for just v, and then it is placed into the equation to obtain the total energy of the electron:

$$E = E_{kinetic} + E_{potential}$$

$$= \frac{1}{2}m_e v^2 - \frac{Ze^2}{4\pi\varepsilon_0 r} = \frac{Ze^2}{8\pi\varepsilon_0 r} - \frac{Ze^2}{4\pi\varepsilon_0 r} = -\frac{Ze^2}{8\pi\varepsilon_0 r} = -\frac{1}{2}m_e v^2$$

Substituting, one obtains the energy of the different levels of hydrogen:

$$E_n = -Z^2 \frac{me^4}{8h^2\varepsilon_0^2} \cdot \frac{1}{n^2}$$

By plugging in values for the constants, we obtain for hydrogen atom

$$E_n = \frac{-13.6 \text{ eV}}{n^2}$$

Thus, the lowest energy level of hydrogen ($n = 1$) is -13.6 eV. The next energy level ($n = 2$) is -3.4 eV. The third ($n = 3$) is -1.51 eV, and so on. Note that these energies are less than zero, meaning that the electron is in a bound state with the proton. Positive energy states correspond to the ionized atom where the electron is no longer bound, but is in a scattering state.

From (2.5) The velocity of the electron in any orbit n is

$$v = \frac{Ze^2}{4\pi\varepsilon_0} \cdot \frac{2\pi}{nh} = \frac{Ze^2}{2\varepsilon_0 h} \frac{1}{n} = \frac{Ze^2}{2\varepsilon_0 h} \frac{c}{c} \frac{1}{n} = \frac{Ze^2}{2\varepsilon_0 hc} \frac{c}{n} = \frac{c}{n}\alpha$$

Thus the velocity of the electron in any orbit n is given by $v = \frac{c}{n}\alpha$

where α is fine-structure constant $\alpha = \left(\frac{1}{4\pi\varepsilon_0}\right) \frac{Ze^2}{(h/2\pi)c} = \frac{1}{137}$

Hence, $$v = \frac{1}{137}\frac{c}{n}$$

2.10 FRANCK–HERTZ EXPERIMENT

A striking demonstration of quantized energy levels as proposed by Bohr has been provided by the experiments by James Franck and Gustav Hertz.[24] In 1914 James Franck and Gustav Hertz published the results of an experiment, which demonstrated the existence of excited states in mercury atoms, helping to confirm the quantum theory, which predicted that electrons occupied only discrete, quantized energy states. It provided strong evidence that Bohr's model of atoms with quantized energy levels was correct. It should be noted that they performed their experiments quite unaware of Bohr's theory.

In order to explain the Franck–Hertz experiments, let us consider a heavy atom like mercury. The electrons in the innermost shells of $^{202}_{80}$Hg are difficult to be removed due to the strong electrostatic attraction of the nucleus. The binding energies of these electrons are generally in the range of a few keV. The outer electrons called the valence electrons are shielded from the nucleus by the screening effect of the electrons in the inner shells. The binding energies of the valence electrons are in the range of a few eV. In the Franck–Hertz[24] experiment only the valence electrons are involved. The energy levels of $^{202}_{80}\text{Hg}$ are as shown in Figure 2.17. These are generally called optical levels as the transition among these levels involves photons of wavelengths in the visible or ultraviolet regions of the electromagnetic spectrum. The energy of the valence electron in the ground state $n = 1$ is -10.42 eV and of the first excited state $n = 2$ is -5.54 eV. Thus the energy needed to raise the electron from the ground state to the first excited state is

$$E_1 = E_{n=2} - E_{n=1} = -5.54 - (-10.42) = 4.88 \text{ eV} \tag{2.90}$$

E_1 is called the first excitation potential of mercury.

If by some reasons the valence electron is excited to the first excited state $n = 2$, the electron returns to the ground state in a very short time generally of the order of 10^{-8} sec. to the ground state $n = 1$. In this case a photon of energy 4.88 eV is emitted which corresponds to the wavelength $\lambda = \dfrac{hc}{E_1} = 2536\,\text{Å}$.

$$E = h\nu, \quad \nu = \frac{E}{h}, \quad \lambda = \frac{c}{\nu} = \frac{hc}{E}$$

$$E_1 = 4.88 \times 1.6 \times 10^{-19}\text{J}, \quad \lambda = \frac{hc}{E_1} = \frac{6.625 \times 10^{-34} \times 3 \times 10^8}{4.88 \times 1.6 \times 10^{-19}} = 2536\,\text{Å}$$

The ionization energy which is the minimum energy required to knock out the valence electron from the atom is 10.42 eV. (Remember the difference between excitation potential and the ionization potential). As mentioned above the first excited state is having energy $E_H = -5.44$ eV, the second excited state having energy $E_1 = -7.8$ eV, the third excited state J having energy $E_J = -8.5$ eV and so on. The energy values of neutral mercury atom are given in Table 2.5.

Let us consider the case where a beam of slow electrons travelling through mercury vapour at low pressure (*See* Figure 2.18). If the kinetic energy of the electron becomes less than 4.9 eV the collision is elastic; that means the translational kinetic energy is conserved. The electron will lose some kinetic energy by the relation

$$\Delta K = \frac{4mM}{(m+M)^2} K = \frac{4m}{M} K \tag{2.91}$$

where M represents the mass of the mercury atom, m that of electron and $K = \dfrac{1}{2}mv^2$, the kinetic energy of the electron. The loss in kinetic energy ΔK is small because $m \ll M$. The energy ΔK gets transferred to mercury atom and appears as recoil energy. This can be represented schematically as follows:

Slow electron + Atom at rest $\rightarrow$ Atom with some recoil energy + Slower electrons

$$e^-\,(\,K_1 < 4.9\ \text{eV}\,) + A \rightarrow A'(\Delta K) + e^-\,(K_2 = K_1 - \Delta K)$$

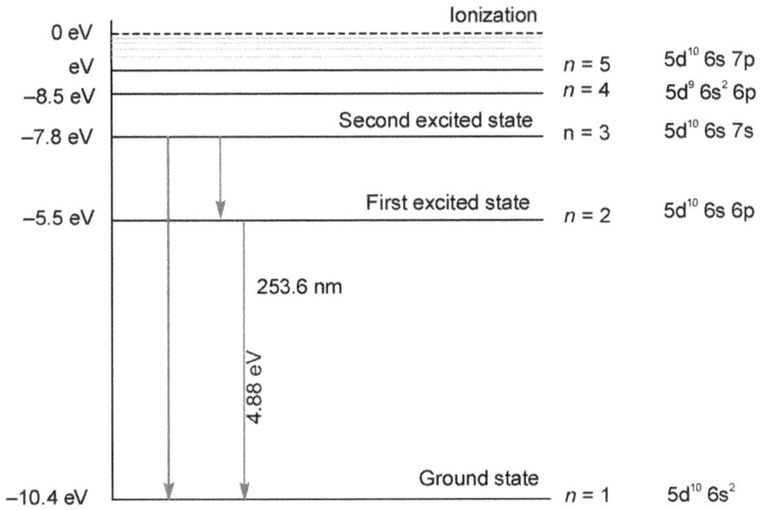

Figure 2.17   The energy level of $^{202}_{80}\text{Hg}$

As ΔK is small the electron experiences many collisions along zigzag path before it comes to rest.

However, when the kinetic energy of the electron becomes larger than 4.9 eV, an inelastic collision may take place in which a part of the kinetic energy is transferred into the atom in the form of internal energy thereby raising the electron from the ground state to the first excited state. The kinetic energy of the electron after the inelastic collision is

$$K_2 = K_1 - (E_H - E_0) = K_1 - 4.9\ \text{eV} \tag{2.92}$$

The series limit is at 84184.1 cm^{-1}. At this limit the mercury atom is ionized. That means one electron is knocked out. In HgII, all levels are doublets or quartets whereas in Hg neutral all levels have odd multiplicity as seen Table 2.5.

This may be represented schematically (*See* Figure 2.18).

Slow electron + Atom in the ground state $\rightarrow$ Atom in the excited state + Slower electrons

$$e^-\,(\,K_1 > 4.9\ \text{eV}\,) + A \rightarrow A^*(\Delta K) + e^-\,(K_2 = K_1 - 4.9)$$

A$^* \rightarrow$ ground state + $h\nu$ (emitted photon 2536 Å)

The second process A$^* \rightarrow$ ground state + $h\nu$ (emitted photon 2536 Å) is almost immediate as the life time of the excited state is about 10^{-8} seconds. When the energy of the incident electron K_1 becomes slightly greater than 4.9 eV, $K_2 < 4.9$ eV and no more inelastic collision will take place. All other collisions will be then elastic. If $K_1 >> 4.9$ eV, the $K_2 > 4.9$ eV and then other inelastic collisions take place.

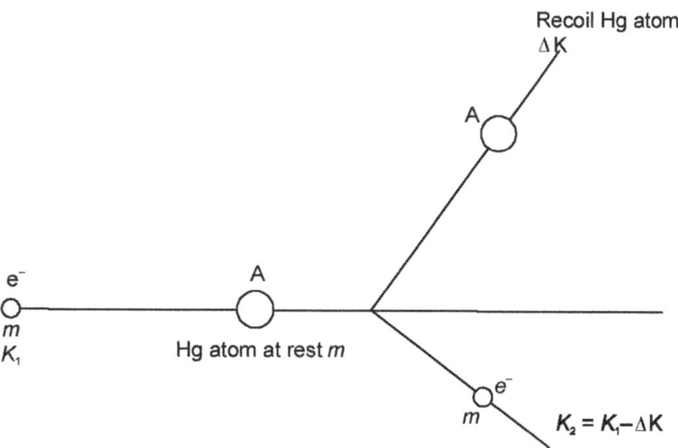

Figure 2.18 Collision of slow electron with Hg atom

Table 2.5 Energy levels of neutral mercury (Hg I)

Configuration	Term	J	Energy difference cm^{-1}	eV
$5d^{10}\,6s^2$	$^1S_0$	0	0.000	
$5d^{10}\,6s6p$	$^3P_0$	0	37645.080	4.67
	$^3P_1$	1	39412.300	4.89
	$^3P_2$	2	44042.977	5.46
$5d^{10}\,6s6p$	$^1P_1$	1	54068.781	6.70
$5d^{10}\,6s7s$	$^3S_1$	1	62350.456	7.73
$5d^{10}\,6s7s$	$^1S_0$	0	63928.243	7.93
$5d^9 6s^2 6p$	$^3P_2$	2	68886.60	8.54
$5d^{10}\,6s7p$	$^3P_0$	0	69516.66	8.62
	$^3P_1$	1	69661.89	8.64
	$^3P_2$	2	71207.51	8.83
$5d^{10}\,6s7p$	$^1P_1$	1	71295.15	8.84
$5d^{10}\,6s6d$	$(^1/_2,^3/_2)_2$	2	71333.182	8.84
	$(^1/_2,^3/_2)_1$	1	71336.164	8.84
$5d^{10}\,6s6d$	$(^1/_2,^5/_2)_2$	2	71396.220	8.85
	$(^1/_2,^5/_2)_3$	3	71431.311	8.86
Hg II	Limit		84184.1	10.44
	$^2S_{1/2}$	1/2		

In the Franck and Hertz experiment the glass tube *T* contains mercury vapour at low pressure and is maintained at a temperature of about 150°C. The tube contains a filament F, a grid G and a collecting plate P. The Franck–Hertz apparatus is shown in Figure 2.19. Between the filament and the grid there is an accelerating voltage V_a which varies from 0 to 60 V. Between the plate P and the grid G is a small retarding potential V_r of around 0.5 volt. Electrons were accelerated by a voltage towards a positively charged grid in a glass envelope filled with mercury vapour. Past the grid was the collecting plate, which is held at a small negative voltage with respect to the grid. When the V_a is increased, the plate current increases. The grid voltage is increased every time at approximately 5 volts. Thus electrons are accelerated and the collected current rises with accelerated voltage. And when the accelerating voltage reaches 4.9 volts, the current sharply drops, indicating a new phenomenon, which takes enough energy away from the electrons and due to this they cannot reach the collector. Thus the electrons with energies slightly greater than 4.9 volt undergoes inelastic collisions and is left with small energy which will not be sufficient to reach the plate. Thus this drop is attributed to inelastic collisions between the accelerated electrons and atomic electrons in the mercury atoms. If V_a is increased by an additional 4.9 volt, some of the electrons that were left with almost to kinetic energy will undergo another inelastic collision and will not reach the plate. This will explain the second dip obtained when plate current is plotted against accelerating potential V_a.

The accelerating voltage where the current dropped gave a measure of the energy necessary to raise an electron to the first excited state of the mercury atom. The transition corresponds to 4.9 volt which is a transition observed in the ultraviolet emission spectrum of mercury at 254 nm (= 4.9eV photon). It has been found that this line is a strong intense line.

$$4.9 \text{ eV} = 4.9 \times 1.6 \times 10^{-19} \text{ Joule} = 7.84 \times 10^{-19} \text{ J}$$

$$v = \frac{E}{h}, \; \lambda = \frac{c}{v}, \; \lambda = \frac{hc}{E} = \lambda = \frac{6.625 \times 10^{-34} \times 3 \times 10^{8}}{7.84 \times 10^{-19}} = 254 \times 10^{-9} \text{m} = 2540 \text{ Å}$$

It corresponds to a transition from the first excited state of mercury to the ground state. Drops in the collected current occur at multiples of 4.9 volts since an accelerated electron which has 4.9 eV of energy removed in a collision can be re-accelerated to produce other such collisions at multiples of 4.9 volts. Franck–Hertz Experiment gives strong confirmation of the idea of quantized atomic energy levels.

This original Franck–Hertz data shows electrons losing 4.9 eV per collision with mercury atoms. It was possible to observe ten sequential bumps at intervals of 4.9 volts. Franck and Hertz were awarded Nobel Prize in 1925 for this work.

For Neon gas, the process of absorbing energy from electron collisions produces visible evidence. When the accelerated electrons excite the electrons in neon to upper states, they de-excite by emission and produce a visible glow in the gas region in which the excitation has taken place. There are about ten excited levels in the range 18.3 to 19.5 eV. They de-excite by dropping to lower states at 16.57 and 16.79 eV. This energy difference gives light in the visible range. Since the accelerated electrons undergo inelastic collisions with the neon and are then accelerated again, they can undergo a series of such collisions if the accelerating voltage is high enough. The accelerating voltage from the Franck–Hertz apparatus used to produce the picture was capable of producing and accelerating voltage of about 80 volts, so one could get up to four collisions. This can be seen under proper conditions as four bands of light from the de-excitation in the collision regions.

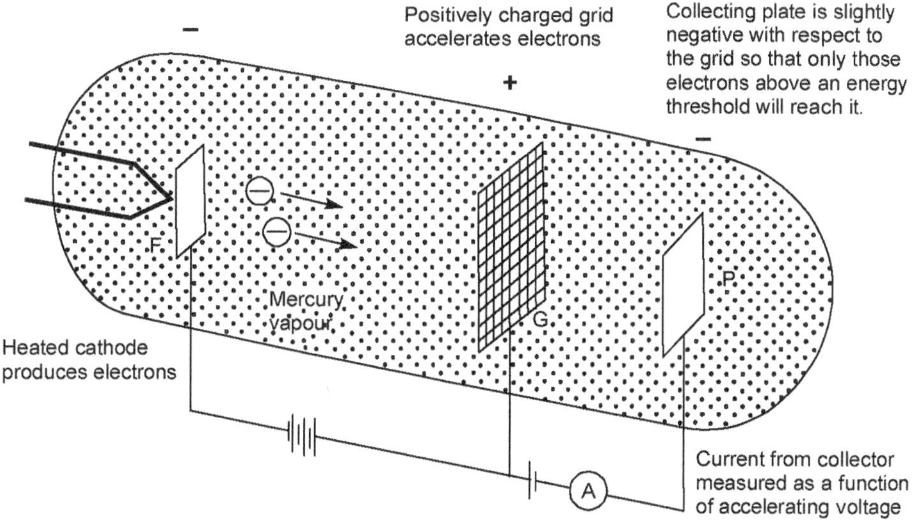

Figure 2.19   Sketch of Franck–Hertz apparatus

The Franck–Hertz display for neon shown was formed by sweeping the accelerating voltage and recording current vs voltage on an X-Y plot (*See* Figure 2.20). The measured separation of the peaks corresponds to about the midpoint of the range of excitation energies of the involved neon transitions.

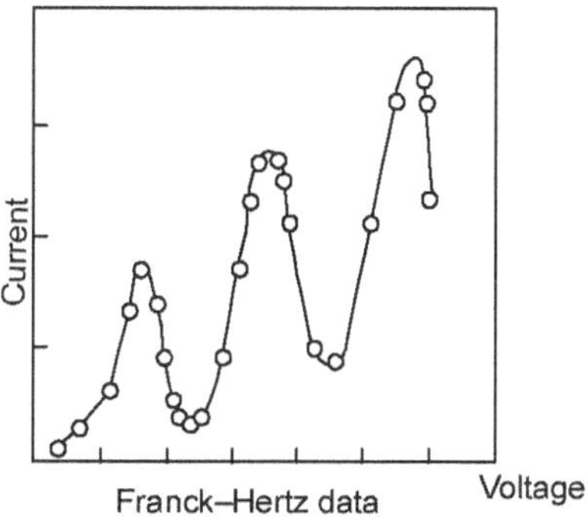

Figure 2.20   Franck–Hertz display for neon

Thus inelastic collisions are observed experimentally by a sudden decrease in the plate current and if a plot is made with plate current as a function of grid potential, we obtain a plot as shown in Figure 2.19.

2.10.1 Shortcomings of Bohr's Theory

The Bohr atom model contains both classical, mechanical treatment and also some primitive quantization conditions. It is known as semiclassical model of the atom. The Bohr model is certainly not a full proof description of the atom. It has many drawbacks.

1. The assumption that the laws of classical mechanics do not apply when there is a quantum jump, does not propose any other law which could replace the classical mechanics.
2. The assumption that the angular momentum is quantized does not explain why it is quantized.
3. The Bohr model could not explain the fine structure of spectral lines even in the hydrogen atom.
4. The angular momentum in the ground state is known to be zero whereas the Bohr model gives an incorrect value $L = \dfrac{h}{2\pi}$ for the ground state orbital angular momentum.
5. The Bohr model fails to explain the spectra of larger atoms. At best it could make approximate predictions only in the case of hydrogen like atoms having single outer-shell electron as in the cases of He^+, Li^{++} etc.
6. The relative intensities of the spectral lines could not be fully explained by Bohr model, though it was possible to give reasonable estimates of the line intensities.
7. Bohr theory could not explain the line splittings in a magnetic (Zeeman effect) or in an electric field (Stark effect).
8. The theory could not explain the distribution of electrons in atoms.

A refinement of theory was essential in order to explain the above experimental results.

However, as we have seen in the earlier discussions that the Bohr model successfully predicted the energies for the hydrogen atom, but had significant failures that were to be corrected by solving the Schrödinger equation for the hydrogen atom. However, because of its simplicity, the Bohr model is still commonly taught to introduce students to quantum mechanics and hence described in this book in detail.

2.11 REFINEMENTS: THE FINE STRUCTURE OF SPECTRAL LINES AND THE SOMMERFELD–WILSON THEORY

2.11.1 Sommerfeld Atomic Theory

In the year 1891 Michelson did experiments with spectrometers of high resolving power and observed that the Balmer series in hydrogen atom was not composed of single lines but in fact contains two or more lines.

These were later called fine structure. In 1916 Paschen performed experiments on the spectra of He^+. Though he was able to predict spectral lines in close conformity with Bohr's theory, it also could not account for the finer details of the spectral lines.

To explain the fine structure of spectral lines Wilson in 1915 and Sommerfeld in 1916 extended independently Bohr's theory by assuming existence of elliptical orbits for the electron, nucleus being at one of the foci of the ellipse. And circular orbits were manifestations of elliptical orbits under certain orbital conditions and electron distribution. Arnold Johannes Wilhelm Sommerfeld (1868–1951) was a German physicist who introduced the fine-structure constant in 1919.

Sommerfeld's elliptical orbits are shown in Figure 2.21.

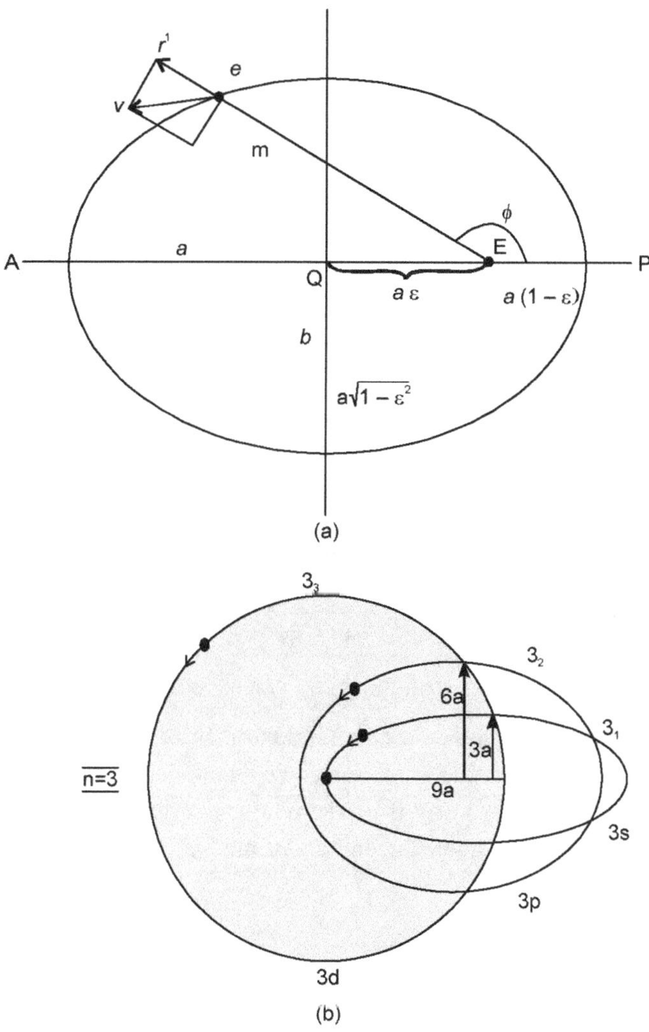

Figure 2.21   (a) Elliptical orbit (b) Relative dimensions of elliptical orbits

For an ellipse we know that

$$1-\varepsilon^2 = \frac{b^2}{a^2} \tag{2.92}$$

where ε is the eccentricity of the ellipse, a and b are the semi-major and semi-minor axes of the ellipse. Thus Sommerfeld modified Bohr's concept by assuming the motion of an electron in an elliptical orbit with the nucleus at one of the foci. In the new concept he envisaged two degrees of freedom for the electron and hence two quantum conditions were applied. Sommerfeld realized that the quantization of angular momenta could now be expressed in terms of phase coordinates or the so called phase integrals by the condition that

$$\oint p_\phi d\phi = kh \tag{2.93}$$

where p_ϕ is the angular momenta corresponding to the azimuthal angle φ and integral is extended to the period of ϕ. He postulated that the stationary states of a periodic system with f degrees of freedom are determined by the conditions

$$\oint p_q dq_q = rh, \; r = \; 1,2,3,...f \tag{2.94}$$

where p_i and q_i are generalized momentum and position coordinates and k is a positive integer extended for a period of q. The type of integrals are called phase integrals or action integrals as the product $p_i \, dq_i$ has the dimension of action.

For an elliptical orbit of this kind, we can apply Kepler's laws of planetary motion with the exception that the coulombs force takes place in the place of gravitational force. The motion of electron with mass m moving about the nucleus of mass M with charge e and Ze respectively can be considered as the motion of a single mass $\mu = \dfrac{mM}{m+M}$ called the reduced mass. As the mass of the nucleus is far more larger than the electron mass, we can assume that $\mu = \dfrac{mM}{M} = m$ where the small value of electron mass m is neglected in the denominator. In polar coordinates the position of the electron is given by the azimuthal angle ϕ and the electron–nuclear distance r. Assuming a potential energy zero when the electron is far away from the nucleus (at infinity), the potential energy of the electron under the influence of the nucleus can be assumed as

$$V = -\frac{eE}{r} = -\left(\frac{1}{4\pi\varepsilon_0}\right)\frac{Ze^2}{r} \tag{2.95}$$

where e and E are electron and nuclear charges respectively, Z is the atomic number.

The kinetic energy of the electron is

$$T = \frac{1}{2}mv^2 = \frac{1}{2}m\,(\dot{r}^2 + r^2\dot{\phi}^2) = \frac{1}{2}m\dot{r}^2 + \frac{1}{2}mr^2\dot{\phi}^2 \tag{2.96}$$

The equation involves two displacement coordinates ϕ and r and two momentum coordinates p_ϕ and p_r.

$$p_\phi = m r^2 \frac{\partial \phi}{\partial t} = m r^2 \dot{\phi} \quad \text{and}$$

$$p_r = m \frac{dr}{dt} = m\dot{r} \tag{2.97}$$

Thus we can build two-phase space diagrams corresponding to the two degrees of freedom in such a way that the motion of electron is represented by two graph points. According to Sommerfeld, the electron motion has two degrees of freedom with only one period and the quantum conditions are

$$\oint p_\phi d\phi = kh$$

$$\oint p_r dr = rh \tag{2.98}$$

where k and r take integral values only. k is called azimuthal quantum number and r as radial quantum number.

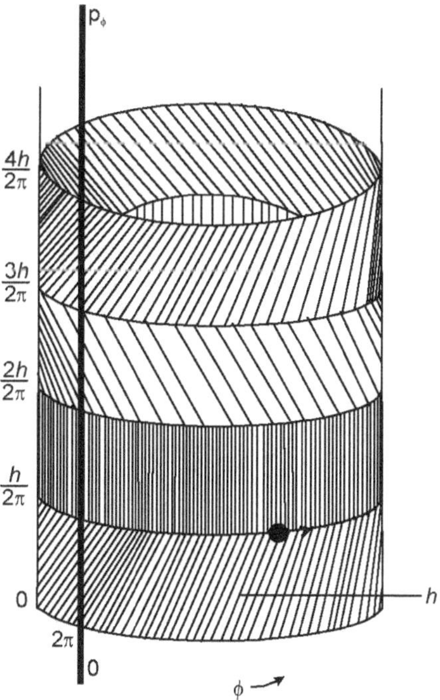

Figure 2.22   Phase-space diagram for Bohr's rotator

According to Kepler's second law of planetary motions, the angular momentum p_φ remains a constant throughout the motion and hence integrating over a complete cycle

$$\int_0^{2\pi} p_\phi d\phi = 2\pi p_\phi = 2\pi mr^2 \dot{\phi} = kh \tag{2.99}$$

In one complete revolution the radius vector turns through an angle 2π. The phase-space diagram can hence be represented by a cylinder of circumference 2π and the motion can be represented by a point moving with a constant speed on one of the circles at a distance $n\dfrac{h}{2\pi}$. The area of each shaded circle in h (*See* Figure 2.22).

2.11.2 Radial Quantum Number

The phase integral for the radial part is[25]

$$\oint p_r dr = rh$$

In order to evaluate this integral let us start with the general equation of an ellipse in polar coordinates

$$\frac{1}{r} = C_1 + C_2 \cos\phi \tag{2.100}$$

where C_1 and C_2 are constants.

The perihelion distance $EP = a(1-\varepsilon)$

The aphelion distance $EA = a(1+\varepsilon)$ $\tag{2.101}$

Semi-minor axis $b = a\sqrt{1-\varepsilon^2}$

And $EQ = a\varepsilon$

When the electron is at perihelion $\phi = 0$ and hence

$$\frac{1}{r} = C_1 + C_2 \cos 0 = C_1 + C_2$$

i.e., $$\frac{1}{a(1-\varepsilon)} = C_1 + C_2 \tag{2.102}$$

When the electron is at the aphelion $\phi = \pi$ and

$$\frac{1}{r} = C_1 + C_2 \cos\pi = C_1 - C_2 \tag{2.103}$$

i.e., $$\frac{1}{a(1+\varepsilon)} = C_1 - C_2 \tag{2.104}$$

Hence $\quad C_1 = \dfrac{1}{a(1-\varepsilon^2)}$ and $C_2 = \dfrac{\varepsilon}{a(1-\varepsilon^2)}$ $\hspace{2cm}$ (2.105)

The equation of the ellipse is then

$$\frac{1}{r} = \frac{1+\varepsilon\cos\phi}{a(1-\varepsilon^2)} \tag{2.106}$$

where a is the semi-major axis and the semi-minor axis $b = a\sqrt{1-\varepsilon^2}$

Taking the log of equation (2.106) and differentiating we obtain

$$\frac{1}{r}\frac{dr}{d\phi} = \frac{\varepsilon\sin\phi}{1+\varepsilon\cos\phi} \tag{2.107}$$

From (2.97) $\quad P_\phi = mr^2\dfrac{\partial\phi}{\partial t} = mr^2\dot{\phi}$

And hence $\quad m\dot{\phi} = \dfrac{P_\phi}{r^2}$

$$p_r = m\frac{dr}{dt} = m\frac{dr}{d\phi}\frac{d\phi}{dt} = m\frac{dr}{d\phi}\dot{\phi} = m\dot{\phi}\frac{dr}{d\phi} = \frac{P_\phi}{r^2}\frac{dr}{d\phi} \tag{2.108}$$

$$P_\phi = mr^2\frac{\partial\phi}{\partial t} = mr^2\dot{\phi}, \ m\dot{\phi} = \frac{p\phi}{r^2}$$

also $\hspace{3cm} dr = \dfrac{dr}{d\phi}d\phi$

Hence $\hspace{2cm} p_r dr = p_\phi \cdot \dfrac{1}{r^2}\dfrac{dr}{d\phi}\left(\dfrac{dr}{d\phi}\right)d\phi = p_\phi\varepsilon^2\dfrac{\sin^2\phi}{(1+\varepsilon\cos\phi)^2}d\phi$ $\hspace{1cm}$ (2.109)

Thus the phase integral $\oint p_r dr = p_\phi\varepsilon^2\displaystyle\int_0^{2\pi}\dfrac{\sin^2\phi}{(1+\varepsilon\cos\phi)^2}d\phi = rh$ $\hspace{1cm}$ (2.110)

$$\int_0^{2\pi} p_\phi d\phi = 2\pi\phi = kh\int_0^{2\pi} p_\phi d\phi = 2\pi\,p_\phi = kh, \quad p_\phi = k\frac{h}{2\pi} \tag{2.111}$$

Thus from equation (2.110) $\dfrac{\varepsilon^2}{2\pi}\displaystyle\int_0^{2\pi}\dfrac{\sin^2\phi\,d\phi}{(1+\varepsilon\cos\phi)^2} = \dfrac{r}{k}$ $\hspace{1.5cm}$ (2.112)

Integrating by part
$$\frac{1}{\sqrt{1-\varepsilon^2}} - 1 = \frac{r}{k} \tag{2.113}$$

$$1 - \varepsilon^2 = \frac{k^2}{(r+k)^2} \tag{2.114}$$

From equation $b = a\sqrt{1-\varepsilon^2}$ and the equation above

$$\frac{a}{b} - 1 = \frac{r}{k}, \ \frac{a}{b} = \frac{r+k}{k} \tag{2.115}$$

Total energy is given by

$$E = T + V = \frac{1}{2}mv^2 + V = \frac{1}{2}m(\dot{r}^2 + r^2\dot{\phi}^2) - \frac{1}{4\pi\varepsilon_0}\frac{Ze^2}{r} = \frac{1}{2}m\dot{r}^2 + \frac{1}{2}mr^2\dot{\phi}^2 - \frac{1}{4\pi\varepsilon_0}\frac{Ze^2}{r} \tag{2.116}$$

$$E = \frac{1}{2m}\left(p_r^{\,2} + \frac{p_\phi^2}{r^2} - \frac{1}{4\pi\varepsilon_0}\frac{Ze^2}{r} \right) \tag{2.117}$$

Using equations
$$\frac{1}{r} = \frac{1+\varepsilon\cos\phi}{a(1-\varepsilon^2)} \cdot \frac{1}{r}\frac{dr}{d\phi} = \frac{\varepsilon\sin\phi}{a(1-\varepsilon^2)} \tag{2.118}$$

$$p_r = m\dot{r} = \frac{p_\phi}{r^2}\frac{dr}{d\phi} \tag{2.119}$$

$$E = \frac{p_\phi^2}{2mr^2}[(\frac{1}{r}\frac{dr}{d\phi})^2 + 1] - \frac{1}{4\pi\varepsilon_0}\frac{Ze^2}{r} \tag{2.120}$$

Substituting for $\dfrac{1}{r}$ and $\dfrac{1}{r}\dfrac{dr}{d\phi}$

$$E = \frac{p_\phi^2}{ma^2(1-\varepsilon^2)^2}\left(\frac{1+\varepsilon^2}{2} + \varepsilon\cos\phi \right) - \frac{1}{4\pi\varepsilon_0}Ze^2\frac{1+\varepsilon\cos\phi}{a(1-\varepsilon^2)} \tag{2.121}$$

The total energy is a constant and is independent of time and angle ϕ. As $\cos\phi$ varies and in order that E is a constant, the coefficient of $\varepsilon\cos\phi$ must vanish. Hence by collecting the terms in $\varepsilon\cos\phi$ and equating to zero

$$\left(\frac{p_\phi^2}{ma^2(1-\varepsilon^2)^2} - \frac{1}{4\pi\varepsilon_0}\frac{Ze^2}{a(1-\varepsilon^2)} \right)\varepsilon\cos\phi = 0 \tag{2.122}$$

i.e.,
$$\left(\frac{p_\phi^2}{ma^2(1-\varepsilon^2)^2} - \frac{1}{4\pi\varepsilon_0}\frac{Ze^2}{a(1-\varepsilon^2)} \right) = 0 \tag{2.123}$$

from which the semi-major axis $a = \dfrac{p_\phi^2}{mZe^2(1-\varepsilon^2)}4\pi\varepsilon_0$ $\tag{2.124}$

Substituting the values of p_φ from (2.111) and $1-\varepsilon^2$ from (2.113)

$$a = 4\pi\varepsilon_0 \frac{h^2}{4\pi^2 mZe^2}(k+r)^2 = a_0\frac{n^2}{Z} \tag{2.125}$$

where $k+r = n$ and a_0 is the Bohr radius $a_0 = \dfrac{h^2\varepsilon_0}{\pi me^2}$

From expression $\left(b = a\sqrt{1-\varepsilon^2}\right)$ and from equation (2.114),

$$\text{i.e.,}\quad 1-\varepsilon^2 = \frac{k^2}{(k+r)^2}$$

the semi-minor axis $b = 4\pi\varepsilon_0 \dfrac{h^2}{4\pi^2 mZe^2}k(k+r) = a_0\dfrac{nk}{Z}$ $\tag{2.126}$

where a_0 is the radius of first Bohr orbit and n is the total quantum number.

Substituting the value of p_ϕ^2 from equation (2.124)

$$p_\phi^2 = \frac{mZe^2(1-\varepsilon^2)a}{4\pi\varepsilon_0}$$

and from equation (2.121) we get

$$E = \frac{mZe^2(1-\varepsilon^2)a}{4\pi\varepsilon_0 ma^2(1-\varepsilon^2)^2}\left[\frac{(1+\varepsilon^2)}{2} + \varepsilon\cos\phi\right] - \frac{1}{4\pi\varepsilon_0}Ze^2\frac{1+\varepsilon\cos\phi}{a(1-\varepsilon^2)}$$

$$= \frac{Ze^2}{4\pi\varepsilon_0}\left\{ \frac{m(1-\varepsilon^2)a\left[\dfrac{(1+\varepsilon^2)}{2} + \varepsilon\cos\phi\right]}{ma^2(1-\varepsilon^2)^2} - \frac{1+\varepsilon\cos\phi}{a(1-\varepsilon^2)} \right\}$$

$$= \frac{Ze^2}{4\pi\varepsilon_0 a(1-\varepsilon^2)}\left\{ \left(\frac{1+\varepsilon^2}{2} + \varepsilon\cos\phi\right) - (1+\varepsilon\cos\phi) \right\}$$

Thus
$$E = \frac{1}{4\pi\varepsilon_0} \frac{Ze^2}{a(1-\varepsilon^2)} \left[\frac{1+\varepsilon^2}{2} - 1 \right] = -\frac{1}{4\pi\varepsilon_0} \frac{Ze^2}{2a} \tag{2.127}$$

If we approximate a to be the Bohr radius, we can

Substitute the value of a from (2.7) $a_n = -\dfrac{n^2 h^2 \varepsilon_0}{\pi m Z e^2}$

$$E = -\frac{1}{4\pi\varepsilon_0} \times \frac{Ze^2}{2} \times \frac{\pi m Z e^2}{n^2 h^2 \varepsilon_0} = -\frac{Z^2 m e^4}{8\varepsilon_0^2 n^2} \tag{2.128}$$

n is substituted for $k + r$

In wave number units $E = -\dfrac{Z^2 m e^4}{8\varepsilon_0^2 n^2 ch^3}$

The energy is the same as in Bohr's theory.

$$E_{n,k} = -\frac{RhcZ^2}{n^2} \quad \text{where} \quad R = \frac{me^4}{8\varepsilon_0^2 ch^3} = \text{Rydberg constant}$$

2.11.3 Sommerfeld's Relativistic Correction

The Sommerfeld theory, though include elliptical orbits, does not add any new energy levels. For a given n, called principal quantum number there are n different quantized elliptical orbits the electron can occupy. Thus for example $n = 4$, the azimuthal quantum number k may have four values, viz., $k = 1, 2, 3, 4$ while the radial quantum number r can have 3,2,1,0 values respectively.

$$E_{n,k} = -\frac{Z^2 m e^4}{8\varepsilon_0^2 n^2 ch^3} = -\frac{RhcZ^2}{n^2} J = -\frac{Rz^2}{n^2} \text{cm}^{-1} \tag{2.129}$$

where R is the Rydberg constant $R = \dfrac{Z^2 m e^4}{8\varepsilon_0^2 ch^3} \text{cm}^{-1}$

$R = 109677.759$ cm^{-1} in hydrogen and it is 109722.403 cm^{-1} in He$^+$.

In short $R = R_\alpha \dfrac{M}{M + m}$

The atomic mass of helium is four times that of hydrogen and hence

$$R_\alpha = R_H \frac{M + m}{m} = R_H \left(1 + \frac{m}{M} \right) = R_{He+} \left(1 + \frac{m}{M} \right)$$

The most probable extrapolated value of R_α obtained by Houston is 109737.424 cm^{-1}.

If R_A is the Rydberg constant of an atom A then

$$\frac{R_\alpha - R_1}{R_\alpha - R_A} = \frac{AM + m}{M + m} = A \tag{2.130}$$

Sommerfeld also considered the variation of mass of the electron in accordance with the special theory of relativity. The mass of the electron moving with a velocity v is given by

$$m = m_o \sqrt{1 - \frac{v^2}{c^2}}$$ where m_o is the rest mass of the electron and v is the orbital velocity. The maximum velocity of the electron occurs at the perihelion of the elliptical orbit and hence the relativistic mass is the maximum at the perihelion. Because of this increase in velocity as the electron approaches the perihelion, the electron tends to overshoot the orbit after every revolution, and the net result is that, there is an advance of the perihelion, resulting in a precession of the electron orbit with a rosette motion about the nucleus. Thus the velocities of the electron in different orbits with the same n, but different k, give relativistic change in mass of the electrons and in the energy. Sommerfeld derived the change in energy of the electron in hydrogen like atom as

$$\Delta T = \frac{R\alpha^2 Z^4}{n^4}\left(\frac{n}{k} - \frac{3}{4}\right) \tag{2.131}$$

where k is the Sommerfeld's azimuthal quantum number. $k = 1,2,3\ldots$ For $s, p, d\ldots$ orbits.

The quantum mechanical treatment has shown that k should be replaced by $l+1/2$. Thus

$$\Delta T_r = \frac{R\alpha^2 Z^4}{n^4}\left(\frac{n}{1 + \dfrac{1}{2}} - \frac{3}{4}\right) \tag{2.132}$$

The term form the spin–orbit interaction should be added to get the total energy shift. The spin–orbit interaction term is

$$\Delta T_{s,1} = -\frac{R\alpha^2 Z^4}{n^3 l\left(1 + \dfrac{1}{2}\right)(l+1)} \cdot \frac{j^{*2} - l^{*2} - s^{*2}}{2} \tag{2.133}$$

where $j^{*2} = j(j+1)$, $l^{*2} = l(l+1)$ and $s^{*2} = s(s+1)$

It has been seen that better agreement between Bohr–Sommerfeld orbital model and quantum mechanical model is obtained when the orbital anglar momentum is replaced by $\sqrt{l(l+1)} \cdot \dfrac{h}{2\pi}$.

So also j by $\sqrt{j(j+1)}\,\dfrac{h}{2\pi}$ and s by $\sqrt{s(s+1)}\,\dfrac{h}{2\pi}$. Thus $j* = \sqrt{j(j+1)}$, $l* = \sqrt{l(l+1)}$ and $s* = \sqrt{s(s+1)}$.

Thus the total shift is $\Delta T + \Delta T_{s,l}$

The total energy in wave numbers is now

$$E_{n,l} = -\frac{RZ^2}{n^2} \pm \left[\frac{R\alpha^2 Z^4}{n^4}\left(\frac{n}{l+\frac{1}{2}} - \frac{3}{4} \right) + \frac{R\alpha^2 Z^4}{n^3 l\left(l+\frac{1}{2}\right)(l+1)} \cdot \frac{j*^2 - l*^2 - s*^2}{2} \right] \tag{2.134}$$

The second term is called Sommerfeld's relativistic correction term.

For $j = l + 1/2$

$$\Delta T_{s,l} = -\frac{R\alpha^2 Z^4}{n^3 l\left(l+\frac{1}{2}\right)(l+1)} \cdot \frac{\left(l+\frac{1}{2}\right)\left(l+\frac{3}{2}\right) - l(l+1) - \frac{1}{2}\cdot\frac{3}{2}}{2}$$

$$= -\frac{R\alpha^2 Z^4}{n^3 2\left(l+\frac{1}{2}\right)(l+1)} \tag{2.135}$$

For $j = l - 1/2$ $\quad \Delta T_{s,l} = \dfrac{R\alpha^2 Z_4}{n_3 l\left(l+\frac{1}{2}\right)(l+1)} \cdot \dfrac{\left(l-\frac{1}{2}\right)\left(l+\frac{1}{2}\right) - l(l+1) - \frac{1}{2}\cdot\frac{3}{2}}{2}$

$$\Delta T_{s,l} = +\frac{R\alpha^2 Z^4}{n^3 2l\left(l+\frac{1}{2}\right)} \tag{2.136}$$

$$\Delta T = \Delta T_{s,l} + \Delta T_r = +\frac{R\alpha^2 Z^4}{n^3}\left[\frac{1}{l+1} - \frac{3}{4n} \right] \quad \text{for } j = l+1/2$$

$$\Delta T = \Delta T_{s,l} + \Delta T_r = +\frac{R\alpha^2 Z^4}{n^3}\left[\frac{1}{l} - \frac{3}{4n} \right] \quad \text{for } j = l-1/2 \tag{2.137}$$

Sommerfeld theory of hydrogen atom thus gives the energy (neglecting spin–orbit interaction)

$$E_{n,l} = -\frac{RZ^2}{n^2} \pm \left[\frac{R\alpha^2 Z^4}{n^4} \left(\frac{n}{j+\frac{1}{2}} - \frac{3}{4} \right) \right] \mathrm{cm}^{-1} \qquad (2.138)$$

where the second term is the relativistic correction.

$$R = 2\pi^2 m_0 e^4 / h^2 (4\pi\varepsilon_0)^2 \; \frac{m_0 e^4}{8h^2 \varepsilon_0^{\,2}} = \text{Rydberg constant}$$

In wavenumber $R = \dfrac{m_0 e^4}{8\varepsilon_0^{\,2} ch^3}$

In terms of term value

$$T_n = RZ^2 / n^2 \pm R\alpha^2 Z^4 (n/l + 1/2)/n^4 \;\; \text{where} \; R = 2\pi^2 m_0 e^4 / ch^3 (4\pi\varepsilon_0)^2$$

The quantum mechanical calculations have shown that the relativistic correction must be written as

$$\frac{R\alpha^2 Z^4}{n^4} \left(\frac{n}{j+\frac{1}{2}} - \frac{3}{4} \right) \qquad (2.139)$$

In general quantum mechanics replaces k, k^2, k^3 by $(l+1/2)$, $l(l+1)$, $(l+1/2)\,l(l+1)$ respectively.

The relativistic correction (mass variation) splits up each level into n components where n is the principal quantum number of the level. The correction due to the spin–orbit interaction splits up each levels into two components (in one electron system) with the exception of S term for which $\Delta T_{ls} = 0$. The spacing between the levels $j = l+1/2$ and $j = l-1/2$ is small compared to the electron levels with different l. Therefore a magnetic interaction contributes to small splitting of a level with a given l. This spitting is called the fine structure of the level. The fine structure of spectral lines is described in Chapter 7.

2.11.4 Fine Structure of the H$_\alpha$ line

Two types of experimental evidence, which arose in the 1920s, suggested an additional property of the electron. One was the closely spaced splitting of the hydrogen spectral lines. They are called fine structure. The other was the Stern–Gerlach experiment that showed in 1922 that a beam of silver atoms directed through an inhomogeneous magnetic field would be forced into two beams. Both these experimental situations were consistent with the possession of an intrinsic angular momentum and a magnetic moment by individual electrons. Classically this could occur if the electrons were considered as spinning ball of charge, and this property was called an electron spin.

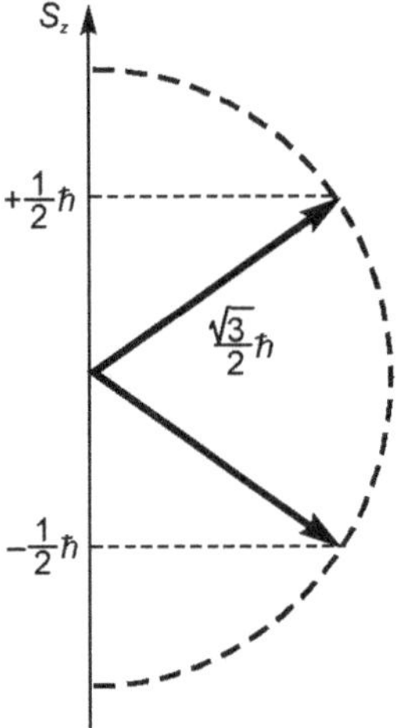

Figure 2.23   Space quantization diagram of electron spin

Spin "up" and "down" allow two electrons for each set of spatial quantum numbers n,l,m_l.

An electron spin is an intrinsic property of electrons. It's value is $s = \frac{1}{2}$. Thus the electrons have intrinsic angular momentum characterized by quantum number 1/2 (*See* Figure 2.23). Similar to other quantized angular momenta, it gives total angular momentum

$$S = \sqrt{\frac{1}{2}\left(\frac{1}{2}+1\right)}\ \hbar = \frac{\sqrt{3}}{2}\hbar$$

The fine structure corresponds to two possibilities for the z-component of the angular momentum.

$$S_z = \pm\frac{1}{2}\hbar$$

Because of this there exists spin magnetic moment for the electron over and above the magnetic moment due to orbital motion. The interaction of the spin magnetic moment of the electron

$$\mu_s = -\frac{e}{2m}gS$$

is the cause of the fine-structure splittings.

We have already seen that the H_α line originates from the transition between

$$n = 2 \quad \rightarrow \quad n = 3$$
$$(l = 0,1) \qquad (l = 0, 1, 2)$$
$$s, p \qquad\qquad s, p, d$$

The resulting spectrum is shown in Figure 2.24.

The $n = 2$, $l = 0$ gives the state $^2S_{1/2}$ and $l = 1$ gives the states $^2P_{1/2}$ and $^2P_{3/2}$. Similar is the case with $n = 3$. $l = 2$ give $^2D_{3/2}$ and $^2D_{5/2}$ states.

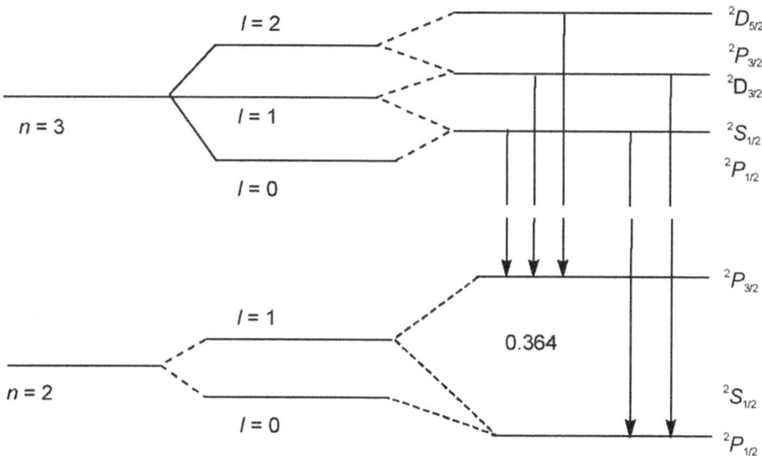

Figure 2.24   Fine structure of hydrogen spectral lines

Note that the $^2P_{1/2}$ and $^2S_{1/2}$ are degenerate in the picture (having the same energy). Later it has been found by Lamb that there is a small splitting in that level which is now known as Lamb shift.

Lamb shift in $1S$ state is measured as (*See* section 5.3).

$$L(1S) = 8172840(22) \text{ kHz}$$

2.11.5 Degenerate State or Degeneracy

In most cases, each energy level represents a single eigen state and will have only one eigen function. However if the state is degenerate, several eigen functions may be associated with the same eigen value or energy. This is called degeneracy. The number of eigen values for the particular energy state or level is equal to the degeneracy of the state. It should be noted that the degeneracy might be removed for example by applying an electric or magnetic field. If we denote g_i as the degeneracy of a state having energy E_i, then g_i is the number of eigen-states corresponding to that energy. For non-degenerate case $g_i = 1$.

2.12 RADIATION FROM ATOMS AND THE SELECTION RULES

Bohr, in his theory on atomic structure has envisaged that radiation occurs when atom undergoes a "quantum jump" from one energy state to the other. Though quantum jumps have now been established through experiments in laser cooled and trapped atoms, the artificial mechanism of quantum jumps could not give any ideas on selection rules as only certain kinds of "quantum jumps" were possible. In order to circumvent these difficulties, Bohr proposed the Correspondence Principle.

2.13 BOHR'S CORRESPONDENCE PRINCIPLE

The principle was proposed by Niels Bohr in 1920 to describe many features of atomic spectra. Quantum mechanics was highly successful in describing microscopic objects, such as atoms and elementary particles. On the other hand, classical mechanics and classical electrodynamics were found sufficient to describe macroscopic systems (springs, capacitors and so forth). Thus ordinarily the quantum theory is used to describe the behaviour of bodies, which are so small that they cannot be seen under an optical microscope, while the theories of classical physics are used to analyse the behaviour of large-scale bodies. However, it is quite reasonable to believe that the ultimate laws of physics must be independent of the size of the physical objects being described. A formal correspondence between classical mechanics and the quantum theory as far as the transitions between the states are concerned is the subject of the Bohr's Correspondence Principle. The principle states that classical physics must emerge as an approximation to quantum physics as systems become "larger".

The basic idea behind the correspondence principle is based on the fundamental assumption that classical mechanics can be understood as limiting case of quantum mechanics. Or conversely many quantum mechanical phenomena can be approximated on the basis of classical mechanics, provided we propose a proper reinterpretation. Thus in brief the Bohr's correspondence principle says that predictions of quantum theory approach those of classical physics in the limit of large quantum numbers. Technically the principle means that the results of a quantum theory analysis of a problem that involves the use of very large quantum numbers must agree with the results of a classical physics analysis. Such correspondence is known as the classical limit of the quantum theory. The correspondence principle provided an important theoretical basis for the development of a detailed correlation between the newer quantum theory and the classical physics that preceded it.

Even before Bohr proposed his atomic theory, Planck had already shown that in the limiting case when $h \rightarrow 0$, the quantum theory converges to classical results. We know that in the limiting case Planck's radiation formula goes over to the Rayleigh–Jean's formula. Planck thus suggested that when quantum of action is infinitesimal, it becomes the case of a classical nature. We can consider, for example, the quantum of energy $E = h\nu$ approaches zero, if h is kept constant and the frequency (of transition) approaches zero. It implies that the energy levels are very close together. The energy difference is infinitely small and in this limit the energies form a continuum.

We have seen in equation (2.13) that the energy according to Bohr's theory is

$$E_n = -\frac{Z^2 e^4 m}{8\varepsilon_0^2 n^2 h^2} = -Z^2 \frac{me^4}{8h^2 \varepsilon_0^2} \cdot \frac{1}{n^2}$$

In terms of Bohr radius $a_0 = \dfrac{h^2 \varepsilon_0}{\pi m Z e^2}$, $E_n = \dfrac{Z^2 e^2}{8h^2} \dfrac{me^2}{\varepsilon_0^2} \dfrac{1}{n^2} = -\dfrac{Ze^2}{8\pi\varepsilon_0 a_0} \dfrac{1}{n^2}$

$$E_n = -\frac{Ze^2}{8\varepsilon_0 n^2 a_0} = -\frac{Z^2 me^4}{8\varepsilon_0^2 n^2 h^2}$$

For $(n+1)$th state $E_n = -\dfrac{Z^2 me^4}{8\varepsilon_0^2 (n+1)^2 h^2}$

Thus $E_{n+1} - E_n = \dfrac{Z^2 me^4}{8\varepsilon_0^2 h^2} \left[\dfrac{(n+1)^2 - n^2}{n^2 (n+1)^2} \right]$

For $n \gg 1$, $E_{n+1} - E_n = \dfrac{Z^2 me^4}{8\varepsilon_0^2 h^2} \dfrac{2}{n^3} = h\nu$

$$\nu = \frac{Z^2 me^4}{8{\varepsilon_0}^2 h^3} \frac{2}{n^3} = \frac{Z^2 me^4}{4{\varepsilon_0}^2 n^3 h^3}$$

Period of electron in the first orbit $T = \dfrac{2\pi a_0}{8\nu}$

Frequency of revolution of the electron at the nth orbit is

$$\frac{1}{T_n} = \frac{\nu_n}{2\pi a_n}$$

where ν_n is the velocity in the nth orbit and a_n is the radius.

We have already seen in equation 2.5 that $\nu_n = \dfrac{1}{4\pi\varepsilon_0} \dfrac{Ze^2}{nh/2\pi} = \dfrac{Ze^2}{2\varepsilon_0 nh}$ from Bohr's theory.

And hence $\dfrac{1}{T_n} = \dfrac{Ze^2}{2\varepsilon_0 nh} \dfrac{1}{2\pi} \dfrac{\pi m Z e^2}{n^2 h^2 \varepsilon_0} = \dfrac{Z^2 me^4}{4\varepsilon_0^2 n^3 h^3}$

which is the same as $E_{n+1} - E_n = \dfrac{Z^2 me^4}{8\varepsilon_0^2 h^3} \dfrac{2}{n^3}$ of that of $n+1$ orbit.

Thus when n is very large Bohr's frequency condition $E_n - E_m = h\nu_{nm}$ is the same as in classical theory, i.e., the frequency of revolution of the electron in the orbit is equal to the radiation frequency. By approximating that the condition of equivalence for the high n values is applicable to moderate n values as well, Bohr put forward the correspondence principle.

We can also derive it in the following way.

If n is very large ΔE can be found by differentiating the following energy equation in MKS unit.

$$E_n = -\frac{me^4 Z^2}{8\varepsilon_0^2 n^2 h^2}$$

Differentiating w.r.t. n, $\Delta E = -\frac{me^4 Z^2}{8\varepsilon_0^2 h^2}\left(-\frac{2}{n^3}\right)\Delta n = \frac{2me^4 Z^2}{8\varepsilon_0^2 h^2 n^3}\Delta n$ \hfill (2.140)

The frequency of light emitted in this case $v = \frac{\Delta E}{h} = \frac{1}{4}\frac{me^4 Z^2}{\varepsilon_0^2 n^3 h^3}\Delta n$

i.e., $v = \frac{\Delta E}{h} = \frac{1}{32}\frac{me^4 Z^2}{\pi^3 \varepsilon_0^2}\left(\frac{2\pi}{nh}\right)^3 \Delta n$ \hfill (2.141)

According to Bohr's first postulate $\dfrac{nh}{2\pi}$ is equal to the angular momentum associated with nth orbit.

Thus $\dfrac{nh}{2\pi} = mvr$

Substituting in equation (2.141) $v = \frac{1}{32}\frac{me^4 Z^2}{\pi^3 m^3 v^3 r^3 \varepsilon_0^2}\Delta n$ \hfill (2.142)

But the centripetal force of attraction and the centrifugal force of motion are equal and hence

$$\frac{mv^2}{r} = \frac{1}{4\pi\varepsilon_0}\frac{Ze^2}{r^2}$$

Hence $mv^2 r = \left(\dfrac{1}{4\pi\varepsilon_0}Ze^2\right)$ \hfill (2.143)

The equation (2.142) can be written as

$$v = \frac{1}{32}\frac{me^4 Z^2}{\pi^3 m^2 v^4 r^2\,\dfrac{mr}{v}}\Delta n$$

By substituting the value of $mv^2 r$ in equation (2.142)

$$v = \frac{1}{32} \frac{me^4 Z^2}{\pi^3 \dfrac{mr}{v}} \left(\frac{4\pi\varepsilon_0}{Ze^2}\right)^2 \Delta n$$

$$v = \frac{1}{32} \frac{me^4 Z^2}{\pi^3 (mr/v)(Ze^2/4\pi\varepsilon_0)^2} \Delta n \tag{2.144}$$

Considering $v/r = \omega = 2\pi f$

$$v = \frac{1}{32} \frac{me^4 Z^2}{\pi^3 (mr/v)(Ze^2/4\pi\varepsilon_0)^2} \Delta n = \frac{1}{2}\frac{\omega}{\pi}\Delta n = f \Delta n \tag{2.145}$$

When $\Delta n = 1$, the frequency of the emitted light is the same as the frequency of rotation in the orbit. When $\Delta n = 1, 2, 3$, etc., radiations are emitted with frequencies that are multiples of f the fundamental frequency. When it is observed carefully it can be observed that the correspondence is not exact, however it reveals the quantum nature of radiation.

Example Show that for large values of the principal quantum number n, the frequencies of revolution of an electron in adjacent energy levels of a hydrogen atom and the radiated frequency for a transition between these levels all approach the same value.

Solution As per Bohr condition $mvr = \dfrac{nh}{2\pi}$

$$v = \frac{nh}{2\pi mr}$$

The frequency of revolution of the electron in the nth orbit

$$v = \frac{v}{2\pi r} = \frac{nh}{4\pi^2 mr^2} = \frac{me^4}{4\varepsilon_0^2 n^2 h^3}$$

The Rydberg constant $R = \dfrac{me^4}{8\varepsilon_0^2 ch^3}$ and hence $v = \dfrac{2Rc}{n^3}$.

The expression for the frequency of radiation emitted in a transition between n_1 and n is

$$v = Rc\left[\frac{1}{n_1^2} - \frac{1}{n^2}\right]$$

If $n_1 = (n-1)$ then $v = Rc\left[\dfrac{1}{(n-1)^2} - \dfrac{1}{n^2}\right] = Rc\left[\dfrac{2n-1}{n^2(n^2-1)}\right]$

If $n \gg 1$ $v \to Rc\dfrac{2n}{n^4} = \dfrac{2Rc}{n^3}$

But $v_n = \dfrac{2Rc}{n^3}$ and $v_{n-1} = \dfrac{2Rc}{(n-1)^3} \sim \dfrac{2Rc}{n^3}$

Thus for large n the values of frequencies v_{n-1}, v_n, v_{n+1} all tend to be the same, viz, $\dfrac{2Rc}{n^3}$.

Thus at large quantum numbers the classical and quantum mechanical values are the same and hence the correspondence principle.

The correspondence principle, however, has assumed a more profound significance. Experimental techniques in atomic, molecular, mesoscopic, and nuclear physics have improved dramatically. High-precision data for many thousands of energy levels are available where the traditional methods of quantum mechanics are not very useful or informative. However, the basic idea behind the correspondence principle must still be valid: Quantum mechanics must be understandable in terms of classical mechanics for the highly excited states, even in difficult cases like the three-body problem, where the overall behaviour seems unpredictable and chaotic. The wider application of Bohr's correspondence principle allows many basic but difficult problems to be seen in a new light.

The classical motions in simple dynamical systems can be understood as composed of independent partial motions, each with its own degree of freedom. Each degree of freedom accumulates its own classical action-integral. The frequency of the classical motion for any particular degree of freedom is given by the partial derivative of the energy function with respect to the corresponding action. Bohr noticed that this classical result yields the correct quantum-theoretical result for the light frequency in a transition from one energy level to another, provided the derivative is replaced by the difference in the energies. Moreover, precise information about the possibility of such transitions and their intensities is obtained by analysing the related classical motion. This information becomes better as the quantum numbers involved become larger. The apparent inconsistencies in Bohr's quantum theory are thereby overcome by a set of rules that came to be called the correspondence principle.

Although the process of radiation cannot be described on the basis of classical electrodynamics, there exists a correspondence between the transition among stationary states and the harmonic components of the classical motion. Classical theory and Maxwell's electrodynamics allowed calculations of the frequencies and also the intensities and polarization of lines. However Bohr's theory disclaimed any information on the mechanism of transitions between the states. The only way was with the correspondence principle. The first major attempt to understand the transitions between the stationary states came from Albert Einstein in the year 1917 by his proposal of transition probabilities.

The quantum theory of Planck and the theories in atomic structure and transitions proposed by Bohr, Sommerfeld and others had proved to be adequate to explain many phenomena, they were not sufficient in explaining many finer details in the spectra. Hence other theories like vector atom model and quantum mechanical descriptions for atoms have been proposed and developed.

The outstanding success of Bohr Theory as far as hydrogen like atoms are concerned is its methodical approach in order to understand the theory and its success in combining classical and quantum ideas.

SOLVED PROBLEMS

1. Why does the mercury light from public lighting appear blue, even though there are yellow and green lines also appearing in the spectrum?

Solution

Among the many atomic lines due to mercury the emission line at 435.8 nm is the most intense among the other emission lines in the visible region. In the design of the mercury lamps the optimum conditions are optimized so that the maximum output is obtained for this transition.

2. i. Find the mass in grams of one atom of the following elements.

(a) Bi atomic weight 208.9804 amu

(b) Xe atomic weight 131.29 amu

(c) He atomic weight 4.0026 amu

Solution

(a) $\dfrac{208.9804}{6.022 \times 10^{23}} = 208.9804 \times 1.660539 \times 10^{24} = 3.47 \times 10^{-22}\,g$

(b) $\dfrac{131.29}{6.022 \times 10^{23}} = 2.18 \times 10^{-22}\,g$

(c) $\dfrac{4.0026}{6.022 \times 10^{23}} = 6.64 \times 10^{-24}\,g$

ii. Find the mass in atomic mass units?

(a) 1 Oxygen atom of mass 2.66×10^{-23} g

(b) 1 Bromine atom of mass 1.31×10^{22} g

Solution

(a) $2.66 \times 10^{-23} \times 6.022 \times 10^{23} = 16$ amu

(b) $1.31 \times 10^{-22} \times 6.022 \times 10^{23} = 78.9$ amu

iii. What is the mass in grams of 6.022×10^{23} nitrogen atoms of mass 14.01 amu?

Solution

Mass of 1 atom $= \dfrac{14.01}{6.022 \times 10^{23}}\,g$

Mass of 6.022×10^{23} atoms $= \dfrac{14.01}{6.022 \times 10^{23}} \times 6.022 \times 1023 = 14.01$ g

iv. What is the mass in grams of 6.022×10^{23} carbon atoms of mass 12.00 amu?

Solution

$$\text{Mass of 1 atom} = \frac{12.0}{6.022 \times 10^{23}} \, \text{g}$$

$$\text{Mass of } 6.022 \times 10^{23} \text{ atoms} = \frac{12.0}{6.022 \times 10^{23}} \times 6.022 \times 10^{23} = 12.00 \, \text{g}$$

v.   How many O atoms of mass 15.99 amu are in 15.99 g of oxygen?

Solution

$$\text{Mass of 1 atom} \quad \frac{15.99}{6.022 \times 10^{23}} \, \text{g}$$

$$\text{Number of atoms in 15.99 gram} = 15.99 \div \frac{15.99}{6.022 \times 10^{23}} = 6.022 \times 10^{23}$$

3.   Calculate the time taken by the electron to traverse the first orbit of hydrogen atom. The radius of the nth orbit $r_n = \dfrac{n^2}{m}\left(\dfrac{h}{2\pi}\right)\dfrac{4\pi\varepsilon_0}{Ze^2} = n^2 \times 5.29 \times 10^{-11} \, \text{m}$

Solution

$$\text{Velocity of the electron in the } n\text{th orbit} \quad \mathbf{v} = \frac{1}{137} \times \frac{c}{n}$$

$$\text{Velocity in the first orbit} \quad \mathbf{v} = \frac{1}{137} \times c$$

$$\text{Period of electron in the first orbit} = \frac{2\pi r}{\mathbf{v}} = \frac{2 \times 3.14 \times 5.29 \times 10^{-11}}{3 \times 10^{8}} \times 137 = 1.517 \times 10^{-16} \, \text{s}$$

4.   The energy of an excited hydrogen atom is − 3.4 eV. Calculate the angular momentum of the electron from Bohr's theory.

Solution

$$\text{The energy of the } n\text{th orbit} \quad E = -\frac{13.6}{n^2} \, \text{eV}$$

$$\text{Therefore,} \quad -3.4 = -\frac{13.6}{n^2}$$

Hence $n = 2$

$$\text{Angular momentum} \quad L = n\frac{h}{2\pi} = \frac{2 \times 6.625 \times 10^{-34}}{2 \times 3.14} = 2.11 \times 10^{-34} \, \text{J.s}$$

5. Hydrogen atom in the ground state is excited by means of monochromatic radiation of wavelength 975 Å. How many different transitions are possible in the resulting spectrum? Calculate the longest wavelength among them.

Solution

$$E = h\nu, \quad \nu = \frac{c}{\lambda}$$

Hence $E = h\dfrac{c}{\lambda}$

Energy equivalent to 975 $\overset{\circ}{A} = h\dfrac{c}{\lambda} = \dfrac{6.625 \times 10^{-34} \times 3 \times 10^8}{975 \times 10^{-10}} \text{J}$

$$= \frac{6.625 \times 10^{-34} \times 3 \times 10^8}{975 \times 10^{-10} \times 1.6 \times 10^{-19}} \text{eV} = 12.75 \text{ eV}$$

Energy of the electron in the ground state is -13.6 eV

Therefore energy in the excited state $-13.6 + 12.75 = -0.85$ eV

$$E_n = -\frac{13.6}{n^2} = 0.85$$

$\therefore n = 4$

Possible transitions are from $n = 4$ $\qquad 4 \to 3, \ 3 \to 2, \ 2 \to 1$

$\qquad\qquad\qquad\qquad$ from $n = 3$ $\qquad 3 \to 2, \ 2 \to 1$

$\qquad\qquad\qquad\qquad$ from $n = 2$ $\qquad 2 \to 1$

$\qquad\qquad\qquad\qquad$ Total 6 transitions

Line of the longest wavelength is the transition from $n = 4 \to 3$

$E_4 - E_3 = -0.85 - (-1.5) = 0.65$ eV

Wavelength $\lambda = \dfrac{hc}{E} = \dfrac{6.625 \times 10^{-34} \times 3 \times 10^8}{0.65 \times 1.6 \times 10^{-19}} = 19110$ Å

6. The series limit of Balmer series is at 3646 Å. Calculate the wavelength of the first member of the series.

Solution

$$\frac{1}{\lambda} = R\left(\frac{1}{2^2} - \frac{1}{n^2}\right)$$

For series limit $n = \alpha$ and hence $\dfrac{1}{\lambda_{\text{limit}}} = \dfrac{R}{4} = \dfrac{1}{3646 \times 10^{-10}}$

$$R = \frac{4 \times 10^{10}}{3646}$$

First member $n = 3$ and hence $\dfrac{1}{\lambda} = R\left(\dfrac{1}{2^2} - \dfrac{1}{3^2}\right) = \dfrac{4 \times 10^{10}}{3646}\left(\dfrac{1}{4} - \dfrac{1}{9}\right)$

$$\lambda = \frac{3646}{4 \times 10^{10}} \times \frac{36}{4} = 6563 \times 10^{-10} \text{ m} = 6563 \text{ Å}$$

7. The wavelength of the H line in hydrogen is 4341 Å. Find the wavelength of the second line in the Paschen series.

Solution

$$\frac{1}{\lambda} = R\left(\frac{1}{n_1^2} - \frac{1}{n_2^2}\right)$$

H_γ is the third line of Balmer series $n = 5 \to 2$

$$\frac{1}{\lambda_1} = R\left(\frac{1}{2^2} - \frac{1}{5^2}\right) = R\left(\frac{1}{4} - \frac{1}{25}\right) = R\frac{21}{100}$$

The second line in Paschen series is $n = 5 \to 3$

$$\frac{1}{\lambda_2} = R\left(\frac{1}{3^2} - \frac{1}{5^2}\right) = R\left(\frac{1}{9} - \frac{1}{25}\right) = R\frac{16}{225}$$

$$\frac{\lambda_2}{\lambda_1} = \frac{225}{16R} \times \frac{21R}{100} = \frac{189}{64}$$

$$\lambda_2 = \frac{189}{64} \times 4341 = 12819 \text{ Å}$$

$$\lambda_2 = 12819 \text{ Å}$$

8. Using Bohr's formula (a) determine the longest wavelength of Lyman series (b) the excitation energy of $n = 3$ level of He$^+$ atom (c) the ionization potential of the ground state of Li^{++} atom ($R = 1.097 \times 10^7$ m^{-1}).

Solution

$$\frac{1}{\lambda} = R\left(\frac{1}{n_1^2} - \frac{1}{n_2^2}\right)$$

For Lyman series $n_1 = 1$

$$\frac{1}{\lambda} = R\left(\frac{1}{1^2} - \frac{1}{n_2^2}\right)$$

For the longest wavelength $n_2 = 2$

$$\frac{1}{\lambda} = R\left(\frac{1}{1^2} - \frac{1}{4}\right) = 1.097 \times 107 \times \frac{3}{4}$$

$$\lambda = \frac{4}{1.087 \times 10^7 \times 3} = 1.215 \times 10^{-7}\,\text{m} = 1215\,\text{Å}$$

Energy of the nth state of ionized atoms

$$E_n = -\frac{13.6}{n^2} Z^2 \text{ eV}$$

For He$^+$ atom $Z = 2$

$$E_1 = -4 \times \frac{13.6}{1} = -54.4 \text{ eV for the first orbit}$$

$$E_3 = -4 \times \frac{13.6}{3^2} = -6.04 \text{ eV for the third orbit}$$

Excitation energy from $n = 1$ to $n = 3$ $= -6.04 + 54.4 = 48.36$ eV

For Lithium $Z = 3$ and

$$E_n = -\frac{13.6}{n^2} 32 \text{ eV} = -\frac{122.4}{n^2} \text{eV}$$

Energy in the ground state $(n = 1) = -122.4$ eV

Ionization energy $= 122.4$ eV and the ionization potential $= 122.4$ Volt.

9. Which state of triply ionizes beryllium (Be^{+++}) has the same orbital radius as that of the ground state of hydrogen atom. Compare the energies of the two states.

Solution

Expression for the nth orbit is $r_n = \frac{n^3}{Z}\frac{\hbar^2 \varepsilon_0}{\pi m e^2} = 0.529 \times 10^{-10}\frac{n^2}{Z}$

The nth orbit of the hydrogen atom is $r_n = n^2 \times 0.529 \times 10^{-10}$ m

For Beryllium $Z = 4$, and $r_n = \frac{n^3}{Z}\frac{\hbar^2 \varepsilon_0}{\pi m e^2} = 0.529 \times 10^{-10}\frac{n^2}{4}\,m$

$$0.529 \times 10^{-10}\frac{n^2}{4} = 0.529 \times 10^{-10}\text{ m (of hydrogen)}$$

$$n^2 = 4 \text{ and } n = 2$$

Energy of the nth orbit of hydrogen $E_n = -\dfrac{13.6}{n^2}\text{eV}$

Energy of the ground state of hydrogen $E_n = -\dfrac{13.6}{1^2}\text{eV}$

Energy of the nth orbit of Be^{+++} $E_n = -\dfrac{13.6}{n^2}Z^2 \text{ eV}$

For $n = 2$, $E_n = \dfrac{13.6}{2^2} \times 4^2\text{eV} = -4 \times 13.6 \text{ eV}$

$$\frac{E_2(Be^{+++})}{E_1(H)} = 4$$

10. Which level of doubly ionized lithium (Li^{++}) has the same energy as the ground state of hydrogen atom? Compare the orbital radii of the two levels.

Solution

For hydrogen $E_n = -\dfrac{13.6}{n^2}\text{eV}$ and $E_1 = -13.6 \text{ eV}$

For other atoms $E_n = -\dfrac{13.6}{n^2}Z^2 \text{ eV}$

Let the ground state energy E_1 of hydrogen be equal to the nth orbit of L^{++} ($Z = 3$)

Therefore, $E_n = \dfrac{13.6}{n^2}32 \text{ eV} = -13.6 \text{ eV}$ (of hydrogen ground state)

$$n^2 = 9, \quad n = 3$$

Radius of the nth orbit of hydrogen $r_n = n^2 \times 5.29 \times 10^{-11} \text{ m}$

for other atoms $r_n = \dfrac{n^3}{Z}\dfrac{\hbar^2 \varepsilon_0}{\pi m e^2} = 0.529 \times 10^{-10}\dfrac{n^2}{Z}$

r_1 (hydrogen) $= 5.29 \times 10^{-11} \text{ m}$

for $n = 3$ orbit of L^{++} $r_3 = \dfrac{5.29 \times 10^{-11}}{3} \times 9 = 3 \times 5.29 \times 10^{-11}\text{m}$

$$\frac{r_3(Li^{++})}{r_1(H)} = 3$$

11. The total energy of the electron in the first excited state of hydrogen is about -3.4 eV.

    (a) What is the K.E. of the electron in this state?

    (b) What is the P.E. of the electron in this state?

Solution

$$\text{K.E.} = \frac{1}{2}\frac{1}{4\pi\varepsilon_0}\frac{Ze^2}{r_n}$$

$$\text{P.E.} = -\frac{1}{4\pi\varepsilon_0}\frac{Ze^2}{r_n} = -2\ \text{K.E.}$$

Total energy T.E. = K.E. + P.E. = K.E. $- 2\ K.E.$ $= -$ K.E.

$$\text{K.E.} = -\text{T.E.} = -(-3.4) = 3.4\ \text{eV}$$

$$\text{P.E.} = -2 \times 3.4 = -6.8\ \text{eV}$$

12. The wavelength of the first line of Balmer series of hydrogen atom is 6563 Å. Calculate the ionization potential and the first excitation potential of hydrogen atom ($h = 6.63 \times 10^{-34}$ Js and $c = 3 \times 10^{8}$ m/s).

Solution

$$\frac{1}{\lambda} = R\left(\frac{1}{2^2} - \frac{1}{n^2}\right)$$

First line $\dfrac{1}{\lambda} = R\left(\dfrac{1}{2^2} - \dfrac{1}{3^2}\right) = R\dfrac{5}{36} = \dfrac{1}{6563 \times 10^{-10}}$

$$R = \frac{36 \times 10^{10}}{5 \times 6563} = 1.097 \times 10^{7}\ \text{m}^{-1}$$

The energy of the electron in the nth orbit

$$E_n = -\frac{Rhc}{n^2} = -\frac{1.097 \times 10^{7} \times 6.63 \times 10^{-34} \times 3 \times 10^{8}}{n^2 \times 1.6 \times 10^{-19}} = -\frac{13.6}{n^2}\text{eV}$$

$$E_1 = -13.6\ \text{eV}$$

$$E_2 = -\frac{13.6}{2^2} = -3.4\ \text{eV}$$

Excitation potential $= 13.6 - 3.4 = 10.2$ eV

13. What is the distance of the closest approach to the nucleus of gold atom by an alpha particle of energy 5, 5 MeV which undergoes scattering by 180°?

Solution

When alpha particle is closest to the nucleus it comes to rest and its kinetic energy is completely converted into potential energy. So the distance of closest approach is when K.E. = P.E.

$$E = \frac{1}{4\pi\varepsilon_0}\frac{Ze \times 2e}{r_0}$$

$$r_0 = \frac{1}{4\pi\varepsilon_0}\frac{2Ze^2}{E}$$

$$E = 5.5 \text{ MeV} = 5.5 \times 10^6 \times 1.6 \times 10^{-19} \text{J} = 8.8 \times 10^{-13} \text{J}$$

For gold $Z = 70$, $\dfrac{1}{4\pi\varepsilon_0} = 9 \times 10^9$ and $e = 1.6 \times 10^{-19}$ Coulomb

$$r_0 = \frac{9 \times 10^9 \times 2 \times 79 \times (1.6 \times 10^{-19})^2}{8.8 \times 10^{-13}} = 41.36 \times 10^{-15} \text{m} = 41.36 \text{ fm}$$

ENDNOTES

1   John Dalton, 6 Sept 1766–27 July 1844.

2   J.J. Thomson (18 December 1856–30 August 1940), was awarded Nobel Prize in 1906.

3   Ernest Rutherford (30 August 1871–19 Oct 1937), obtained Nobel Prize in Chemistry in 1908.

4   Niels Bohr (7 October 1885–18 November 1962), received Nobel Prize in 1922.

5   Hänsch T.W., Nobel Prize 2005.

6   Schmidt-Kaler, A., Leibfried, D., Seel, S., Zimmermann, C., Koenig, W., Weitz, M. and Hänsch. T.W. (1995). High Resolution Spectroscopy of the 1S–2S Transition of Atomic Hydrogen and Deuterium, *Physical Review* A, 51 2789.

7   Johann Jakob Balmer (May 1, 1825–March 12, 1898).

8   Louis Karl Heinrich Friedrich Paschen (January 22, 1865–February 25, 1947).

9   Frederick Sumner Brackett (August 1, 1896–January 28, 1988).

10 August Herman Pfund (December 28, 1879–January 4, 1949).

11 Houston, W.V. (1927). *Phys.Rev.* 30, 608.

12 NIST reference on constants, units and uncertainty, NIST 2006.

13 Dirac, P.A.M. (8 August 1902–20 October 1984), Nobel Prize in Physics 1933.

14   Willis Eugene Lamb Jr., (12 July 1913–15 May 2008), Nobel Prize in Physics 1955.

15   Uzan, John Philippe. (2003) The fundamental constants and their variation: observational status and theoretical motivations." *Reviews of Modern Physics.* 75

16   Olive, Keith; Quian, Young-Zhong (2003). "Were Fundamental Constants Different in the Past?" *Physics Today.* (American Institute of Physics) 57 (10): 40–45.

17   Barrow, John D., *The constants of Nature: From alpha to omega-the numbers that encode the* deepest secrets of the universe. Vintage, London.

18   Uzan, Jean-Philippe; Bénédicte Leclercq (2008). *The natural laws of the universe-Understanding fundamental constants.* Springer. Praxis, Berlin Heidelberg New York.

19   Fujii, Yasunori (2004). "Oklo Constant on the Time-Variability of the Fine-Structure Constant." *Astrophysics, Clocks and Fundamental Constants .Lecture Notes in Physics.* Springer. Heidelberg Berlin. pp. 167–185.

20   Uzan, John-Philippe. (2004) " Variation of the constants in the late and early universe." *Astrophys.* 0409424.

21   Feynman, Richard, P. (1985). QED : *The Strange Theory of Light and Matter.* Princeton University Press. pp. 129.

22. Kaufmann W.J. (1988). *Universe.* W.H. Freeman and Co.

23   G. W. F. Drake, (Ed.). 1996 *Atomic, Molecular and Optical Physics Handbook.* AIP Press, Woodbury, New York.

24   James Franck and Gustav Hertz, Verhand. (1914). *Deut.Physik Ges.* 16, 457–467.

25   White, H.E. (1934). *Introduction to Atomic Spectra.* McGraw-Hill New York.

REFERENCES

Barrow, D. John. (2002). The Constatns of Nature: From Alpha to Omega-The Numbers that Encode the Deepest Secrets of the Universe. Vintage, London.

Biraben, M., *et al.* (2001). In the hydrogen atom, precision physics of simple atomic systems. (edts.). Karshenboim, S.G., Pavone, F.S., Basssani. G.F., Inguscio, M. and Hänsch, T.W. Springer, Berlin. Heidelberg. 18.

Bourzeix, S., de Beauvoir, B., Nez, F., Plimmer, M.D., de Tomasi, F., Julien, L. and Biraben, F. (1996). *Phys. Rev. Lett.* 76, 384.

Condon, E.U. and Shortley, G.H. (1935). *The Theory of Atomic Spectra.* Cambridge University Press, Cambridge.

Cowan, R.D. (1981). *The Theory of Atomic Structure and Spectra.* University of California Press, Berkeley.

de Beauvoir, B., Nez, F., Julien, L., Cagnac, B., Biraben, F., Touahri, D., Hilico, L., Acef, O., Clairon, A. and Zondy, J.J. (1997). *Phys. Rev. Lett.* 78, 440.

Feynman, Richard P. (1985). QED: The Strange Theory of Light and Matter. Princeton University Press. p. 129.

Fujii, Yasunori (2004)."Oklo Constant on the Time-Variability of the Fine-Structure Constant". Astrophysics, Clocks and Fundamental Constants. Lecture Notes in Physics. Heidelberg: Springer Berlin. pp. 167–185.

Hänsch, T.W., Alnis, J., Fendel, P. Fischer, M., Gohle, C., Herrmann, M., Holzwarth, R., Kolachevsky, N., Udem, T.H. and Zimmermann, M. (2005). Precision spectroscopy of hydrogen and femtosecond laser frequency combs. *Philosophical Transactions of Royal Society A.* 363 2155–2163.

Hänsch, T.W., Andreae, T., Huber, A., Koenig, W., Leibfried, D., Pachucki, K., Prevedelli, M., Schmidt-Kaler, F., Weitz, M., Wynands, R. and Zimmermann, C. (1994). Precision Laser Spectrosocpy of Atomic Hydrogen, Proceedings of the 11th International Conference on Laser Spectroscopy, Hot Springs, Virginia, (1993). Bloomfield, L., Gallagher, T. and Larson, D. (eds.). AIP Conference Proceeding, V 290, p3.

Huber, A., Udem, T.H., Gross, B., Reichert, J., Kourogi, M., Pachucki, K., Weitz, M. and Hänsch, T.W. (1998). *Phys. Rev. Lett.* 80, 468.

Kolachevsky, N., Fischer, M., Karshenboim, S. and Hänsch, T.W. (2004). High precision optical measurement of the 2S hyperfine interval in atomic hydrogen. *Physical Review Letters.* 92, 033.

Kuhn, H.G. *Atomic Spectra.* (1962). Longmans, London.

Niering, M. *et al.* (2000). Measurement of the hydrogen 1S-2S transition frequency of phase coherent. Comparison with microwave cesium fountain clock. *Physical Review Letters.* 84, 5496–5499.

Olive, Keith, Quian, Young-Zhong. "Were Fundamental Constants Different in the Past?" Physics Today (Americal Institute of Physics) 57 (10). 40–45 (2003).

Pachucki, K., Leibfried, D., Weitz, M., Huber, A., König, W. and Hänsch, T.W. (1996). *J. Phys. B* 29, 177.

Schmidt-Kaler, A., Leibfried, D., Seel, S., Zimmermann, C., Koenig, W., Weitz M., and Hänsch, T.W. (1995). High resolution spectroscopy of the 1S–2S transition of atomic hydrogen and deuterium. *Physical Review* A. 51, 2789.

Udem, T.H., Huber, A., Gross, B., Reichert, J., Prevedelli, M., Weitz, M. and Hänsch, T.W. (1997). *Phys. Rev. Lett.* 79, 2646.

Uzan, Jean-Philippe, Bénédicte Leclercq (2008). The Natural Laws of the Universe-Understanding Fundamental Constants. Berlin Heidelberg New-York: Springer Praxis.

Uzan, John Philippe. (2003). The fundamental constants and their variation: observational status and theoretical motivations". *Reviews of Modern Physics.* 75.

Uzan, John-Philippe. (2004). Variation of the constants in the late and early Universe *Astrophys*. 0409424.

Weitz, M., Huber, A., Schmidt-Kaler, F., Leibfried, D., Vassen, W., Zimmermann, C., Pachucki K., Häensch, T.W., Julien, L. and Biraben, F. (1995). Precision measurement of the 1S ground state lamb shift in atomic hydrogen and deuterium by frequency comparison; *Physical Review* A. 52. 2664.

Weitz, M., Huber, A., Schmidt-Kaler, F., Leibfried, D. and Hänsch, T.W. (1994). Precision Measurement of the hydrogen and deuterium 1S ground state lamb shift. *Physical Review Letters*. 72, 328.

HISTORICAL

Niels Bohr (1913). "On the constitution of atoms and molecules." *Philosophical Magazine*. 26: 1–25.

Niels Bohr (1913). "On the constitution of atoms and molecules, part II systems containing only a single nucleus." *Philosophical Magazine*. 26: 476–502.

Niels Bohr (1913). "On the constitution of atoms and molecules, Part III." *Philosophical Magazine*. 26: 857–875.

Niels Bohr (1914). "The spectra of helium and hydrogen." *Nature*. 92: 231–232.

Niels Bohr (1921). "Atomic Structure." March 24, *Nature*.

3

THE INTERACTION OF RADIATION WITH MATTER

The interaction of radiation with matter is huge and fascinating and involves both experimental and theoretical approach and combines the ideas of quantum mechanics and relativity.

The two theories on light, viz., the particle and wave concepts are necessary for a detailed description of the interaction of light with matter and the decoding of messages within the radiation. The simple picture of light is as particle of both electric and magnetic field that oscillates in time perpendicular to the direction of propagation. There is another simple picture of photon, which is as a wave with a known wavelength. The wave–particle duality is an interesting problem.

Either as a wave or particle, there are a number of possible interactions of photon with matter and for the moment we consider the atom, although any matter will do. The list of possible interactions is as follows although it is not complete.

i. Elastic scattering
ii. Non-elastic scattering
iii. Stimulated absorption or merely absorption
iv. Stimulated emission
v. Spontaneous emission

Elastic scattering involves the interaction of a photon with an atom that does not change the energy of the photon and conversely non-elastic scatter allows the photon to be absorbed by the atom and re-emitted at a different wavelength so that the atom absorbs some energy or gives additional energy to the photon. Absorption allows for the photon to be absorbed completely and destroyed, or formally annihilated, and the atom then absorbs both energy and the momentum of the photon. Stimulated emission requires the atom to be in an excited state and when a photon of the light energy/

frequency passes by, the atom is stimulated into the emission of a photon of exactly the same colours of the passing photon. Finally the atom may be in an excited state and spontaneously creates a photon which is emitted, relaxing the atom to the lower energy state.

3.1 EINSTEIN'S TRANSITION PROBABILITIES

The fundamental concept on the foundation of quantum mechanical amplifier of electromagnetic radiation (viz., the concept of maser and laser) was laid down by Albert Einstein way back in 1917 when he proposed that a quantized atomic system in thermodynamic equilibrium exhibits not two but three radiative processes, viz., absorption, spontaneous emission and induced or stimulated emission[1]. Under normal conditions the first two processes dominate and the third one, viz., stimulated emission is comparatively negligible. It took another 37 years to reveal the wide ranging consequences of these processes by the invention of the first maser in 1954 by Gorden, Zeiger and Townes.[2, 3]

Let us consider two quantized energy levels of an atom, a lower E_1 state and an upper E_2 state as shown in Figure 3.1. Let there be N_1 number of atoms in the energy level E_1 and N_2 number of atoms in energy level E_2. An atom in the lower energy state E_1 can absorb radiation or a photon from an outside source and if it gets sufficient energy (equal to the energy difference of the two states), the atom is brought to the upper state E_2. The process is called absorption. Thus an excitation can take place only in the presence of radiation from an outside source. Thus the absorption process will depend on the external radiation field at the particular transition frequency (resonance frequency) between the two levels and also on the number of atoms in the lower state. The transition is followed by the Planck's relation

$$\Delta E = E_2 - E_1 = h\nu \tag{3.1}$$

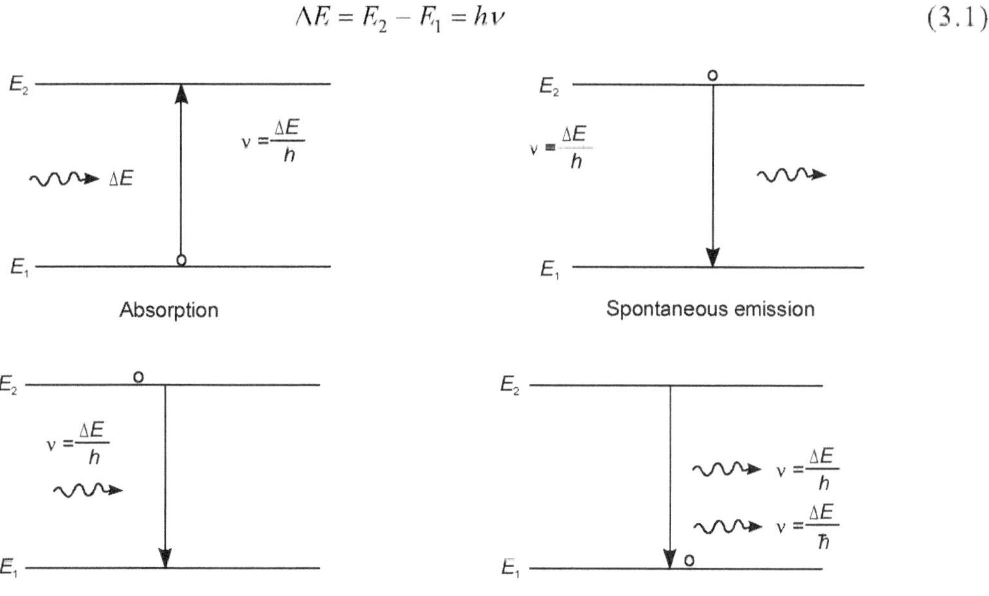

Figure 3.1   Radiative processes

As $\omega = 2\pi\nu$ we can write $E_2 - E_1 = h\omega / 2\pi$

$$\omega = \frac{E_2 - E_1}{\hbar} \quad \text{where} \quad \hbar = \frac{h}{2\pi} \tag{3.2}$$

If the energy density at the absorption frequency ω is $u(\omega)$, the number of absorptions per unit area per unit time W_{12} is given by

$$W_{12} \propto N_1\, u(\omega)$$

or
$$W_{12} = B_{12}\, N_1 u(\omega) \tag{3.3}$$

where B_{12} is the proportionality constant which is known as Einstein's coefficient of absorption and is characteristic of the energy level.

Once the atom is raised to the excited state, it doesn't stay in the excited state for a long time. The time the atom stays in the excited state, is called the lifetime of the state. The lifetime of majority of the states is of the order of 10^{-8} seconds. However there are states in which atom stays for a longer period. In that case the lifetime of the state is said to be large and such states are called metastable states. Let us now consider that the lifetime of the excited state is small. In that case the atom tries to reverse back to the original state. Einstein postulated two kinds of possibilities of this transition from the upper state to the lower state. One is called spontaneous emission and the other is induced or stimulated emission. However the frequency of the emitted radiation in either case is the same as in the case of absorption $\omega = \dfrac{E_2 - E_1}{\hbar}$. The spontaneous emission occurs without any external field as the name suggests. However the stimulated emission is induced by an external field. If we consider the number of transitions per second in spontaneous emission, it is evident that the number of emissions per unit area per unit time depends only on the number of atoms in the excited state whereas in the case of stimulated emission it depends on the intensity of the external field as well.

Thus the number of spontaneous emissions per unit area per unit time $W_{21} = A_{21}\, N_2$ and the number of stimulated emissions per unit area per unit time $W_{21} = B_{21}\, N_2 u(\omega)$ where the proportionality factor A_{21} is called Einstein's coefficient of spontaneous emission and the proportionality factor B_{21} as Einstein's coefficient of stimulated emission. Thus there are three types of Einstein's coefficients, viz., Einstein's coefficient of absorption, Einstein's coefficient of spontaneous emission and Einstein's coefficient of stimulated emission.

In thermal equilibrium the number of upward transitions is equal to number of downward transitions. Thus we have,

$$B_{12}\, N_1\, u(\omega) = A_{21}\, N_2 + B_{21}\, N_2\, u(\omega)$$

$$u(\omega)\, [B_{12}\, N_1 - B_{21}\, N_2] = A_{21}\, N_2 \tag{3.4}$$

$$u(\omega) = \frac{N_2 A_{21}}{[N_1 B_{12} - N_2 B_{21}]} = \frac{A_{21}}{\left[\left(\dfrac{N_1}{N_2}\right) B_{12} - B_{21}\right]} \tag{3.5}$$

The distribution of atoms under thermal equilibrium follows Boltzmann distribution law and hence the number of atoms in the excited state is given by

$$N_2 = N_1 \exp\left[-\left(\frac{E_2 - E_1}{kT}\right)\right] = N_1 \exp\left(-\frac{\hbar\omega}{kT}\right) \tag{3.6}$$

$$\frac{N_1}{N_2} = \exp\left(\frac{\hbar\omega}{kT}\right) \tag{3.7}$$

Thus we can write the radiation field $u(\omega)$ as

$$u(\omega) = \frac{A_{21}}{\left[\left(\dfrac{N_1}{N_2}\right) B_{12} - B_{21}\right]}$$

$$= \frac{A_{21}}{\left[\left(B_{12} \exp\left(\dfrac{\hbar\omega}{kT}\right) - B_{21}\right)\right]} = \frac{\left(\dfrac{A_{21}}{B_{21}}\right)}{\left[\dfrac{B_{12}}{B_{21}} \exp\left(\dfrac{\hbar\omega}{kT}\right) - 1\right]} \tag{3.8}$$

Planck's distribution law states that

$$u(\omega) = \frac{8\pi h\nu^3}{c^3} \frac{1}{\left[\exp\left(\dfrac{h\nu}{kT}\right) - 1\right]} = \frac{\dfrac{8\pi h\nu^3}{c^3}}{\left[\exp\left(\dfrac{h\nu}{kT}\right) - 1\right]} = \frac{\left(\dfrac{2\hbar\omega^3}{\pi c^3}\right)}{\left[\exp\left(\dfrac{\hbar\omega}{kT}\right) - 1\right]} = \frac{\left(\dfrac{\hbar\omega^3}{\pi^2 c^3}\right)}{\left[\exp\left(\dfrac{\hbar\omega}{kT}\right) - 1\right]} \tag{3.9}$$

By comparing (3.8) and (3.9) we obtain $\dfrac{B_{12}}{B_{21}} = 1$

hence $B_{12} = B_{21} = B$

$$\text{Also } \frac{A_{21}}{B_{21}} = \frac{2\hbar\omega^3}{\pi c^3} = \frac{\hbar\omega^3}{\pi^2 c^3} \tag{3.10}$$

The ratio of spontaneous emission to stimulated emission in the case of a thermal source is

$$\frac{N_2 A_{21}}{N_2 B u(\omega)} = e^{\frac{\hbar\omega}{kT}} - 1 \tag{3.11}$$

$$\frac{A_{21}}{Bu(\omega)} = e^{\frac{\hbar\omega}{kT}} - 1 \tag{3.12}$$

If $\omega > \left(\dfrac{kT}{\hbar}\right)$, the right-hand side of the expression (3.12) becomes large and hence the spontaneous

emission will supersede stimulated emission. And if $\omega < \left(\dfrac{kT}{\hbar}\right)$, the right-hand side of the equation

(3.12) becomes small and in this case stimulated emission is large. At optical frequencies where

$\omega < \left(\dfrac{kT}{\hbar}\right)$, spontaneous emission is the most probable kind of transition.

Bohr incorporated the idea proposed by Einstein. Einstein stated that an atomic system may transfer from an energy state to an energetically lower state "without excitation by an external cause". This was interpreted by Bohr as to mean that the system will spontaneously pass to a stationary state of lower energy. Although Einstein did not use the word in his paper, the A_{mn} is known now as probability factor of spontaneous emission or coefficient of spontaneous emission. Extending the correspondence principle to the classical representation of Fourier expansion of the dipole moment

$$A_{mn} \leftrightarrow \frac{16\pi^4 \nu^3}{3c^3 h} \left|\overline{P}_{nm}\right|^2 \tag{3.13}$$

where P_{nm} is the transition dipole moment.

Let us consider two plane mirrors M_1 and M_2 each having an area A placed parallel to each other at x and $x + dx$ so that the distance between them is dx. We consider a beam of light of frequency ω propagates in the x-direction. The mirrors are kept perpendicular to the line of propagation of the beam of light. We shall now calculate the rate of change of intensity as the beam passes through the medium.

Now let us consider an electric dipole $\vec{P}$ fixed in direction and oscillates harmonically with a frequency ν_0 and with amplitude $\vec{P}_0$ so that

$$\vec{P} = \vec{P}_0 \cos 2\pi \nu_0 t \tag{3.14}$$

We consider a point R from the dipole where R is large compared to the wavelength of the radiation emitted by the oscillating dipole. The electric and magnetic fields at the point at the distance R can be written as

$$\left|\vec{E}\right| = \left|\vec{H}\right| = \frac{1}{c^2 R}\left|\frac{d^2\vec{P}}{dt^2}\right|\sin\theta \tag{3.15}$$

where higher powers of $1/R$ in the expansion are neglected. We already know that the electric vector and magnetic vector are in mutually perpendicular directions and they are right angles to the direction of propagation of the wave. The angle θ is the angle between the direction of the propagation of radiation and the direction in which the dipole is oscillating. Let us consider that the electric vector is in the plane of the paper which is the plane of R and $\vec{P}$.

Combining (3.14) and (3.15) we obtain

$$\left|\vec{E}\right| = \left|\vec{H}\right| = \frac{4\pi^2 v_0^2}{c^2 R}\,\overline{P}\sin\theta \tag{3.16}$$

If the dipole oscillation is subjected to electromagnetic forces due to its own motion, then there will be less energy of radiation and this energy is emitted out. The rate of emitted radiation energy is related to the Poynting's vector and is obtained by integration of Poynting vector over the surface of the sphere of radius R.

The Poynting vector is written as

$$\left|\vec{S}\right| = \frac{c}{4\pi}\left|\vec{E}\times\vec{H}\right| = \frac{c}{4\pi}\frac{16\pi^4 v_0^4}{c^4 R^2}\left|\vec{P}\right|^2 \sin^2\theta \tag{3.17}$$

The instantaneous rate of flow energy between the angle θ and $\theta + d\theta$ is

$$\left|\vec{S}\right| \times 2\pi\,R^2 \sin\theta\,d\theta \tag{3.18}$$

Hence the rate of emission is

$$-\frac{dE}{dt} = \frac{8\pi^4 v_0^4}{c^3}\left|\vec{P}\right|^2 \int_0^\pi \sin^2\theta\,d\theta = \frac{32\pi^4 v_0^4}{3c^3}\left|\vec{P}\right|^2 \tag{3.19}$$

The above expression turns out to be $\quad -\dfrac{dE}{dt} = \dfrac{16\pi^4 v_0^4}{3c^3}\left|\vec{P}_0\right|^2 \tag{3.20}$

The average rate of radiation can be shown to be $\dfrac{16\pi^4 v_0^4}{3c^3}\,\vec{M}.\vec{M}^*$ where $\vec{M} = \int \rho(x,y,z)\vec{r}dxdydz$ is the electric moment of the oscillating charges $\vec{M}^*$ is its complex conjugate.

Quantum mechanically it can be shown that it is equal to

$$\frac{64\pi^4 e^2 v_{mn}^4}{3c^3 h}\left[\left|\langle x_{mn}\rangle^2\right| + \left|\langle y_{mn}\rangle^2\right| + \left|\langle z_{mn}\rangle^2\right|\right] \tag{3.21}$$

where,

$$\left|\langle x_{mn}\rangle^2\right| = \left|\int \Psi_n^* x \Psi_m \, d\tau\right|^2$$

$$\left|\langle y_{mn}\rangle^2\right| = \left|\int \Psi_n^* y \Psi_m \, d\tau\right|^2$$

$$\left|\langle z_{mn}\rangle^2\right| = \left|\int \Psi_n^* z \Psi_m \, d\tau\right|^2$$

and

$$A_{mn} = \frac{64\pi^4 e^2 v_{mn}^4}{3hc^3}\left[\left|\langle x_{mn}\rangle^2\right| + \left|\langle y_{mn}\rangle^2\right| + \left|\langle z_{mn}\rangle^2\right|\right] \tag{3.22}$$

3.2 SELECTION RULES

It had been proposed by Niels Bohr that only certain transitions between the energy states were possible which he called as selection rules. It means that a major pair of energy levels are not connected with the observed line spectrum. We have already seen that $\Delta l = \pm 1$ and $\Delta m = 0$ or ± 1 are the selection rules for dipole transitions. Though Bohr proposed them as selection rules without any mathematical proof, solutions of Schrödinger equation and the determination of transition matrix elements give mathematical proof for earlier conclusions. Thus we can derive selection rules on the basis of quantum mechanics which also gives information on the relative intensities of the different allowed transitions.

The dipole selection rules are only to a first approximation. And hence deviations in selection rules can occur. Although a transition may be forbidden by dipole approximation, they can however, be allowed by magnetic dipole or electric quadrupole interactions. Various kinds of selection rules are listed in Table 3.1.

Table 3.1 Electric dipole, magnetic dipole and electric quadrupole selection rules under LS coupling scheme.

Quantum number	Electric dipole	Magnetic dipole	Electric quadrupole
Total angular momentum quantum number (J)	$\Delta J = 0, \pm 1$ $J' = 0 \neq J'' = 0$	$\Delta J = 0, \pm 1$ $J' = 0 \neq J'' = 0$	$\Delta J = 0, \pm 1$ $J' - J = \pm 2$
Total orbital angular momentum quantum number (L)	$\Delta L \pm 1$	$\Delta L = 0, \pm 1$	$\Delta L = 0, \pm 1, \pm 2$ $L' = 0 \neq L = 0$
Total spin angular momentum quantum number (S)	$\Delta S = 0$	$\Delta S = 0$	$\Delta S = 0$

3.3 LIGHT AMPLIFICATION

Consider a near monochromatic beam of light of frequency ω propagating in the x-direction. In order to study the rate of change of intensity as it propagates through the medium and to understand light amplification we consider two mirrors M_1 and M_2 of area A perpendicular to the line of propagation at x and $x + dx$ as shown in Figure 3.2.

From the geometrical figure the volume of the medium in between the two mirrors is $A dx$. If the number of absorptions per unit time per unit area is W_{12}, the number of absorption per unit time in the volume element $A dx$ is $W_{12} A dx$. As each photon has an energy $\hbar\omega$, the energy absorbed per unit time is $W_{12} A dx \, \hbar\omega$. The energy of stimulated emission per unit time is $W_{21} A dx . \hbar\omega$. The net energy absorbed per unit time is then equal to

$$(W_{12} - W_{21}) \, \hbar\omega \, A \, dx \tag{3.23}$$

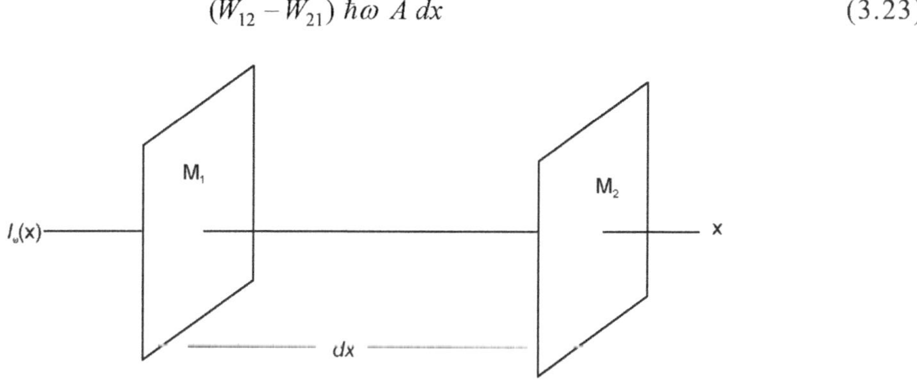

Figure 3.2 Light amplification

In the discussion we neglect the spontaneous emission as they are at random and cancels out and may be lost and has no effect.

Now let us consider it in another perceptive. If $I_\omega (x)$ is the intensity of the beam entering the mirror M_1, then the energy of the photons entering the volume $A dx$ per unit time is $I_\omega(x) A$. If $I_\omega (x + dx)$ is the intensity leaving the mirror M_2, the total energy leaving the volume per unit time is

$$I_\omega(x + dx)A = I_\omega(x)A + \left(\frac{\partial I_\omega}{\partial x}\right) dxA \tag{3.24}$$

The net amount of energy leaving the volume is then $\left(\dfrac{\partial I_\omega}{\partial x}\right) dxA$

This should be equal to the negative of the amount of energy expressed in equation (3.23).

Hence, $\left(\dfrac{\partial I_\omega}{\partial x}\right) dxA = -(W_{12} - W_{21}) \, \hbar\omega \, A \, dx \tag{3.25}$

We have already seen in equation (3.10) that

$$\frac{A_{21}}{B_{21}} = \frac{2\hbar\omega^3}{\pi c^3} = \frac{h\omega^3}{\pi^2 c^3}$$

or
$$\frac{A_{21}}{B} = \frac{h\omega^3}{\pi^2 c^3}, \quad \text{i.e.,} \quad B = \frac{\pi^2 c^3}{h\omega^3} A_{21} \tag{3.26}$$

A_{21} is related to the life time t_{sp} of the excited state by the relation $A_{21} = \dfrac{1}{t_{sp}}$

Hence
$$B = \frac{\pi^2 c^3}{h t_{sp}\omega^3} \tag{3.27}$$

$$W_{12} = Bu(\omega)\, N_1 \quad \text{and} \quad W_{21} = Bu(\omega)\, N_2 \tag{3.28}$$

From (3.25) and (3.27) and (3.28) we have

$$\left(\frac{\partial I_\omega}{\partial x}\right) dxA = -\left(\frac{\pi^2 c^3}{h t_{sp}\omega^3}\right) u(\omega)\, h\omega\, A\, dx\, (N_1 - N_2) \tag{3.29}$$

$$\frac{\partial I_\omega}{\partial x} = -\left(\frac{\pi c^3}{2 t_{sp}\omega^2}\right) u(\omega)\, (N_1 - N_2) \tag{3.30}$$

We can also write $u(\omega)$ in terms of the intensity I_ω and the velocity of light in the medium as

$$I_\omega = vu(\omega) = \left(\frac{c}{n_0}\right) u(\omega) \tag{3.31}$$

Here c is the velocity of light in vacuum and n_0 is the refractive index of the medium.

Thus
$$\frac{\partial I_\omega}{\partial x} = -\left(\frac{\pi c^3}{2 t_{sp}\omega^2}\right) \frac{n_0}{c} I_\omega (N_1 - N_2) \tag{3.32}$$

or
$$\frac{1}{I_\omega}\frac{\partial I_\omega}{\partial x} = -\alpha \tag{3.33}$$

where $\alpha = \left(\dfrac{\pi c^3}{2 t_{sp}\omega^2}\right) \dfrac{n_0}{c}(N_1 - N_2).$

The solution of equation (3.33) is $I_\omega(x) = I(0)\exp(-\alpha x)$ $\tag{3.34}$

The above solution shows that if the value of α is positive, the intensity of the beam will decrease exponentially. And if the value of α is negative, the intensity of the beam will increase exponentially. The light is then amplified. The condition α negative implies that (N_1-N_2) is negative or $N_2 > N_1$. The necessary condition for the amplification of light is that N_2 should be larger than N_1. This is called the condition of population inversion. That means the population in the upper state should be larger than the population in the lower state and is the necessary condition for light amplification, in other words, for the so-called laser action. We know that at normal condition of temperature and pressure the population in the lower state is larger than in upper state. So we should create a condition that the number of atoms in the excited state becomes larger than the lower state. Scientists have found out many methods to achieve population inversion. So the necessary and essential condition for laser action is the population inversion in the atomic or molecular systems. The methods available for creating population inversion are i) electrical discharge as in the case of He-Ne or Argon ion laser, ii) optical pumping as in the case of dye lasers, iii) Chemical reactions as in the case HF laser and iv) State selection as in the case of Ammonia maser.

SOLVED PROBLEMS

1. Show that the total energy of all radiations from a black body is i) proportional to the fourth power of the temperature of the body, ii) the wavelength corresponding to the maximum energy density is inversely proportional to the temperature.

Solution

Planck's law of radiation is

$$E(v)\,dv = \left(\frac{8\pi h v^3}{c^3}\right)\frac{dv}{\exp(hv/kT)-1}$$

As

$$v=\frac{c}{\lambda},\quad dv=\frac{c}{\lambda^2}d\lambda$$

$$E(\lambda)\,d\lambda = -\left(\frac{8\pi hc}{\lambda^5}\right)\frac{d\lambda}{\exp(hc/\lambda kT)-1}$$

We assume $\quad x=\dfrac{hc}{\lambda kT},\ dx=-\dfrac{hc}{\lambda^2 kT}d\lambda$

or $\quad \lambda=\dfrac{hc}{xkT},\ d\lambda=-\dfrac{hc}{kT}\dfrac{dx}{x^2}$

$$E(x)\, dx = -\frac{8\pi hc}{(hc\,/\,xkT)^5(e^x-1)}\left(-\frac{hc}{kT}\right)\frac{dx}{x^2}$$

$$= \frac{8\pi(kT)^4}{(hc)^3}\frac{x^3}{e^x-1}dx$$

$$E = \int_0^\infty E(\lambda)d\lambda = \int_0^\infty E(x)dx = \frac{8\pi(kT)^4}{(hc)^3}\int_0^\infty \frac{x^3}{e^x-1}dx$$

$$\int_0^\infty \frac{x^3}{e^x-1}dx = \frac{\pi^4}{15}$$

Thus

$$E = \frac{\pi^4}{15}\frac{8\pi(kT)^4}{(hc)^3}\quad\text{or}\quad E\propto T^4$$

At maximum energy density $\dfrac{dE}{d\lambda}=0$

$$\frac{dE}{d\lambda} = -\frac{5}{\lambda^6}\left(\frac{1}{\exp(hc\,/\,\lambda kT)-1}\right)+\frac{1}{\lambda^7}\left[\frac{(hc\,/\,kT)\exp(hc\,/\,\lambda kT)}{\exp(hc\,/\,kT)-1)^2}\right]=0$$

$$\text{or}\quad \lambda_{\text{max}} = \frac{hc}{5kT}\left(\frac{\exp(hc\,/\,\lambda kT)}{\exp(hc\,/\,\lambda kT)-1}\right)$$

Assuming $\dfrac{hv}{kT}\geq 1,\quad$ i.e., $\quad\dfrac{hc}{\lambda kT}\geq 1\quad \lambda_{\text{max}}=\left(\dfrac{hc}{5k}\right)\dfrac{1}{T}$

i.e.,

$$\lambda_{\text{max}}\propto\frac{1}{T}\qquad\text{(Wien's law)}$$

ENDNOTES

1. Einstein, A. (1917). "Zur Quantentheorie der Strahlung." *Z. Phys.* 18, 121.

2. Gordon, J.P., Zeiger, H.J. and Townes, C.H. (1954). "Molecular microwave oscillator and new hyperfine structure in the microwave spectrum of NH_3." *Phys.Rev.* 95, 282.

3. Gordon, J.P., Zeiger, H.J. and Townes, C.H. (1955). The maser: a new type of microwave amplifier, frequency standard and spectrometer. *Phys. Rev.* 99, 1264.

REFERENCES

Nair, K.P.R. (2006, 2008). *Atoms, Molecules and Lasers*. Narosa, Delhi and Alpha Science International Oxford.

Svelto, O. (1998). *Principles of Lasers*. Plenum Press, NewYork.

Thyagarajan, K. and Ghatak, A.K. (1984). *Lasers: Theory and Applications.*, Macmillan India Ltd.

Yariv, A. (1989). *Quantum Electronics*. John Wiley and Sons Inc. New York.

4

SCHRÖDINGER EQUATION APPLIED TO ATOMS

4.1 QUANTUM MECHANICS AND SCHRÖDINGER WAVE EQUATION

We have already seen that the quantum conditions which Bohr introduced in his theory of atomic model led to successful interpretation of many features of atomic spectra. Though quantum theory introduced by Max Planck, Bohr's atomic structure, Sommerfeld's improvements and others had proved to be successful in interpreting the atomic spectra, it was felt that a more consistent theory of atomic structure was necessary to account for many finer experimental details in the observations for which hitherto theories were not sufficient. Bohr's correspondence principle had shown a way to relate the old quantum theory with classical theory. In classical theory light was regarded as a wave whereas electron was regarded as a particle. Later it was considered that in order to explain phenomena like interference and refraction light must be thought to be behaving as waves and to explain phenomena like photoelectric effect, Compton effect, etc., light must be considered as behaving like particles. Though it would look paradoxical how light could behave as wave-like and particle-like it was later considered that the nature of experiment makes us to be believed so. The dual nature of light as wave and as particle led de Broglie to develop his theory of matter waves. In 1924 de Broglie postulated the dual nature of matter. In the hands of Schrödinger the theory was greatly improved which led to the development of quantum mechanics. The theory is free from the drawbacks of the old quantum theory. Almost at the same time of the development of Quantum Mechanics by Schrödinger, Heisenberg put forward an independent theory of quantum dynamics which is later known as Matrix Mechanics. Though both these theories are different in their mathematical formulation, Schrödinger showed later that both theories are in effect the same and arrive at the same results.

The quantum mechanics proposed by Schrödinger became a successful tool in explaining atomic spectra. It is a mathematical formulation and replaced the old quantum theory. However, in order to give a physical picture of atoms, even now Bohr's orbital picture is many times used.

An imaginary idealized diffraction experiment is generally used to understand the basic differences between the ideas of motion in classical mechanics and those in quantum mechanics. Let us consider a beam of electrons incident on a screen in which two slits are cut. If we close one slit, the electrons pass through the other and we could observe an intensity pattern on a screen. Similarly repeating the experiment closing the other slit and allowing the electrons to pass through the first one, we obtain another intensity pattern. By opening both the slits at the same time we allow the electrons to pass through both the slits simultaneously. Thus by opening both the slits for the passage of the electrons we expect intensity pattern classically which is the superposition of the two patterns obtained earlier. We expect such a result by classical mechanics, because that the electrons moving in its path and passing through one of the slits do not have any effect on the path of the electrons passing through the other. However, contrary to all classical expectations we obtain a diffraction pattern which does not correspond only to the superposition of patterns given by individual slits. This certainly contradicts the classical ideas that the electrons move in well-defined paths.

There is no concept of the path of the electron in quantum mechanics. This forms the basic idea behind the uncertainty principle which was proposed by Werner Heisenberg in 1927. According to classical mechanics, the position and momentum of moving electron can be found with great accuracy. When an electron is considered as a wave, it is difficult to exactly locate the moving electron as it extends throughout the region of space. Then the question arises whether it is possible to know the exact position of the electron at a given instant of time especially when the electron exhibits dual nature (wave and particle). The answer to this question is given by Werner Heisenberg who states that "it is impossible to determine simultaneously the position and momentum of the electron with any desired accuracy." Or in other words the Heisenberg principle states that "There exists no predictive law which gives information about the simultaneous position and momentum of a particle".

A mathematical formulation of uncertainty principle can be obtained by considering different patterns of a beam of electrons while passing through a single slit (*See* Figure 4.1).

Let us consider a slit of width Δx and measure the diffraction pattern of the electron beam. By observing the diffraction pattern we make simultaneously measurement of position and momentum. By considering the electron as a particle, the width of the slit Δx gives the uncertainty in position perpendicular to the direction of propagation of the beam. This is because the exact position in the slit through which the electrons pass, the slit remain indefinite and can be anywhere inside the space Δx. Thus Δx remains the uncertainty. The electrons suffer a deflection upwards or downwards, the electrons thus acquire a momentum Δp as shown in Figure 4.1, the momentum p remaining the same.

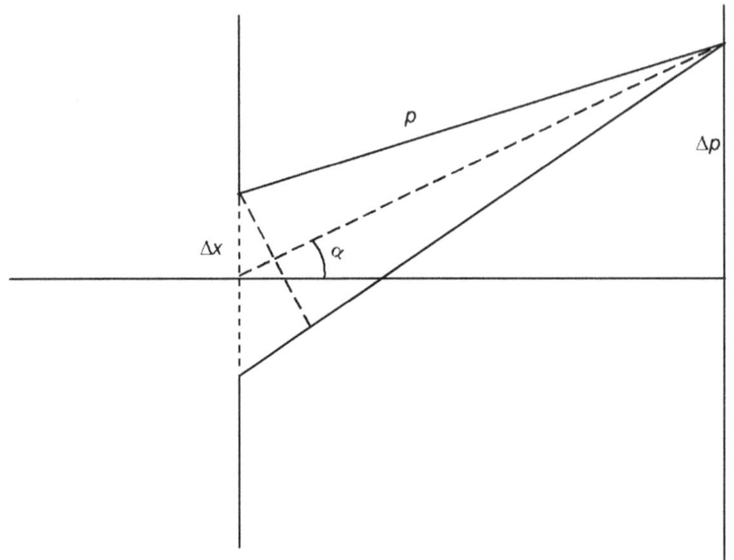

Figure 4.1   Diffraction of electron at a slit

$$\Delta p = p \, \sin \, \alpha \qquad (4.1)$$

where α is the mean angle of deflection.

The experimental observation can be explained by associating a wavelength λ to the electron such that

$$\lambda = \frac{h}{mv} = \frac{h}{p} \qquad (4.2)$$

The diffraction equation in terms of Δx, α and λ is written as

$$\Delta x \sin \alpha \approx \lambda = \frac{h}{p} \qquad (4.3)$$

$$\Delta x \frac{\Delta p}{p} \approx \frac{h}{p}$$

$$\Delta x \cdot \Delta p \approx \hbar \qquad (4.4)$$

Practically this relation can also be written as $\Delta x \cdot \Delta p \approx \dfrac{h}{2\pi}$ or $\geq \dfrac{h}{4\pi}$

That means the product of the uncertainties in determining the position and momentum of the particle can never be smaller than the number of the order $\dfrac{1}{2}\hbar$. Thus $\Delta x \cdot \Delta p \geq \dfrac{\hbar}{2}$.

$$\Delta E \cdot \Delta t \geq \frac{\hbar}{2}$$

$$\Delta J \cdot \Delta \phi \geq \frac{\hbar}{2} \tag{4.5}$$

The relation shows that $\hbar$ is an absolute limit to the accuracy of simultaneous measurement of position coordinates and momentum which we can never get beneath.

Generally Heisenberg principle is written mathematically as $\Delta x \cdot \Delta p \approx \hbar$ where Δx is the uncertainty in the determination of position and δp is the uncertainty in the determination of momentum. Mathematical equation can be stated in words as "the product of the uncertainty in the simultaneous determination of the position and momentum of a particle is equal to or greater than $\hbar$. Thus it sets a limit to the accurate and simultaneous determination of position and momentum of a particle.

From the above relation it is also clear that when the accuracy of the measurement of momentum increased, the accuracy on the measurement of the position coordinates decrease and vice versa.

If the velocity of the electron is $\dfrac{dx}{dt}$, then there exists an uncertainty in time given by $\Delta t = \Delta x / \left(\dfrac{dx}{dt}\right)$.

Similarly an uncertainty in energy is $\Delta E = \left(\dfrac{dx}{dt}\right)\Delta p$ or $\Delta E \cdot \Delta t \approx \hbar$.

It means that if the time for which the system remains in a particular energy state is short, then its energy will be more defined and for longer stay in state the energy uncertainty is less. In the Bohr atom model we have already described that when the electron gets sufficient energy it goes to the excited state and from there it jumps back to the ground state with the emission of a photon. If Δt is the time it remains in the excited state, it gives the error in the measurement of time. The error in the measurement of energy then cannot be less than $\Delta E = \frac{\hbar}{\Delta t}$. Thus there occurs always an error in the measurement of the energy of photon, and hence in the measurement of frequency or wavelength. This will give a natural breadth to the line.

The width of a spectral line and the monochromatic nature of α-decay are explained by the above relation in the uncertainty in energy and time.

The theory of matter waves indicated a more general theory of atomic structure. However, the Bohr–Sommerfeld orbital theory does not lose its orbital picture entirely. It was the Austrian physicist Schrödinger who for the first time showed that the assumptions made by Bohr could be replaced by a set of postulates. There is no apriori assumption of quantum numbers. These postulates remain the basic building block of quantum mechanics. Schrödinger modified the classical wave equation in such a way that its solutions included the features of the new mechanics as envisaged by de Broglie equation and uncertainty principle.

Starting with de Broglie equation we can derive the Schrödinger equation. The well known wave equation is

$$\nabla^2 \psi = \frac{1}{c^2} \cdot \frac{\partial^2 \psi}{\partial t^2} \qquad (4.6)$$

where c is the wave velocity and $\psi = \psi\,(x,t)$ is an unknown function representing the displacement of the wave. $\nabla^2 \psi$ stands for the Laplacian of ψ which in cartesian coordinates x, y, z is given by

$$\nabla^2 \psi = \frac{\partial^2 \psi}{\partial x^2} + \frac{\partial^2 \psi}{\partial y^2} + \frac{\partial^2 \psi}{\partial z^2} \qquad (4.7)$$

A solution of equation (4.7) is expected to be a periodic function of time such that

$$\psi = \psi\, e^{i\omega t} \qquad (4.8)$$

In this trial solution, the amplitude ψ is a function of coordinates only but not of time. $\omega = 2\pi \nu$ where ν is the frequency.

$$\frac{\partial \psi}{\partial t} = i\omega \psi\, e^{i\omega t}$$

$$\frac{\partial^2 \psi}{\partial t^2} = -\omega^2 \psi\, e^{i\omega t} \qquad (4.9)$$

and so on.

On substituting in equation (4.6)

$$\nabla^2 \psi = \frac{1}{c^2} \cdot \frac{\partial^2 \psi}{\partial t^2} = -\frac{1}{c^2} \omega^2 \psi e^{i\omega t} = -\frac{1}{c^2} \omega^2 \psi$$

$$\nabla^2 \psi + \frac{\omega^2}{c^2} \psi = 0 \qquad (4.10)$$

we know that $\nu\lambda = c$ and hence $\lambda = \dfrac{c}{\nu} = \dfrac{2\pi c}{\omega}$, $\dfrac{c^2}{\omega^2} = \dfrac{\lambda^2}{4\pi^2}$

$$\nabla^2 \psi + \frac{4\pi^2}{\lambda^2} \psi = 0 \qquad (4.11)$$

Introducing de Broglie wavelength for moving particle

$$\lambda = \frac{h}{p} = \frac{h}{m\text{v}} \qquad \lambda^2 = \frac{h^2}{m^2\text{v}^2}$$

$$\nabla^2 \psi + \frac{4\pi^2 m^2 \text{v}^2}{h^2}\psi = 0 \tag{4.12}$$

Thus with the introduction of the de Broglie's wavelength ψ is interpreted as the amplitude of the wave associated with a particle moving with a velocity.

If we express the total energy of the particle

$$W = \frac{m\text{v}^2}{2} + V$$

$$\frac{m\text{v}^2}{2} = W - V \text{ or } m\text{v}^2 = 2(W - V)$$

The equation (4.12) becomes

$$\nabla^2 \psi + \frac{4\pi^2 m\, 2(W - V)}{h^2}\psi = 0$$

$$\nabla^2 \psi + \frac{8\pi^2 m(W - V)}{h^2}\psi = 0 \tag{4.13}$$

Thus we arrive at the famous Schrödinger equation or sometimes called the wave equation.

4.2   QUANTUM MECHANICAL DESCRIPTION OF HYDROGEN ATOM: SCHRÖDINGER EQUATION APPLIED TO HYDROGEN ATOM

In the old quantum theory, the discreteness of the energy was introduced quite arbitrarily, but we arrive at the results from a solution of the Schrödinger equation. The Bohr–Sommerfeld electron orbit is now replaced by quantum mechanical probability distribution of electrons. Or the presence of the electron in an orbit is decided by the probability of finding the electron in a volume element dv, i.e., by the function $\psi\psi^* \, dv$.

By introducing suitable values of the potential energy and then the total energy and applying reasonable boundary conditions we can solve the Schrödinger equation and the solutions give the energy states of the atom. The hydrogen atom can be considered as a problem involving a single mass point with reduced mass μ moving about a central part. As the proton mass is very large compared to the electron mass, the reduced mass is the electron mass itself

$$\mu = \frac{Mm}{M + m} \approx \frac{Mm}{M} = m \tag{4.14}$$

where M is the proton mass which is high compared to the electron mass.

For an electron mass m with charge e moving in the central force field of a nuclear mass M and charge $E = Ze$, the potential energy is $V = -\dfrac{eE}{r} = -\dfrac{Ze^2}{r}$.

And if the reduced mass is $\mu = m$ as described by equation 4.14, the wave equation can be written as

$$\nabla^2 \Psi + \frac{8\pi^2 m(W - V)}{h^2} \Psi = 0$$

It is generally written in the form

$$\nabla^2 \Psi + \frac{8\pi^2 m(E - V)}{h^2} \Psi = 0 \tag{4.15}$$

$$\nabla^2 \Psi + \left(\frac{8\pi^2 m \left[E - \left(-\dfrac{Ze^2}{r} \right) \right]}{h^2} \right) \Psi = 0$$

$$\nabla^2 \Psi + \left[8\pi^2 m \frac{Ze^2}{h^2 r} \right] \psi = -\frac{8\pi^2 m}{h^2} E\Psi \tag{4.16}$$

$$\left(-\frac{h^2}{2m} \nabla^2 - \frac{Ze^2}{r} \right) \Psi = E\psi \tag{4.17}$$

$$H\psi = E\psi \tag{4.18}$$

In order to solve the Schrödinger equation we change from cartesian coordinates (x, y, z) to polar coordinates (r, θ, ϕ). The motion of the electron in three dimension is shown in Figure 4.2 and the relation between the cartesian and spherical coordinates is shown in Figure 4.3, and applying boundary conditions we can arrive at suitable solutions. The boundary conditions imposed on ψ are that ψ and its first derivative $d\psi/dt$ are everywhere continuous, single-valued and finite. The wave function, which satisfies these boundary conditions, is called eigen function.

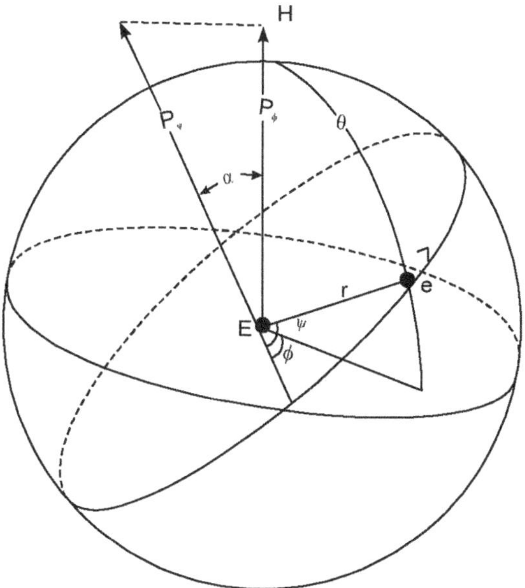

Figure 4.2   The motion of the electron with three degrees of freedom

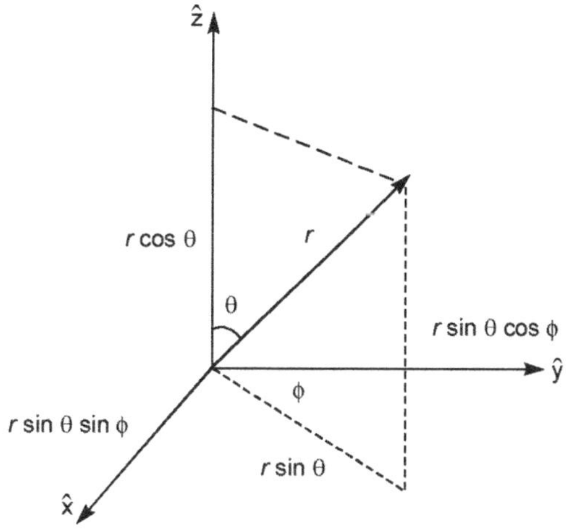

Figure 4.3   Relation between cartesian and spherical coordinates.

In Figure 4.3 polar coordinates $x = r \sin \theta \sin \phi$

$$y = r \sin \theta \cos \phi$$

$$z = r \cos \theta$$

The method of changing from cartesian coordinates to spherical coordinates is shown in Figure 4.4.

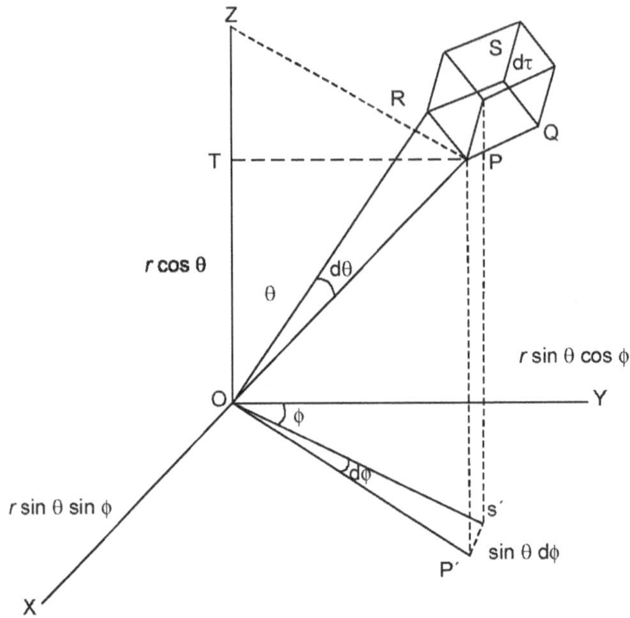

Figure 4.4   Method of changing from cartesian to polar coordinates

$$PT = OP' = r\sin\theta$$

The coordinate $x = OP'\sin\phi = r\sin\theta\sin\phi$

The coordinate $y = OP'\cos\phi = r\sin\theta\cos\phi$

The coordinate $z = OZ = r\cos\theta$

The volume element in spherical coordinate $d\tau = r\sin\theta\ dr\ d\theta\ d\phi$

The limits of variation of coordinates are

$$-\infty < x, y, z;\ 0 < r < \infty;\ 0 < \theta < \pi \ \text{ and } \ 0 < \phi < 2\pi$$

$$r = \sqrt{x^2 + y^2 + z^2}$$

In some books, x and y coordinates are interchanged. In that case,

$$x = r\sin\theta\ \cos\phi$$
$$y = r\sin\theta\ \sin\phi$$
$$z = r\cos\theta$$

In polar coordinates the equation (4.17) becomes

$$-\frac{h^2}{8\pi^2 m}\left[\left(\frac{1}{r^2}\right)\frac{\partial}{\partial r}\left(r^2\frac{\partial\psi}{\partial r}\right) + \frac{1}{r^2\sin\theta}\frac{\partial}{\partial\theta}\left(\sin\theta\frac{\partial}{\partial\theta} + \frac{1}{r^2\sin^2\theta}\frac{\partial}{\partial\phi}\right)\psi\right] - \frac{Ze^2}{r}\psi = E\psi \qquad (4.19)$$

The acceptable solutions of ψ are those where ψ and its first derivative $\delta\psi / \delta t$ are finite, single-valued and continuous.

Interpretation of wave function   The interpretation of the wave function in terms of the location of the particle is based on a suggestion made by Max Born. He made use of the analogy with the wave theory of light in which the square of the amplitude of an electromagnetic wave in a region is interpreted as its intensity and therefore in quantum terms as a measure of the probability of finding a photon present in the region. The Born interpretation of the wave function focuses on the square of the wave function (or square modulus $|\psi|^2 = \psi^*\psi$ if ψ is complex). It states that the value of $|\psi|^2$ at a point is proportional to the probability of finding the particle at that point.

The wave function in the lowest energy state of hydrogen atom is proportional to e^{-r/a_0} where a_0 is a constant and r is the distance from the nucleus. Note that this wave function depends only on r and not on angular positions relative to nucleus. We can now calculate the relative probability of finding the electron inside a region of volume 1.0 pm³ located at the nucleus and also at a distance a_0 from the nucleus.

$$P \propto \psi^2 \delta V$$

$\delta V = 1.0$ pm³ and at the nucleus $r = 0$ and hence $P \propto e^0 \times (1.0\ \mathrm{pm}^3) = 1.0 \times 1.0\ \mathrm{pm}^3$

At a distance $r = a_0$, $P \propto e^2 \times (1.0\ \mathrm{pm}^3) = 0.14 \times 1.0\ \mathrm{pm}^3$

The ratio of the probabilities $1/0.14 = 7.1$

Repeat the calculation for He$^+$ where the wave function is e^{-2r/a_0}

Normalization   In Schrödinger equation solution if Ψ is a solution, then $N\Psi$ is also a solution where N is a constant. It is thus possible to find a constant factor that is always possible and is called the normalization factor. We find the normalization factor by considering the fact that the probability that a particle is in the region dx is $(N\Psi^*)(N\Psi)dx$. The sum over all space of this individual probability is 1. Expressing it mathematically $N^2 \int \Psi^*\Psi dx = 1$ where the integral is over all space, i.e., from $-\infty$ to $+\infty$.

Thus $N = \dfrac{1}{\left(\int (\psi^*\psi\, dx) \right)^{\frac{1}{2}}}$

Thus we always consider the fact that the wave function is normalized by taking $\int \Psi^* \Psi dx = 1$. In three dimension, it is $\int \Psi^*\Psi dxdydz = 1$ or $\int \Psi^* \Psi d\tau = 1$ where $d\tau = dxdydz$. In spherical polar coordinates $x = r\sin\theta\cos\varphi, y = r\sin\theta\sin\varphi, z = r\cos\theta$.

The volume element $d\tau = r^2 \sin\theta\, dr\, d\theta\, d\varphi$. r ranges from 0 to infinity, θ from 0 to π and φ from 0 to 2π.

$$\int \Psi^* \Psi \, d\tau = N^2 \left(\int_0^\infty r^2 e^{-2r/a_0} \, dr \int_0^\pi \sin\theta d\theta \int_0^{2\pi} d\phi \right)$$

In hydrogen atom, $\int \Psi^* \Psi \, d\tau = N^2 \left(\int_0^\infty r^2 e^{-2r/a_0} \, dr \int_0^\pi \sin\theta d\theta \int_0^{2\pi} d\phi \right)$.

$$1 = N^2 \frac{a_0^3}{2} \times 2\pi = \pi N^2 a_0^3. \text{ Thus } N = \frac{1}{\pi a_0^3}$$

$$\Psi = \left(\frac{1}{\pi a_0^3} \right)^{1/2} e^{-\frac{r}{a_0}}$$

The Born interpretation puts several restrictions on the acceptability of wave functions. The principal constraint is that the wave function must not be infinite anywhere. If it were, the integral in equation would be infinite and the normalization constant would be zero. The normalized function would be zero everywhere except where it is infinite, which would be unacceptable. Thus ψ should be finite everywhere. These conditions rules out many possible solutions of the Schrödinger equation. And also the wave function must be single-valued.

Thus the physical interpretation of ψ is that $|\psi|^2$ represents the probability density of finding the electron at the point specified by r, θ, ϕ. $|\psi|^2 d\tau$ is the probability of finding the electron in an infinitesimal volume element $d\tau$. $d\tau$ can be defined such that the distance of the electron from the nucleus is between r and $r + dr$, the zenith angle is between θ and $\theta + d\theta$ and the azimuthal angle is between ϕ and $\phi + d\phi$. In Figure 4.4, the element length PQ is dr. For the small angle $d\theta$, RP = $rd\theta$ and the $\lfloor$RPQ is 90°. The distance PT = OP $\sin\theta = r \sin\theta$ = OP′, the projection of PT on the XY plane. For $d\phi$, the PS or the projection $P'S'$ in the XV plane is $r \sin\theta \, d\phi$. PS is almost perpendicular to RP. Thus the rectangle volume $d\tau = (dr)(rd\theta)(r\sin\theta \, d\phi) = r^2 \sin\theta \, dr \, d\theta \, d\phi$.

The normalization condition is $\int \psi^2 \, dr = \int_{r=0}^\infty \int_{\theta=0}^\pi \int_{\phi=0}^{2\pi} |\Psi(r,\theta,\phi)|^2 r^2 \sin\theta \, dr \, d\theta \, d\phi$

The above equation can be separated into three differential equations by assuming

$$\Psi(r,\theta,\phi) = R(r) \cdot \Theta(\theta) \cdot \Phi(\phi) \tag{4.20}$$

where r, θ and ϕ represent three independent position coordinates of the electron. Substituting equation (4.19), we obtain

$$-\frac{h^2}{8\pi^2 m} \left[\Theta \cdot \Phi \frac{1}{r^2} \frac{d}{dr}\left(r^2 \frac{dR}{dr} \right) + R \cdot \Phi \frac{1}{r^2 \sin\theta} \frac{d}{d\theta}\left(\sin\theta \frac{d\Theta}{d\theta} \right) + R\cdot\Theta \frac{1}{r^2 \sin^2\theta} \frac{d^2\Phi}{d\phi^2} \right]$$

$$-\frac{Ze^2}{r} R\Theta\Phi = E \, R\Theta\Phi \tag{4.21}$$

Multiplying throughout by $-\dfrac{8\pi^2 m}{h^2}\dfrac{r^2}{R\Theta\Phi}$ we get

$$\left(\frac{1}{R}\right)\frac{d}{dr}\left(r^2\frac{\partial R}{\partial r}\right)+\left(\frac{1}{\Theta}\right)\left(\frac{1}{\sin\theta}\frac{d}{d\theta}\left(\sin\theta\frac{\partial\Theta}{\partial\theta}\right)\right)+\left(\frac{1}{\Phi}\right)\frac{1}{\sin^2\theta}\frac{d^2\Phi}{d\phi^2}+\frac{8\pi^2 m}{h^2}\left(E+\frac{Ze^2}{r}\right)r^2=0 \qquad (4.22)$$

Multiplying again by $\sin^2\theta$

$$\frac{1}{R}\sin^2\theta\frac{d}{dr}\left(r^2\frac{\partial R}{\partial r}\right)+\frac{\sin\theta}{\Theta}\frac{\partial}{\partial\theta}\left(\sin\theta\frac{\partial\Theta}{\partial\theta}\right)+\frac{1}{\Phi}\frac{\partial^2\Phi}{\partial\phi^2}+\frac{8\pi^2 mr^2\sin^2\theta}{h^2}\left(E+\frac{Ze^2}{r}\right)=0 \qquad (4.23)$$

This enables us to solve each separately. We can first separate out the factors involving Φ only and set them equal to a constant. By doing so, we assume that when Φ changes the first three terms do not change so that

$$\frac{1}{\Phi}\frac{d^2\Phi}{d\phi^2}=\text{Constant}=-m^2 \qquad (4.24)$$

$$\frac{d^2\Phi}{d\phi^2}=-m^2\Phi \qquad (4.25)$$

$$\frac{d^2\Phi}{d\phi^2}+m^2\Phi=0 \qquad (4.26)$$

$$\left(\frac{d^2}{d\phi^2}+im\right)\left(\frac{d^2}{d\phi^2}-im\right)\Phi=0 \qquad (4.27)$$

the general solution is $\qquad\qquad \Phi=A\,e^{\pm im\phi} \qquad\qquad\qquad\qquad (4.28)$

A is an arbitrary constant. It is possible to evaluate A by normalizing Φ.

$$\frac{\partial\Phi}{\partial\phi}=A\,im\,e^{\pm im\phi}$$

$$\frac{d^2\Phi}{d\phi^2}=-A\,m^2\,e^{\pm im\phi}=-m^2\Phi$$

Thus we arrive at the original equation (4.26) $\dfrac{1}{\Phi}\dfrac{d^2\Phi}{d\phi^2}=-m^2$

The solution of (4.26) is $\Phi=A\,e^{\pm im\phi}$

It is possible to evaluate A by normalizing Φ

Normalizing we get $\displaystyle\int_0^{2\pi}\Phi^*\Phi\,d\phi=1,\quad$ i.e., $\displaystyle\int_0^{2\pi}Ae^{im\phi}.Ae^{-im\phi}\,d\phi$

$$\int_0^{2\pi} \Phi_m \Phi^*_{m'} \partial\phi = A^2 \int_0^{2\pi} e^{im\phi} . e^{-im'\phi} d\phi = A^2 \int_0^{2\pi} e^{i(m-m')\phi} . \partial\varphi = \int_0^{2\pi} A^2 d\phi \text{ for } m = m'$$

Thus

$$= 0 \text{ for } m \neq m'$$

Thus

$$\int_0^{2\pi} \Phi^* \Phi \, d\phi = A^2 \int_0^{2\pi} d\phi = A^2 \cdot 2\pi = 1 \text{ or } A^2 = \frac{1}{2\pi}, A = \frac{1}{\sqrt{2\pi}} \qquad (4.29)$$

ϕ is a wave function that has to be single-valued. It means that it should have the same value at $\phi = 0$ and $\phi = 2\pi$.

$$\text{i.e., } \Phi = \frac{1}{\sqrt{2\pi}} e^{\pm im\phi} = \frac{1}{\sqrt{2\pi}} (\cos m\phi \pm i \sin m\phi) \qquad (4.30)$$

The above equation is valid only if $m = 0$ or an integer (positive or negative).

If m were not an integer, an increase in ϕ by 2π would not repeat the same value to Φ, i.e., the wave function in that case would not be single-valued. Hence m must be an integer. m is generally associated with the magnetic quantum number.

$$\Phi = A = A \, e^{\pm 2\pi im}$$

$$e^{\pm 2\pi im} = \cos 2\pi m \pm \sin 2\pi m$$

$$\Phi = A = A \, (\cos 2\pi m \pm \sin 2\pi m)$$

This equation is valid only if $m = 0$ or an integer, positive or negative.

Thus $\Phi = \dfrac{1}{\sqrt{2\pi}} e^{im\phi}$, $m = 0, \pm 1, \pm 2, \pm 3, \cdots$

The positive or negative values correspond to distinct solutions while for $m = 0$ there is only one solution.

Substituting equation (4.24) in (4.23) and dividing by $\sin^2 \theta$ we have

$$\left(\frac{1}{R}\right)\frac{d}{dr}\left(r^2\frac{\partial R}{\partial r}\right) + \left(\frac{1}{\Theta}\right)\frac{1}{\sin\theta}\frac{d}{d\theta}\left(\sin\theta\frac{\partial\Theta}{\partial\theta}\right) + \frac{8\pi^2 mr^2}{h^2}\left(E + \frac{Ze^2}{r}\right) - \frac{m^2}{\sin^2\theta} = 0 \qquad (4.31)$$

On rearranging, we get

$$\left(\frac{1}{R}\right)\frac{d}{dr}\left(r^2\frac{\partial R}{\partial r}\right) + \frac{8\pi^2 mr^2}{h^2}\left(E + \frac{Ze^2}{r}\right) + \frac{1}{\Theta}\frac{1}{\sin\theta}\frac{d}{d\theta}\left(\sin\theta\frac{d\Theta}{d\theta}\right) - \frac{m^2}{\sin^2\theta} = 0 \qquad (4.32)$$

This separates r, θ parts. The first and the second terms depend on r and the third and the fourth terms depend on θ only.

Hence the first two terms may be set equal to a constant, say β, and hence the sum of the third and fourth term becomes equal to $-\beta$.

Thus,
$$\left(\frac{1}{R}\right)\frac{d}{dr}\left(r^2\frac{\partial R}{\partial r}\right)+\frac{8\pi^2 mr^2}{h^2}\left(E+\frac{Ze^2}{r}\right)=\beta$$

and
$$\left(\frac{1}{\Theta}\right)\frac{1}{\sin\theta}\frac{d}{d\theta}\left(\sin\theta\frac{d\Theta}{d\theta}\right)-\frac{m^2}{\sin^2\theta}=-\beta \tag{4.33}$$

Thus,
$$\left(\frac{1}{R}\right)\frac{d}{dr}\left(r^2\frac{\partial R}{\partial r}\right)+\left[\frac{8\pi^2 mr^2}{h^2}\left(E+\frac{Ze^2}{r}\right)-\beta\right]=0$$

and
$$\left(\frac{1}{\sin\theta}\right)\frac{d}{d\theta}\left(\sin\theta\frac{d\Theta}{d\theta}\right)-\frac{m^2\Theta}{\sin^2\theta}=-\beta\Theta \tag{4.34}$$

4.2.1 The Part Involving θ

The equation involving Θ only is

$$\left(\frac{1}{\Theta}\right)\frac{1}{\sin\theta}\frac{d}{d\theta}\left(\sin\theta\frac{d\Theta}{d\theta}\right)-\frac{m^2}{\sin^2\theta}=-\beta \tag{4.35}$$

$$\frac{1}{\sin\theta}\frac{d}{d\theta}\left(\sin\theta\frac{d\Theta}{d\theta}\right)-\frac{m^2}{\sin^2\theta}\Theta=-\beta\Theta \tag{4.36}$$

$$\frac{1}{\Theta}\frac{1}{\sin\theta}\frac{\partial}{\partial\theta}\left(\sin\theta\frac{\partial\Theta}{\partial\theta}\right)-\frac{m^2}{\sin^2\theta}=-\beta=-l(l+1) \tag{4.37}$$

We can show that above the equation is equivalent to associate Legendre differential equation. It is tedious and lengthy to solve this equation. It has been solved by applying different methods. We shall here describe some significant properties of the solutions.

$$\frac{1}{\sin\theta}\frac{d}{d\theta}\left(\sin\theta\frac{d\Theta}{d\theta}\right)+\left[l(l+1)-\frac{m^2}{\sin^2\theta}\right]\Theta=0 \tag{4.38}$$

It is possible to solve the equation by assuming $x=\cos\theta$.

$$-\sin\theta\,d\theta=dx$$

$$\frac{d\Theta}{d\theta}=\frac{dx}{d\theta}\frac{d\Theta}{dx}=-\sin\theta\frac{d\Theta}{dx}$$

$$\frac{d}{d\theta}=-\sin\theta\frac{d}{dx}$$

$$\frac{1}{\sin\theta}\frac{d}{d\theta}=-\frac{d}{dx}$$

$$\sin\theta\frac{d\Theta}{d\theta}=-\sin^2\theta\frac{d\Theta}{dx}$$

Therefore $\quad \sin\theta \dfrac{d\Theta}{d\theta} = -\sin^2\theta \dfrac{d\Theta}{dx} = -(1-x^2)\dfrac{d\Theta}{dx}$

Thus by substituting $x = \cos\theta$, the equation (4.38) can be written as

$$\frac{d}{dx}\left\{(1-x^2)\frac{d\Theta}{dx}\right\} + \left[l(l+1) - \frac{m^2}{1-x^2}\right]\Theta = 0$$

$$(1-x^2)\frac{d^2\Theta}{dx^2} - 2x.\frac{d\Theta}{dx} + \left[l(l+1) - \frac{m^2}{1-x^2}\right]\Theta = 0 \tag{4.39}$$

This equation is similar to the general equation, viz., associated Legendre equation

$$(1-x^2)\frac{d^2}{dx^2}P_l^m(x) - 2x.\frac{d}{dx}P_l^m(x) + \left[l(l+1) - \frac{m^2}{1-x^2}\right]P_l^m(x) = 0$$

whose solution is

$$x = B\ P_l^{|m|}(x)$$

Thus the solution of the equation (4.39) is

$$\Theta_{lm}(\theta) = B\ P_l^{|m|}(\cos\theta) \tag{4.40}$$

where B is the normalization factor. It is given by

$$B = \left\{\frac{2l+1}{2}\frac{(l-|m|)!}{(l+|m|)!}\right\}^{\frac{1}{2}} \tag{4.41}$$

B is determined by evaluation of $\displaystyle\int \Theta^*\Theta \sin\theta d\theta = 1$

$$\int_{-1}^{+1} B^2 P_l^m(x)P_l^m(x)dx = 1$$

Hence the solution in the normalized form is

$$\Theta_{lm}(\theta) = \left\{\frac{2l+1}{2}\frac{(l-|m|)!}{(l+|m|)!}\right\}^{\frac{1}{2}} P_l^{|m|}(\cos\theta) \tag{4.42}$$

$$P_l^{|m|}(x) = \left(1-x^2\right)^{\frac{|m|}{2}}, \frac{d^{|m|}}{dx^{|m|}}.P_l(x) \tag{4.43}$$

where, $\quad P_l(x) = \dfrac{1}{2^l\,l!}\dfrac{d^l}{dx^l}(1-x^2)^l$

$$P_l^{|m|}(x) = (1-x^2)^{\frac{|m|}{2}} \cdot \frac{d^{|m|}}{dx^{|m|}}\left[\frac{1}{2^l\,l!}\frac{d^l}{dx^l}(1-x^2)^l\right]$$

where, $x = \cos\theta$ and m is the magnetic quantum number.

For a given value of l, $m = 0, \pm1, \pm2, \pm3, \ldots \pm l$

The most important property of $P_l^{|m|}(x)$ for our purpose is its orthogonality.

$$\text{i.e.,} \int P^m{}_l(x)\int P^m{}_l{}'(x)\, dx = 0 \quad \text{if } l \neq l'$$

$$= \frac{2}{2l+1}\frac{(l+m)!}{(l-m)!} \quad \text{if } l = l' \tag{4.44}$$

$Y_{lm} + \Theta_{lm}(\theta)\, e^{im\phi}$ is known as the spherical harmonics.

We can see that by the solution of the equation involving Θ only, the following results are obtained, viz., $\beta = l\,(l{+}1)$ where l is a new quantum number and can have values $0, 1, 2, 3, \cdots, l \geq m$

This can have a single-valued solution when $-\beta = -l\,(l+1)$ where l takes positive whole number values $\geq m$ only.

The associated Legendre polynomials for a complete set of solutions

$$\Theta_{lm}(\theta) = \left\{ \frac{2l+1}{2}\frac{(l-|m|)!}{(l+|m|)!} \right\}^{\frac{1}{2}} P_l^{|m|}(\cos\theta) \tag{4.45}$$

$$P_l^{|m|}(\cos\theta) = (\sin\theta)^{|m|}\frac{d^{|m|}}{d(\cos\theta)^{|m|}}P_l(\cos\theta)$$

$$P_l(\cos\theta) = \frac{1}{2^l\, l!}\frac{d^l}{d(\cos\theta)^l}\left(1 - \cos^2\theta\right)^l$$

However, it is easier to use the expressions (4.42) and (4.43) to obtain the probability density functions. For example

$$P_1^1(x) = \frac{1}{2}\left(1 - x^2\right)^{\frac{1}{2}}\frac{d^2}{dx^2}\left(x^2 - 1\right) = \frac{1}{2}\left(1 - x^2\right)^{\frac{1}{2}}2 = \left(1 - x^2\right)^{\frac{1}{2}}$$

when $\qquad x = \cos\theta \quad P_1^1\left(\cos\theta\right) = \left(1 - \cos^2\theta\right)^{\frac{1}{2}} = \sin\theta$

Similarly $\qquad P_2^1(x) = \left(1 - x^2\right)^{\frac{1}{2}}\frac{d}{dx}\left[\frac{1}{4.2!}\frac{d^2}{dx^2}\left(1 - x^2\right)^2\right]$

$$= \left(1 - x^2\right)^{\frac{1}{2}}\frac{d}{dx}\left[\frac{1}{8}\left(3x^2 - 1\right)\right] = \left(1 - x^2\right)^{\frac{1}{2}}.3x$$

Hence $\quad P_2^1\left(\cos\theta\right) = 3\sin\theta\cos\theta$

The associated Legendre polynomials and probability density factor for s, p, d electrons are shown in Table 4.1.

Table 4.1   The associated Legendre polynomials and probability density factor for s,p,d electrons

Electron	l	m	$P_l^{\lvert m \rvert}(\cos\theta)$	$(\Theta_{l,m})^2$
s	0	0	1	$\dfrac{1}{2}$
p	1	1	$\sin\theta$	$\dfrac{3}{4}\sin^2\theta$
	1	0	$\cos\theta$	$\dfrac{3}{2}\cos^2\theta$
d	2	2	$3\sin^2\theta$	$\dfrac{15}{16}\sin^4\theta$
	2	1	$3\cos\theta\sin\theta$	$\dfrac{15}{4}\sin^2\theta\cos^2\theta$
	2	0	$\dfrac{1}{2}(3\cos^2\theta-1)$	$\dfrac{10}{16}(9\cos^4\theta-6\cos^2\theta+1)$
f	3	3	$15\sin^3\theta$	$\dfrac{7}{1440}\sin^6\theta$
	3	2	$15\cos\theta\sin^2\theta$	$\dfrac{105}{16}\left(\sin^4\theta\cos^2\theta\right)$
	3	1	$\dfrac{3}{2}(5\cos^2\theta-1)\sin\theta$	$\dfrac{21}{32}\sin^2\theta(25\cos^4\theta-10\cos^2\theta+1)$
	3	0	$\dfrac{1}{2}(5\cos^3\theta-3\cos\theta)$	$\dfrac{7}{8}(25\cos^2\theta-30\cos^4\theta+9\cos^2\theta)$
g	4	4	$105\sin^4\theta$	$\dfrac{315}{256}\sin^8\theta$
	4	3	$105\cos\theta\sin^3\theta$	$\dfrac{315}{32}\sin^6\theta\cos^2\theta$
	4	2	$\dfrac{15}{2}(7\cos^2\theta-1)$	$\dfrac{225}{320}\sin^4\theta(49\cos^4\theta-14\cos^2\theta+1)$
	4	1	$\dfrac{5}{2}(7\cos^2\theta-3\cos\theta)\sin\theta$	$\dfrac{225}{160}\sin^2\theta(49\cos^6\theta-42\cos^4\theta+9\cos^2\theta)$
	4	0	$\dfrac{1}{8}(35\cos^4\theta-3\cos^2\theta+3)$	$\dfrac{9}{328}(1225\cos^8\theta-2100\cos^6\theta-1110\cos^4\theta-180\cos^2\theta+9)$

We can consider now that $\phi_m\, \phi_m^*\, d\phi$ is the probability of finding the electron between the angle φ and $\phi + d\phi$ and $\phi_{ml}\, \phi_{ml}^*\, \sin\theta\, d\theta$ is the probability of finding the electron between θ and $\theta + d\theta$.

For a given value of m the value $\phi_m\phi_m^* = \dfrac{1}{2\pi} = $ constant which is independent of ϕ.

This means that the probability of finding an electron in a small element of $d\phi_1$ is the same as for any of the small element of angle $d\phi_2$, thus there is equal probability of finding the electrons in all direction of ϕ from 0 to 2π which can be represented by a circle.

By the classical picture of atoms, the nucleus is surrounded by various shells and subshells of electrons in orbits and the quantum mechanical picture, on the other hand, deal with the probability density distribution of electrons.

The θ-dependent probability for s electron where $l = 0$ and $m = 0$ is $(\Theta_{m,l})^2 = \tfrac{1}{2}$ and hence is independent of angle θ. Thus there is equal probability between 0 and π in regard to θ and can be represented by a half circle. Taking angles θ and ϕ together the half circle rotated through the range of ϕ which is 2π is a sphere. Therefore for an s electron the charge density is spherically symmetric. Comparing the classical orbital model this result means that there is no motion in a direction changing ϕ or θ. The probability of charge clouds in the case of $1s$, $2s$ and $3s$ electrons are given in Figure 4.5.

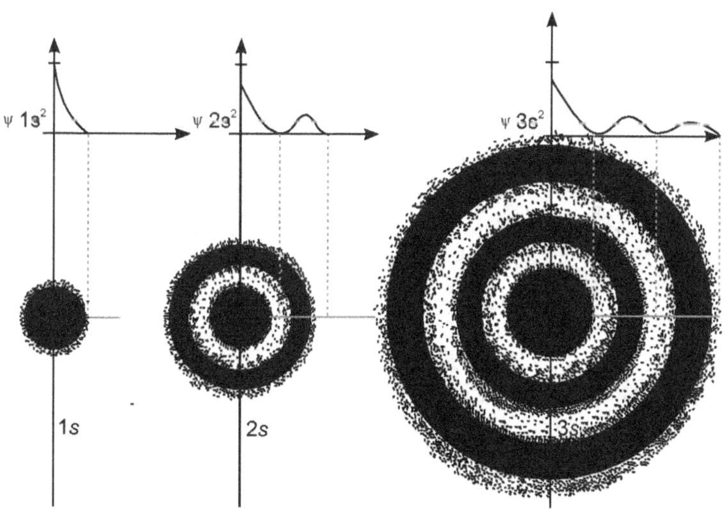

Figure 4.5   Charge clouds associated with $1s$, $2s$ and $3s$ orbitals

For a p electron $l = 1$ and m has values 1, 0 and -1. The probability distribution is shown in Figure 4.6. It can be seen that they are not spherical but are lobed.

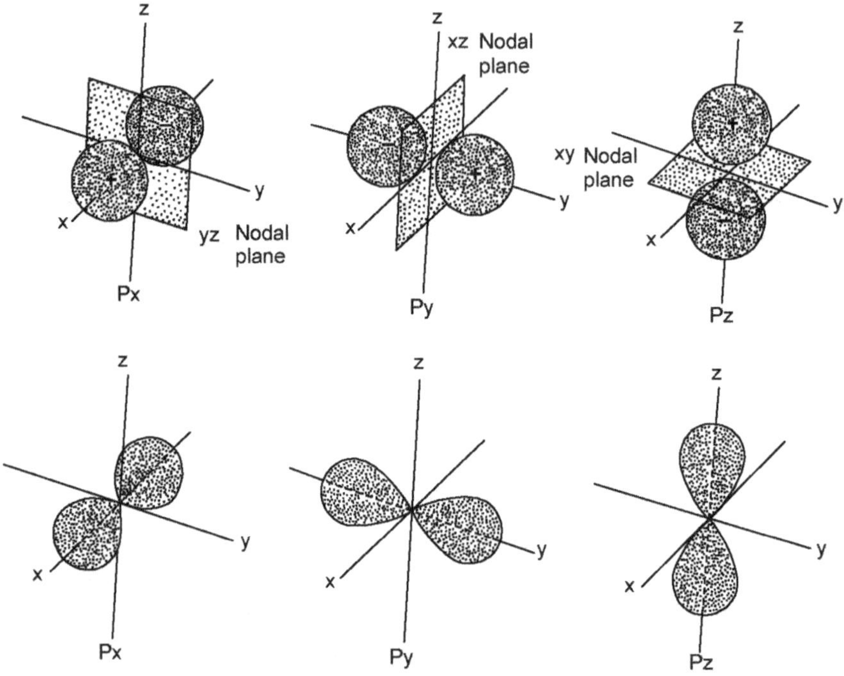

Figure 4.6 The probability distribution for $2p\,(l=1)$ orbitals

For $l=2$, $m=2,1,0,-1$ and -2, the probability distribution is shown in Figure 4.7.

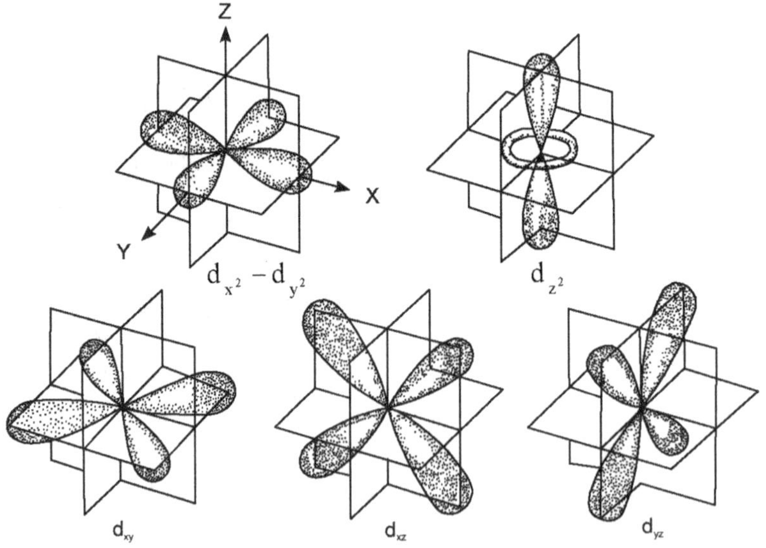

Figure 4.7 The probability distribution for $3d\,(l=2)$ orbitals

In all cases, with the exception of s electrons, the charge density is maximum in the poles, i.e., in the direction $\theta = 0$ and $\theta = \pi$. $e^{im\phi} = 0$ means that there is no motion in the ϕ coordinates. Thus the motion of the electron in the plane of the orbit is in a meridian plane through the ϕ axis, all meridian axis being equally probable. This is in agreement with classical theory. The only difference is that the orbital angular momentum should be taken as $\sqrt{(l(l+1)}\,\dfrac{h}{2\pi}$ instead of $k\dfrac{h}{2\pi}\left(\text{or } l\dfrac{h}{2\pi}, l \text{ being } k-1\right)$.

The projection of l on the ϕ axis is $m\dfrac{h}{2\pi}$.

4.2.2 The Radial Equation Part

The solution of the radial part gives the energy of the system.

We have seen in equation (4.33)

$$\left(\frac{1}{R}\right)\frac{d}{dr}\left(r^2\frac{\partial R}{\partial r}\right)+\frac{8\pi^2 m r^2}{h^2}\left(E+\frac{Ze^2}{r}\right)=\beta=l(l+1)$$

By multiplying by R and dividing by r^2 we get

$$\frac{1}{r^2}\frac{d}{dr}\left(r^2\frac{dR}{dr}\right)-\frac{l(l+1)}{r^2}R+\frac{8\pi^2 m}{h^2}\left(E+\frac{Ze^2}{r}\right)R=0 \tag{4.46}$$

The radial equation

$$\frac{1}{r^2}\frac{d}{dr}\left(r^2\frac{dR}{dr}\right)-\frac{l(l+1)}{r^2}R+\frac{8\pi^2 m}{h^2}\left(E+\frac{Ze^2}{r}\right)R=0$$

can be written as

$$\frac{1}{r^2}\frac{d}{dr}\left(r^2\frac{dR}{dr}\right)+\frac{8\pi^2 m}{h^2}\left(E+\frac{Ze^2}{r}\right)R-\frac{l(l+1)}{r^2}R=0 \tag{4.47}$$

$$\frac{d^2 R}{dr^2}+\frac{2}{r}\frac{dR}{dr}+\frac{8\pi^2 m}{h^2}\left(E+\frac{Ze^2}{r}\right)R-\frac{l(l+1)}{r^2}R=0 \tag{4.48}$$

where $R = R(r)$

If we still simplify the equation (4.48) by substituting $R=\dfrac{1}{r}R'$ we get

$$\frac{dR}{dr}=-\frac{1}{r^2}\frac{dR'}{dr} \quad \text{and} \quad \frac{d^2 R}{dr^2}=-\frac{1}{r^2}\frac{dR'^2}{dr^2}+\frac{2}{r^3}\frac{dR'}{dr}.$$

Then the equation (4.48) becomes

$$\frac{d^2 R'}{dr^2} + \frac{8\pi^2 m}{h^2}\left(E + \frac{Ze^2}{r}\right)R' - \frac{l(l+1)}{r^2}R' = 0$$

The solution of the equation is tedious, but we can make an attempt with some assumptions.

Let us make the following substitutions

$$n^2 = -\frac{2\pi^2 mZ^2 e^4}{h^2 E} \quad \text{i.e., } E = -\frac{2\pi^2 mZ^2 e^4}{n^2 h^2} \tag{4.49}$$

and
$$\rho = \frac{2Zr}{na_0} \quad \text{i.e., } r = \frac{na_0\rho}{2Z} = \frac{nh^2\rho}{8\pi^2 me^2 Z} \tag{4.50}$$

where a_0 is the first Bohr radius we have already seen.

$$a_0 = \frac{h^2}{4\pi^2 me^2}$$

ρ is dimensionless. It may look that the choice of substitution is peculiar but we know from

Bohr's theory omitting the term $\left(\dfrac{1}{4\pi\varepsilon_0}\right)^2$, $E = \dfrac{2\pi^2 mZ^2 e^4}{n^2 h^2}$. (*See* equation 2.27).

$$\frac{dR'}{dr} = \frac{dR'}{d\rho}\frac{d\rho}{dr} = \frac{dR'}{d\rho}\left(\frac{2Z}{na_0}\right)$$

Similarly $\dfrac{d^2 R'}{dr^2} = \left(\dfrac{2Z}{na_0}\right)^2 \dfrac{d^2 R'}{d\rho^2}$

Substituting in equation (4.48) we may obtain

$$\frac{d^2 R'}{d\rho^2} + \frac{2}{\rho}\frac{dR'}{d\rho} + \left[-\frac{1}{4} + \frac{n}{\rho} - \frac{l(l+1)}{\rho^2}\right]R = 0 \tag{4.51}$$

where R is now $R(\rho)$ and the limits are from 0 to infinity. We have now purely an algebraic equation. Even to solve this equation is tedious and hence we consider some extreme cases.

1. for very large ρ

    In this case $\dfrac{n}{\rho}$ and $\dfrac{l(l+1)}{\rho^2}$ are very small and the equation (4.48) can be approximated as

$$\frac{d^2 R}{d\rho^2} - \frac{R}{4} = 0 \tag{4.52}$$

we have here reverted back the general nomenclature R instead of R'.

The above equation has the general solution $R = A\ \exp(-\rho/2) + B\exp(\rho/2)$

where

A and B are arbitrary constants. Taking the boundary condition that R must be finite for $\rho \to \alpha$, we must set $B = 0$. Hence the acceptable solution is

$$R = F(\rho)\exp(-\rho/2) \tag{4.53}$$

$F(\rho)$ is a polynomial of finite order in ρ

By substituting of equation (4.53) into equation (4.51) we can get an equation for $F(\rho)$

$$F'' + \left(\frac{2}{\rho} - 1\right)F' + \left[\frac{n-1}{\rho} - \frac{l(l+1)}{\rho^2}\right]F = 0 \tag{4.54}$$

The differentiation in the above equation is with respect to ρ

We can find a solution of F in the form of a polynomial

$$F(\rho) = \rho^s\left(a_0 + a_1\rho + a_2\rho^2 + \dots\right) = \rho^s L(\rho) \quad a_0 \neq 0 \text{ and } s \geq 0 \tag{4.55}$$

The above polynomial is finite for $\rho = 0$

Substituting equation (4.55) into equation (4.54) we obtain an equation for L.

$$\rho^2 L'' + \rho\left[2(s+1) - \rho\right]L' + \left[\rho(n-s-1) + s(s+1) - l(l+1)\right]L = 0 \tag{4.56}$$

If ρ is made zero it follows from 4.55 that $s(s+1) - l(l+1) = 0$

This gives two roots for s. $s = l$ or $s = -(l+1)$

The requirement in the boundary condition implied on $R(\tilde{n})$ forces us to choose $s = l$.

The equation in L is then

$$\rho^2 L'' + \rho\left[2(l+1) - \rho\right]L' + \left[\rho(n-l-1)\right]L = 0$$
$$\text{or} \qquad \rho L'' + \left[2(l+1) - \rho\right]L' + \left[(n-l-1)\right]L = 0 \tag{4.57}$$

The above equation can be solved by substitution of a power series and the recursion relation between the coefficients of successive terms of the series is

$$a_{k+1} = \frac{k+l+1-n}{(k+1)(k+2l+2)}a_k \tag{4.58}$$

It can be shown that in order to obey the boundary condition L must terminate. If the highest power of ρ in L is ρ^n, we must choose n as a positive integer such that $n = n' + l + 1$. Here n' is called radial quantum number and n as total quantum number. n can take values 1,2,3… and n' and l can have values 0,1,2,… The energy eigen value is

$$E_n = -\left|E_n\right| = \frac{mZ^2 e^4}{2n^2 \hbar^2} \tag{4.59}$$

2. very small value of ρ

Here
$$\frac{l(l+1)}{\rho^2} >> \frac{n}{\rho} \text{ and } \frac{1}{4}$$

The equation (4.51) reduces to

$$\frac{d^2 R}{d\rho^2} + \frac{2}{\rho}\frac{dR}{d\rho} - \frac{l(l+1)}{\rho^2}R = 0 \tag{4.60}$$

We try to solve the above equation by assuming $R = \rho^k$ \hfill (4.61)

Substituting in (4.60) we obtain $[k(k-1) + 2k - l(l+1)]\,\rho^{k-2} = 0$

Equation (4.60) will be satisfied by (4.61) if the coefficient

$$k(k-1) + 2k - l(l+1) = 0$$

i.e.,
$$k^2 + k = l^2 + l$$

Thus $k = l$ or $-(l+1)$

The solution of (4.60) is then $R = \rho^l$ or $\rho^{-(l+1)}$.

The first solution is possible as the second solution approaches to infinity as $\rho \to \alpha$.

Thus the solution allowed is $R = \rho^l$ \hfill (4.62)

Intermediate solution   We have seen the solutions of the equation for large values of ρ and for small values of ρ. We can expect that the general solution of the equation may oscillate between these two extreme values. Therefore we can write the general solution of the equation for all values of ρ as

$$R = \rho^l L(\rho)\exp\left(-\frac{\rho}{2}\right) \tag{4.63}$$

The $L(\rho)$ is a power series in ρ and hence is a polynomial. This polynomial is called Laguerre polynomial.

$$\frac{dR}{d\rho} = l\rho^{l-1} L \exp\left(-\frac{\rho}{2}\right) + \rho^l \frac{dL}{d\rho}\exp\left(-\frac{\rho}{2}\right) - \frac{1}{2}\rho^l L \exp\left(-\frac{\rho}{2}\right) \tag{4.64}$$

$$\frac{d^2 R}{d\rho^2} = \left\{\rho^l \frac{d^2 L}{d\rho^2} + \left(2l\rho^{l-1} - \rho^l\right)\frac{dL}{d\rho} + [l(l-1)\rho^{l-2} - l\rho^{l-1} + \frac{1}{4}\rho^l]\right\}\exp\left(-\frac{\rho}{2}\right) \tag{4.65}$$

Substituting for R, and $\dfrac{dR}{dr}$ and $\dfrac{d^2R}{dr^2}$ in equation (4.65)

we get after taking out the factor $\exp(-\rho/2)$ as it is common

$$\left\{\rho^l\frac{d^2L}{d\rho^2}+\left(2l\rho^{l-1}-2\rho^{l-1}-\rho^l\right)\frac{dL}{d\rho}\right\}+$$

$$\left[[l(l-1)+2l]\rho^{l-2}-(l+1)\rho^{l-1}+\frac{n}{\rho}\rho^l-\frac{l(l+1)\rho^l}{\rho^2}\right]L=0 \qquad (4.66)$$

Dividing (4.60) by ρ^{l-1} we get

$$\rho\frac{d^2L}{d\rho^2}+[2l(l+1)-\rho]\frac{dL}{d\rho}+(n-l-1)L=0 \qquad (4.67)$$

This is associated Laguerre equation and the solutions are in terms of Laguerre polynomials.

The solution of equation is

$$R(\rho)=\rho^l L_{n+l}^{2l+1}(\rho)\exp\left(-\frac{\rho}{2}\right) \qquad (4.68)$$

Here we have again used the general nomenclature R instead of R'.

The normalization is done by taking the integral

$$N^2\int_0^\infty\left[\rho^l L_{n+l}^{2l+1}(\rho)\exp\left(-\frac{\rho}{2}\right)\right]^2\rho^2 d\rho=1 \qquad (4.69)$$

$\rho^2 d\rho$ is the volume element. The solution is tedious and the result of the normalization factor

$$N=-\sqrt{\left(\frac{\rho}{r}\right)^3\frac{(n-l-1)!}{2n\{(n+l)!\}^3}} \qquad (4.70)$$

$$\rho=\frac{2Zr}{na_0}$$

Thus the radial eigen function is obtained as

$$R_{n,l}(\rho)=\sqrt{\left(\frac{2Z}{na_0}\right)^3\frac{(n-l-1)!}{2n\{(n+l)!\}^3}}(\rho)^l.e^{-\frac{Zr}{na_1}}\cdot L_{n+l}^{2l+1}(\rho)$$

$$R_{n,l}(\rho)=\sqrt{\left(\frac{2Z}{na_0}\right)^3\frac{(n-l-1)!}{2n\{(n+l)!\}^3}}\left(\frac{2Zr}{na_0}\right)^l.e^{-\frac{Zr}{na_1}}\cdot L_{n+l}^{2l+1}\left(\frac{2Zr}{na_0}\right) \qquad (4.71)$$

The radial eigen function for $n = 1$, $l = 0$ can be evaluated as follows

$$R_{1,0}(\rho) = N\rho^0 L_{1+0}^{0+1}(\rho)\exp\left(-\frac{\rho}{2}\right) = NL_1^1(\rho)\exp\left(-\frac{\rho}{2}\right) \tag{4.72}$$

$$L_1^1(\rho) = \frac{d}{d\rho}\left[e^\rho \frac{d}{d\rho}(\rho\exp(-\rho))\right] = \frac{d}{d\rho}\left[e^\rho\left\{e^{-\rho}(1-\rho)\right\}\right] = \frac{d}{d\rho}(1-\rho) = -1 \tag{4.73}$$

$$N = -\sqrt{\left(\frac{2Z}{a_0}\right)^3 \frac{(1-0-1)!}{2\times 1\times\{(1+0)!\}^3}} = -\sqrt{\frac{8Z^3}{a_0^3}}\times\frac{1}{2} = -2\left(\frac{Z}{a_0}\right)^{3/2} \tag{4.74}$$

Hence $R_{n,l}(\rho) = (-2)(-1)\left(\frac{Z}{a_0}\right)^{3/2}\cdot\exp\left(-\frac{\rho}{2}\right) = 2\left(\frac{Z}{a_0}\right)^{3/2}\cdot\exp\left(-\frac{\rho}{2}\right)$

$$R_{1,0}(r) = 2\left(\frac{Z}{a_0}\right)^{3/2}\cdot\exp\left(-\frac{Zr}{a_0}\right) = 2\left(\frac{Z}{a_0}\right)^{3/2}\cdot\exp\left(-\frac{Zr}{a_0}\right) \tag{4.75}$$

The general solution of (4.47) gives

$$E = -\frac{2\pi^2 m e^4 Z^2}{n^2 h^2} = -\frac{m^2 e^4 Z^2}{2n^2 \hbar^2} \tag{4.76}$$

where n takes integral values $n = 1,2,3,4\ldots$

m in (4.76) is the mass of the electron.

The n is the total quantum number in the Bohr–Sommerfeld theory. n is called the principal quantum number and for a given value of n, the azimuthal quantum number l takes values $l = 0, 1, 2, 3,\ldots(n-1)$. L is related to the azimuthal quantum number k of the Sommerfeld theory and in this case $l = k - 1$. Thus the quantum number proposed in Bohr–Sommerfeld theory appears as the outcome of the solutions of the Schrödinger equation. Thus m, l and n are quantum numbers, which can take values

$$n = 1, 2, 3, \ldots \alpha$$
$$l = 0, 1, 2, 3\ldots(n-1)$$

and the magnetic quantum number

$$m = 0, \pm 1, \pm 2, \ldots \pm l$$

The wave function $\psi_{n,l,m}(r,\theta,\phi) = R_{n,l}(r)\Theta_{m,l}(\theta)\Phi_m(\phi)$ is referred to as atomic orbital.

The theoretical results obtained by solving the Schrödinger equation were found to predict the observed phenomena correctly. The product of ψ with its complex conjugate over a volume element dv is the probability density of the electron. In spherical polar coordinates $\psi\psi^* dv = \int \psi\psi^* \, r^2 \sin\theta \, dr \, d\theta \, d\phi$ is thus the probability of the electron that will be found in a small volume element $r^2 \sin\theta \, dr \, d\theta \, d\phi$ specified by the coordinates r, θ, ϕ. When the probability is summed over all space

$$\int \psi\psi^* dv = \int \psi\psi^* \sin\theta dr \, d\theta d\phi = 1$$

The above result is accomplished by adjusting the constants occurring in the solution and is known as normalization.

$$\psi\psi^* = RR^* \, \Theta\Theta^* \, \phi\phi^*$$

RR^* gives the probability density as a function of r alone

$\Theta\Theta^*$ is the probability density as a function of θ alone

$\phi\phi^*$ is the probability density as a function of ϕ alone

The radial factor $R_{n,l}$ of the eigen function   The radial equation arising from the solution

$$\psi = R_{n,l} \, \Theta_{m,l} \, \Phi_m$$

can be written as

$$\frac{1}{r^2}\frac{d}{dr}\left(r^2\frac{dR}{dr}\right) - \frac{l(l+1)}{r^2}R + \frac{8\pi^2\mu}{h^2}(E-V)R = 0 \tag{4.77}$$

The potential energy V of the electron moving in the influence of the nucleus is

$$V = -\frac{Ze^2}{r} \tag{4.78}$$

In fact we should take it as $V = -\dfrac{1}{4\pi\varepsilon_0}\dfrac{Ze^2}{r}$. But for most of our discussions in this chapter $\dfrac{1}{4\pi\varepsilon_0}$ is neglected.

The boundary condition is that R is finite and continuous everywhere.

Solution

$$E = -\frac{m^2 Z^2 e^4}{2n^2\hbar^2} \tag{4.79}$$

$$R_{n,l}(r) = \sqrt{\left(\frac{2Z}{na_0}\right)^3 \frac{(n-l-1)!}{2n\{(n+l)!\}^3}} \left(\frac{2Zr}{na_0}\right)^l \cdot e^{-\frac{Zr}{na_1}} \cdot L_{n+l}^{2l+1}\left(\frac{2Zr}{na_0}\right) \tag{4.80}$$

where, $a_0 = \dfrac{h^2}{4\pi^2 me^2} = 0.53 \times 10^{-8}\,\text{cm}$

If we consider $\left(\dfrac{2Zr}{na_0}\right) = \rho$, the expression (4.71) can be written as

$$R_{n,l}(\rho) = \sqrt{\left(\frac{\rho}{r}\right)^3 \frac{(n-l-1)!}{2n\{(n+l)!\}^3}}\,(\rho)^l \cdot e^{-\rho/2} \cdot L_{n+l}^{2l+1}(\rho)$$

$L_{n+l}^{2l+1}\left(\dfrac{2Zr}{na_0}\right)$ is the associated Laguerre polynomials.

$$L_k^p(x) = \frac{d^p}{dx^p}\left[L_k(x)\right] = \frac{d^p}{dx^p}\left[e^x\left\{\frac{d^k}{dx^k} x^k e^{-x}\right\}\right] \tag{4.81}$$

$$L_{n+l}^{2l+1}\left(\frac{2Zr}{na_0}\right) = \sum_{k=0}^{n-l-1}(-1)^k \frac{[(n+l)!]^2}{(n-l-1-k)!(2l+1+k)!}\left(\frac{2Zr}{na_0}\right)^k \tag{4.82}$$

We consider an expression

$$y = x^k e^{-x}$$

the kth derivative of $y = x^k e^{-x}$ will be $\dfrac{d^k y}{dx^k} = \dfrac{d^k(x^k e^{-x})}{dx^k} = e^{-x}L_k(x)$

where $L_k(x)$ is a polynomial in x.

Thus $L_k(x) = e^{-x}\dfrac{d^k y}{dx^k} = e^{-x}\left[\dfrac{d^k(x^k e^{-x})}{dx^k}\right]$ called Laguerre polynomial of degree k. The pth dervative of $L_k(x)$ is denoted as $L_k^p(x)$.

$$L_k^p(x) = \frac{d^p}{dx^p}[L_k(x)] = \frac{d^p}{dx^p}[e^x\left\{\frac{d^k}{dx^k}(x^k e^{-x})\right\}]$$

This polynomial is of degree $k-p$ and the order is p.

We have already derived R_{10}

$$N = \sqrt{\left[\frac{(1-0-1)!}{2\times1\times\{(1+0)!\}^3}\left(\frac{2Z}{a_1}\right)^3\right]} = 2\left(\frac{Z}{a_0}\right)^{3/2} \tag{4.83}$$

$$L_1^1(r) = \frac{d}{dr}\left[e^r \frac{d}{dr}\left(r\exp\left(-\frac{r}{2}\right)\right)\right] = \frac{d}{dr}(1-r) = -1 \tag{4.84}$$

$$R_{10} = (-2)(-1)\left(\frac{Z}{a_0}\right)^{3/2}\exp\left(-\frac{Zr}{a_0}\right) = 2\left(\frac{Z}{a_0}\right)^{3/2}\exp\left(-\frac{Zr}{a_0}\right) \tag{4.85}$$

For $n = 2$, $l = 0$

$$R_{n,l}(r) = \sqrt{\left(\frac{2Z}{na_0}\right)^3 \frac{(n-l-1)!}{2n\{(n+l)!\}^3}} \left(\frac{2Zr}{na_0}\right)^l \cdot e^{-\frac{Zr}{na_1}} \cdot L_{n+l}^{2l+1}\left(\frac{2Zr}{na_0}\right)$$

$$R_{2,0}(r) = \sqrt{\left(\frac{2Z}{2a_0}\right)^3 \frac{(2-0-1)!}{2\times 2\{(2+0)!\}^3}} \left(\frac{2Zr}{2xa_0}\right)^l \cdot e^{-\frac{Zr}{2xa_1}} \cdot L_2^1\left(\frac{2Zr}{na_0}\right) = -\frac{1}{4\sqrt{2}}\left(\frac{Z}{a_0}\right)^{3/2} \cdot L_2^1\left(\frac{2Zr}{na_0}\right)$$

$$L_2^1\left(\frac{2Zr}{na_0}\right) = \frac{d}{dr}\left[e^{2Zr/na_0} \frac{d^2}{dr^2}\left(\frac{2Zr}{na_0}\right)^2 \exp\left(-\frac{2Zr}{a_0}\right)\right]$$

$$L_2^1(\rho) = \frac{d}{d\rho}[e^\rho \frac{d^2}{dr^2}(\rho)^2 \exp(-\rho) = \frac{d}{d\rho}[2-4\rho+\rho^2] = 2\rho-4 = 2\frac{2Zr}{2a_0}-4 = 2\left(\frac{Zr}{a_0}-2\right)$$

$$R_{2,0}(r) = -\frac{1}{4\sqrt{2}}\left(\frac{Z}{a_0}\right)^{3/2}\left[2\left(\frac{Zr}{a_0}-2\right)\right] = -\frac{1}{2\sqrt{2}}\left(\frac{Z}{a_0}\right)^{3/2} \times \left(2-\frac{Zr}{a_0}\right)\exp\left(-\frac{Zr}{2a_0}\right) \tag{4.86}$$

$$R_{2,1}(r) = \sqrt{\left(\frac{2Z}{2a_0}\right)^3 \frac{(2-1-1)!}{2\times 2\{(2+1)!\}^3}} \left(\frac{2Zr}{2a_0}\right)^l \cdot e^{-\frac{Zr}{2a_1}} \cdot L_2^3\left(\frac{2Zr}{na_0}\right) = -\frac{1}{4\sqrt{2}}\left(\frac{Z}{a_0}\right)^{3/2} \cdot L_2^3\left(\frac{2Zr}{na_0}\right)$$

$$R_{n,l}(\rho) = \sqrt{\left(\frac{\rho}{r}\right)^3 \frac{(n-l-1)!}{2n\{(n+l)!\}^3}} (\rho)^l \cdot e^{-\rho/2} \cdot L_{n+l}^{2l+1}(\rho)$$

$$L_2^3(\rho) = \frac{d^3}{d\rho^3}\left[e^\rho \frac{d^3}{d\rho^3}(\rho^3 \exp(-\rho))\right] = \frac{d^3}{d\rho^3}\left[e^\rho \frac{d^2}{d\rho^2}(3\rho^2-\rho^3)\exp(-\rho)\right]$$

$$= \frac{d^3}{d\rho^3}\left[e^\rho \frac{d}{d\rho}(6\rho-6\rho^2+\rho^3)\exp(-\rho)\right] = \frac{d^3}{d\rho^3}[e^\rho(6-18\rho+9\rho^2-\rho^3)e^{-\rho}]$$

$$= \frac{d^3}{d\rho^3}[(6-18\rho+9\rho^2-\rho^3)] = \frac{d^2}{d\rho^2}[(-18+18\rho-3\rho^2)]$$

$$= \frac{d}{d\rho}[(18-6\rho)] = -6$$

$$R_{n,l}(\rho) = -\frac{1}{12\sqrt{6}}\left(\frac{Z}{a_0}\right)^{3/2}\rho(-6)e^{-\rho/2} = -\frac{1}{2\sqrt{6}}\left(\frac{Z}{a_0}\right)^{3/2}\rho e^{-\rho/2}$$

$$R_{2,1}(\rho) = -\frac{1}{2\sqrt{6}}\left(\frac{Z}{a_0}\right)^{3/2} r\exp\left(-\frac{Zr}{2a_0}\right) \tag{4.87}$$

The first few radial wave-functions for the hydrogen atom are listed below:

$$R_{1,0}(r) = \frac{2}{a_0^{3/2}}\exp\left(-\frac{r}{a_0}\right)$$

$$R_{2,0}(r) = \frac{2}{(2a_0)^{3/2}}\left(1-\frac{r}{2a_0}\right)\exp\left(-\frac{r}{2a_0}\right)$$

$$R_{2,1}(r) = \frac{1}{\sqrt{3}(2a_0)^{3/2}}\frac{r}{a_0}\exp\left(-\frac{r}{2a_0}\right)$$

$$R_{3,0}(r) = \frac{2}{(3a_0)^{3/2}}\left(1-\frac{2r}{3a_0}+\frac{2r^2}{27a_0^2}\right)\exp\left(-\frac{r}{3a_0}\right)$$

$$R_{3,1}(r) = \frac{4\sqrt{2}}{9(3a_0)^{3/2}}\frac{r}{a_0}\left(1-\frac{r}{6a_0}\right)\exp\left(-\frac{r}{3a_0}\right)$$

$$R_{3,2}(r) = \frac{2\sqrt{2}}{27\sqrt{5}(3a_0)^{3/2}}\left(\frac{r}{a_0}\right)^2\exp\left(-\frac{r}{3a_0}\right)$$

Normalized values of $R_{n,l}(r)$

$$R_{1,0}(r) = 2\left(\frac{Z}{a_0}\right)^{3/2}\exp\left(-\frac{Zr}{a_0}\right)$$

$$R_{2,0}(r) = \frac{1}{2\sqrt{2}}\left(\frac{Z}{a_0}\right)^{\frac{3}{2}}\left(2-\frac{Zr}{a_0}\right)\exp\left(-\frac{Zr}{2a_0}\right)$$

$$R_{2,1}(r) = \frac{1}{2\sqrt{6}}\left(\frac{Z}{a_0}\right)^{\frac{3}{2}}\left(\frac{Zr}{a_0}\right)\exp\left(-\frac{Zr}{2a_0}\right)$$

$$R_{3,0}(r) = \frac{1}{81\sqrt{3}}\left(\frac{Z}{a_0}\right)^{\frac{3}{2}}\left[27-18\frac{Zr}{a_0}+\frac{2Z^2r^2}{a_0^2}\right]\exp\left(-\frac{Zr}{3a_0}\right)$$

$$= \frac{1}{9\sqrt{3}}\left(\frac{Z}{a_0}\right)^{\frac{3}{2}}\left[6-4\frac{Zr}{a_0}+\frac{4}{9}\frac{Z^2r^2}{a_0^2}\right]\exp\left(-\frac{Zr}{3a_0}\right)$$

$$R_{3,1}(r) = \frac{4}{81\sqrt{6}}\left(\frac{Z}{a_0}\right)^{\frac{3}{2}}\left[6\frac{Zr}{a_0} - \frac{Z^2 r^2}{a_0^2}\right]\exp\left(-\frac{Zr}{3a_0}\right)$$

$$= \frac{4}{27\sqrt{6}}\left(\frac{Z}{a_0}\right)^{\frac{3}{2}}\left[4 - \frac{2}{3}\frac{Zr}{a_0}\right]\frac{2Zr}{a_0}\exp\left(-\frac{Zr}{3a_0}\right)$$

$$R_{3,2}(r) = \frac{1}{81\sqrt{30}}\left(\frac{Z}{a_0}\right)^{\frac{3}{2}}\frac{Z^2 r^2}{a_0^2}\exp\left(-\frac{Zr}{3a_0}\right)$$

$$= \frac{1}{81\sqrt{30}}\left(\frac{Z}{a_0}\right)^{\frac{3}{2}}\frac{4Z^2 r^2}{a_0^2}\exp\left(-\frac{Zr}{3a_0}\right)$$

Normalized values of $\Theta_{l,m}(\theta)$

$$\Theta_{0,0}(\theta) = \frac{1}{\sqrt{2}}$$

$$\Theta_{1,0}(\theta) = \frac{\sqrt{3}}{2}\cos\theta$$

$$\Theta_{1,\pm1}(\theta) = \frac{\sqrt{3}}{2}\sin\theta$$

$$\Theta_{2,0}(\theta) = \frac{\sqrt{5}}{8}(3\cos^2\theta - 1)$$

$$\Theta_{2,\pm1}(\theta) = \frac{\sqrt{15}}{4}\sin\theta\cos\theta$$

$$\Theta_{2,\pm2}(\theta) = \frac{\sqrt{15}}{16}\sin^2\theta \tag{4.88}$$

The wave function for the hydrogen atom is thus given by

$$\Psi_{n,l,m}(r,\theta,\phi) = R_{n,l}(r)\,\Theta_{l,m}(\theta)\,\Phi_m(\phi)$$

and is referred to as atomic orbital. Thus the complete eigen function or the description of the atomic orbital is

$$\Psi_{n,l,m} = \frac{e^{im\varphi}}{\sqrt{2\pi}}\frac{\sqrt{(2l+1)(l-1)}}{2(l+m)!}\sin^m\theta\, P_l^m(\cos\theta)\frac{\sqrt{4(n-l-1)!Z^2}}{[(n+l)!]^3 n^4 a_0^3}\left(\frac{2Zr}{na_0}\right)^l \cdot e^{-\frac{2Zr}{na_1}} \cdot L_{n+l}^{2l+1}\left(\frac{2Zr}{na_0}\right) \tag{4.89}$$

$\Psi_{n,l,m}\,\Psi_{n',l',m'}\,dv = 1$ for $m = m'$, $l = l'$, $n = n'$ and 0 otherwise.

The dependence of $R_{n,l}$ on r can be obtained by knowing the value of Laguerre polynomial. Laguerre polynomial for some orbitals are given in Table 4.2.

Table 4.2   Laguerre polynomial for some orbitals

Electron	n	l	$L_{n+l}^{2l+1}\left(\dfrac{2Zr}{na_0}\right)$
$1s$	1	0	$-1!$
$2p$	2	1	$-3!$
$3d$	3	2	$-5!$
$4f$	4	3	$-7!$
$2s$	2	0	$2x - 4$
$3p$	3	1	$24x - 96$
$4d$	4	2	$720x - 5760$
$3s$	3	0	$-3x^2 + 18x - 18$
$4p$	4	1	$-60x^2 + 600x - 1200$
$4s$	4	0	$4x^2 - 48x^2 + 144x - 96$

Figure 4.8 shows the graphical representation of Radial function $R_{n,l}$ and $(R_{n,l})^2$ for hydrogen with z = 1, n = 3. (Note that the number of nodes n $-$ l $-$1).

The dotted curve represents the radial function $R_{n,l}\,R_{n,l}^{\;*} = (R_{n,l})^2$ of the probability density $\psi\psi^*$.

In the Figure 4.8, the electron–nuclear distance r is given in units of Bohr orbit, i.e., $a_1 = 0.53$ Å.

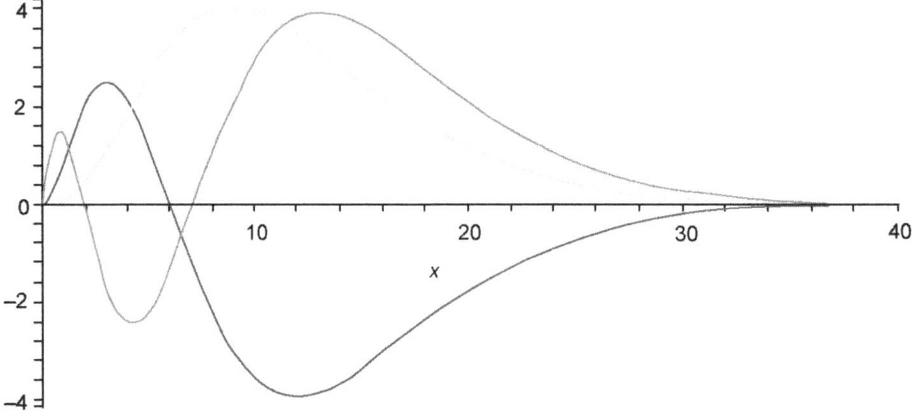

Figure 4.8   Radial function $R_{n,l}$ and $(R_{n,l})^2$ for hydrogen with $Z = 1$, $n = 3$

The probability of finding the electron in the volume element $4\pi r^2\,dr$ enclosed between spheres of radii r and $r + dr$ depend only on the radial part of the state function and is given by

$$D = 4\pi r^2 \left(R_{n,l} \right)^2 \tag{4.90}$$

R_{max} goes on increasing as n increases but for the same value of n it decreases with l. This is in good agreement with the elliptical orbital density as in Sommerfeld. The eccentricity is largest when $l = 0$ and the orbit is called penetrating orbit. The orbits with $n = l + 1$ is called non-penetrating orbit.

In each orbit the angular momentum is $\sqrt{l(l+1)}\,\dfrac{h}{2\pi}$

A still more striking comparison with the older theory can be obtained by comparing the average value of r in each case. It is shown in Table 4.3.

Table 4.3   The average value of r

Quantum Mechanics: probability density distribution	Quantum theory: Bohr–Sommerfeld orbits
$\bar{r} = \dfrac{a_1 n^2}{Z}\left\{1 + \dfrac{1}{2}(1 - \dfrac{l(l+1)}{n^2})\right\}$	$\bar{r} = \dfrac{a_1 n^2}{Z}\left\{1 + \dfrac{1}{2}\left(1 - \dfrac{k^2}{n^2}\right)\right\}$
$\overline{r^2} = \dfrac{a_1^2 n^4}{Z^2}\left\{1 + \dfrac{3}{2}\left(1 - \dfrac{l(l+1) - \frac{1}{3}}{n^2}\right)\right\}$	$\overline{r^2} = \dfrac{a_1^2 n^2}{Z^2}\left\{1 + \dfrac{3}{2}\left(1 - \dfrac{k^2}{n^2}\right)\right\}$
$\overline{\dfrac{1}{r^3}} = \dfrac{Z^3}{a_1^3 n^3 l\left(l + \frac{1}{2}\right)(l+1)}$	$\overline{\dfrac{1}{r^3}} = \dfrac{Z^3}{a_1^3 n^3 k^3}$

Thus it can be seen that k^2 in the old quantum theory should be replaced by $l(l+1)$ to be identical with quantum mechanics. Thus $k = \sqrt{l(l+1)}\,\hbar$.

Thus the angular momentum in old quantum theory is replaced by $l^* = \sqrt{l(l+1)}$ and the component along the z axis is $m\dfrac{h}{2\pi}$.

In Figure 4.9, the graphs of the radial wave functions $(R_{nl})^2$ in the case of first three principal quantum numbers of hydrogen are given and Figure 4.10 again shows different energy levels and transitions in hydrogen atom.

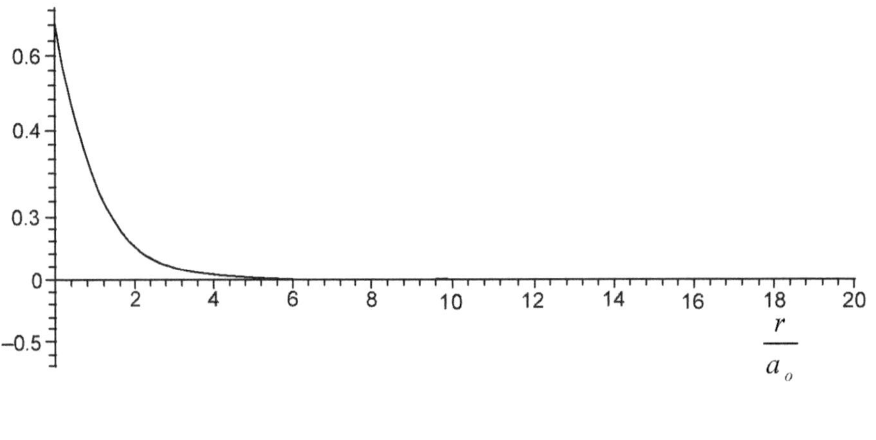

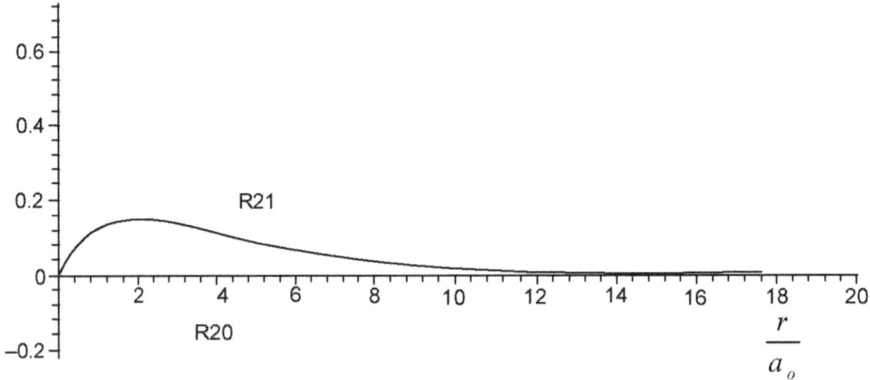

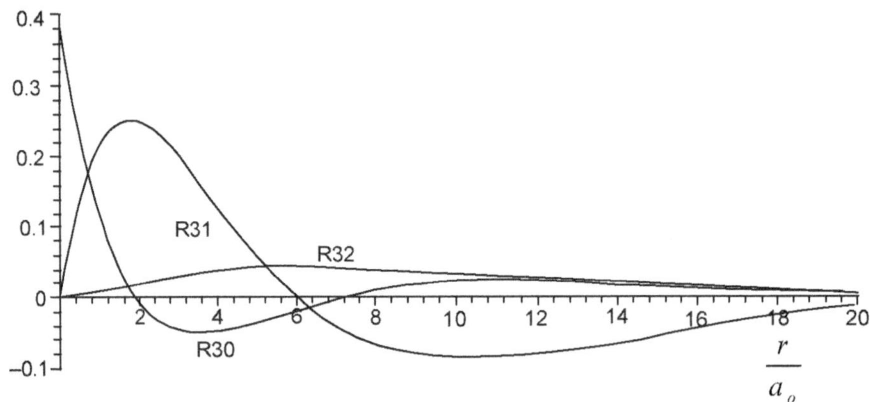

Figure 4.9   Graphs of the radial wave functions $(R_{nl})^2$ for the first three principal quantum numbers of hydrogen

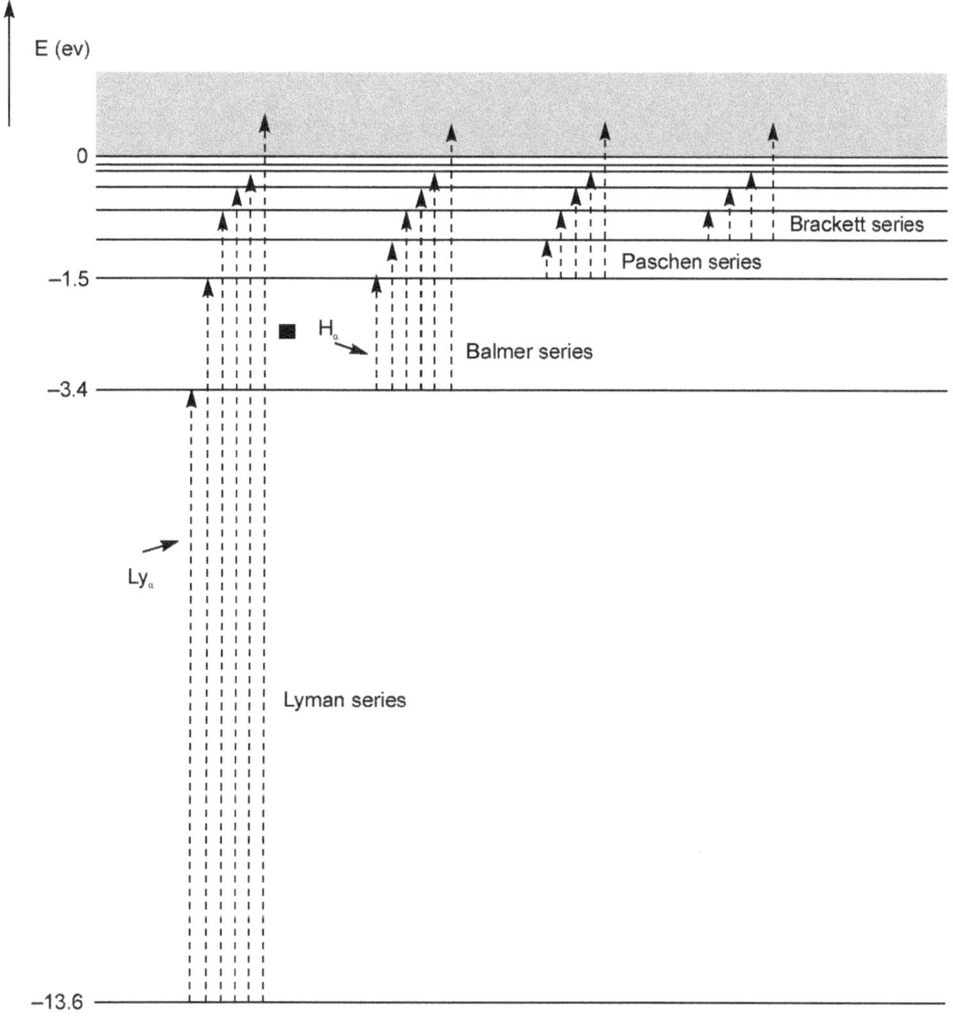

Figure 4.10   Hydrogen spectrum

4.3 THE HYDROGEN WAVE FUNCTIONS

$$\Psi_{n,l,m} = \frac{e^{im\phi}}{\sqrt{2\pi}} \frac{\sqrt{(2l+1)(l-1)}}{2(l+m)!} \sin^m \theta \, P_l^m(\cos\theta) \frac{\sqrt{4(n-l-1)!Z^2}}{[(n+l)!]^3 n^4 a_0^{\,3}} \left(\frac{2Zr}{na_0}\right)^l \cdot e^{-\frac{2Zr}{na_0}} \cdot L_{n+l}^{2l+1}\left(\frac{2Zr}{na_0}\right)$$

$$\Psi_{1,0,0} = \Psi_{1s} - \frac{1}{\sqrt{\pi}} \left(\frac{Z}{a_0}\right)^{3/2} e^{-Zr/a_0}$$

$$\Psi_{2,0,0} = \Psi_{2s} = \frac{1}{4\sqrt{2\pi}} \left(\frac{Z}{a_0}\right)^{3/2} \left(2 - \frac{Zr}{a_0}\right) e^{-Zr/2a_0}$$

$$\psi_{2,1,0} = \psi_{2pz} = \frac{1}{4\sqrt{2\pi}} \left(\frac{Z}{a_0}\right)^{5/2} re^{-Zr/2a_0} \cos\theta$$

$$\psi_{2,1,+1} = \psi_{2px} = \frac{1}{4\sqrt{2\pi}} \left(\frac{Z}{a_0}\right)^{5/2} re^{-Zr/2a_0} \sin\theta\cos\phi$$

$$\psi_{2,1,-1} = \psi_{2py} = \frac{1}{4\sqrt{2\pi}} \left(\frac{Z}{a_0}\right)^{5/2} re^{-Zr/2a_0} \sin\theta\sin\phi$$

$$\psi_{3,0,0} = \psi_{3s} = \frac{1}{81\sqrt{3\pi}} \left(\frac{Z}{a_0}\right)^{3/2} \left(27 - 18\frac{Zr}{a_0} + 2\frac{Z^2 r^2}{a_0^2}\right) e^{-Zr/3a_0}$$

$$\psi_{3,1,0} = \psi_{3pz} = \frac{\sqrt{2}}{81\sqrt{\pi}} \left(\frac{Z}{a_0}\right)^{5/2} \left(6 - \frac{Zr}{a_0}\right) re^{-Zr/3a_0} \cos\theta$$

$$\psi_{3,1,+1} = \psi_{3px} = \frac{\sqrt{2}}{81\sqrt{\pi}} \left(\frac{Z}{a_0}\right)^{5/2} \left(6 - \frac{Zr}{a_0}\right) re^{-Zr/3a_0} \sin\theta\cos\phi$$

$$\psi_{3,1,-1} = \psi_{3py} = \frac{\sqrt{2}}{81\sqrt{\pi}} \left(\frac{Z}{a_0}\right)^{5/2} \left(6 - \frac{Zr}{a_0}\right) re^{-Zr/3a_0} \sin\theta\sin\phi$$

$$\psi_{3,2,0} = \psi_{3dzz} = \frac{1}{81\sqrt{6\pi}} \left(\frac{Z}{a_0}\right)^{7/2} r^2 e^{-Zr/3a_0} (\cos^2\theta - 1)$$

$$\psi_{3,2,+1} = \psi_{3dxz} = \frac{\sqrt{2}}{81\sqrt{\pi}} \left(\frac{Z}{a_0}\right)^{7/2} r^2 e^{-Zr/3a_0} \sin\theta\cos^2\theta\cos\phi$$

$$\psi_{3,2,-1} = \psi_{3dyz} = \frac{\sqrt{2}}{81\sqrt{\pi}} \left(\frac{Z}{a_0}\right)^{7/2} r^2 e^{-Zr/3a_0} \sin\theta\cos\theta\sin\phi$$

$$\psi_{3,2,+2} = \psi_{3dx^2-y^2} = \frac{1}{81\sqrt{2\pi}} \left(\frac{Z}{a_0}\right)^{7/2} r^2 e^{-Zr/3a_0} \sin^2\theta\cos 2\phi$$

$$\psi_{3,2,-2} = \psi_{3dxy} = \frac{1}{81\sqrt{2\pi}} \left(\frac{Z}{a_0}\right)^{7/2} r^2 e^{-Zr/3a_0} \sin^2\theta\sin 2\phi \tag{4.91}$$

4.4 ATOMIC UNITS

It is sometimes convenient to write down the Hamiltonian and energy eigen values in atomic units. Some authors prefer to use atomic units in their calculations. The atomic units are explained in Table 4.4.

Table 4.4 Atomic units

Quantity	S.I. unit	Atomic unit
Mass of the electron	9.1×10^{-31} kg	1 a.u.
Charge of the electron	1.6×10^{-19} m	−1 a.u
Radius of the Bohr orbit of hydrogen atom a_0	0.529×10^{-10} m	1 a.u.
Twice the energy of the electron in the first Bohr orbit	-4.36×10^{-13} J	−1 a.u. (1 Hartree)
Angular momentum unit $(h/2\pi)$	1.05×10^{-34} Js	1 a.u.

The Hamiltonian of hydrogen atom $\hat{H} = -\dfrac{1}{2}\nabla_1^2 - \dfrac{1}{r}$

In atomic unit it is, $\hat{H} = -\dfrac{\hbar^2}{2m}\nabla_1^2 - \dfrac{Ze^2}{r_1}$

$$\tag{4.92}$$

And the 1s wave function is

$$\psi_{1s} = \frac{1}{\sqrt{\pi}} e^{-r} \tag{4.93}$$

4.4.1 Helium Atom and the Hartree–Fock Theory

Helium has the atomic structure $1s^2$. It has thus two electrons and the nuclear charge is +2e. The ground state of helium is obtained by the following way. As the two electrons are in the s state $l_1 = 0$ and $l_2 = 0$ and the total orbital angular momentum $L = l_1 + l_2 = 0$ and hence an S state. The electron's spin can have their directions either parallel or in opposite directions. Hence $S = s_1 + s_1 = 1$ or $s_1 - s_1 = 0$. The first one gives a triplet state and the second one a singlet. The ground state of He is $^1S_0$; All states in helium are either singlets or triplets. And hence only singlet and triplet transitions are possible in helium and intercombination transitions are not possible. This creates two types of helium one with all states singlets called parahelium and the other with all states triplets called orthohelium. The ortho species are three time numerous than the para species. The calculation of energy terms are not as simple as in single electron systems and hence new theories are to be sought. The wave mechanical theories give better results in multielectron systems including helium atom.

We shall consider a coordinate system with nucleus as origin and the coordinates of electron 1 as (x_1, y_1, z_1) and electron 2 as (x_2, y_2, z_2). We consider the nucleus to be at rest. As a general consideration we can take the nuclear charge as Ze where Z is the atomic number. We can treat the helium like ions such as Li^+, Be^{++}, etc., in this manner. The Hamiltonian operator is then

$$H = -\frac{\hbar^2}{2m}\nabla_1^2 - \frac{\hbar^2}{2m}\nabla_2^2 - \frac{Ze^2}{r_1} - \frac{Ze^2}{r_2} + \frac{e^2}{r_{12}} \qquad (4.94)$$

where m is the mass of the electron, r_1 and r_2 are the distances of electrons 1 and 2 from the nucleus.

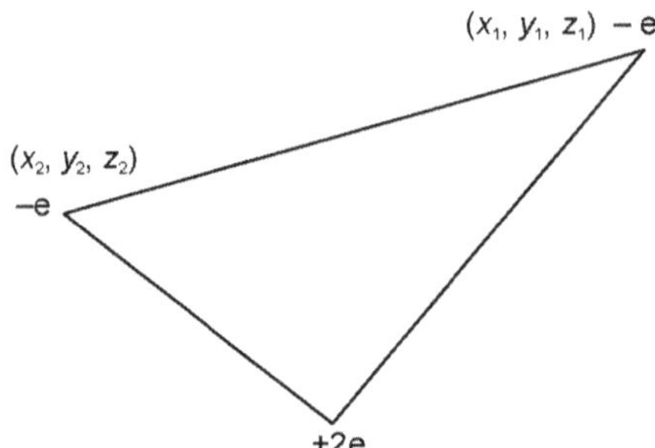

The first two terms in the expression are the operators for the electron kinetic energy, the third and the fourth terms are the potential energies of attraction between the electrons and the nucleus and the last term is the electron repulsion term.

The Schrödinger equation is

$$H\psi = E\psi \qquad (4.95)$$

We can also write with a different nomenclature, i.e., $H(1,2)\ \psi(1,2) = E(1,2)\ \psi(1,2)$

where $\psi(1,2)$ or ψ is the eigen function and $E\ (1,2)$ is the eigen value of the two electrons taken together in the atom.

The function $\psi(1,2) = \psi$ has six coordinates which can be written as $\psi(x_1, y_1, z_1, x_2, y_2, z_2)$ and

$$r_{12} = [(x_1 - x_2)^2 + (y_1 - y_2)^2 - (z_1 - z_2)^2]^{1/2}$$
$$\psi^2 = [\psi(x_1, y_1, z_1, x_2, y_2, z_2)]^2 \text{ and}$$
$$\psi^2 d\tau = [\psi(x_1, y_1, z_1, x_2, y_2, z_2)]^2 dx_1\, dy_1\, dz_1 \cdot dx_2\, dy_2\, dz_2$$

ψ^2 gives the probability per unit volume that the electron 1 is at the point (x_1, y_1, z_1) and the electron 2 is at the point (x_2, y_2, z_2) at any given instant. $\psi^2 d\tau$ gives the probability that at any given instant the electron 1 is in the volume element $d\tau_1\ (dx_1\, dy_1\, dz_1)$ and the electron 2 is in the volume element $d\tau_2$ $(dx_2\, dy_2\, dz_2)$.

In spherical polar coordinate

$$\psi(1,2) = \psi(r_1,\ \theta_1, \phi_1, r_2, \theta_2, \phi_2) \qquad (4.96)$$

The normalization condition is $\int [\psi(x_1, y_1, z_1, x_2, y_2, z_2)]^2 d\tau_1\, d\tau_2 = 1$

In spherical polar coordinates $d\tau_1 = r_1^2 dr_1 \sin\theta_1\, d\theta_1\, d\phi_1$
$$d\tau_2 = r_2^2\, dr_2 \sin\theta_2\, d\theta_2\, d\phi_2$$

4.4.2 Independent Electron approximation

The helium atom Hamiltonian can be rearranged in the form

$$H = \left(-\frac{\hbar^2}{2m}\nabla_1^2 - \frac{Ze^2}{r_1} \right) + \left(-\frac{\hbar^2}{2m}\nabla_2^2 - \frac{Ze^2}{r_2} \right) + \frac{e^2}{r_{12}} \tag{4.97}$$

$$= H_1^0 + H_2^0 + \frac{e^2}{r_{12}}$$

where
$$H_1^0 = -\frac{\hbar^2}{2m}\nabla_1^2 - \frac{Ze^2}{r_1} \quad \text{and}$$

$$H_2^0 = -\frac{\hbar^2}{2m}\nabla_2^2 - \frac{Ze^2}{r_2} \tag{4.98}$$

are considered to be the unperturbed Hamiltonian, say H^0 of the helium atom. The unperturbed system is thus a helium atom in which the two electrons exert no forces on each other. Although there is no physical reality in considering so, it becomes easy to apply perturbation theory to such a system to get approximation to the actual helium atom. The unperturbed Hamiltonian thus is the sum of the Hamiltonians of two independent hydrogen-like particles. Hence H_1^0 and H_2^0 have hydrogen like eigen functions (orbitals) say $\phi(1)$ and $\phi(2)$.

Thus
$$H_1^0\phi(1) = E(1)\phi(1)$$

$$H_2^0\phi(2) = E(2)\phi(2) \tag{4.99}$$

$E(1)$ and $E(2)$ are the eigen values.

We can write $H^0\psi^0 = \left[H(1) + H(2) \right] \left[\phi(1)\phi(2) \right]$

$$= H(1)\,[\phi(1)\,\phi(2)] + H(2)\,[\phi(1)\,\phi(2)]$$
$$= [H(1)\,\phi(1)]\,\phi(2) + [H(2)\,\phi(2)]\,\phi(1)$$
$$= E^0(1)\,\phi(1)\,\phi(2) + E^0(2)\,\phi(1)\,\phi(2)$$
$$= [E^0(1) + E^0(2)]\,[\phi(1)\,\phi(2)] \tag{4.100}$$

Thus the unperturbed eigen values and eigen functions can be approximated as

$$E^0 = E^0(1) + E^0(2)$$
$$\psi^0 = \phi(1)\,\phi(2) \tag{4.101}$$

H(1) operates on $\phi(1)$ and H(2) on $\phi(2)$ only, and hence $\phi(1)\,\phi(2) = \phi(2)\,\phi(1)$.

In independent electron approximation the function $\psi^0 = \phi(1)\,\phi(2)$.

Thus the function $\psi^0 = \phi(1)\,\phi(2)$ which is the true eigen function of H^0 and can be used as an eigen function of the Hamiltonian $H = H(1) + H(2) + \dfrac{e^2}{r_{12}}$ which can be considered as an approximate eigen function of H^0 and thus the eigen functions of the one electron Hamiltonians $H(1)$ and $H(2)$ may serve as a basis wave function for two electron system. The two electron problem can thus be considered as that of two one electron hydrogen problems.

The energies are then

$$E^0(1) = -\frac{Z^2}{n_1^2}\frac{e^2}{2a_0}$$

$$E^0(2) = -\frac{Z^2}{n_2^2}\frac{e^2}{2a_0} \quad \text{and}$$

$$E^0 = -Z^2\left(\frac{1}{n_1^2} + \frac{1}{n_2^2}\right)\frac{e^2}{2a_0}$$

(4.102)

where,
$n_1 = 1,2,3,$ etc. and $n_2 = 1,2,3,$ etc.

where a_0 is the Bohr radius.

For helium atom the two electrons are in the 1s orbital and thus both of them have the same orbital function which can be written as

$$1s(1) = \sqrt{\frac{Z^3}{\pi a_0^3}}\, e^{\left(-\frac{Zr_1}{a_0}\right)} \quad \text{and}$$

$$1s(2) = \sqrt{\frac{Z^3}{\pi a_0^3}}\, e^{\left(-\frac{Zr_2}{a_0}\right)}$$

(4.103)

Thus
$$\psi = 1s(1)1s(2) = \sqrt{\frac{Z^3}{\pi a_0^3}}\, e^{\left(-\frac{Z(r_1+r_2)}{a_0}\right)}$$

(4.104)

For hydrogen like atom $E = -\dfrac{1}{2}\dfrac{Z^2 e^2}{n^2 a_0}$

and for He^+ $Z = 2$ and electron in 1s state $n = 1$.

and
$$E^{He} = E^{He^+}(1) + E^{He^+}(2) = -Z^2\left[\frac{1}{1^2} + \frac{1}{1^2}\right]\frac{e^2}{2a_0} = -2Z^2 \cdot \frac{e^2}{2a_0}$$

(4.105)

The expression $-\dfrac{e^2}{2a_0}$ is the ground state energy of the hydrogen atom which is equal to -13.606 eV.

For helium $Z = 2$ and hence $E = -2 \times 4 \times \dfrac{e^2}{2a_0} = -8 \times 13.606 = -108.8$ eV $= -1.74 \times 10^{-17}$ J.

However the experimental value of the helium atom is -79 eV. Thus the energy calculated is much below the observed value and error is almost 38%. This clearly indicates that we cannot neglect the $\dfrac{1}{r_{12}}$ term in the Hamiltonian.

Hence the next step is to include the term $\dfrac{e^2}{r_{12}}$ through a perturbation treatment. The first order perturbation correction can be written as

$$E^{(1)} = \int \psi^{(0)*} \hat{H}' \psi^{(0)} d\tau$$

$$E^{(1)} = \frac{Z^6 e^2}{\pi^2 a_0^6} \int\limits_0^{2\pi}\int\limits_0^{2\pi}\int\limits_0^{\pi}\int\limits_0^{\pi}\int\limits_0^{\infty}\int\limits_0^{\infty} e^{-2Zr_1/a_0} e^{-2Zr_2/a_0} \frac{1}{r_{12}} r_1^2 \sin\theta_1 r_2^2 \sin\theta_2 \, dr_1 dr_2 d\theta_1 d\theta_2 d\phi_1 d\phi_2 \tag{4.106}$$

$$\frac{1}{r_{12}} = \left[(x_1 - x_2)^2 + (y_1 - y_2)^2 + (z_1 - z_2)^2 \right]$$

An expansion of $\dfrac{1}{r_{12}}$ in terms of spherical harmonics gives

$$\frac{1}{r_{12}} = \frac{1}{r_1} \sum_{l=0}^{\infty} \sum_{m=-l}^{l} \frac{4\pi}{2l+1} \left(\frac{r_2}{r_1} \right)^l [Y_l^m(\theta_1\phi_1)]^* Y_l^m(\theta_2\phi_2) \text{ for } r_1 \geq r_2$$

$$= \frac{1}{r_2} \sum_{l=0}^{\infty} \sum_{m=-l}^{l} \frac{4\pi}{2l+1} \left(\frac{r_1}{r_2} \right)^l [Y_l^m(\theta_1\phi_1)]^* Y_l^m(\theta_2\phi_2) \text{ for } r_2 \geq r_1 \tag{4.107}$$

We can simplify it into

$$E^{(1)} = \frac{16 Z^6 e^2}{a_0^6} \sum_{0}^{\infty} \sum_{m=-l}^{l} \frac{1}{2l+1} \int\limits_0^{\infty}\int\limits_0^{\infty} e^{-2Zr_1/a_0} e^{-2Zr_2/a_0} \frac{r_<^l}{r_>^{l+1}} r_1^2 r_2^2 \, dr_1 dr_2 \times$$

$$\int\limits_0^{2\pi}\int\limits_0^{\pi} [Y_l^m(\theta_1\phi_1)]^* Y_0^0(\theta_1\phi_1) \sin\theta_1 d\theta_1 d\phi_1 \times \int\limits_0^{2\pi}\int\limits_0^{\pi} [Y_0^0(\theta_2\phi_2)]^* Y_l^m(\theta_2\phi_2) \sin\theta_2 d\theta_2 d\phi_2 \tag{4.108}$$

$r_<$ means the smaller of r_1 and r_2, $r_>$ means the larger of r_1 and r_2.

We already know that spherical harmonics are orthogonal to each other, i.e.,

$$\int_0^{2\pi}\int_0^{\pi}[Y_l^m(\theta,\varphi)]^* Y_l^{m'}(\theta,\varphi)\sin\theta\, d\theta\, d\varphi = \partial_{l,l'}\partial_{m,m'} \tag{4.109}$$

where ∂ is Kronecker delta.

Hence $\displaystyle E^{(1)} = \frac{16Z^6 e^2}{a_0^6}\sum_0^{\infty}\sum_{m=-l}^{l}\frac{1}{2l+1}\int_0^{\infty}\int_0^{\infty}e^{2Zr_1/a_0}e^{-2Zr_2/a_0}\frac{r_<^l}{r_>^{l+1}}r_1^2 r_2^2\, dr_1 dr_2 \partial_{l,0}\partial_{m,0}\partial_{l,0}\partial_{m,0}$

Therefore $\displaystyle E^{(1)} = \frac{16Z^6 e^2}{a_0^6}\int_0^{\infty}\int_0^{\infty}e^{-2Zr_1/a_0}e^{-2Zr_2/a_0}\frac{1}{r_>}r_1^2 r_2^2\, dr_1 dr_2$

When we integrate over r_1 in the range $0 \le r_1 \le r_2$ we have $r_> = r_2$ and in the range $r_1 \le r_2 \le 0$ we have $r_> = r_1$. Therefore

$$E^{(1)} = \frac{16Z^6 e^2}{a_0^6}\int_0^{\infty}e^{-2Zr_2/a_0}r_2^2[\int_0^{r_2}e^{-2Zr_1/a_0}\frac{r_1^2}{r_2}\, dr_1 + \int_{r_2}^{\infty}e^{-2Zr_1/a_0}\frac{r_1^2}{r_2}]dr_2$$

$$E^{(1)} = \frac{16Z^6 e^2}{a_0^6}\int_0^{\infty}e^{-2Zr_2/a_0}r_2^2[\int_0^{r_2}e^{-2Zr_1/a_0}r_1^2\, dr_1]dr_2 + \frac{16Z^6 e^2}{a_0^6}\int_0^{\infty}e^{-2Zr_2/a_0}r_2^2[\int_{r_2}^{\infty}e^{-2Zr_1/a_0}r_1\, dr_1]dr_2 \tag{4.110}$$

We know that $\displaystyle \int xe^{bx}dx = \frac{e^{bx}}{b^2}(bx-1)$ and

$$\int x^2 e^{bx}dx = e^{bx}\left(\frac{x^2}{b}-\frac{2x}{b^2}+\frac{2}{b^3}\right)$$

$$\int x^n e^{qx}dx = \frac{n!}{q^{n+1}} \tag{4.111}$$

By carrying out the integration we obtain the result as

$$E^{(1)} = \frac{5Z}{8}\left(\frac{e^2}{a_0}\right) \tag{4.112}$$

Thus the first order correction is $\displaystyle E^{(1)} = \frac{5\times 2}{8}(27.212) = 34\text{ eV}$ \tag{4.113}

Thus $E = E^{(1)} + E^{(2)} = -108.8\text{ eV} + 34\text{ eV} = -74.8\text{ eV}$. The experimental value is -79.0 eV.

SOLVED PROBLEMS

1. Calculate the time taken by the electron to traverse the first orbit of hydrogen atom. The radius of the nth orbit $r_n = \dfrac{n^2}{m}\left(\dfrac{h}{2\pi}\right)\dfrac{4\pi\varepsilon_0}{Ze^2} = n^2 \times 5.29 \times 10^{-11}$ metre.

Solution

Velocity of the electron in the nth orbit $\quad v = \dfrac{1}{137} \times \dfrac{c}{n}$

Velocity in the first orbit $v = \dfrac{1}{137} \times c$

Period of electron in the first orbit $= \dfrac{2\pi r}{v} = \dfrac{2 \times 3.14 \times 5.29^{-11}}{3 \times 10^8} \times 137 = 1.517 \times 10^{-16}\,\text{s}$

2. The energy of an excited hydrogen atom is -3.4 eV. Calculate the angular momentum of the electron from Bohr's theory.

Solution

The energy of the nth orbit $E = -\dfrac{13.6}{n^2}\text{eV}$

Therefore, $-3.4 = -\dfrac{13.6}{n^2}$

hence $\qquad n = 2$

Angular momentum $L = n\dfrac{h}{2\pi} = \dfrac{2 \times 6.625 \times 10^{-34}}{2 \times 3.14} = 2.11 \times 10^{-34}\,\text{J.s}$

3. Hydrogen atom in the ground state is excited by means of monochromatic radiation of wavelength 975 Å. How many different transitions are possible in the resulting spectrum? Calculate the longest wavelength among them.

Solution

$$E = h\nu,\; v = \dfrac{c}{\lambda}$$

Hence $E = h\dfrac{c}{\lambda}$

Energy equivalent to 975 $\mathring{A} = h\dfrac{c}{\lambda} = \dfrac{6.625 \times 10^{-34} \times 3 \times 10^8}{975 \times 10^{-10}}\,\text{J}$

$$= \dfrac{6.625 \times 10^{-34} \times 3 \times 10^8}{975 \times 10^{-10} \times 1.6 \times 10^{-19}}\,\text{eV} = 12.75 \text{ eV}$$

Energy of the electron in the ground state is -13.6 eV

Therefore energy in the excited state $-13.6 + 12.75 = -0.85$ eV

$$E_n = -\frac{13.6}{n^2} = -0.85$$

$$n = 4$$

Possible transitions are from $n = 4$ $4 \to 3,\ 3 \to 2,\ 2 \to 1$

from $n = 3$ $3 \to 2,\ 2 \to 1$

from $n = 2$ $2 \to 1$

Total 6 transitions

Line of the longest wavelength is the transition from $n = 4 \to 3$

$$E_4 - E_3 = -0.85 - (-1.5) = 0.65 \text{ eV}$$

Wavelength $\lambda = \dfrac{hc}{E} = \dfrac{6.625 \times 10^{-34} \times 3 \times 10^8}{0.65 \times 1.6 \times 10^{-19}} = 19110 \,\text{Å}$

4. The series limit of Balmer series is at 3646 Å. Calculate the wavelength of the first member of the series.

Solution

$$\frac{1}{\lambda} = R\left(\frac{1}{2^2} - \frac{1}{n^2}\right)$$

For series limit $n = \alpha$ and hence $\dfrac{1}{\lambda_{\text{limit}}} = \dfrac{R}{4} = \dfrac{1}{3646 \times 10^{-10}}$

$$R = \frac{4 \times 10^{10}}{3646}$$

First member $n = 3$ and hence $\dfrac{1}{\lambda} = R\left(\dfrac{1}{2^2} - \dfrac{1}{3^2}\right) = \dfrac{4 \times 10^{10}}{3646}\left(\dfrac{1}{4} - \dfrac{1}{9}\right)$

$$\lambda = \frac{3646}{4 \times 10^{10}} \times \frac{36}{4} = 6563 \times 10^{-10}\,\text{m} = 6563 \,\text{Å}$$

5. The wavelength of the H gama line in hydrogen is 4341 Å. Find the wavelength of the second line in the Paschen series.

Solution

$$\frac{1}{\lambda} = R\left(\frac{1}{n_1^2} - \frac{1}{n_2^2}\right)$$

H_γ is the third line of Balmer series $n = 5 \to 2$

$$\frac{1}{\lambda_1} = R\left(\frac{1}{2^2} - \frac{1}{5^2}\right) = R\left(\frac{1}{4} - \frac{1}{25}\right) = R\frac{21}{100}$$

Second line in Paschen series is $n = 5 \rightarrow 3$

$$\frac{1}{\lambda_2} = R\left(\frac{1}{3^2} - \frac{1}{5^2}\right) = R\left(\frac{1}{9} - \frac{1}{25}\right) = R\frac{16}{225}$$

$$\frac{\lambda_2}{\lambda_1} = \frac{-225}{16R} \times \frac{21R}{100} = \frac{189}{64}$$

$$\lambda_2 = \frac{189}{64} \times 4341 = 12819 \text{ Å}$$

6. Using Bohr's formula i) determine the longest wavelength of Lyman series ii) the excitation energy of $n = 3$ level of He$^+$ atom iii) the ionization potential of the ground state of Li^{++} atom ($R = 1.097 \times 10^7$ m^{-1}).

Solution

$$\frac{1}{\lambda} = R\left(\frac{1}{n_1^2} - \frac{1}{n_2^2}\right)$$

For Lyman series $n_1 = 1$

$$\frac{1}{\lambda} = R\left(\frac{1}{1^2} - \frac{1}{n_2^2}\right)$$

for the longest wavelength $n_2 = 2$

$$\frac{1}{\lambda} = R\left(\frac{1}{1^2} - \frac{1}{4}\right) = 1.097 \times 10^7 \times \frac{3}{4}$$

$$\lambda = \frac{4}{1.097 \times 10^7 \times 3} = 1.215 \times 10^{-7} \text{ m} = 1215 \text{ Å}$$

Energy of the *n*th state of ionized atoms

$$E_n = -\frac{13.6}{n^2} Z^2 \text{ eV}$$

For He$^+$ atom $Z = 2$

$$E_1 = 4 \times \frac{13.6}{1} = -54.4 \text{ eV for the first orbit}$$

$$E_3 = 4 \times \frac{13.6}{3^2} = -6.04 \text{ eV for the third orbit}$$

Excitation energy from $n = 1$ to $n = 3 = -6.04 + 54.4 = 48.36$ eV

For lithium $Z = 3$ and $E_n = -\dfrac{13.6}{n^2} 3^2 \text{ eV} = -\dfrac{122.4}{n^2} \text{eV}$

Energy in the ground state $(n = 1) = -122.4$ eV

Ionization energy $= 122.4$ eV and ionization potential-122.4 volt

7. Which state of triply ionized beryllium (Be^{+++}) has the same orbital radius as that of the ground state of hydrogen atom. Compare the energies of the two states.

Solution

Expression for the nth orbit is $r_n = \dfrac{n^3}{Z}\dfrac{\hbar^2 \varepsilon_0}{\pi m e^2} = 0.529 \times 10^{-10} \dfrac{n^2}{Z}$

The nth orbit of the hydrogen atom is $r_n = n^2 \times 0.529 \times 10^{-10}$ m

For Beryllium $Z = 4$, and $r_n = \dfrac{n^3}{Z}\dfrac{\hbar^2 \varepsilon_0}{\pi m e^2} = 0.529 \times 10^{-10} \dfrac{n^2}{4}$ m

$0.529 \times 10^{-10}\dfrac{n^2}{4} = 0.529 \times 10^{-10}$ m (of hydrogen)

$$n^2 = 4 \quad \text{and} \quad n = 2$$

Energy of the nth orbit of hydrogen $E_n = -\dfrac{13.6}{n^2}$ eV

Energy of the ground state of hydrogen $E_n = -\dfrac{13.6}{1^2}$ eV

Energy of the nth orbit of Be^{+++} $E_n = -\dfrac{13.6}{n^2}Z^2$ eV

For $n = 2$, $E_n = -\dfrac{13.6}{2^2} \times 4^2 \text{ eV} = -4 \times 13.6$ eV

$$\frac{E_2(Be^{+++})}{E_1(H)} = 4$$

8. Which level of doubly ionized lithium (Li^{++}) has the same energy as the ground state of hydrogen atom? Compare the orbital radii of the two levels.

Solution

For hydrogen $E_n = -\dfrac{13.6}{n^2}$ eV and $E_1 = -13.6$ eV

For other atoms $E_n = -\dfrac{13.6}{n^2}Z^2$ eV

Let the ground state energy E_1 of hydrogen be equal to the nth orbit of L^{++} $(Z = 3)$

Therefore, $E_n = -\dfrac{13.6}{n^2}3^2$ eV $=-13.6$ eV (of hydrogen ground state)

$$n^2 = 9, \quad n = 3$$

Radius of the nth orbit of hydrogen $r_n = n^2 \times 5.29 \times 10^{-11}$ m

for other atoms $r_n = \dfrac{n^3}{Z}\dfrac{\hbar^2 \varepsilon_0}{\pi m e^2} = 0.529 \times 10^{-10}\dfrac{n^2}{Z}$

r_1 (hydrogen) $= 5.29 \times 10^{-11}$ m

for $n = 3$ orbit of L^{++} $r_3 = \dfrac{5.29 \times 10^{-11}}{3} \times 9 = 3 \times 5.29 \times 10^{-11}$ m

$$\dfrac{r_3(\mathrm{Li}^{++})}{r_1(\mathrm{H})} = 3$$

9. The total energy of the electron in the first excited state of hydrogen is about -3.4 eV.

    i.   What is the K.E. of the electron in this state?

    ii. What is the P.E. of the electron in this state?

Solution

$$\mathrm{K.E.} = \dfrac{1}{2}\dfrac{1}{4\pi\varepsilon_0}\dfrac{Ze^2}{r_n}$$

$$\mathrm{P.E.} = -\dfrac{1}{4\pi\varepsilon_0}\dfrac{Ze^2}{r_n} = -2\,\mathrm{K.E.}$$

Total energy T.E. $=$ K.E. $+$ P.E. $=$ K.E. -2 K.E. $=$ K.E.

K.E. $= -$ T.E. $= -(-3.4) = 3.4$ eV

P.E. $= -2 \times 3.4 = -6.8$ eV

10. The wavelength of the first line of Balmer series of hydrogen atom is 6563 Å. Calculate the ionization potential and the first excitation potential of hydrogen atom ($h = 8.83 \times 10^{-34}$ Js and $c = 3 \times 10^8$ m/s).

Solution

$$\dfrac{1}{\lambda} = R\left(\dfrac{1}{2^2} - \dfrac{1}{n^2}\right)$$

First line $\dfrac{1}{\lambda} = R\left(\dfrac{1}{2^2} - \dfrac{1}{3^2}\right) = R\dfrac{5}{36} = \dfrac{1}{6563 \times 10^{-10}}$

$$R = \dfrac{36 \times 10^{10}}{5 \times 6563} = 1.097 \times 10^7\,\mathrm{m}^{-1}$$

The energy of the electron in the nth orbit

$$E_n = -\frac{Rhc}{n^2} = -\frac{1.097 \times 10^7 \times 6.63 \times 10^{-34} \times 3 \times 10^8}{n^2 \times 1.6 \times 10^{-19}} = -\frac{13.6}{n^2}\,\text{eV}$$

$$E_1 = -13.6 \text{ eV}$$

$$E_2 = -\frac{13.6}{2^2} = -3.4 \text{ eV}$$

Excitation potential $= 13.6 - 3.4 = 10.2$ eV

11. What is the distance of the closest approach to the nucleus of Gold atom by an alpha particle of energy 5.5 MeV which undergoes scattering by 180°.

Solution

When alpha particle is closest to the nucleus it comes to rest and its kinetic energy is completely converted into potential energy. So the distance of closest approach is when K.E. = P.E.

$$E = \frac{1}{4\pi\varepsilon_0} \frac{Ze \times 2e}{r_0}$$

$$r_0 = \frac{1}{4\pi\varepsilon_0} \frac{2Ze^2}{E}$$

$$E = 5.5 \text{ MeV} = 5.5 \times 10^6 \times 1.6 \times 10^{-19} \text{ J} = 8.8 \times 10^{-13} \text{ J}$$

For gold $Z = 79$, $\dfrac{1}{4\pi\varepsilon_0} = 9 \times 10^9$ and $e = 1.6 \times 10^{-19}$ Coulomb

$$r_0 = \frac{9 \times 10^9 \times 2 \times 79 \times (1.6 \times 10^{-19})^2}{8.8 \times 10^{-13}} = 41.36 \times 10^{-15}\, m = 41.36 \text{ fm}$$

12. The Hamiltonian for hydrogen atom is $\hat{H} = -\dfrac{1}{2}\nabla^2 - \dfrac{1}{r}$ and the 1s wave function in atomic units are $\psi_{1s} = \dfrac{1}{\sqrt{\pi}}e^{-r}$. Calculate the ground state energy of the hydrogen atom.

Solution

We know the del function $\nabla^2 = -\dfrac{1}{2r^2}\left[\dfrac{\partial}{\partial r}\left(r^2\dfrac{\partial}{\partial r}\right) + \dfrac{1}{\sin\theta}\dfrac{\partial}{\partial\theta}\left(\sin\theta\dfrac{\partial}{\partial\theta}\right) + \dfrac{1}{\sin^2\theta}\dfrac{\partial^2}{\partial\phi^2}\right] - \dfrac{1}{r}$

Radial part is $-\dfrac{1}{2r^2}\left[\dfrac{\partial}{\partial r}\left(r^2\dfrac{\partial}{\partial r}\right)\right] - \dfrac{1}{r}$

$$E_{1s} = \int \Psi_{1s} \hat{H} \Psi_{1s} d\tau = \frac{1}{\pi} \int_0^\infty e^{-r} \left[-\frac{1}{2r^2} \left[\frac{d}{dr} \left(r^2 \frac{d}{dr} \right) \right] . - \frac{1}{r} \right] - e^{-r} r^2 dr \int_0^\pi \sin\theta d\theta \int_0^{2\pi} d\phi$$

$$= -\frac{4\pi}{2\pi} \int_0^\infty r^2 e^{-2r} dr = -\frac{1}{2} \text{a.u.} = -2.18 \times 10^{-18} \, \text{Js} = -13.6 \, \text{eV}$$

13. Work out the angular momentum function for the five $3d$ orbitals.

Solution

The angular momentum function is $\Theta(\theta)\, \Phi(\phi)$

$$= \frac{1}{\sqrt{2\pi}} e^{iM\varphi} N P_l^{|M|}(\cos\theta)$$

The normalization factor $N = [\dfrac{2l+1}{2} \dfrac{(l-|M|)!}{(l+|M|)!}]^{1/2}$ and

$$P_l^{|M|}(\cos\theta) = \frac{1}{2^l l!} (1-x^2)^{|M|/2} \frac{d^{l+|M|}}{dx^{l+|M|}} (x^2-1)^l$$

For $l=2, |M|=2, \quad N = \left[\frac{(4+1)}{2} \frac{(2-2)!}{(2+2)!} \right]^{1/2} = \sqrt{\frac{5}{48}}$

$$P_2^2(x) = \frac{1}{2^2 2!} (1-x^2) \frac{d^4}{dx^4} (x^4 - 2x^2 + 1) = \frac{1}{8}(1-x^2)24 = 3\sin^2\theta$$

(we had assumed $\cos\theta$ as x)

Therefore $P_l^{|M|}(\theta) = P_{2,\pm2}(\theta) = \sqrt{\frac{5}{48}} \times 3 \, \sin^2\theta = \sqrt{\frac{15}{16}} \sin^2\theta$

$$A(\theta,\phi) = \sqrt{\frac{15}{32\pi}} \sin^2\theta e^{2i\phi}$$

Similarly $P_{2,\pm1}(\theta) = \sqrt{\frac{15}{4}} \sin\theta \cos\theta$

$$A(\theta,\phi) = \sqrt{\frac{15}{8\pi}} \sin\theta \cos\theta \, e^{2i\phi}$$

$$P_{2,0}(\theta) = \sqrt{\frac{5}{8}} (3 \, \cos^2\theta - 1); \; A(\theta,\phi) = \sqrt{\frac{15}{16\pi}} (3\cos^2\theta - 1) \backslash$$

REFERENCES

Condon, E.U. and Shortley, G.H. *The Theory of Atomic Spectra*. Cambridge University Press, Cambridge, 1935, reprinted 1951.

Cowan, R.D. (1981). *The Theory of Atomic Structure and Spectra*. University of California Press, Berkeley.

David Halliday. (2007). *Fundamentals of Physics*, 8th (edn.). Wiley. New Jersey.

David, J. Griffiths. (2004). *Introduction to Quantum Mechanics*, 2nd (edn.). Benjamin Cummings, San Francisco.

Edmonds, A.R. (1960). *Angular Momentum in Quantum Mechanics*. Princeton University Press, Princeton.

Erwin Schrödinger. (1926). *Annalen der Physik, (Leipzig)*.

Hannabuss, K. (1997). *An Introduction to Quantum Theory*. Oxford University Press, Oxford.

Kuhn, H. *Atomic Spectra*. (1962). Longmans, London.

Moss, R.E. (1973). *Advanced Molecular Quantum Mechanics*. Chapman and Hall, London.

Paul Adrien Maurice Dirac. (1958). *The Principles of Quantum Mechanics*, 4th (edn.). Oxford University Press.

Prasad, R.K. (1992). *Quantum Chemistry*. Wiley Eastern Ltd., New Delhi.

Richard Liboff. (2002). *Introductory Quantum Mechanics*, 4th (edn.). Addision Wesley. San Francisco.

Rose, M.E. (1957). *Elementary Theory of Angular Momentum*. John Wiley and Sons, New York.

Schiff, L.I. (1968). *Quantum Mechanics*, McGraw-Hill Book Co. New York.

Schrödinger (1992). *Life and Thought by Walter John Moore*. Cambridge University Press.

Schrödinger, Erwin. (1926). "An undulatory theory of the mechanics of atoms and molecules." *Phys. Rev.* 28(6): 1049–1070.

Serway R.A, Moses, C.J. and Moyer, C.A. (2004). *Modern Physics*, 3rd (edn.). Brooks Cole. California.

White, H.E. *Introduction to Atomic Spectra*. McGraw-Hill, New York, 1934.

5

SPECTROSCOPIC STUDIES ON THE SHELL STRUCTURE OF ATOMS

5.1 X-RAY SPECTRA

In 1895 Röntgen[1] discovered a new kind of radiation known as X-rays or Rontgen rays. Since its invention, X-rays are used for a number of applications ranging from medical use to determination of crystal structure. It was Laue who developed an X-ray spectrometer, which led to the determination of the wavelengths of the characteristic lines of a large number of elements.

Considering a crystal made up of regularly spaced atoms, Laue suggested the possibility of considering a crystal like a diffraction grating. By passing a narrow beam of X-rays through a crystal and recording the diffraction pattern on a photographic plate, it was possible to observe small symmetrically spaced spots which were characteristics of the element constituting the crystal. These spots are known as Laue spots or Laue pattern[2]. Later Bragg[3] showed that the X-ray diffraction pattern is due to the regularly spaced atom in the crystal, which can be related to the distance between the successive layers of the atoms in the crystal. The well known Bragg equation in X-ray diffraction is

$$2d \sin \theta = n\lambda \tag{5.1}$$

where λ is the wavelength of the X-ray, θ is the angle of incidence and d is the distance between the successive layers of the atoms in the crystal.

Prior to Laue's discovery of X-ray diffraction, Barkla and Sadler studied absorption of X-rays in aluminum and by measuring their hardness or penetrability they have identified two groups of radiations, one is more penetrating and the other less penetrating. The more penetrating ones are hard radiations called K radiations and the less penetrating ones as soft radiations or L radiations.

By measurement of the X-ray diffraction by crystals it is now known that K-radiations of an element are characteristic of short wavelength and L radiations are in long wavelengths. Barkla and Sadler detected K series for lighter elements and L series in heavier elements. Soft radiations called M, N, O are observed in heavy elements at higher wavelengths.

Moseley[4] had used the Laue's technique of X-ray diffraction of crystals to determine the characteristic lines of a large number of elements. By a systematic study of K radiations in a number of elements ranging from Ca to Zn the following conclusions had been arrived.

1. As the atomic number increases, the corresponding lines in the spectra shift regularly and successively to shorter wavelengths.

2. The frequency of each corresponding X-ray line is approximately proportional to the square of the atomic number.

3. For each element the spectra showed two lines of which one is intense and is in the longer wavelength region and the other is weaker and is in the shorter wavelength.

4. It means that if we follow the lines in successive elements the shift in wavelength is gradual and regular and this characteristic can be used to identify the change in atomic weights of the elements. For example if one finds a sudden and abrupt change in otherwise expected gradual change in wavelength, it may be concluded that there is missing element, which can be investigated further, and the missing element may be searched. For example in Moseley's first experiments it was found that there was a large difference in wavelengths passing from Ca to Ti indicating that they are not consecutive elements. Prior to Moseley's experiments, nickel and cobalt were listed in order of nickel, cobalt because of their atomic weights 58.7 and 58.9 respectively. Ni is expected to come before Co in the periodic table according to atomic weight, but the chemical properties indicate that Co to be placed before Ni. Moseley's X-ray photographs showed undoubtedly that the order is cobalt, nickel. It is also supported by the X-ray diffraction experiments. In 1913 from studies on different elements, Moseley found that there is a relation between the wave number and the atomic number of the element. Thus he put forward his relation, which is now known as Moseley law. In brief it states that the frequency of each corresponding X-ray line is approximately proportional to the square of the atomic number of the emitting element. From a graph of the K lines measured by Moseley for different elements, he had arrived at the following expression

$$v(\mathrm{cm}^{-1}) = K'R(Z-\sigma)^2 \tag{5.2}$$

where R is the Rydberg constant $= 109737$ cm^{-1} and σ and K' are constants. It has been found that in the case of K_α radiations, the value of $\sigma = 1$ and $K' = \frac{3}{4}$. By an analogy with the Balmer series expression for the hydrogen atom Moseley's expression is

$$v(cm^{-1}) = R(Z-1)^2 \left(\frac{1}{1^2} - \frac{1}{2^2} \right) \tag{5.3}$$

Moseley first plotted $\sqrt{\nu}$ against Z and later it was modified by Sommerfeld and plotted $\sqrt{\dfrac{\nu}{R}}$ with Z. This is known as Moseley law or Moseley's pictures.

Plot of K lines with $\sqrt{\dfrac{\nu}{R}}$ had shown that K radiation consisted of two strong spectral lines which were called K_α and K_β. Continuing the work of Moseley, other investigators have noticed that in the heavier elements, each of the K_α and K_β are themselves closed doublets. These four lines are called $K_{\alpha 1}$, $K_{\alpha 2}$, $K_{\beta 1}$ and $K_{\beta 2}$ lines. Later it was found that Mosleys law holds good for the spectra of all elements in the periodic table. Mosley's law holds good for all L series lines also. L radiations were also found to be consisted of a group of spectral lines called $L_{\alpha 1}$, $L_{\alpha 2}$, $L_{\beta 1}$ and $L_{\beta 2}$ lines. In heavy elements the L series is composed of as many as twenty lines. Siegbahn in 1916 and later Dolejsek observed many other series of lines in heavier elements, which are now called M and N radiations. Moseley's relation also could represent these lines.

5.1.1 Absorption Spectra

In the experiments conducted in 1916 by de Broglie, the X-ray spectrogram showed blackening in two distinct positions on the photographic plates. The wavelengths of these absorption edges were independent of the X-ray source and were found to be due to selective absorption of silver and bromine used for photographic emulsion. However, later experiments have revealed that when the layers of metals like Cu, Ag, Fe, Sn, etc. were placed in front of the photographic plate, certain additional absorption edges appear on the photographic plate, which were characteristics of the metals placed on the way of the X-ray. In later experiments several different absorption edges were noticed especially in heavier elements. The absorption edge occurring on the shortest wavelength for a given element is called the K-absorption limit corresponding to the K radiation mentioned earlier. Beyond K limit to longer wavelengths other absorption limits are obtained in heavier elements, which were called L limits consisting of edges L_I, L_{II}, L_{III} and M limits M_I, M_{II}, M_{III}, M_{IV}, M_V. The observation of these absorption limits was very helpful in explaining the origin of X-rays from within the atoms. Kossel in 1916 proposed a detailed theory on the origin of X-ray spectra. A sharp absorption edge for a given element indicates that when the incident X-rays have an energy (vis-à-vis frequency) larger than a certain critical energy (vis-à-vis frequency) they are largely absorbed whereas those of lower energy (or lower frequency) simply pass through without change in intensity. When the electrons in an atom encounter the critical energy $h\nu_c$, this energy of the X-ray beam is used by the electron to get ejected out of the atom. Thus the critical energy ejects an electron from the atom. They are the photoelectrons. The atom thus gets ionized. The different absorption limits observed as K, L and M limits thus correspond to energies necessary to remove electrons from different shells or subshells in the atom. This led to the concept of shell structure in atoms. The absorption limits suggests that the electrons occupy different shells in the atom. Kossel suggested that the electrons in atoms can be thought to be arranged in groups or shells and analogy with *K, L, M* limits he called these shells as *K, L, M,* etc. shells. The absorption limit observed thus gave a direct concept to describe the energy levels of an atom.

Prior to the emission of X-rays by an atom, the following rearrangements of electrons happen inside the atoms.

A complete removal of electron from one of the completed inner shells takes place. This is accomplished by an inelastic impact of the atom by a high speed particle (like an electron, proton or another atom) or by absorption of a photon of energy $h\nu_c$ or greater. By such an impact an electron from an inner shell is knocked out from the atom. Thus a vacancy is created in the inner shell and an electron from a nearby outer shell tries to occupy the vacant place and jump into the unoccupied orbit giving rise to an X-radiation the energy of which is the difference in energy between the two stationary states of the atom in which the electron exchange occurred. A vacancy is now created in the orbit from which the electron jumped to the earlier vacant place. Another electron from a shell farther out may now jump to the place where the new vacancy created. This process is repeated till the positive ion (positively charged atom due to the loss of the electron) captures an electron (from outside). Thus all these facts gave the idea to Kossel who suggested that the electrons in atom could be arranged in groups of shells. Comparison with hydrogen atom, the *K* shell is assumed to be associated with $n = 1$. The other shell *L* with $n = 2$, *M* with $n = 3$, *N* with $n = 4$ and so on. Kossel suggested that lines of *K* series are emitted when an electron from any of the shells *L, M, N* ... makes a transition to the *K* shell. Thus the existence of a vacant positioning the shell is imperative to observe a *K* line. This vacancy is created if the energy of an incident electron or photon is sufficient to remove the electron from *K* shell, which do not make a line. Later experiments by de Broglie, Siegbahn and others and theoretical work by Smekel, Corten, Wentzel and others refined Kossel's theory of shell structure of electron in atoms. Note that with the so called optical spectra we deal with radiations arising from electronic transitions in the outermost part of the atom. The valence electrons play a major role in the optical spectra whereas in X-ray spectra the inner shell electrons are mainly involved.

To arrive at a comprehensive picture of what we have described above, let us consider the example of cadmium atom. Cadmium has an atom number 48. That means there are 48 electrons revolving round Cd nucleus. Cadmium has 48 protons, 64 neutrons, 48 electrons (atomic number is 48), atomic mass 112.411 amu. Cd is chosen as an example to describe the shell arrangement as it contains completed shells with electrons like the inert gas He, Ne, Ar, Kr, etc. The electron configuration of cadmium is

$$1s^2 \quad 2s^2\,2p^6 \quad 3s^2 3p^6 3d^{10} \quad 4s^2 4p^6 4d^{10} \quad 5s^2 \qquad {}^1S_0$$
$$\quad K \qquad L \qquad\quad M \qquad\qquad N \qquad\quad O$$

The Table 5.1 shows that there are five completed *s* sub-shells followed by three *p* sub-shells and two completed *d* shells in cadmium.

Table 5.1   The electronic configuration of cadmium

Shell subshell	K	L	M	N	O
s	$1s^2$	$2s^2$	$3s^2$	$4s^2$	$5s^2$
p		$2p^6$	$3p^6$	$4p^6$	
d			$3d^{10}$	$4d^{10}$	

The X-ray notation *K, L, M, N* correspond to $n = 1, 2, 3, 4, \ldots$ and are the main shells of the electrons. However, each electron orbit in cadmium interpenetrate all other electron orbits, but for our explanations, we can consider the time average of the position and represent them in circular orbits as shown in Figure 5.1.

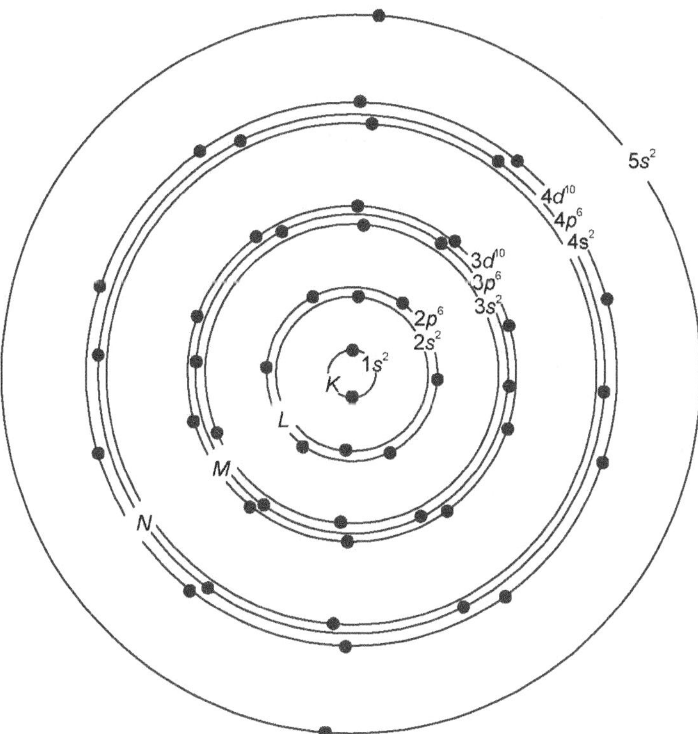

Figure 5.1   Shells and subshells of electrons (e.g., Cadmium atom)

The two $1s$ electrons are the most tightly bound electrons in cadmium followed by $2s$, $2p$ ... electrons in their binding energies. The binding energies of these electrons can be measured by the absorption edges. Thus we can create an energy level diagram to represent the energy level of the cadmium atom.

The ground state of the cadmium atom is represented as $^1S_0$. In this case of definition of an atomic state S corresponds to the state where $L = 0$. Superscript on the left is the multiplicity of the state. In this case it is singlet. The suffix zero on the right hand side represents the total angular momentum quantum number J and in this case $J = 0$. The total angular momentum quantum number J is obtained by the relation $J = L \pm S$ where L is the orbital quantum number and S is the total electron spin quantum number. Note that S also is used for the nomenclature of the state with $L = 0$. Both have the same nomenclature in spectroscopy. In principle J can take values $J = L + S$, $L + S - 1,\ldots|L - S|$.

The energy level diagram of cadmium atom is shown in Figure 5.2. The bottom of the diagram is the ground state of the cadmium atom with all its electrons intact. If we remove a 5s electron from the atom the atom, gets ionized. The state of the ionized cadmium by the removal of the 5s electron is represented by $5s^{-1}$ state. Its characteristic is $5^2S_{1/2}$. Note that in the normal atom the multiplicities of the states were singlets or triplets (odd multiplicity), however in the ionized atom the states are doublets or quartets (even multiplicity). $^2S_{1/2}$ state means $L = 0$ (S state), total spin quantum number $S = \frac{1}{2}$ and the total quantum number $J = L \pm S = \frac{1}{2}$. The total spin quantum number can be obtained from the multiplicity of the state. Here the superscript on the left represents the multiplicity and in this case it is 2. The multiplicity depends upon the electron spin and it is $2S + 1$ where S is the spin. Thus $2S + 1 = 2$ and hence $S = \frac{1}{2}$. As stated above S is also used for the nomenclature of the state with $L = 0$. The highest energy state is obtained by the removal of an electron from the innermost $1s$ orbital and the state is $1s^{-1}$ in the figure. $1s^{-1}$ represents the fact that there is one vacancy in the $1s$ orbital. It is again a $^2S_{1/2}$ state denoted as $5s^{-1}\ ^2S_{1/2}$. Similar way, the removal of a 2s electron raises the atom to the $2s^{-1}$ state which again gives a $^2S_{1/2}$ state. Similarly if we could remove an electron from the p orbital, it is designated as $2p^{-1}$ and gives rise to a ^2P state, thus giving rise to two states $^2P_{1/2}$ and $^2P_{3/2}$ of which the latter lies lowest because of the inverted nature of the $2P$ term. In the case of a P state $L = 1$ and if it is doublet as in this case $S = \frac{1}{2}$. The configuration gives rise to two states with $J = L \pm S = \frac{3}{2}$ or $\frac{1}{2}$ thereby giving rise to $^2P_{1/2}$ and $^2P_{3/2}$ states. The absorption limits L_{III} and L_{II} are obtained by transition to these levels. It should be noted that when the electron shell is half or less filled, the configuration gives rise to normal states whereas the shell is filled with more than half it leads to inverted states. In deriving the state with incompletely filled shells and if the shell is more than half filled, the number of unfilled electrons are counted for the derivation of electronic states.

Thus the energy states of the cadmium atom can be easily constructed. The transitions are allowed if $\Delta L = \pm 1$ and $\Delta J = 0,\ \pm 1$. Only few transitions are shown in the Figure 5.2. The levels with the same value of L and different values of J give close doublets in X-ray spectra.

In deriving the electronic states of an atom the electrons in the valence shells are used to derive the states unlike in this case. A subshell p occupied by a single electron or a subshell p lacking an electron to complete the shell give rise to the same type of states, the only difference being that the former is a normal multiplet, in this case doublet whereas the latter is an inverted state.

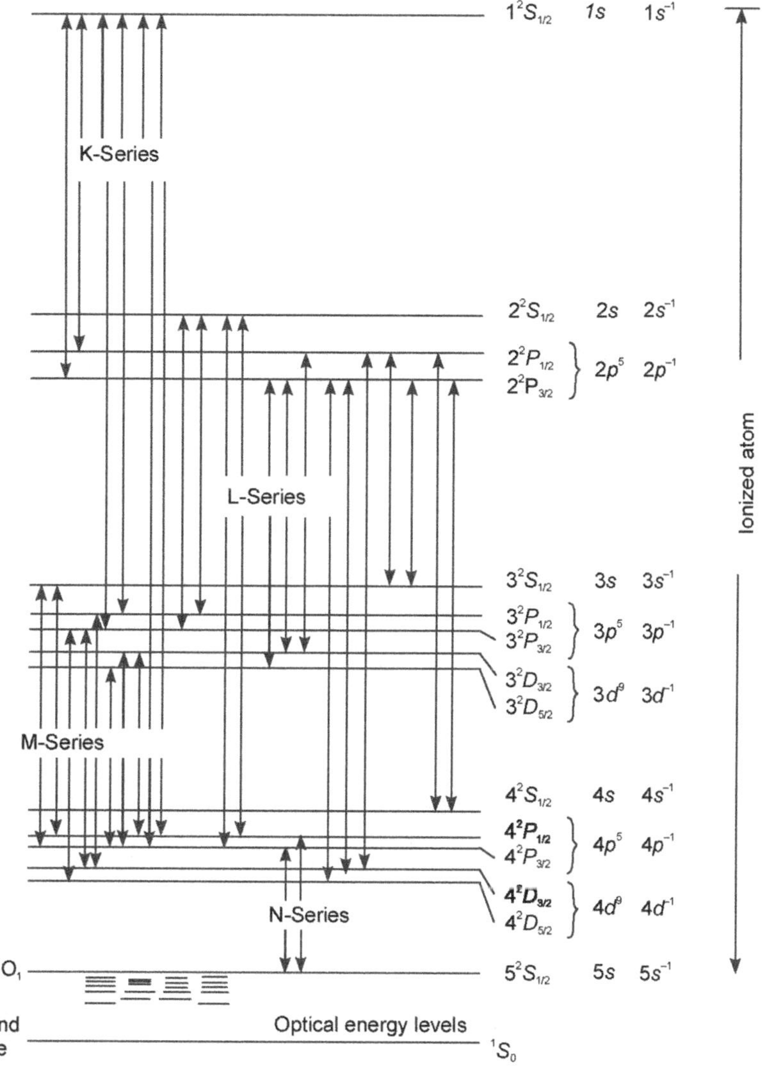

Figure 5.2   X-ray energy levels of Cd

5.2   *AUFBAU PRINZIP*—BUILDING PRINCIPLE

A number of experimental evidences obtained from the X-ray spectra throw light on the shell structure of atoms. The *Aufbau Prinzip* is a German word meaning "building principle." The information obtained from the X-ray spectra along with their chemical properties were coined together to give a complete picture of the building up principle for the elements. The scheme proposed by Bohr and later modified by Stoner could explain the position of elements in the periodic table of Mendeleev[5].

Periodic Table of the Elements

Legend:
- hydrogen
- alkali metals
- alkali earth metals
- transition metals
- poor metals
- nonmetals
- noble gases
- rare earth metals

Group	IA 1	IIA 2	IIIB 3	IVB 4	VB 5	VIB 6	VIIB 7	VIII 8	VIII 9	VIII 10	IB 11	IIB 12	IIIA 13	IVA 14	VA 15	VIA 16	VIIA 17	VIIIA 18
1s / 1	1 H																	2 He
2s / 2	3 Li	4 Be											5 B	6 C	7 N	8 O	9 F	10 Ne
3s / 3	11 Na	12 Mg											13 Al	14 Si	15 P	16 S	17 Cl	18 Ar
4s / 4	19 K	20 Ca	21 Sc (3d)	22 Ti	23 V	24 Cr	25 Mn	26 Fe	27 Co	28 Ni	29 Cu	30 Zn	31 Ga	32 Ge	33 As	34 Se	35 Br	36 Kr
5s / 5	37 Rb	38 Sr	39 Y (4d)	40 Zr	41 Nb	42 Mo	43 Tc	44 Ru	45 Rh	46 Pd	47 Ag	48 Cd	49 In	50 Sn	51 Sb	52 Te	53 I	54 Xe
6s / 6	55 Cs	56 Ba	57 La (5d)	72 Hf ($5d^2$)	73 Ta	74 W	75 Re	76 Os	77 Ir	78 Pt	79 Au	80 Hg	81 Ti	82 Pb	83 Bi	84 Po	85 At	86 Rn
7s / 7	87 Fr	88 Ra	89 Ac (5f)	104 Unq ($6d^2$)	105 Unp	106 Unh	107 Uns	108 Uno	109 Une	110 Unn					114 Uuq	116 Uuh		118 Uuo

Subshell labels: s, p, d, f; 2p, 3p, 4p, 5p, 6p, 5d, 6d; 4f, 5f

Series															
Lanthanids series (4f)	58 Ce (4f)	59 Pr	60 Nd	61 Pm	62 Sm	63 Eu	64 Gd	65 Tb	66 Dy	67 Ho	68 Er	69 Tm	70 Yb	71 Lu	
Actinide series (5f)	90 Th ($5f^2$)	91 Pa	92 U	93 Np	94 Pu	95 Am	96 Cm	97 Bk	98 Cf	99 Es	100 Fm	101 Md	102 No	103 Lr	

Figure 5.3   The periodic table of the elements

Bohr proposed a scheme of building up of the elements in the periodic table, which was later, improved by Stoner and is now known as Bohr-Stoner Scheme. The periodic table is given in Figure 5.3. The number and distribution of electrons in various shell under Bohr–stoner scheme is given in Table 5.1.

Bohr proposed his scheme based on the following assumptions. The number of electrons around the nucleus of a neutral atom is equal to the atomic number Z. Each of these elements takes on a definite quantized state described by a set of quantum numbers. The number of electrons in state with a particular principal quantum number is limited. The maximum number of electrons in each shell in given by $2n^2$ where n is the principal quantum number. This was also suggested later by Pauli's[6] exclusion principle and already proposed by Kossel[7] in explaining the X-ray spectrum of elements. After considering and assigning quantum numbers to each electron in an atom with atomic number Z, the next atom in the periodic table is obtained by adding one more electron. The quantum numbers to the added electron are assigned such that the electron in the state is in the most tightly bound state. With such a scheme the atom will have the least amount of energy possible.

Earlier Rydberg suggested that the atomic number of the inert gases He, Ne, A, Kr, Xe and Rn can be represented by the formula

$$
\begin{array}{cccccc}
\text{He} & \text{Ne} & \text{A} & \text{Kr} & \text{Xe} & \text{Rn} \\[2mm]
Z = 2(1^2 & +\ 2^2 & +2^2 & +3^2 & +3^2 & +4^2) \\[2mm]
2 & 10 & 18 & 36 & 54 & 86
\end{array}
$$

Bohr immediately used this idea and correlated this information of electron shells for the arrangements of electrons in other elements. The electrons are thought to be arranged in electron shells and various electrons are classified under so called shells. All electrons belonging to the same shell are characterized by the same principal quantum number n. The first shell is with $n = 1$ followed by $n = 2$, $n = 3$, etc. The lower shells are filled one by one with the electrons available with 2, 8, 18, 32, etc., electrons respectively. The number of electrons in a shell with principal quantum number n is $2n^2$. The shells with $n = 1,2,3,4$, etc., are called K, L, M, N, etc. shells respectively. The nomenclature is taken from X-ray spectroscopy. The electrons in a particular shell with principal quantum number n is further divided into subshells so that electrons belonging to the same subshell have the same azimuthal quantum number l. Electrons with $l = 0, 1, 2, 3,...$ are called s, p, d, f, etc., electrons respectively. $l = k - 1$ where k is the azimuthal quantum number in the Bohr–Sommerfeld theory.

In Bohr's representation the electrons are distributed equally in subshells, but it was immediately realized it as wrong on the basis of X-ray evidence. Stoner suggested a modified scheme of accommodating electrons in subshells. This is in conformity with the Pauli's exclusion principle and is also in agreement with experimental observations. Bohr and Stoner schemes are given in Table 5.2.

Table 5.2 Distribution and number of electrons in various shells in Bohr–Stoner scheme

Shell	$K(n=1)$	$L\ (n=2)$		$M(n=3)$			$N(s=4)$			
Sub shells	$1s$	$2s$	$2p$	$3s$	$3p$	$3d$	$4s$	$4p$	$4d$	$4f$
Bohr's scheme	2	4	4	6	6	6	8	8	8	8
Stoner's scheme	2	2	6	2	6	10	2	6	10	14

5.2.1 Electron Configuration

Electron configuration in an atom tells us how the electrons are arranged in atomic orbitals. Pauli's exclusion principle plays a major role in describing the distribution of electrons among various atomic orbitals. "It should be forbidden for more than one electron with the same value of the principal quantum number n to have the same value of the other three quantum numbers l, m_1 and m_s." It basically says that two electrons can occupy the same orbital only if they have opposite spins. This means that we can have only two electrons in an orbital. In a p orbital we can have two electrons each in p_x, p_y and p_z orbitals. Electron spin quantum number is an important factor, though it has no effect on the energy, size, shape or spatial orientation of an orbital. The electron spin quantum number has the values $m_s = +\dfrac{1}{2}$ or $m_s = -\dfrac{1}{2}$. According to the Aufbau principle, electrons are filled beginning with the orbital of the lowest energy in accordance with the Pauli's exclusion principle. Aufbau principle says that " a maximum number of two electrons are put into orbitals in the order of increasing orbital energies: the lowest energy orbitals are filled before electrons are placed in higher energy orbitals". The principle works well with the first 18 elements, then less well for the rest of elements. The modern form of Aufbau Principle gives an order of energy given by Erwin Madelung[8] .

1. The orbitals are filled in the order of increasing $n + l$.

2. Where two orbitals have the same value $n + l$, they are filled in order of increasing n.

The following order have been followed in filling the orbitals

$$1s,\ 2s,\ 2p,\ 3s,\ 3p,\ 4s,\ 3d,\ 4p,\ 5s,\ 4d,\ 5p,\ 6s,\ 4f,\ 5d,\ 6p,\ 7s,\ 5f,\ 6d,\ 7p$$

There are other methods to determine the order of orbital energies like Hartree–Fock calculations which are tedious and cumbersome and often need approximations.

5.2.2 First Period

Let us now consider the periodic system of elements on the basis of Bohr–Stoner hypothesis. Hydrogen is the smallest and the simplest atom with a single electron. The single electron is bound to its normal state $1s$ with an energy binding of 109677 cm^{-1}. The normal energy state in hydrogen is designated as $^2S_{1/2}$ which is derived from the electron configuration $1s$. Helium contains 2 electrons, both of them are bound to $1s$, thus giving rise to the electronic configuration $(1s)^2$. The energy state derived from this electronic configuration is $^1S_0$.

Thus the configuration of hydrogen and helium are

$$\text{H} \qquad 1s \qquad\qquad ^2S_{1/2}$$

$$\text{He} \qquad 1s^2 \qquad\qquad ^1S_0$$

Principal quantum number $(n) = 1$

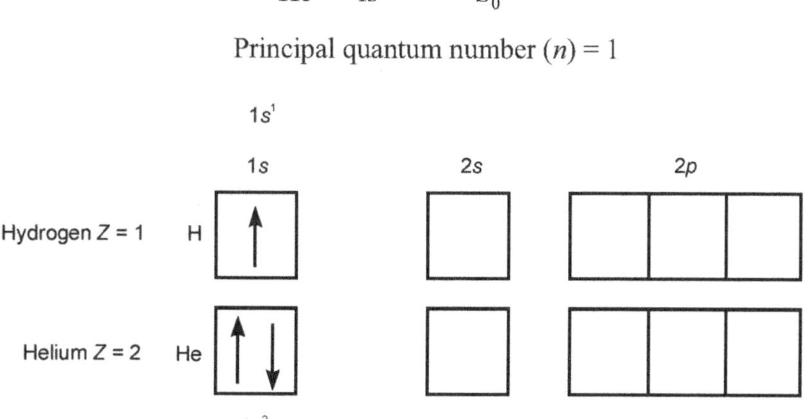

Diamagnetic and paramagnetic properties   A diamagnetic substance is unaffected by a magnetic field whereas a paramagnetic substance is affected by a magnetic field. In a diamagnetic substance all electron spins are paired, e.g., 'He' whereas a paramagnetic substance contains atleast an electron spin unpaired.

Paired He [↑↓]   Unpaired H [↑]

5.2.3 Second Period

Lithium has an atomic number $Z = 3$ and has 3 electrons. The two electrons are used to fill the $1s$ orbit thus completing the K shell. The third electron occupies the 2s subshell. The electronic configuration of Li is $1s^2\ 2s$ and its normal energy state is designated as $^2S_{1/2}$. Beryllium has 4 electrons with atomic number $Z = 4$ and after using 2 electrons to fill the $1s$ orbital (K shell), the remaining 2 electrons are used to complete the $2s$ subshells. The fifth element is boron and has an atomic number $Z = 5$ with 5 electrons. The $1s$ and $2s$ subshells are filled with 2 electrons each in this case and the remaining electron goes to the next subshell $2p$ with $n = 2$ and $l = 1$. Thus the electronic configuration of Boron is $1s^2\ 2s^2\ 2p$ with the normal electronic state designated as $^2P_{1/2}$. In C, N, O, F and Ne the electrons are added to the $2p$ subshell successively from carbon till neon after the occupation of the $1s$ and $2s$ with 2 electrons each. Thus carbon with $Z = 6$ marks the addition of 2 electrons in the $2p$ shell making the electronic configuration of carbon as $1s^2 2s^2\ 2p^2$. The maximum number of electrons the p shell can hold, is 6 and this is achieved in Ne with the electronic configuration $1s^2 2s^2\, 2p^6$ and the normal energy state of Neon is $^1S_0$.

Principal quantum number $n = 2$

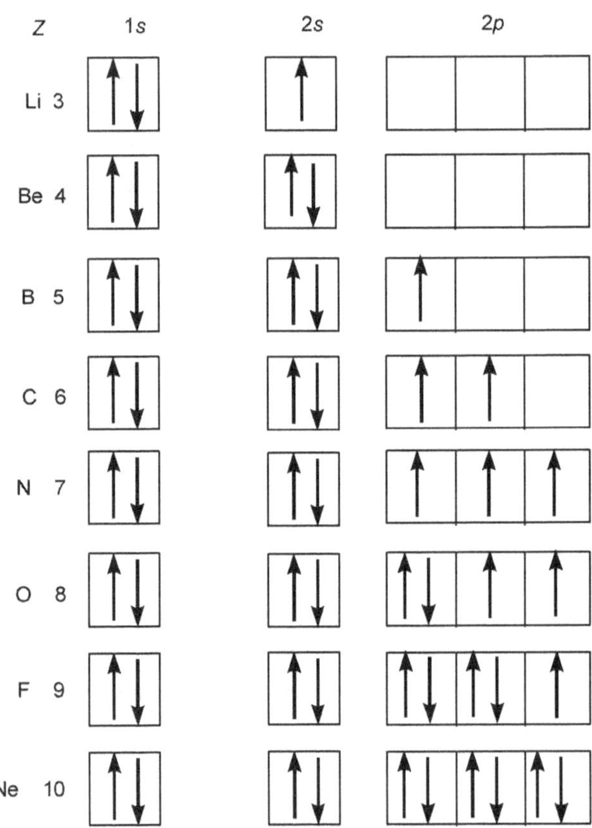

It should be noted that while filling up the orbitals the Hund's rule is followed just like the Pauli's exclusion principle. The Hund's rule states that when there are two or more orbitals of the same energy, each one is singly occupied before any one is doubly occupied.

In the above example in the p shell the p_x, p_y, p_z are having the same energies and are occupied singly first. The most stable arrangement of electrons in a subshell is the one in which there are greatest number of paired spins.

5.2.4 Third Period

There are 8 elements in this period, viz., Na, Mg, Al, Si, P, S, Cl and Ar. The building principle is the same as in the case of eight elements in the second period as described above. The electronic configurations are confirmed by spectroscopic observations of the energy transitions vis-à-vis energy states. The first element in the third period is sodium and spectroscopically it has been found that Na has $^2S_{1/2}$ as its ground energy state. Thus the 11th electron is bound to $3s$ subshell. Magnesium has 12 electrons and after filling the K and L shells with 10 electrons the remaining 2 electrons are arranged in $3s$ subshell thus giving rise to the electron configuration $1s^2\,2s^2\,2p^6\,3s^2$. And for aluminium to argon one electron each is added to the $3p$ shell successively till it becomes six in argon giving rise to the electronic configuration for argon as $1s^2\,2s^2\,2p^6\,3s^2\,3p^6$. This is conveniently written as $KL\ 3s^2\,3p^6$ as the 10 electrons occupy K and L shells. It is also represented as [Ne] $3s^2\,3p^6$. Only the last bound electrons or the so called valence electrons are responsible for optical spectra.

$$
\begin{array}{llll}
\text{Na} & 11 & [\text{Ne}] & 3s^1 \\
\text{Mg} & 12 & [\text{Ne}] & 3s^2 \\
\text{Al} & 13 & [\text{Ne}] & 3s^2 3p^1 \\
\text{Si} & 14 & [\text{Ne}] & 3s^2 3p^2 \\
\text{P} & 15 & [\text{Ne}] & 3s^2 3p^3 \\
\text{S} & 16 & [\text{Ne}] & 3s^2 3p^4 \\
\text{Cl} & 17 & [\text{Ne}] & 3s^2 3p^5 \\
\text{Ar} & 18 & [\text{Ne}] & 3s^2 3p^6 \\
\end{array}
$$

where, the third period Ne core is $\left[Ne\right] = 1s^2\,2s^2\,2p^6$

5.2.5 Fourth Period

Potassium with $Z = 19$ is the first member of this long period of elements and it ends with krypton with $Z = 36$. This period contains 18 elements. Since Argon with $Z = 18$ has filled with $3s$ and $3p$ electrons, by addition of one more electron in potassium, it is expected to occupy the $3d$ subshell. This configuration is expected to give a D electronic state, i.e., one due to the valence $3d$ electron. However, the spectroscopic evidence shows that potassium has an S state which is possible only when the electron occupies the $4s$ subshell instead of the $3d$ shell. Thus the electron added to argon does not occupy the $3d$ shell instead prefer to occupy the $4s$ subshell. This is due to the fact that $4s$ electron is more tightly bound than the $3d$ electron. The well known spectrum of potassium is similar to the spectra of sodium with S state at the bottom. Potassium is thus an alkali metal atom similar to Li and Na. In calcium one more electron is added to potassium and this electron is added to the 4s shell thereby the $4s$ shell is complete. Thus the electronic configuration of calcium is $1s^2\,2s^2\,2p^6\,3s^2\,3p^6\,4s^2$. Note that $3d$ subshells are not filled yet. Calcium is an alkaline earth atom similar to Be and Mg.

$$
\begin{array}{llll}
\text{K} & 19 & [\text{Ar}] & 4s^1 \\
\text{Ca} & 20 & [\text{Ar}] & 4s^2 \\
\text{Sc} & 21 & [\text{Ar}] & 4s^2 3d^1 \\
\end{array}
$$

where, $\left[Ar\right] = 1s^2\,2s^2\,2p^6\,3s^2\,3p^6$

Following calcium the next 10 elements Sc, Ti, V, Cr, Mn, Fe, Co, Ni, Cu and Zn with atomic numbers ranging from $Z = 21$ to $Z = 30$ are built up by filling the $3d$ subshell successively from Sc to Zn. The electrons in a $3d$ subshell is more tightly bound than a $4p$ orbital. Electronic configuration for zinc $(Z = 30)$ is

$$1s^2\ 2s^2\ 2p^6\ 3s^2\ 3p^6\ \ 3d^{10}\ 4s^2$$

Here again the valence electron is $4s^2$ which contributes to the chemical and spectroscopic properties of an element. The ground state of zinc is $^1S_0$. However, copper with $Z = 29$ has ten electrons in the $3d$ subshell and one electron in $4s$ which explains its behaviour as an alkali. The electronic configuration of copper is $1s^2\ 2s^2\ 2p^6\ 3s^2\ 3p^6\ 3d^{10}\ 4s$ and the normal state is $^2S_{1/2}$. Zinc which has 2 electrons in the 4s orbit resembles an alkaline earth.

The process of filling up of subshells goes on successively till nickel with $Z = 28$ except in the case of Cr where the electronic configuration is $1s^2\ 2s^2\ 2p^6\ 3s^2\ 3p^6\ 3d^{\,5}\ 4s$ with the electronic ground state as $^7S_3$.

Vanadium having $Z = 23$ has an electronic configuration

$$V(Z = 23)\ 1s^2\ 2s^2\ 2p^6\ 3s^2\ 3p^6\ 3d^3\ 4s^2\qquad ^4F_{3/2}$$

The next element Chromium Cr with $Z = 24$, instead of expected $1s^2\ 2s^2\ 2p^6\ 3s^2\ 3p^6\ 3d^4\ 4s^2$ configuration, has the electronic configuration

$$Cr(Z = 24)\qquad 1s^2\ 2s^2\ 2p^6\ 3s^2\ 3p^6\ 3d^5\ 4s,\qquad ^7S_3$$

$1s^2\ 2s^2\ 2p^6\ 3s^2\ 3p^6\ 3d^5\ 4s$ is the one which is confirmed by spectroscopic observations. The elements from Sc till Ni have thus partially filled 3d electrons.

The electron configuration of $Sc(Z = 21)$ is $1s^2\ 2s^2\ 2p^6\ 3s^2\ 3p^6\ 3d\ 4s^2$ and that of Ni $(Z = 28)$ is $1s^2\ 2s^2\ 2p^6\ 3s^2\ 3p^6\ 3d^8\ 4s^2$. These elements are called elements of iron group or $3d$ transition group. They have incompletely filled d subshells and readily gives rise to cations that have incompletely filled d subshells. The atoms are characterized by their paramagnetic and ferromagnetic behaviour and their ionic solutions show colours, which can be explained by in terms of electronic transitions of $3d$ electrons. Copper is expected to have an electronic configuration $[Ar]3d^{\,9}\ 4s^2$, but the spectroscopic investigations reveal a configuration $[Ar]3d^{10}\ 4s$.

Note the electronic configuration of the elements Co, Ni, Cu and Zn.

$$Co\ (Z = 27)\ 1s^2\ 2s^2\ 2p^6\ 3s^2\ 3p^6\ 3d^7\ 4s^2,\quad i.e.,\quad [Ar]\ 4s^2\ 3d^7$$
$$Ni\ (Z = 28)\ 1s^2\ 2s^2\ 2p^6\ 3s^2\ 3p^6\ 3d^8\ 4s^2,\quad i.e.,\quad [Ar]\ 4s^2\ 3d^8$$
$$Cu\ (Z = 29)\ 1s^2\ 2s^2\ 2p^6\ 3s^2\ 3p^6\ 3d^{10}\ 4s^1,\quad i.e.,\quad [Ar]\ 4s\ \ 3d^{10}$$
$$Zn\ (Z = 30)\ 1s^2\ 2s^2\ 2p^6\ 3s^2\ 3p^6\ 3d^{10}\ 4s^2,\quad i.e.,\quad [Ar]\ 4s^2\ 3d^{10}$$

These are attributed to the fact that half filled and completely filled levels which include d^5, d^{10}, f^7, f^{14} have lower energies. Thus inorder to acquire more stability, one of the $4s$ electrons in Cu goes over to $3d$ orbital so that it gets half filled. The extra stability of half filled and completely filled orbitals are believed to be due to their symmetrical arrangements and large exchange energies.

First transition metal series are Sc (21), Ti (22), V (23), Cr (24), Mn (25), Fe (26), Co (27), Ni(28), Cu (29), Zn (30).

The electronic configuration of all elements in this group are

$$\begin{array}{lll}
\text{Sc} & 21 & [\text{Ar}] \ 4s^2 3d^1 \\
\text{Ti} & 22 & [\text{Ar}] \ 4s^2 3d^2 \\
\text{V} & 23 & [\text{Ar}] \ 4s^2 3d^3 \\
\text{Cr} & 24 & [\text{Ar}] \ 4s^1 3d^5 \\
\text{Mn} & 25 & [\text{Ar}] \ 4s^2 3d^5 \\
\text{Fe} & 26 & [\text{Ar}] \ 4s^2 3d^6 \\
\text{Co} & 27 & [\text{Ar}] \ 4s^2 3d^7 \\
\text{Ni} & 28 & [\text{Ar}] \ 4s^2 3d^8 \\
\text{Cu} & 29 & [\text{Ar}] \ 4s^1 3d^{10} \\
\text{Zn} & 30 & [\text{Ar}] \ 4s^2 3d^{10}
\end{array}$$

Note the periodic anomalies in the following transition metals:

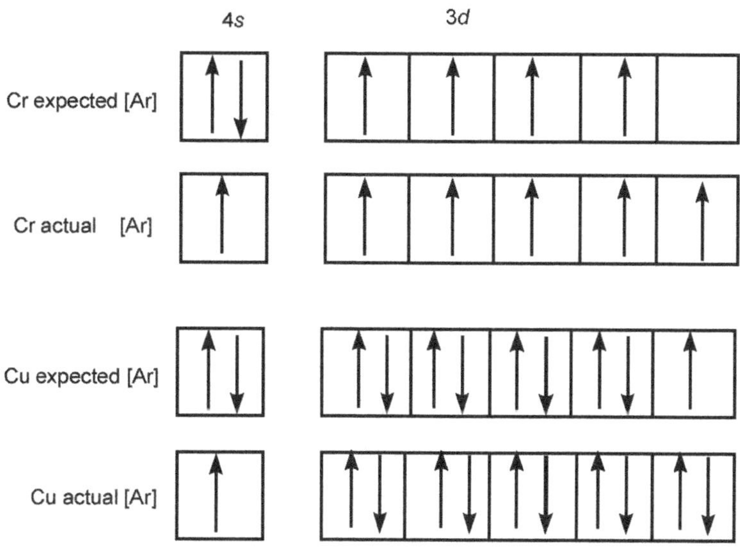

The next elements are Ga, Ge, As, Se, Br and Kr, the atomic numbers ranging from $Z = 31$ to $Z = 36$. They are built up by successive addition of 6 electrons in the $4p$ subshell and the electronic configuration of the last one in this long group, viz., of Kr is

$$\text{Kr}\quad 1s^2\ 2s^2\ 2p^6\ 3s^2\ 3p^6\ 3d^{10}\ 4s^2\ 4p^6,\ {}^1S_0$$

Gallium ($Z = 31$) has one $4p$ electron and is similar to aluminium. The atoms in this group have configuration

$$\text{Ga}\quad 1s^2\ 2s^2\ 2p^6\ 3s^2\ 3p^6\ 3d^{10}\ 4s^2\ 4p$$
$$\text{Ge}\quad 1s^2\ 2s^2\ 2p^6\ 3s^2\ 3p^6\ 3d^{10}\ 4s^2\ 4p^2$$
$$\text{As}\quad 1s^2\ 2s^2\ 2p^6\ 3s^2\ 3p^6\ 3d^{10}\ 4s^2\ 4p^3$$
$$\text{Se}\quad 1s^2\ 2s^2\ 2p^6\ 3s^2\ 3p^6\ 3d^{10}\ 4s^2\ 4p^4$$
$$\text{Br}\quad 1s^2\ 2s^2\ 2p^6\ 3s^2\ 3p^6\ 3d^{10}\ 4s^2\ 4p^5$$
$$\text{Kr}\quad 1s^2\ 2s^2\ 2p^6\ 3s^2\ 3p^6\ 3d^{10}\ 4s^2\ 4p^6$$

5.2.6   Fifth Period

It contains 18 elements starting with an alkali metal and ending with an inert gas. This is the second longest period and consists of Rb, Sr, Y, Zr, Nb, Mo, Tc, Ru, Rh, Pd, Ag, Cd, In, Sn, Sb, Te, I, Xe, atomic number Z ranging from 37 to 54. As Kr in the fourth group is filled with $4p^6$ electrons, one would expect that the next electron in the Rb should be filled in $4d$ or $4f$ orbital as they are yet to be occupied. However, Rubidium with $Z = 37$ instead of allowing the last electron to occupy the $4d$ orbit, occupies the $5s$ orbital. The next element strontium with $Z = 38$ completes the s subshell by filling it with the last electron, thereby the electronic configuration of the two elements rubidium and strontium are

$$\text{Rb}\quad 1s^2\ 2s^2\ 2p^6\ 3s^2\ 3p^6\ 3d^{10}\ 4s^2\ 4p^6\ 5s$$
$$\text{Sr}\quad 1s^2\ 2s^2\ 2p^6\ 3s^2\ 3p^6\ 3d^{10}\ 4s^2\ 4p^6\ 5s^2$$

In the following elements yttrium, zirconium, columbium (Niobium, Cb or Nb), molybdenum, technetrum, ruthenium, rhodium, palladium, silver and cadmium $4d$ orbitals are filled successively. However, spectroscopic observations have revealed certain aberrations; but Bohr–Stoner scheme is still valid. The relative energies are in the order $4s$, $3d$ and $4p$ and similarly $5s$, $4d$ and $5p$. The electronic configuration of yttrium and zirconium are

$$Z = 39\quad \text{Y}\quad 1s^2\ 2s^2\ 2p^6\ 3s^2\ 3p^6\ 3d^{10}\ 4s^2\ 4p^6\ 4d\ 5s^2$$
$$Z = 40\quad \text{Zr}\quad 1s^2\ 2s^2\ 2p^6\ 3s^2\ 3p^6\ 3d^{10}\ 4s^2\ 4p^6\ 4d^2\ 5s^2$$

The electronic configuration of Nb with $Z = 41$ is $1s^2\ 2s^2\ 2p^6\ 3s^2\ 3p^6\ 3d^{10}\ 4s^2\ 4p^6\ 4d^4\ 5s$, i.e., [Kr] $4d^4\ 5s$ and of M_o with $Z = 42$ is $1s^2\ 2s^2\ 2p^6\ 3s^2\ 3p^6\ 3d^{10}\ 4s^2\ 4p^6\ 4d^5\ 5s$, i.e., [Kr] $4d^5\ 5s$. The elements Palladium ($Z = 46$) the configuration is $1s^2\ 2s^2\ 2p^6\ 3s^2\ 3p^6\ 3d^{10}\ 4s^2\ 4p^6\ 4d^{10}$, i.e., [Kr]

$4d^{10}$, silver (Z = 47) with [Kr] $4d^{10}$ $5s$, cadmium (Z = 48) with the configuration is [Kr] $4d^{10}$ $5s^2$ are similar to nickel, copper and zinc respectively. The elements indium, tin, antimony, tellurium, iodine and xenon mark the filling of $5p$ electrons successively. Xenon (Z = 54) has configuration [Kr] $5s^2$ $5p^6$ the normal state being $^1S_0$.

Thus the fifth period contains [Kr] core second series of transition metals in which the $4d$ orbitals are filled. The $4d$ subshells are filled in a manner similar but not in identical manner to those of the $3d$ orbitals in the first transition series.

5.2.7 Sixth Period

This is the third long period. It starts with an alkali metal and ends with an inert gas. The period contains 32 elements including 14 rare earth elements. The first element cesium and the second element barium with Z = 55 and 56 mark the occupation of $6s$ subshell and are similar to rubidium and strontium.

Cs (Z = 55) $1s^2$ $2s^2$ $2p^6$ $3s^2$ $3p^6$ $3d^{10}$ $4s^2$ $4p^6$ $4d^{10}$ $5s^2$ $5p^6$ $6s$,   i.e., [Xe] 6s

Ba (Z = 56) $1s^2$ $2s^2$ $2p^6$ $3s^2$ $3p^6$ $3d^{10}$ $4s^2$ $4p^6$ $4d^{10}$ $5s^2$ $5p^6$ $6s^2$,   i.e., [Xe] 6s²

The $5d$ subshells are not occupied in the above two elements. In Lanthanum with Z = 57, the last electron occupies a $5d$ subshell the configuration being

La (Z = 57) $1s^2$ $2s^2$ $2p^6$ $3s^2$ $3p^6$ $3d^{10}$ $4s^2$ $4p^6$ $4d^{10}$ $5s^2$ $5p^6$ $5d$ $6s^2$,   i.e., [Xe] 5d 6s²

Also written as KLM $4s^2$ $4p^6$ $4d^{10}$ $5s^2$ $5p^6$ $5d$ $6s^2$.

From here onwards the energy of $4f$ orbital becomes comparable to the energy of the $5d$ orbital and the next 14 elements starting with Ce till Lu mark the occupation of $4f$ subshell successively. Thus the configuration of Ce with Z = 58 is $1s^2$ $2s^2$ $2p^6$ $3s^2$ $3p^6$ $3d^{10}$ $4s^2$ $4p^6$ $4d^{10}$ $4f$ $5s^2$ $5p^6$ $5d$ $6s^2$ and Lutetium (Z = 71) with $1s^2$ $2s^2$ $2p^6$ $3s^2$ $3p^6$ $3d^{10}$ $4s^2$ $4p^6$ $4d^{10}$ $5s^2$ $5p^6$ $6s^2$ $4f^{14}$ $5d^1$ $6s^2$ (KLM $4f^{14}$ $5d$ $6s^2$). These 14 elements are the rare earth elements. The last element completes the $4f$ shell thereby the N shell. After Lu, the next element is Hf with Z = 72 with the electronic configuration

Hafnium Hf (Z = 72) [Xe] $6s^2$ $4f^{14}$ $5d^2$ and from this element onwards addition in the $5d$ orbital is resumed until the subshell is completed in gold (in Latin Aurum – Au).

Platinum Pt (Z = 78) $1s^2$ $2s^2$ $2p^6$ $3s^2$ $3p^6$ $3d^{10}$ $4s^2$ $4p^6$ $4d^{10}$ $4f^{14}$ $5s^2$ $5p^6$ $5d^9$ $6s$

Gold Au (Z = 79) $1s^2$ $2s^2$ $2p^6$ $3s^2$ $3p^6$ $3d^{10}$ $4s^2$ $4p^6$ $4d^{10}$ $4f^{14}$ $5s^2$ $5p^6$ $5d^{10}$ $6s$

Gold Au (Z = 79) has one 6s electron and mercury Hg(Z = 80) has two 6s electrons. From thallium with Z = 81 till radon with Z = 86, electrons occupy the $6p$ shells successively. The electronic configuration of thallium and radon are

Tl (Z = 81) $1s^2$ $2s^2$ $2p^6$ $3s^2$ $3p^6$ $3d^{10}$ $4s^2$ $4p^6$ $4d^{10}$ $4f^{14}$ $5s^2$ $5p^6$ $5d^{10}$ $6s^2$ $6p^1$

Rn (Z = 86) $1s^2$ $2s^2$ $2p^6$ $3s^2$ $3p^6$ $3d^{10}$ $4s^2$ $4p^6$ $4d^{10}$ $4f^{14}$ $5s^2$ $5p^6$ $5d^{10}$ $6s^2$ $6p^6$

Note that $5f$ is not occupied till now. Radon completes the 6th period, which is the third long period.

The sixth period contains [Xe] core atoms and $6s$ is filled before $4f$. In the Lanthanide series there are 14 elements corresponding to filling $4f$ orbitals. Note that $4f$ are filled before $5d$.

5.2.8 Seventh Period

This is the last period and starts again with an alkali metal francium (cornellium) followed by radium, actinium, thorium, uranium. In the first two elements $7s$ is occupied. Radium with two electrons in the $7s$ subshell is an alkaline earth. It contains [Rn] core filled elements. These are called Actinide series with 14 elements corresponding to filling $5f$ orbitals. $5fs$ are filled before $6d$. Actinium has two $7s$ electrons and one $6d$ electron, but beyond this $5f$ orbitals are lower in energies than $6d$ orbitals and a new series of rare earth elements starts with thorium, protactinium, uranium and the radioactive transuranic elements. This is known as the actinide group and lawrencium with $Z = 103$ is the last member of this group. The spectroscopic studies on the 7th group are very meagre. The electronic configurations are given below.

Actinium Ac $(Z = 89)$ [Rn] $7s^2\ 6d^1$ but it is [Rn] $7s^2\ 5f^1$ (discovered 1899)

Thorium Th $(Z = 90)$ [Rn] $7s^2\ 6d^2$ but it is [Rn] $7s^2\ 5f^2$ (discovered 1829)

Proctatinium Pa $(Z = 91)$ [Rn] $7s^2\ 5f^2\ 6d^1$ but it is [Rn] $7s^2\ 5f^3$ (discovered 1913)

Uranium U $(Z = 92)$ [Rn] $7s^2\ 5f^3\ 6d^1$ but it is [Rn] $7s^2\ 5f^4$ (discovered 1789)

Neptunium Np $(Z = 93)$ [Rn] $7s^2\ 5f^4\ 6d^1$ but it is [Rn] $7s^2\ 5f^5$ (discovered 1940)

Plutonium Pu $(Z = 94)$ [Rn] $7s^2\ 5f^6$ (discovered 1940)

Americium Am $(Z = 95)$ [Rn] $7s^2\ 5f^7$ (discovered 1944)

Curium Cm $(Z = 96)$ [Rn] $7s^2\ 5f^7 6d^1$ (discovered 1944)

Berkelium Bk $(Z = 97)$ [Rn] $7s^2\ 5f^9$ (discovered 1949)

Californium Cf $(Z = 98)$ (discovered 1950)

Einsteinium Es $(Z = 99)$ (discovered 1952)

Fermium Fm $(Z = 100)$ (discovered 1952)

Mendelevium Md $(Z = 101)$ (discovered 1955)

Nobelium No $(Z = 102)$ (discovered 1958)

Lawrencium Lr $(Z = 103)$ [Rn] $7s^2\ 5f^{14}\ 7p^1$ instead of the expected [Rn] $7s^2\ 5f^{14}\ 6d$
(discovered 1961)

Rutherfordium Rf (104) (discovered 1964)

Dubnium Db $(Z = 105)$ (discovered 1967)

Seaborgium Sg $(Z = 106)$ (discovered 1974)

Bohrium Bh $\quad$ (Z = 107) (discovered 1981)

Hassium Hs $\quad$ (Z = 108) (discovered 1984)

Meitnerium Mt $\quad$ (Z = 109) (discovered 1982)

Figure 5.4 shows subshell filling in the periodic table.

Bohr–Stoner scheme could satisfactorily explain the shell structure in elements and could explain many observed features in the X-ray spectra of elements. The filling of the elements with larger Z should be viewed with the spectroscopic observations. The periodic table is as shown in Figure 5.3 where elements are arranged in groups with similar electronic configuration. The order in which various subshells are filled by electrons is determined by the following rules:

1. np electrons have higher energy than ns electrons but substantially lower than $(n + 1)s$ electrons.

2. nd electrons have almost or slightly higher than the energy of $(n + 1)s$ electrons.

3. nf electrons have almost the same energy as $(n + 1)d$ orbitals.

The wavefunction of an atom with N electrons in the self consistent field (SCF) theory can be written as

$$\psi(1,2,\ldots N) = \phi_1(1)\ \phi_2(2)\ldots\ \phi_N(N)$$

Here $\phi_i(N)$ represent the atomic orbitals associated with a given pair of n and l quantum numbers. The orbital energies are in the order of

$$1s < 2s = 2p < 3s = 3p = 3d < 4s = 4p = 4d = 4f = \ldots$$

However from spectroscopic evidences, this order of the energy levels is not maintained. For example the electronic configuration of Be($Z = 4$) is proved to be $1s^2\ 2s^2$ from observation of the ground state as $^1S_0$. This is because of the fact that the energy due to above configuration is below with the configuration $1s^2\ 2p^2$. Hence the order of the energy levels was corrected to be

$$1s < 2s < 2p < 3s < 3p < 3d < 4s < 4p < 4d < 4f$$

If this order is correct, the electronic configuration in potassium should be $1s^2\ 2s^2\ 2p^6\ 3s^2\ 3p^6 3d$ whereas experimental observation suggested that the configuration of electrons in potassium is $1s^2\ 2s^2\ 2p^6\ 3s^2\ 3p^6\ 4s$ (the normal state being $^2S_{1/2}$).

The present accepted order of the orbital energy is

$$1s < 2s < 2p < 3s < 3p < 4s < 3d < 4p < 5s < 4d < 5p < 6s < 4f < 5d < 6p < 7s$$

though there may be aberrations. It should be considered as an empirically arranged order.

With the above rules in mind we can fill the periodic table of elements. Till Ba there is no deviations from the above. However, Cr with $Z = 24$ has only one electron in the $4s$ shell and five $3d$ electrons thereby having 2 in complete shells. Note the configuration of V with $Z = 23$ and Mn with $Z = 25$.

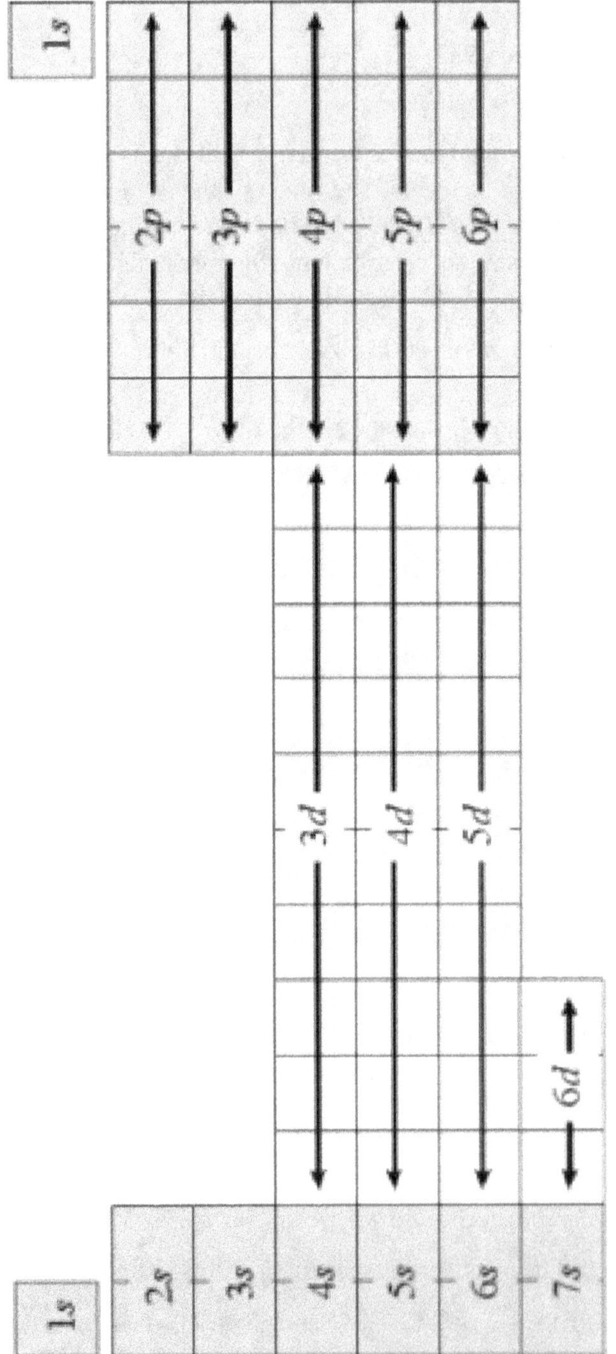

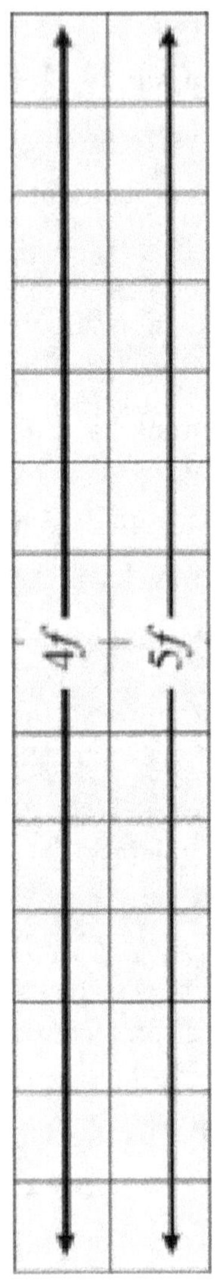

Figure 5.4   Subshell filling in the periodic table

$$\text{V } (Z=23) \; 1s^2 \; 2s^2 \; 2p^6 \; 3s^2 \; 3p^6 \; 3d^3 \; 4s^2 \; {}^4F_{3/2}$$

$$\text{Cr } (Z=24) \; 1s^2 \; 2s^2 \; 2p^6 \; 3s^2 \; 3p^6 \; 3d^5 \; 4s \; {}^4S_{5/2}$$

$$\text{Mn}(Z=25) \; 1s^2 \; 2s^2 \; 2p^6 \; 3s^2 \; 3p^6 \; 3d^5 \; 4s^2 \; {}^7S_3$$

In general only one incomplete shell is found in all elements.

The electronic configuration of Ba is

$$\text{Ba } (Z=56) \; 1s^2 \; 2s^2 \; 2p^6 \; 3s^2 \; 3p^6 \; 3d^{10} \; 4s^2 4p^6 4d^{10} \; 5s^2 \; 5p^6 \; 6s^2$$

In the next element La with $Z = 57$ the last electron goes to the $5d$ subshell whereas in the next 14 elements from Ce to Lu, the outermost electrons go into the $4f$ subshells after filling $6s^2$ and one $5d$ orbital.

$$\text{La } (Z=57) \; 1s^2 \; 2s^2 \; 2p^6 \; 3s^2 \; 3p^6 \; 3d^{10} \; 4s^2 4p^6 4d^{10} \; 5s^2 \; 5p^6 \; 5d \; 6s^2$$

$$\text{Ce } (Z=58) \; 1s^2 \; 2s^2 \; 2p^6 \; 3s^2 \; 3p^6 \; 3d^{10} \; 4s^2 4p^6 4d^{10} \; 4f \; 5s^2 \; 5p^6 \; 5d \; 6s^2$$

.

.

.

$$\text{Lu } (Z=71) \; 1s^2 \; 2s^2 \; 2p^6 \; 3s^2 \; 3p^6 \; 3d^{10} \; 4s^2 4p^6 4d^{10} \; 4f^{14} \; 5s^2 \; 5p^6 \; 5d \; 6s^2$$

Note the electronic configuration in Np, Pu, Am and Cm

$$\text{Np } (Z=93) \; 1s^2 \; 2s^2 \; 2p^6 \; 3s^2 \; 3p^6 \; 3d^{10} \; 4s^2 4p^6 4d^{10} \; 4f^{14} \; 5s^2 \; 5p^6 \; 5d^{10} 5f^2 \; 6s^2 \; 6p^5 \; 6d^5 \; 7s$$

$$\text{Pu } (Z=94) \; 1s^2 \; 2s^2 \; 2p^6 \; 3s^2 \; 3p^6 \; 3d^{10} \; 4s^2 4p^6 4d^{10} \; 4f^{14} \; 5s^2 \; 5p^6 \; 5d^{10} \; 5f^2 \; 6s^2 \; 6p^5 \; 6d^5 \; 7s^2$$

$$\text{Am } (Z=95) \; 1s^2 \; 2s^2 \; 2p^6 \; 3s^2 \; 3p^6 \; 3d^{10} \; 4s^2 4p^6 4d^{10} \; 4f^{14} \; 5s^2 \; 5p^6 \; 5d^{10} \; 5f^3 \; 6s^2 \; 6p^5 \; 6d^5 \; 7s^2$$

$$\text{Cm } (Z=96) \; 1s^2 \; 2s^2 \; 2p^6 \; 3s^2 \; 3p^6 \; 3d^{10} \; 4s^2 4p^6 4d^{10} \; 4f^{14} \; 5s^2 \; 5p^6 \; 5d^{10} \; 5f^7 \; 6s^2 \; 6p^6 \; 6d^1 \; 7s^2$$

The distribution of electrons in the K, L, M, N sheels are given in Table 5.3.

Table 5.3 Electron shells

n	l	m_l	Notation	No. of electrons	Shell
1	0	0	$1s$	2	K
2	0	0	$2s$	2	
2	1	$-1, 0, +1$	$2p$	6	L (8)
3	0	0	$3s$	2	
3	1	$-1, 0, +1$	$3p$	6	M (18)
3	2	$-2, -1, 0, +1, +2$	$3d$	10	
4	0	0	$4s$	2	
4	1	$-1, 0, +1$	$4p$	6	N (32)
4	2	$-2, -1, 0, +1, +2$	$4d$	10	
4	3	$-3, -2, -1, 0, +1, +2, +3$	$4f$	14	

5.3 ELECTRON SPIN

5.3.1 Electron Intrinsic Angular Momentum

Though Schrödinger equation could predict the energy levels of the hydrogen atom with reasonable accuracy, it could not explain certain observed phenomena. The fine structure of the atomic lines and the anomalous Zeeman Effect could not be explained by hitherto known theories. In order to overcome the difficulties in the explanation of the doublet fine structure in the alkali metal it was Uhlenbeck and Goudsmit[9] and independently by Bichowsky and Urey who introduced the concept of a new angular momentum assigned to the valence electron. According to them every electron, whether in an atom or a free electron, possesses an intrinsic angular momentum arising from the intrinsic spin of the electron. Its magnitude parallel or antiparallel to a given direction have the magnitude $\dfrac{1}{2}\dfrac{h}{2\pi}$. Along with these the Stern–Gerlach[10] experiment of 1922 revealed that a beam of silver atoms passing through an inhomogeneous magnetic field showed two deflected beams corresponding to $+\dfrac{1}{2}$ and $-\dfrac{1}{2}$ orientation of the spin angular momentum which is in conformity of an electron spin. Classically the electron can be considered like a spinning cricket ball, only difference in this case is that the spinning ball has a charge.

The z-component of the angular momentum

$$S_z = \pm\frac{1}{2}\hbar$$

It is evident that an angular momentum and a magnetic moment would arise from a spinning sphere of charge and the property of electron spin should be considered as a quantum concept. Thus a spin angular momentum quantum number $\frac{1}{2}$ was required for the electron in order to explain these phenomena. The quantum numbers connected with the electron spin follow the usual pattern, i.e.,

$$s = \sqrt{s(s+1)}\ \hbar, \quad m_s = \pm\frac{1}{2}, \quad s_z = m_s\hbar$$

The quantum mechanical treatment of the hydrogen atom spectra has revealed that atleast three quantum numbers are necessary to explain the spectrum of hydrogen atom. Magnetic quantum numbers are necessary to explain the spectrum of atoms in magnetic or electric fields. The hydrogen atom spectra appear to be explainable, as we have seen so far just by changes in the principal quantum numbers n. However, when the hydrogen atom spectra are examined by a spectrograph of high resolving power, splittings of the lines were observed. The first two lines of the Balmer series are expected to consist of five lines. Out of which two are strong components and the other three are weak. They are shown in Figure 5.5.

Thus shortly before quantum mechanics was developed Goudsmit and Uhlenbeck postulated that electron had an intrinsic angular momentum which they called as "spin angular momentum". It was also found that the electron spin has unusual characteristics and also there was no classical analog of spin. But there were experimental evidences supporting the electron spin hypothesis. Of course electron spin arises in the equation of motion of electrons treated relativistically, which is beyond the scope of this book.

Though the fine structure in the spectrum of hydrogen atom and the Stern–Gerlach experiment could be considered as the first experimental evidence to consider the existence of an electron spin, the experimental evidences for spin hypothesis could be summarized as the following observations:

1. Stern–Gerlach experiment
2. Fine structure splittings in atomic spectra
3. The anomalous Zeeman effect
4. The degeneracy of the excited states of atoms and molecules
5. Electron and magnetic resonance spectroscopy.

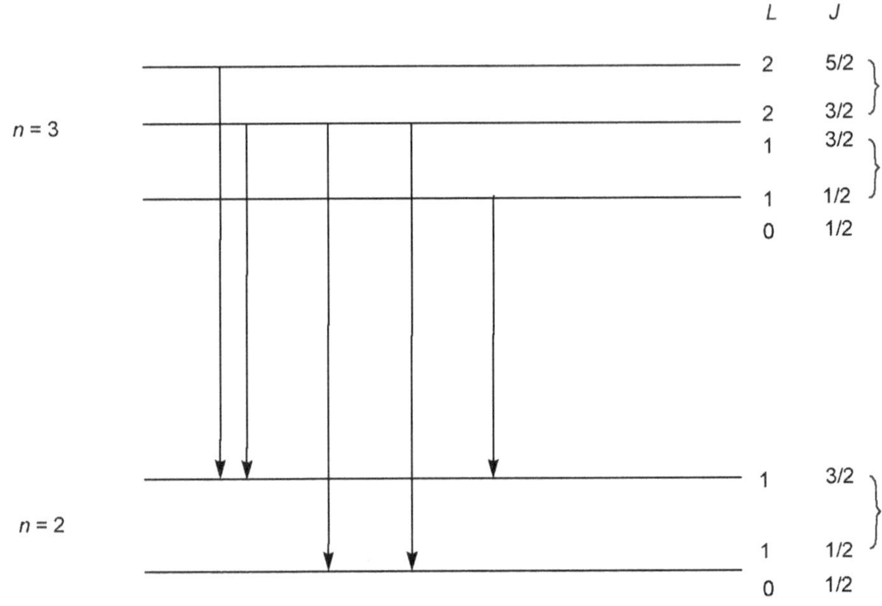

Figure 5.5   Balmer series

5.3.2 Electron Spin Magnetic Moment

We now know that electron is a charged particle and displays an intrinsic spin angular momentum in addition to its orbital angular momentum. Hence we must expect a magnetic moment, over and above the magnetic moment originating from the orbital motion. The z-component of the magnetic moment associated with the orbital motion is

$$\mu_z = m_l\, \mu_B$$

The z-component of magnetic moment associated with the electron spin is then expected to be

$$\mu_z = m_s\, \mu_B = \pm\mu_B\,\frac{1}{2}$$

but the measured value turns out to be about twice that. The measured value is written as

$$\mu_z = \pm\frac{1}{2}g\mu_B$$

where g is called the gyromagnetic ratio or the electron spin g-factor. g has the value of 2, more exactly $g = 2.00232$. For the orbital angular momentum $g = 1$. It means that $\mu_z = \pm m_s g\mu_B = \pm\frac{1}{2}\times 2.00232\mu_B$ for spin motion and and $\mu_z = \pm m_l\, g\,\mu_B = \pm m_l \times 1 \times \mu_B$. The Dirac[11] equation in the relativistic quantum mechanics predicts the precise value of g and was measured in the Lamb shift measurements. The magnetic moment is usually expressed in units of Bohr magneton.

$$\mu_B = \frac{eh}{2m_e} = \frac{eh}{4\pi m_e} = 9.2740154 \times 10^{-24} \ \text{J/T} = 5.7883826 \times 10^{-5} \ \text{eV/T} \ \ \text{Bohr magneton}$$

$$= 9.2740 \times 10^{-21} \ \text{eg} \ G^{-1}$$

The electron spin magnetic moment is an important factor in determining the spin–orbit interaction which determines the fine structure in the spectra of atoms. The energy levels are split by spin–orbit interaction giving rise to fine structure spectra. The electron spin magnetic moment is also a factor in dealing with interaction of atoms with external magnetic fields (Zeeman effect).

According to classical theory, the ratio between the magnetic moment μ and the mechanical moment p due to the orbital motion of the electron is

$$\frac{\mu}{p} = \frac{e}{2mc} \quad \text{or it can be written as} \quad \frac{\mu_l}{p_l} = \frac{e}{2mc}$$

And it has been noticed by observing anomalous Zeeman effect that the corresponding ratio in the case of spinning electron is

$$\frac{\mu_s}{p_s} = 2\frac{e}{2mc}$$

We cannot follow the usual method of writing down the classical expressions and converting them to quantum mechanical operators as there is no classical analog of electron spin. Hence in order to calculate quantum mechanical operators, commutation rules, etc., a series of postulates were involved which gave right answers to the experiments listed above.

Postulate I   The operators for spin angular momentum commute and combine in the same way as in the case of ordinary angular momentum. The operators for electron spin are exactly analogous to the operators of the orbital motion. Thus $\hat{S}^2$, $\hat{S}_x, \hat{S}_y, \hat{S}_+, \hat{S}_-$ are analogous to $\hat{L}^2, \hat{L}_x, \hat{L}_y, \hat{L}_z, \hat{L}_+, \hat{L}_-$.

Postulate II   For a single electron there are only two simultaneous eigen functions of $\hat{S}^2$ and $\hat{S}_z$. These are called α and β and follow the eigenvalue equation. These depend on the orientation of the electron spin either in the $+\frac{1}{2}$ and $-\frac{1}{2}$ orientation.

$$\hat{S}_z\alpha = \frac{1}{2}\hbar\alpha, \qquad\qquad \hat{S}^2\alpha = \frac{1}{2}\left(\frac{1}{2}+1\right)\hbar^2\alpha$$

$$\hat{S}_z\beta = -\frac{1}{2}\hbar\beta \qquad\qquad \hat{S}^2\beta = \frac{1}{2}\left(\frac{1}{2}+1\right)\hbar^2\beta$$

where α and β are the eigen functions of $\hat{S}_z$ having the eigenvalues $+\frac{1}{2}\hbar$ and $-\frac{1}{2}\hbar$ or in the a.u. of $\hbar$, $m_S = +\frac{1}{2}$ or $-\frac{1}{2}$ respectively. The functions α and β are considered to be normalized, i.e., integrals of α^2 and β^2 over all spin space is 1. They are also orthogonal as postulate I. The eigenvalue of S^2 is $\frac{1}{2}\left(\frac{1}{2}+1\right)\hbar^2$ or $S = \sqrt{\frac{1}{2}\left(\frac{1}{2}+1\right)}\ \hbar$.

Postulate III   As the electron has a charge the spinning electron acts like a magnetic dipole or magnet, the magnetic dipole moment of which is

$$\hat{\mu} = -g_0\beta_m\hat{S}$$

where g_0 and β_m are found to be the spectroscopic splitting factor and the Bohr magneton respectively. As mentioned above the value of g_0 is 2.00232 and that of β_m is 9.2740×10^{-21} eg G^{-1}. The minus sign in the above expression indicates that the direction of the dipole moment vector is antiparallel to the spin vector.

Though the value of g_0 was first obtained as an empirical constant, Dirac in 1928 solved the relativistic wave equations for the electron and predicted a magnetic moment for the electron originating from the spin and the value of g_0 was derived thus from relativistic concept.

It can be seen that

$$\hat{\mu}_z = -g_0\beta_m\hat{S}_z$$

The electron spin operators affect only spin coordinates and hence they commute with all operators that are functions only of space coordinates. Thus the operators $\hat{S}^2$ and $\hat{S}_z$ commute with the operators $\hat{H}$, $\hat{L}^2$ and $\hat{L}_z$ as long as $\hat{H}$ contains no spin terms. Atomic wavefunctions can thus be chosen as simultaneous wave functions of all five of these operators, viz., $\hat{S}^2, \hat{S}_z, \hat{H}, \hat{L}^2$ and $\hat{L}_z$. This is equivalent of characterizing atomic wavefunctions by quantum numbers n, l, m and $\hat{S}_z$ or we can think of electron moving in a four dimensional space which are three space coordinates and one spin coordinate.

It should be noticed that the half integral spin of electron is not an angular momentum which takes different values as in the case of n or l, but is an intrinsic property of the electron having the quantum number value $\frac{1}{2}$. The total angular momentum of an atom by an electron is in principle due to the motion of the centre of mass of the electron around the nucleus in an orbit (orbital motion) and the spin of the electron about an axis through the centre of mass of the electron. To obtain the total angular momentum of the atom due to the valence electron we have to combine the orbital angular momentum l and the electron spin momentum s vectorially. Thus the spin angular momentum $s\frac{h}{2\pi}$ is added vectorially to the orbital angular momentum l. $\frac{h}{2\pi}$ to form the resultant $j\frac{h}{2\pi}$. $j = l \pm s$. In another model the spin angular momentum $s * \cdot \frac{h}{2\pi}$ is added vectorially to the orbital angular momentum

$l * \cdot \frac{h}{2\pi}$ to obtain the resultant $j * \cdot \frac{h}{2\pi}$. In this case $s* = \sqrt{s(s+1)}$, $l* = \sqrt{l(l+1)}$, $j* = \sqrt{j(j+1)}$ and $j = l \pm s$. The second model is generally preferred, as it gives quantum mechanical results, though not always. J is called the total quantum number.

The maximum number of j values which can be obtained is 2 for spin 1/2 for states with $l \neq 0$ and for state with $l = 0$ (s electrons), there is only one possibility.

Thus in short the electron spin $s = 1/2$ is an intrinsic property of electrons. The spin angular momentum of electron is characterized by quantum number 1/2 which is an intrinsic angular momentum. This gives total angular momentum which is in similar pattern with other quantized angular momenta.

$$S = \sqrt{\frac{1}{2}\left(\frac{1}{2}+1\right)} \, \hbar = \frac{\sqrt{3}}{2}\hbar$$

The fine structure in the hydrogen spectra originates from the two orientation possibilities for the z-component of the angular momentum, viz.,

$$S_z = \pm\frac{1}{2}\hbar$$

This causes an energy splitting because of the magnetic moment of the electron. The electron spin orientations are shown in Figure 5.6.

From expression $\frac{\mu}{p} = g\frac{e}{2m}$ (for orbital motion $g = 1$ and for spin motion $g \approx 2$).

$$\mu_s = -\frac{e}{2m}gS$$

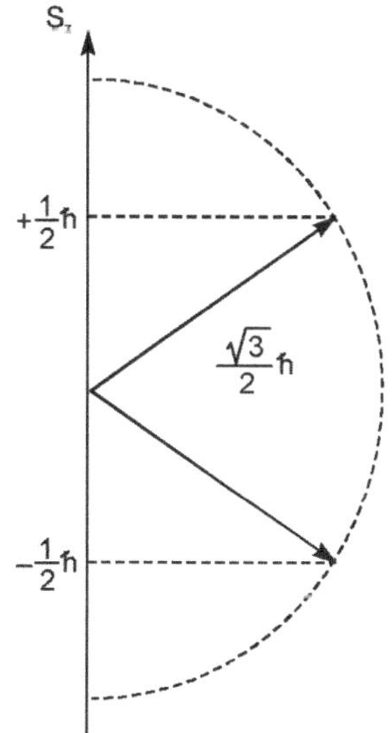

Figure 5.6   The spin angular momentum of an electron

Spin "up" and "down" allows two electrons for each set of spatial quantum numbers n, l, m_l.

Thus a spin angular momentum quantum number $\frac{1}{2}$ was required for the electron in order to explain the phenomena like the fine structure, Stern–Gerlach experiment, Zeeman effect and so on. The quantum numbers connected with the electron spin follow the usual pattern, i.e.,

$$s = \sqrt{s(s+1)} \; \hbar, \quad ms = \pm\frac{1}{2}, \quad s_z = m_s \, \hbar$$

5.4 STERN–GERLACH EXPERIMENT

It was in 1921 that Otto Stern and Walter Gerlach[12] performed experiments on silver atoms and showed the quantization of electron spin into two orientations[13]. These experimental results have made a breakthrough and had made a great contribution to the development of the quantum theory of the atoms. The Stern–Gerlach experiment was performed in 1921 at the Institute of Experimental Physics, Frankfurt Univeristy. The important results which gave a break through in atomic physics was published in 1922.

The experiment was carried out with a beam of silver atoms. Silver atomic beams were produced in a hot oven and allowed to pass through an inhomogeneous magnetic field in a direction perpendicular to the magnetic field. In such an experiment a magnetic dipole will experience a torque while passing through the nonuniform magnetic field. The force is proportional to the field gradient since the two "poles" will be subject to different fields. Classically one would expect all possible orientations of the dipoles so that a continuous smear would be produced on the photographic plate, but in their experiments Stern and Gerlach found that the magnetic field separated the beam into two distinct parts, which indicated that there were only two possible orientations of the magnetic moment of the electron.

We already know that silver with 47 electrons has an electronic configuration

$$\text{Ag} \quad 1s^2 \; 2s^2 \; 2p^6 \; 3s^2 \; 3p^6 \; 3d^{10} \; 4s^2 \; 4p^6 \; 4d^{10} \; 5s$$

The silver has a single electron in the outermost shell ($5s$ electron) which moves in the Coulomb potential caused by the 47 protons of the nucleus shielded by the 46 inner electrons. The s electron has zero orbital angular momentum ($l = 0$), and hence one would expect no interaction with the external magnetic field.

It has become a puzzle how does the electron obtain a magnetic moment if it has zero angular momentum. Because of zero magnetic dipole it can produces no "current loop" to produce a magnetic moment. It was in 1925 that Samuel A. Goudsmit and George E. Uhlenbeck[14] explained by postulating that the electron had an intrinsic angular momentum, independent of its orbital characteristics[15]. In a classical understanding, a spinning ball of charge could have a magnetic moment if it were spinning such that the charge at the edges produced an effective current loop. This kind of reasoning led to the use of "electron spin" to describe the intrinsic angular momentum and the two spots in the Stern–Gerlach experiment. Thus the magnetic moment of a silver atom is that of the spin of an electron.

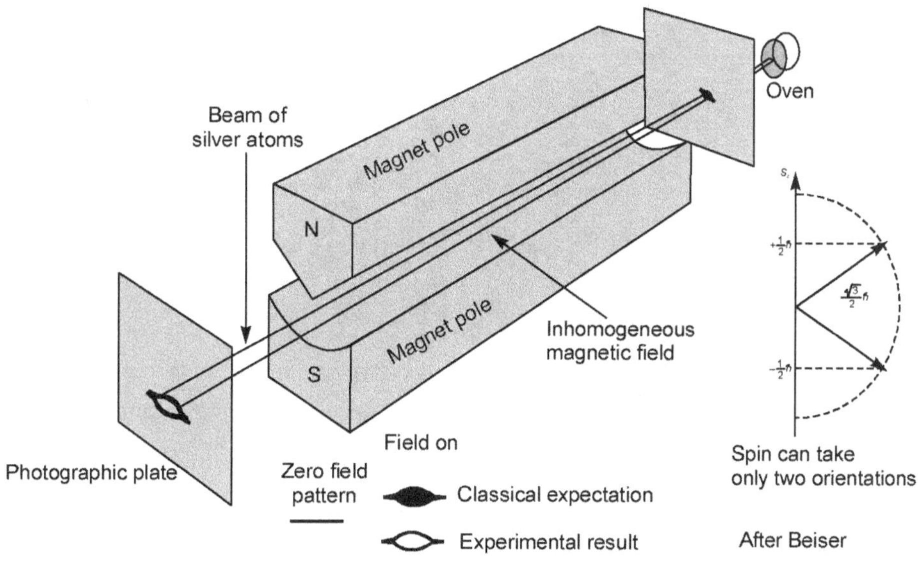

Figure 5.7   Stern–Gerlach experiment

Stern–Gerlach experiment confirmed the quantization of electron spin into two orientations in a nonuniform magnetic field. This has become a major contribution to the development of the quantum theory of the atom. The experimental set-up of stern and Gerlach is shown in Figure 5.7 and the deflection of beam in a nonhomogenius magnetic field is shown in Figure 5.8.

If we apply a magnetic field in the z direction, the potential energy of the electron spin magnetic moment is given by

$$U = -\mu \cdot B = -\mu_B \frac{g}{2} B_z = \pm \mu_B B_z$$

where g is the electron spin g-factor and μ_B is the Bohr magneton.

We can use the relationship between force and potential energy and gives

$$F_z = -\frac{\partial U}{\partial z} = \pm \mu_B \frac{\partial B_z}{\partial z}$$

The deflection can be shown to be proportional to the spin and to the magnitude of the magnetic field gradient.

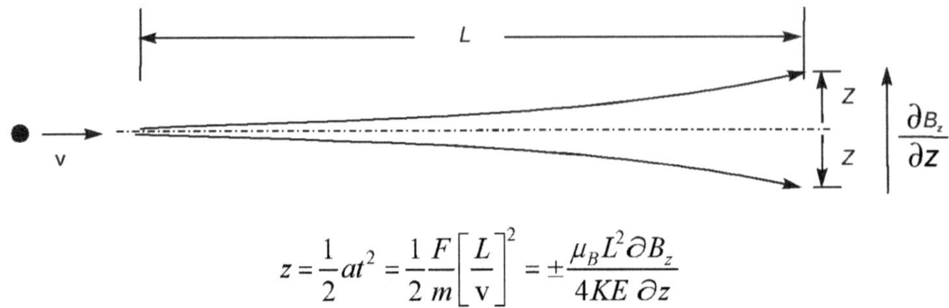

$$z = \frac{1}{2}at^2 = \frac{1}{2}\frac{F}{m}\left[\frac{L}{\mathrm{v}}\right]^2 = \pm\frac{\mu_B L^2}{4KE}\frac{\partial B_z}{\partial z}$$

Figure 5.8   Electron deflection in a magnetic field

5.5 HYDROGEN FINE STRUCTURE

The basic structure of the hydrogen energy levels can be calculated from the Schrödinger equation. The energy levels agree with the earlier Bohr model, and agree with experiment within a small fraction of an electron volt.

However, when the spectral lines of the hydrogen atom are examined at very high resolution, they are found to be closely spaced doublets. This splitting is called fine structure and was one of the first experimental evidences for electron spin. Splitting in the energy levels evidently causes the splitting in the spectrum. This is attributed to an interaction between the electron spin S and the orbital angular momentum L. It is called the spin–orbit interaction, which we will discuss in detail in later chapters. Because of this interaction the $2p$ level is split into a pair of lines. The $2s$ and $2p$ states are found to differ a small amount due to Lamb shift. Note that the interaction of the electron spin and orbital motion is called fine structure and the interaction of electron spin and nuclear spin is called hyperfine structure. The fine structure splittings and the transitions are shown in Figures 5.9 and 5.10.

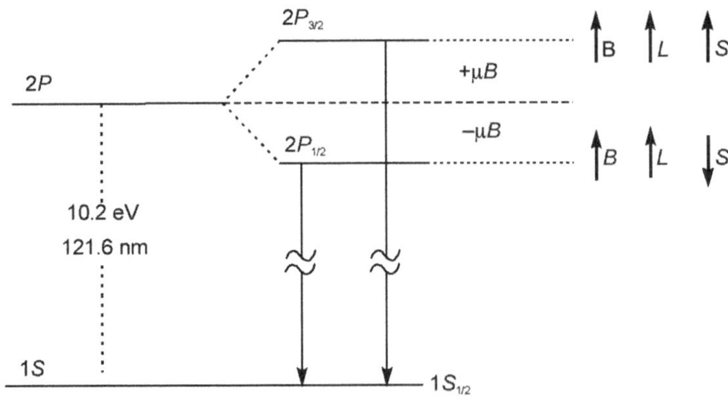

Figure 5.9   Hydrogen fine structure

We have seen that Bohr theory as well as the solution of Schrödinger equation predict the red H-alpha line of hydrogen as a single line. The Bohr theory predicts a transition wavelength 656.11 nm for H-alpha line treating the nucleus as a fixed centre. If we use the reduced mass, we get 656.47 nm for hydrogen and 656.29 nm for the corresponding transition in deuterium. Thus the difference between the hydrogen and deuterium lines is about 0.18 nm and the splitting of each of them is about 0.016 nm, corresponding to an energy difference of about 0.000045 eV. This corresponds to an internal magnetic field on the electron of about 0.4 Tesla.

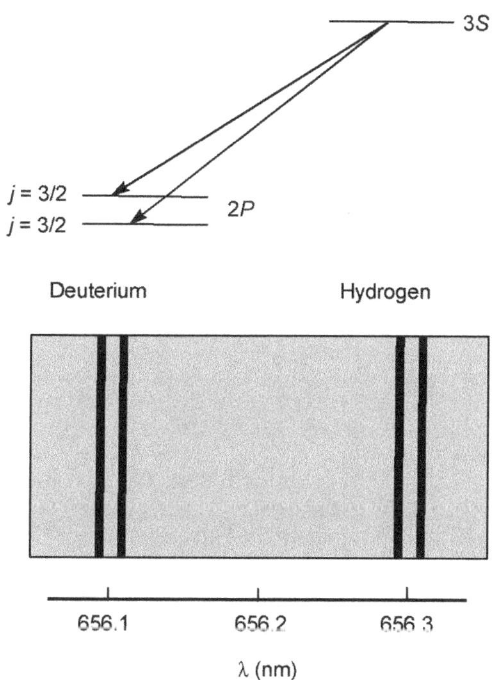

Figure 5.10   Red H-alpha line of hydrogen and deuterium

The hydrogen atom is a one electron system. We have seen that the electronic configuration of hydrogen atom is $1s1$. In this system $L = l$, $S = s = 1/2$ and $J = L \pm S$. Thus we have an S state as $l = 0$ and the multiplicity is (2 × electron spin $s + 1$) and $j = 1/2$. If we consider the excited P state where $l = 1$ we will have $j = 1/2$ or 3/2 thereby giving two states $^2P_{1/2}$ and $^2P_{3/2}$. All quantum states are doublets except the ground state. We have already seen that the H_α line, the first member of the Balmer series is due to a transition between $n = 3$ to $n = 2$. The upper state with $n = 3$ can have $L = l = 0,1,2$. The states are thus S, P and D states. The possible values of J are

For the S state $J = L + S = 1/2$ and thus the state is termed $^2S_{1/2}$.

For the P state $J = L + S ... |L - S| = 3/2$, and 1/2 and the state is termed $^2P_{1/2}$ and $^2P_{3/2}$.

And for the D state $J = 2 + \frac{1}{2}$ and $2 - \frac{1}{2}$, i.e., 5/2 and 3/2 and the states are termed $^2D_{5/2}$ and $^2D_{3/2}$.

The lower state with $n = 2$ has $L = l = 0$ and 1 and thus an S state and P state. The J value for the S state is $J = 0 + \frac{1}{2}$ or $\left|0 - \frac{1}{2}\right| = \frac{1}{2}$ thereby giving the state $^2S_{1/2}$.

The P state has $J = 3/2$ and $1/2$ as derived above thereby having two states $^2P_{1/2}$ and $^2P_{3/2}$. Thus we have eight states. The transitions are possible following the selection rules. The selection rules are $\Delta L = \pm 1$ and $\Delta J = 0$ or ± 1. Thus the permitted transitions are seven and they are

$$^2D_{5/2} \rightarrow {}^2P_{3/2}$$
$$^2D_{3/2} \rightarrow {}^2P_{3/2}$$
$$^2D_{3/2} \rightarrow {}^2P_{1/2}$$
$$^2P_{3/2} \rightarrow {}^2S_{1/2}$$
$$^2P_{1/2} \rightarrow {}^2S_{1/2}$$
$$^2S_{1/2} \rightarrow {}^2P_{3/2}$$
$$^2S_{1/2} \rightarrow {}^2P_{1/2}$$

$^2D_{3/2} \rightarrow {}^2P_{1/2}$ and $^2P_{3/2} \rightarrow {}^2S_{1/2}$ are coincident so also $^2P_{1/2} \rightarrow {}^2S_{1/2}$ and $^2S_{1/2} \rightarrow {}^2P_{1/2}$.

Taking into consideration of the above facts the fine structure of the H_α line has five components.

The strongest components are $^2D_{5/2} \rightarrow {}^2P_{3/2}$ and the combined $^2D_{3/2} \rightarrow {}^2P_{1/2}$ and $^2P_{3/2} \rightarrow {}^2S_{1/2}$. Note that they are of the type $\Delta L = +1$ and $\Delta J = +1$.

$^2D_{3/2} \rightarrow {}^2P_{3/2}$ and $^2P_{1/2} \rightarrow {}^2S_{1/2}$ are less intense. Note that in this case $\Delta L = +1$ and $\Delta J = 0$.

$^2S_{1/2} \rightarrow {}^2P_{3/2}$ still weak $(\Delta L = -1 \text{ and } \Delta J = -1)$. The weakest line is $^2S_{1/2} \rightarrow {}^2P_{1/2}$ $(\Delta L = -1 \text{ and } \Delta J = 0)$.

The energy level diagram and the transitions are shown in Figure 5.11.

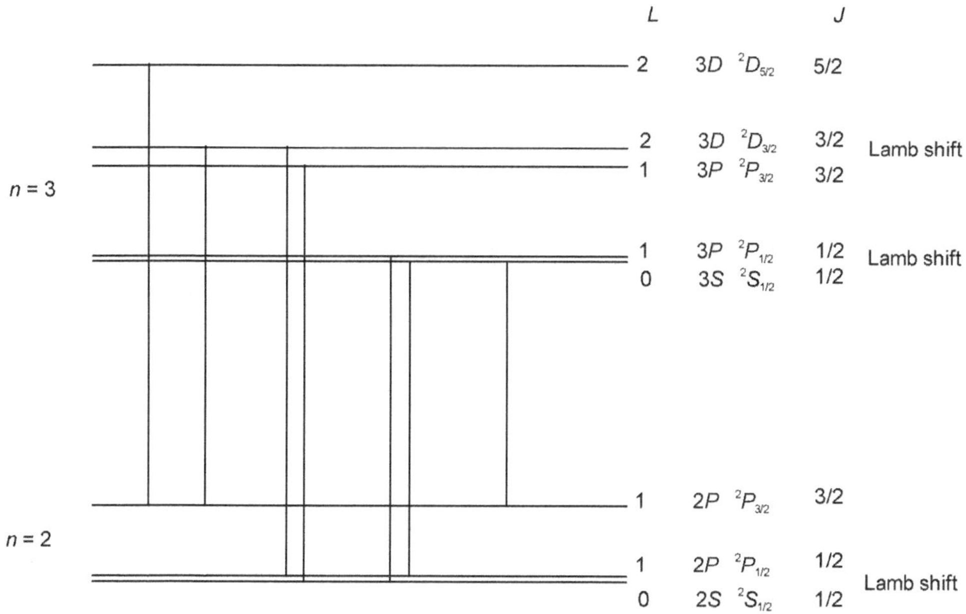

Figure 5.11   The energy level diagram of the hydrogen atom and the transitions

5.6 PAULI'S EXCLUSION PRINCIPLE

The principle was formulated by Pauli even before the electron spin was postulated by Goudmit and Uhlenbeck in the year 1925. Pauli in 1925 put forward the principle, which came to play an important role in the shell structure of atoms and the development of complex spectra. In considering the quantum state of an electron, the directional degeneration must be removed and this is achieved considering the orientation of electrons in an external magnetic field. The electrons orient in the direction of the magnetic field. The individual electrons are then qunatized by four quantum numbers n, l, ml and m_s.

n is the principal quantum number $n = 1,2\ldots$

l is the orbital quantum number $l = 0.1.2.3\ldots(n-1)$

m_l is the magnetic orbital quantum number $m_l = -l, (-l+1)\ldots -l$ (i.e., m_l takes values from $-l$ to $+l$).

m_s is the spin magnetic quantum number $m_s = +1/2$ or $-1/2$.

The simplest form of Pauli's exculsion principle is as follows:

No two electrons in the same state can have all their quantum numbers the same.

This is the well known Pauli's exclusion principle.

For a given value of l the quantum number m_l can have $(2l + 1)$ values from $-l$ to $+l$. Corresponding to these we can attribute two m_s values giving rise to $2(2l + 1)$ quantum states.

We can now explain the ground state of helium atom and those of alkaline earth atoms on the basis of Pauli's exclusion principle. There are two equivalent electrons in these atoms. If a triplet state is to be expected, the total spin should be 1 (total spin $S = 1$). Remember splitting of level is determined by $2S + 1$ where S is the total spin. In order to have a triplet state as per the above law, the total spin should be 1. To obtain a total spin of 1, the individual spins of the two electrons should be in the same direction $(\uparrow\uparrow)$, which will give a total spin $1/2 + 1/2$. This requires that all 4 quantum numbers of the two electrons to be identical. This is against Pauli's exclusion principle. If the electron spins are in the opposite direction $(\uparrow\downarrow)$ it obeys Pauli's exclusion principle. Individual electron spins in opposite spins in opposite direction means the total spin $S = 1/2 - 1/2 = 0$ which give a multiplicity of $2S + 1 = 1$ (singlet). Thus in the helium atom only the singlet state is allowed as the normal state. This is the case with other alkali metal atoms as well.

For two equivalent p electrons

$$m_l = 1 \quad 0 \quad -1 \quad\quad 1 \quad\quad 0 \quad\quad -1$$

$$m_s = \frac{1}{2} \quad \frac{1}{2} \quad \frac{1}{2} \quad -\frac{1}{2} \quad -\frac{1}{2} \quad -\frac{1}{2}$$

Equivalent electrons means n and l are the same for the electrons. Hence in order to follow the Pauli's exclusion principle we have to consider only m_l and m_s.

$$m_l = \;\;\;1 \;\;\text{ and } m_s = +1/2 \;\text{ and } -1/2 \text{ is possible}$$
$$m_l = \;\;\;0 \;\;\text{ and } m_s = +1/2 \;\text{ and } -1/2 \text{ is possible}$$
$$m_l = -1 \;\;\text{ and } m_s = +1/2 \;\text{ and } -1/2 \text{ is possible}$$

If $m_l = 1$ and $m_s = +1/2$ and $+1/2$ it violates the Pauli's exclusion principle as n and l are already the same and thus all 4 quantum numbers are the same which is not possible. Similar is the argument with $m_s = -1/2$ and $-1/2$. For non-equivalent electrons the case is different.

5.7   QUANTIZED ANGULAR MOMENTUM

We have already seen in the solution of the Schrödinger equation for hydrogen atom that the orbital angular momentum is quantized and the relationship derived was

$$l^2 = l(l+1)\hbar^2, \qquad l = \text{angular momentum quantum number}$$

The magnitude of the angular momentum in terms of the orbital quantum number is

$$l = \sqrt{l(l+1)}\;\hbar \qquad \begin{array}{l} l = \text{angular momentum quantum number} \\ l = 0,1,2, \;\dots\; n-1 \end{array}$$

In some books a definition is given for l^* as $l^* = \sqrt{l(l+1)}$

Remember that Bohr proposed the quantization of angular momentum in the form

$$L = mvr = \frac{nh}{2\pi}$$

The z-component of the angular momentum in terms of the magnetic quantum number takes the form

$$L_z = m_z \hbar$$

The above general form applies to orbital angular momentum, spin angular momentum and also to the total angular momentum for the atomic or molecular systems.

Thus quantum numbers associated with electron spin give the following relation:

$$s = \sqrt{s(s+1)}\,\hbar,\; s = \frac{1}{2},\; m_s = \pm\frac{1}{2}$$

The vector model gives a relationship between the magnitude of the angular momentum and its projection along any direction is space.

5.8 TOTAL ANGULAR MOMENTUM

When the orbital angular momentum and the spin angular momentum are coupled the total angular momentum or quantized angular momentum is obtained. The quantized angular momentum has the relation

$$J = \sqrt{j(j+1)}\,\hbar,\; j = \text{total angular momentum quantum number}$$

where the total angular momentum quantum number is

$$j = l \pm s = l \pm \frac{1}{2}$$

The z-component of angular momentum

$$J_z = m_j \hbar,\; m_j = -j,\, -j+1,\, -j+2,\, \ldots j-1,\, j$$

In a magnetic fiel4 m_j levels are split. The above kind of coupling gives an even number of angular momentum levels, which is consistent with the multiplets seen in anomalous Zeeman effects in atoms such as sodium.

As long as external interactions are not extremely strong, the total angular momentum of an electron can be considered to be conserved and j is said to be a "good quantum number". This quantum number is used to characterize the splitting of atomic energy levels, such as the spin–orbit splitting which leads to the sodium doublet.

The spectroscopic notation used for characterizing energy levels of atomic electrons is based upon the orbital quantum number.

5.9 HYDROGEN ENERGY LEVELS

We have seen that the energy levels of the hydrogen atom can be calculated by Bohr atomic model and also by solving the Schrödinger equation. Both of them agree with experimental results within a small fraction of an electron volt.

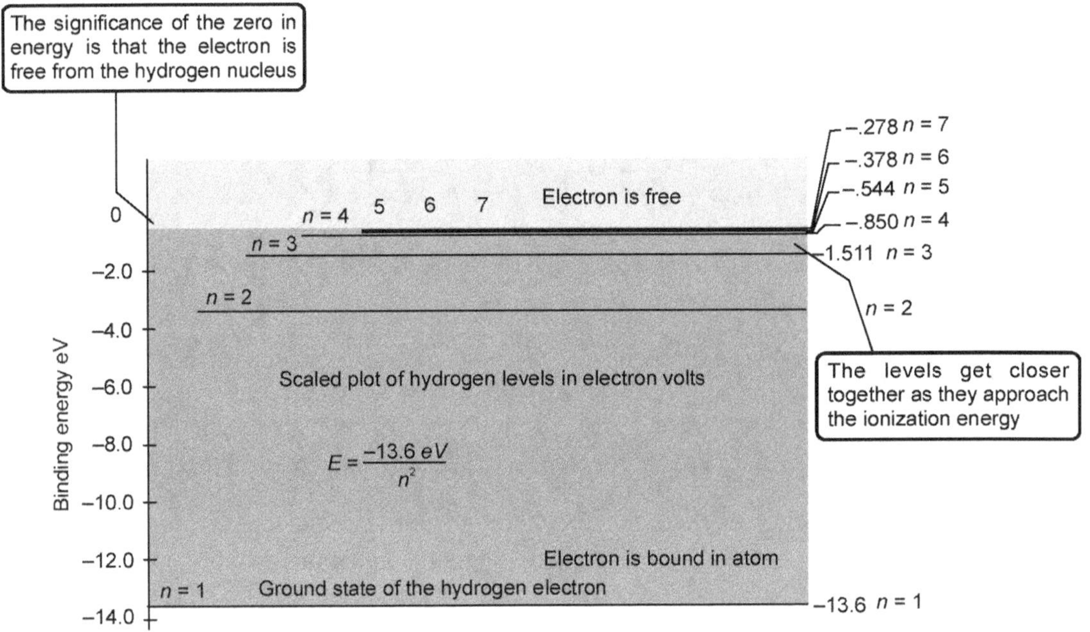

Figure 5.12 Hydrogen energy level plot

The energy levels of hydrogen are again given in Figure 5.12 for immediate reference.

However, studying the spectra at high resolution we have also found evidence of some other small effects on the energy. The ^2P level split into two components, viz., $^2P_{1/2}$ and $^2P_{3/2}$ due to spin–orbit interaction and leading to fine structure in the spectrum. The $2S$ state has only one level even with spin orbit interaction and is named $2^2S_{1/2}$. It has been seen that the levels $(n = 2)$ $^2P_{1/2}$ and $(n = 2)$ $^2S_{1/2}$ have the same energy. However, later experiments have shown that the energies of $2p$ $^2P_{1/2}$ and $2s$ $^2S_{1/2}$ and states differ a small amount in what is called the Lamb shift. We do not here deal with the hyperfine structure for the time being which is caused by the interaction of electron spin and nuclear spin in what is called hyperfine structure. Even the ground $1s$ state is split by hyperfine interaction. The Lamb shift is explained in the following section.

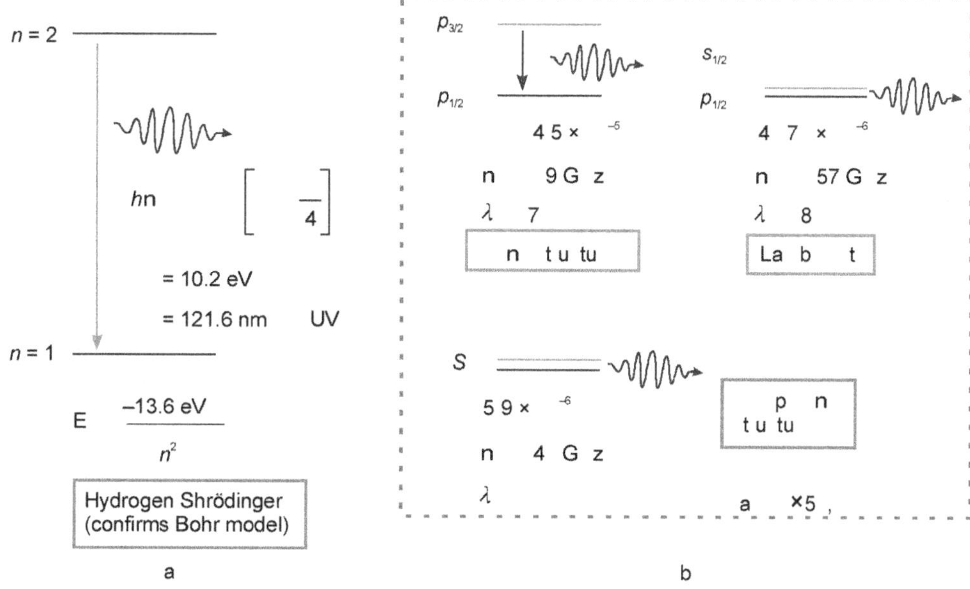

Figure 5.13   (a) Low resolution spectrum   (b) Types of splitting of levels

Note that there are three types of splitting of levels in Figure 5.13 (b). The fine structure originates from spin–orbit interaction (electron spin and orbital motion of electron), the hyperfine structure originates from the nuclear spin–orbital motion interaction (in the above case it is due to the two orientations of the proton spin and the electron spin) and the Lamb shift is due to the interaction of electrons and electromagnetic radiation fluctuation. Figure 5.13(a) gives a low resolution spectrum of $n = 2 \rightarrow n = 1$ transition and Figure 5.13(b)fine structure, hyperfine structure and Lamb shift in $n = 2 - 1$ transition.

5.10 LAMB SHIFT

We have already seen in Chapter 2 that the theories of Bohr, Sommerfeld and Schrödinger could give a reasonably good explanation for the hydrogen spectra. But the fine structure in the spectra could not be adequately explained and a qualitative explanation came with the introduction of electron spin. The inadequacy of Sommerfeld theory was rectified by taking into account

1.   The relativistic dependence of electron mass on velocity

2.   The electron spin.

Both these effects have been taken into account in Dirac's theory. The energy levels obtained from Dirac's theory in wave number units can be expressed by

$$E_{n,l,j} = -\frac{RZ^2}{n^2}\left[1 + \frac{Z^2\alpha^2}{n^2}\left(\frac{n}{\left(j+\dfrac{1}{2}\right)} - \frac{3}{4}\right)\right]$$

where n is the principal quantum number, l is orbital quantum number and j is the total quantum number. The expression derived by Dirac is the same as that derived by Sommerfeld (equation 2.138) except that k in Sommerfeld expression is replaced by $(j + 1/2)$. The energy level and transitions are shown in Figure 5.14. The H_α line of Balmer series $(n = 3 \rightarrow n = 2)$ should consist of seven fine structure components according to the selection rule $\Delta l = \pm 1$, $\Delta j = 0, \pm 1$. However, the transitions ν_4 and ν_5 so also ν_6 and ν_7 coincide and only five lines are observed. In a similar manner the L_α line in Lyman series consists of two lines, viz., $2\,^2P_{1/2} - 1\,^2S_{1/2}$ and $2\,^2P_{3/2} - 1\,^2S_{1/2}$.

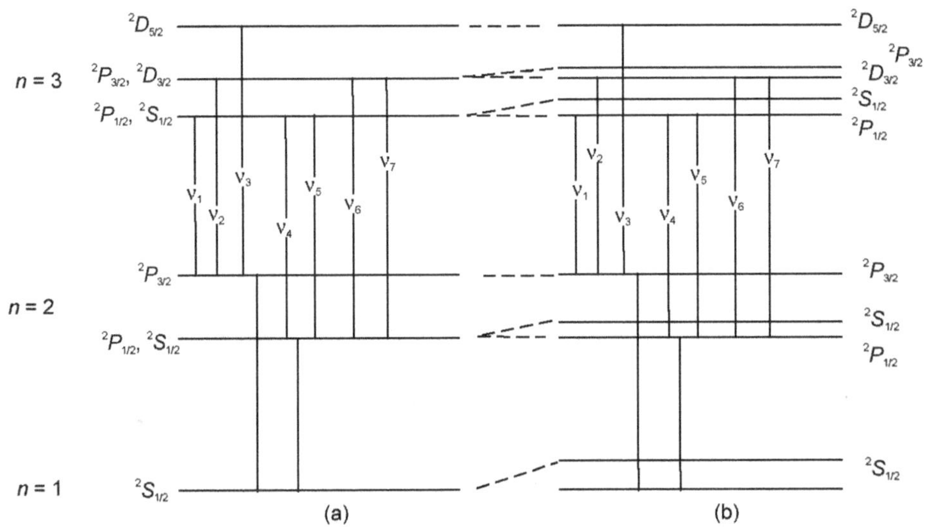

Figure 5.14   Hydrogen atom energy levels (a) Bohr Sommerfeld–Dirac theory (b) Lamb shifts

From equation 2.134 we can write the relativistic total energy neglecting spin–orbit interaction term, for the time being, as

$$E = \frac{RZ^2}{n^2} + \frac{R\alpha^2 Z^4}{n^2}\left[\frac{n}{j+\frac{1}{2}} - \frac{3}{4}\right] = \frac{R}{n^2}\left[1 + \frac{\alpha^2}{n^2}\left(\frac{n}{j+\frac{1}{2}} - \frac{3}{4}\right)\right] \mathrm{cm}^{-1}$$

and hence according to Dirac's relativistic theory the energy levels of the $2\,^2S_{1/2}$, $2\,^2P_{1/2}$ and the $2\,^2P_{3/2}$ are

$$E(2\,^2S_{1/2}) = E(2\,^2P_{1/2}) = \frac{R}{4}\left[1 + \frac{\alpha^2}{4}\left(2 - \frac{3}{4}\right)\right] \text{ and}$$

$$E(2\,^2P_{3/2}) = -\frac{R}{4}\left[1 + \frac{\alpha^2}{4}\left(1 - \frac{3}{4}\right)\right]$$

The transition between the levels $2\,^2P_{3/2} - 2\,^2S_{1/2}$ is an allowed transition as per the selection rule $\Delta l = \pm 1$, $\Delta j = \pm 1$ and the transition frequency lies in the microwave region and is 10950 MHz equivalent to $0.365\ \mathrm{cm}^{-1}\left(\dfrac{R\alpha^2}{16}\right)$. It should be noted from above that $2^2S_{1/2}$ and $2^2P_{1/2}$ have the same energy.

Thus the Dirac's theory could explain the salient features of the Balmer series by taking into account the relativistic correction on the dependence of electron mass on velocity and the spin of the electron. However, more measurements have shown still discrepancies. We have seen that as per Dirac's theory, the levels having same n and same j but different l are degenerate. Thus for example the levels $2\,^2S_{1/2}$ and $2\,^2P_{1/2}$ coincide. However, measurements in the microwave region by Lamb and co-workers in 1947 have shown that the levels $2\,^2S_{1/2}$ and $2\,^2P_{1/2}$ are not degenerate and there is a small splitting in the levels. When the transitions to these states were measured in the optical region (i.e., from different n to $2\,^2S_{1/2}$ and $2\,^2P_{1/2}$), these splittings were not noticeable as they were very small. However, by the rapid development of microwave spectroscopy after World War II, it was possible to measure transitions between $2\,^2S_{1/2}$ and $2\,^2P_{1/2}$ levels.

The transition between $2\,^2S_{1/2} - 1\,^2S_{1/2}$ is not an allowed transition because the selection rule $\Delta l = 1$ is not obeyed here. Thus the $2\,^2S_{1/2}$ is a metastable state. An excited atom in the $2\,^2S_{1/2}$ cannot be spontaneously transferred into the $2\,^2P_{1/2}$ state because of very small energy separation between them. However by applying appropriate microwave frequencies transitions from $2\,^2S_{1/2}$ to $2\,^2P_{1/2}$ and $2\,^2S_{1/2}$ to $2\,^2P_{3/2}$ can be obtained. Thus Lamb and co-workers investigated for the first time the transitions from the $2\,^2S_{1/2}$ state to $2\,^2P_{1/2}$ and to $2\,^2P_{3/2}$ levels. From these results it has been unequivocally proved that the levels having the same n and the same j (with different l) are not degenerate. This splitting is called Lamb shift.

The microwave experimental set-up of Lamb and co-workers is shown in Figure 5.15. A beam of hydrogen atom is obtained by dissociation of hydrogen molecules in a tungsten furnace. The atomic beam is collimated by an aperture and the collimated atomic beam is made to collide with an electron beam during its passage to a metallic target and the atoms are brought to the metastable $2\,^2S_{1/2}$ state. The hydrogen atoms in the metastable state strike the metal target and the excitation energy of this state is transferred to the electrons on the metallic target. These electrons in the metallic target are knocked off and produce a current, which is measured by a sensitive galvanometer. The ground state hydrogen atoms (atoms in the 1s $^2S_{1/2}$) are not capable of ejecting electron from the metallic target. The atomic beam is made to pass through a microwave cavity placed in between the electron beam and the metal target. The experimental set-up is placed in an evacuated chamber.

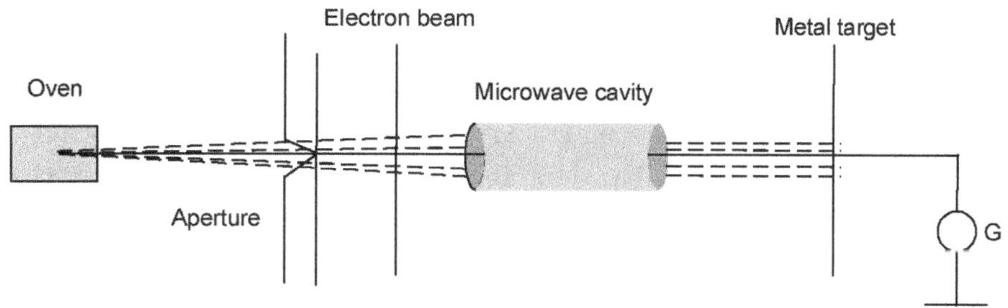

Figure 5.15 The microwave experimental set-up of Lamb and co-workers

If the Dirac theory is correct, a microwave frequency of 10950 MHz should give an absorption signal or in other words the atoms from the metastable $2\ ^2S_{1/2}$ should be transferred to the $2\ ^2P_{3/2}$ level. Thereby we would observe a decrease in the galvanometer current. In the actual experiment after fixing the microwave frequency, a magnetic field was applied in the microwave cavity so that the transition occurs in the Zeeman sublevels arising from the transition $2\ ^2S_{1/2} - 2\ ^2P_{3/2}$ and the magnetic field is slowly varied in order to obtain a resonance. The transition between the Zeeman sublevels of $2\ ^2S_{1/2} - 2\ ^2P_{3/2}$ were also monitored during the experiment. These experiments showed that there were two transitions instead of one as predicted by Dirac's theory. These transitions were found at microwave frequencies 9910 and 1057 MHz and it was found that the former corresponds to $2\ ^2S_{1/2} - 2\ ^2P_{3/2}$ and the latter to $2\ ^2P_{1/2} - 2\ ^2S_{1/2}$ (*See* Figure 5.16). In both cases the atoms in metastable state $2\ ^2S_{1/2}$ is transferred to the $2\ ^2P_{3/2}$ and $2\ ^2P_{1/2}$ states from where spontaneous transitions were occurred to the ground state by emitting ultraviolet photons. This makes a decrease in the galvanometer current. Thus the experiments by Lamb and Robert Rutherford have shown that the $2\ ^2P_{1/2}$ and $2\ ^2S_{1/2}$ levels are not coincident but are separated by 1057 MHz equivalent to 0.0353 cm^{-1} in energy. This energy shift is known as Lamb shift.

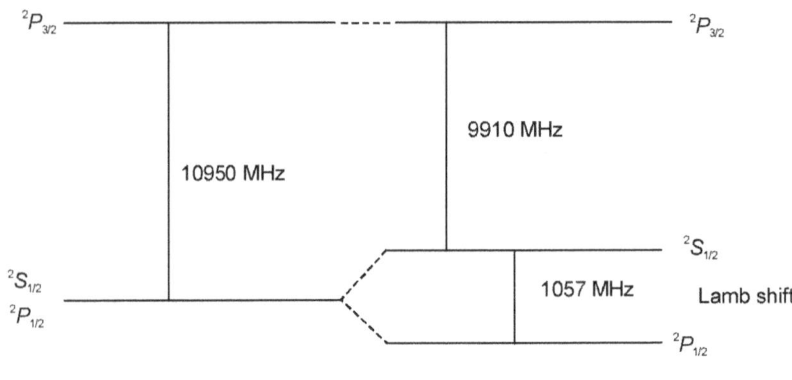

Figure 5.16   Lamb shift

5.10.1 Theory of Lamb Shift

We have already discussed that the lamb shift is the removal of the degeneracy observed in states having same n and same j. For example the earlier notion was that the states $2\ ^2P_{1/2}$ and $2\ ^2S_{1/2}$ coincide. Under the microwave studies the transitions $2\ ^2P_{3/2} \leftrightarrow 2\ ^2S_{1/2}$ was found to be 9910 MHz and the transition $2\ ^2P_{1/2} \leftrightarrow 2\ ^2S_{1/2}$ was measured to be 1057 MHz. This observation clearly proved that the $2\ ^2P_{1/2}$ and $2\ ^2S_{1/2}$ are separate. The splitting is due to the interaction of the electron with the electromagnetic field. We already know that the quantum theory predicts a zero point energy associated with the electromagnetic field. This energy W associated with the electromagnetic field

$$W = \sum_v \left(n_v + \frac{1}{2} \right) h v$$

where n_v is the number of photons of frequency v. It shows that even if there are no photons there exists an energy equals $\sum_v \frac{1}{2} hv$. This is called zero point energy. The electrons are subjected to fluctuations because of this field. We shall now calculate the electric field arising from the zero point energy. The number of standing waves or photons in the frequency range v and $v + dv$ is $\frac{8\pi}{c^3} v^2 dv$. In a radiation cavity of volume V the number of photons are $\frac{8\pi}{c^3} v^2 dv V$. The energy of electric field is then $\frac{8\pi}{c^3} v^2 dv V \times \frac{hv}{2}$.

Thus
$$\frac{E_v^2 V}{8\pi} = \left(\frac{8\pi}{c^3} v^2 dv \right) V \times \frac{hv}{2}$$

The amplitude of the oscillating electric field is then $E_v^2 = \frac{32}{c^3} \pi^2 h v^3 dv$.

The fluctuation in the electric and magnetic fields of the electromagnetic radiation will cause fluctuations in the position of the electron in the hydrogen atom. If we define the displacement of the electron from the Bohr–Sommerfeld orbit as Δs, then the corresponding change in the potential energy U can be expressed in Taylor series and is

$$\Delta U = \frac{\partial U}{\partial x} \Delta x + \frac{\partial U}{\partial y} \Delta y + \frac{\partial U}{\partial z} \Delta z + \frac{1}{2} \frac{\partial^2 U}{\partial x^2} (\Delta x)^2 + \frac{1}{2} \frac{\partial^2 U}{\partial y^2} (\Delta y)^2 + \frac{1}{2} \frac{\partial^2 U}{\partial z^2} (\Delta z)^2$$

We can consider the fluctuations at the zero point energy are isotropic and hence

$$\overline{\Delta x} = \overline{\Delta y} = \overline{\Delta z} = 0$$

$$\overline{(\Delta x)^2} = \overline{(\Delta y)^2} = \overline{(\Delta z)^2} = \frac{1}{3} \overline{(\Delta s)^2}$$

and $\overline{\Delta U} = \frac{1}{6} \overline{(\Delta s)^2} \left(\frac{\partial^2 U}{\partial x^2} + \frac{\partial^2 U}{\partial y^2} + \frac{\partial^2 U}{\partial z^2} \right) = \frac{1}{6} \overline{(\Delta s)^2} \Delta^2 U$

If we consider $\overline{\Delta U}$ as a small perturbation then the first order energy can be written as

$$\Delta E = \int \psi_{n,l,m}^* (r) \overline{\Delta U} \psi_{n,l,m} (r) d\tau = \frac{\overline{(\Delta s^2)}}{6} \int \psi_{n,l,m}^* (r) \overline{\Delta^2 U} \psi_{n,l,m} (r) d\tau$$

$\Delta^2 U = 0$ except at $r = 0$ and at $r = 0$, $\Delta^2 U$ is infinite.

$\int \Delta^2 U d\tau = \int \nabla U . d\overline{s} = 4\pi Z e^2$ where ds is the vector area of any closed surface.

$$\Delta E = \frac{\overline{(\Delta s^2)}}{6} 4\pi Z e^2 \left|\psi(0)_{n,l,m}\right|^2$$

$\psi_{n,l,m} = 0$ except for states with $l = 0$ and

$$\left|\psi(0)_{n,l,m}\right|^2 = \frac{Z^3}{\pi n^3 a_0^3} \quad \text{where } a_0 \text{ is Bohr radius.}$$

The $^2S_{1/2}$ state moves upward by an energy $\Delta E = \dfrac{\overline{(\Delta s^2)}}{6} \dfrac{4Z^4 e^2}{n^3 a_0^3}$.

Thus the degeneracy is removed.

At room temperatures the fine structure in the H^α is not well resolved as the Doppler broadening is large at optical frequencies. However, one could lower the temperature and narrow the Doppler broadening. It was Hänsch and Shahin who resolved all the structures in the optical region by using the technique of saturation spectroscopy.

5.10.2 Measurement of the Lamb Shift

The Lamb shift is extremely small to observe in optical or in UV regions. However, with the availability of lasers it was made possible to measure these small splittings in the optical region by Doppler free studies. Before the availability of high precision lasers, Willis Lamb has made measurements of transitions directly between the sublevels by going to other regions of the electromagnetic spectrum, viz., the microwave region. He was able to produce a beam of hydrogen atoms in the 2s, $^2S_{1/2}$ state. However, because of the selection rule $\Delta l = \pm 1$ it is not possible to observe a transition to the 1s $^2S_{1/2}$ as in this case $\Delta l = 0$. Note that the upper state and lower states are S states. However, Lamb got the idea of putting the atoms in a magnetic field. By doing so, levels are split by the Zeeman effect. It means that the state $^2P_{3/2}$ with $j = 3/2$ will split into $(2j + 1)$ components, i.e, 4 magnetic sublevels with $m_j = +3/2, +1/2, -1/2, -3/2$. Similarly the $^2S_{1/2}$ state will be split into $m_j = +1/2, -1/2$ magnetic sublevels. Figure 5.17 shows the various magnetic sublevels. Lamb exposed the hydrogen atoms in the magnetic field to microwave radiations of 2395 MHz and could observe transition between magnetic sublevels. Note that ordinary microwave oven is of 2450 MHz. The experiment can be performed by varying the magnetic field or the microwave frequencies. In his experiments Lamb varied the magnetic field until that frequency produced the transition from $(n = 2)\ ^2P_{1/2}$ (i.e., same as $^2S_{1/2}$ without Lamb shift)– $(n = 2)\ ^2P_{3/2}$. He could then measure the allowed transition from $(n = 2)\ ^2P_{3/2} - (n = 1)\ ^2S_{1/2}$. He then used the frequency to extrapolate to zero magnetic field in order to obtain the transition and the frequency of 1057 MHz for the separation of $2\ ^2P_{1/2}$ and $2\ ^2S_{1/2}$ levels was obtained. In electron volt

$$E = h\nu = 6.62606896 \times 10^{-34}\ \text{Js} \times 1057 \times 10^6\ \text{Hz}$$

$$= \frac{6.62606896 \times 10^{-34}\ \text{Js} \times 1057 \times 10^6\ \text{Hz}}{1.60217646 \times 10^{-19}\ \text{coulomb}} = 4.371 \times 10^{-6}\ \text{eV}$$

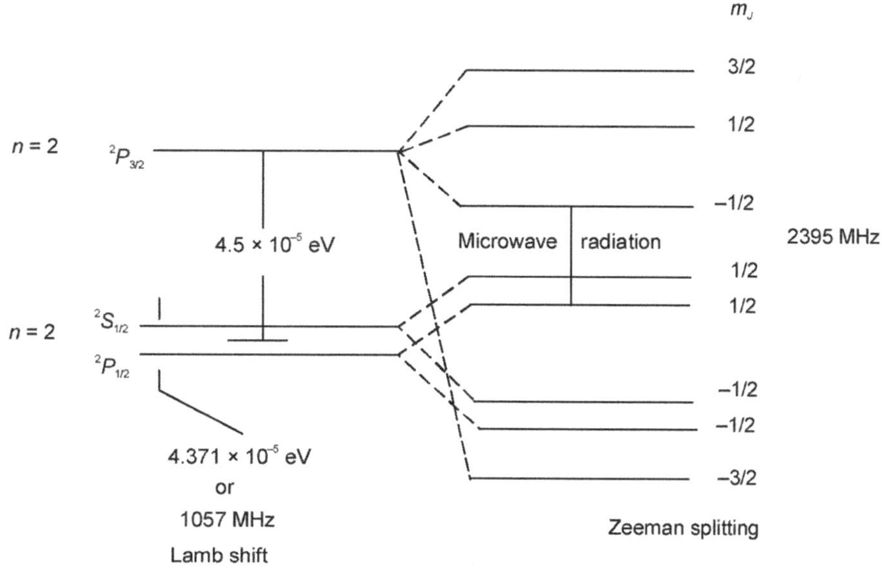

Figure 5.17 Measurement of the Lamb shift

In old quantum theory the $n = 2$, $^2S_{1/2}$ and $^2P_{1/2}$ are degenerate. Thus it was Lamb who showed for the first time experimentally that they are split by an amount of 1057 MHz (4.3×10^{-6} eV).

5.10.3 The Lamb Shift by Saturation Spectroscopy

It was already mentioned that it is not possible to measure these small splittings due to the Lambshift in ordinary spectroscopic work. However with the advent of tunable lasers it is possible to measure the splittings even in optical region. When one tries to observe these small splittings in high resolution by using lasers, the line broadening by Doppler effect always obscures it. Doppler broadening is due to the thermal motion of atoms and it is particularly important in the case of hydrogen atoms as it has low mass and hence high thermal velocity.

The rms velocity of the hydrogen gas at room temperature (300 K) is about 2700 m/s. This would correspond to a Doppler shift of around 4 GHz, which means a line broadening twice the value. We have seen that the Lamb shift is only 1.057 GHz and hence it lies under the Doppler line width and is obscure for the Lamb shift.

Tunable lasers are widely used nowadays to carry out Doppler free spectroscopy. Doppler free saturation spectroscopy minimizes the effects of the Doppler broadening. In Doppler free spectroscopy the light from a laser beam is split into two beams by a beam splitter and sent through the gas in opposite directions as shown in the Figure 5.18. The two beams are arranged in such a manner that they cross in a region of gas cell containing hydrogen. The power of the two beams are arranged in such a way that one beam is powerful and is called saturating beam or pump beam and the other is made less powerful and is called the probe beam. When the laser is tuned to a transition frequency of the hydrogen atom, the saturating beam is intense enough to raise the atoms from the lower state to the upper state. This means that the saturating beam depletes the lower level by "pumping" the atoms to the upper state. If we introduce a rotating chopper in the way of the saturating beam, the beam is made "on" and "off" by the frequency of the chopper. This is called modulation of the frequency (or frequency modulation). Thus the absorption due to the saturating beam is turned on and off at intervals of the frequency of the chopper. The probe beam when passed through the hydrogen gas then encounters the atoms at time intervals. So the absorption due to the probe beam depends on whether the saturating beam is off or on. A sensitive detector locked to the modulation frequency then detects the probe beam. The detector thus detect the absorption due to the probe beam corresponding to the on and off of the absorption of the probe beam.

The advantage of this set-up is that it enables to produce a Doppler free spectrum. The gas in the cell experiences the Doppler effect due to the two beams from the opposite directions. The hydrogen atoms, which move in the direction of the two beams, experience a Doppler shift while for those atoms whose velocity components are in the perpendicular directions will not experience any Doppler shift. Thus the only hydrogen atoms, which are in resonance with both of the beams, are those, which have no component of velocity in the direction of the beams. They absorb at the frequency associated with the rest frame of the atom. Other atoms see the incoming radiation shifted up for the saturation beam and down for probe beam or vice versa, so they are not resonant for both. Tuning the dye laser slowly through the frequency range of the transition and measuring the change in intensity of the probe beam at the modulation or "chopping" frequency is shown in Figure 5.18.

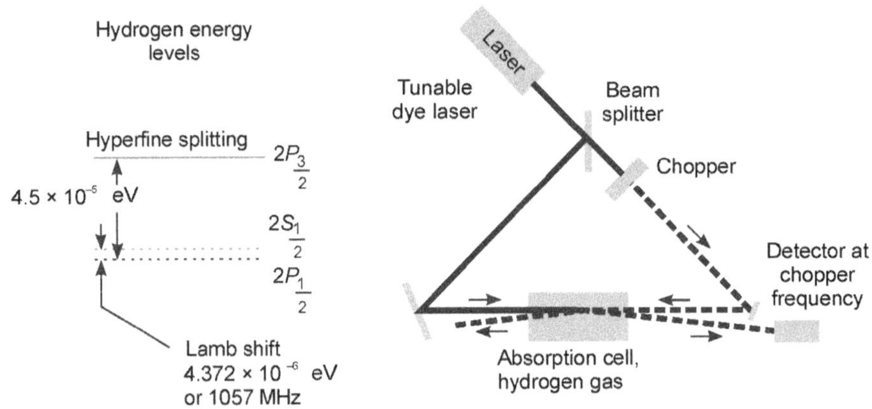

Figure 5.18 Doppler free saturation spectroscopy

The tremendous advantage of the technique by Hänsch and co-workers is in eliminating the effects of Doppler broadening. This is accomplished by directing the saturation beam and probe beam through the gas in opposite directions. The only hydrogen atoms which are in resonance with both of the beams are those which have no component of velocity in the direction of the beams, and therefore absorb at the frequency associated with the rest frame of the atom. Other atoms in their rest frames see the incoming radiation shifted up for the saturation beam and down for probe beam or vice versa, so they are not resonant for both. The spectrum shown was obtained by Hänsch's group and was detected by tuning the dye laser slowly through the frequency range of the transition and measuring the change in intensity of the probe beam at the modulation or "chopping" frequency. Figure 5.19 shows the Doppler free spectrum and Lamb shift obtained by Hänsch and his group.

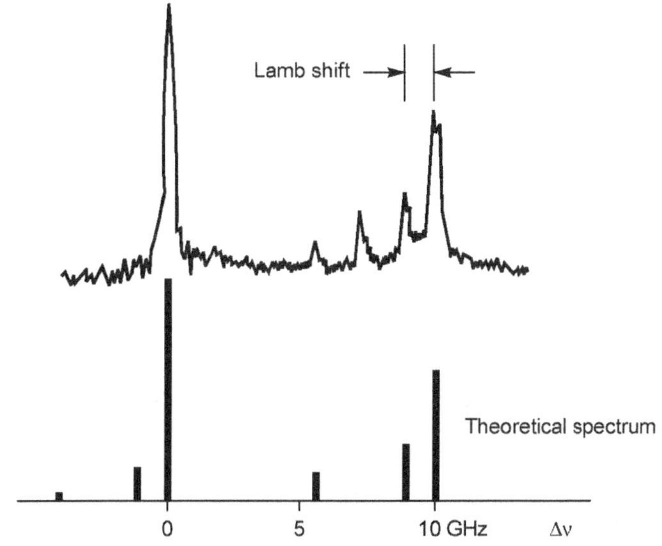

Hydrogen fine structure and hyperfine structure for the n = 3 ′ 2 transition.

Figure 5.19   Doppler free spectrum

5.11 SIGNIFICANCE OF THE LAMB SHIFT

The measurement of Lamb shift gave immediately a verification of theoretical calculations made with the quantum theory of electrodynamics. The theory of electrodynamics predicted that electrons continually exchanged photons, which being the mechanism of electromagnetic radiation.

The small Lamb shift, measured with great precision, agreed to many decimal places with the calculated result from quantum electrodynamics. The measured precision gives us the electron spin g-factor as

$$g = 2.002319304386$$

5.12 THE HYDROGEN 21-CM LINE

The 21-cm line of hydrogen atom plays a major role in mapping our galaxy and interstellar media. As it lies in the microwave region, this radiation can penetrate through the dust clouds and give a complete map of the hydrogen as the visible light do not penetrate the dust clouds. The 21-centimetre line refers to the transition between the energy levels created by changes in the energy states due to spin orientations of the single electron and the single proton of the neutral hydrogen atom. The transition occurs at the microwave frequency 1420.405751 7667(10) MHz equivalent to vacuum wavelength 21 cm. The transition is extensively used in Astronomy and Astrophysics as it penetrates through the clouds which are otherwise opaque to visible radiations.

$$\text{We know that } \nu\lambda = c \text{ or } \nu = \frac{c}{\lambda} = \frac{2.99792458 \times 10^{10}\,\text{cm}}{21\,\text{cm}} = 1.420 \times 10^{9} = 1.420\,\text{GHz}$$

As stated above the 1420 MHz radiation comes from the transition between the two hyperfine levels of the hydrogen 1s ground state, slightly split by the interaction between the electron spin and the nuclear spin. We have already seen that hydrogen atom essentially consists of a nucleus and an electron. The nucleus of the normal hydrogen atom consists of only a proton. It does not have a neutron as in the case of other atoms. Only the isotopes of hydrogen like deuterium and tritium have neutrons in addition to a single proton. We have also seen that an electron possesses a spin whose value is ½. Thus the neutral hydrogen has a proton orbited by an electron. This is analogous to rotational motion of the Earth rotating on its axis as it orbits the Sun, but in atomic case they have slightly different concept as they are quantum particles. The spin of the electron and proton can be in either direction—in the classical analogy they are rotating clockwise or anticlockwise around a given axis. We have already seen that there is an interaction between the electron spin and the orbital motion of the electron, which leads to the fine structure in the spectra. However the 1420 MHz line originates as the levels in the 1s ground state is lightly split by the interaction between the electron spin and the nuclear spin. The splitting is known as hyperfine structure. The electron and the proton may both have their spins oriented in the same direction or in opposite directions. Because of magnetic interactions between the particles electron and proton, a hydrogen atom that has the spins of the electron and proton aligned in the same direction (parallel) is slightly higher in energy (less tightly bound) than one where the spins of the electron and proton are in opposite directions (anti-parallel). This is shown in Figure 5.20

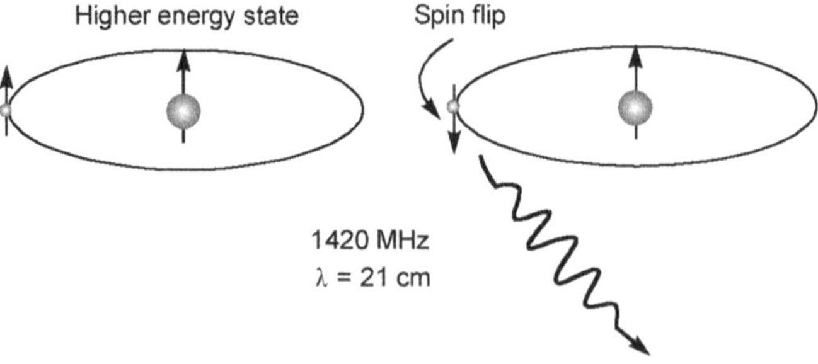

Figure 5.20   The spins of the electron and proton of a hydrogen atom in parallel or anti-parallel alignment

Thus the lowest orbital energy state of the hydrogen atom split into two energy states due to this interaction and the splitting is called hyperfine splitting. This splitting of the hydrogen ground state is extremely small compared to the ground state energy of -13.6 eV, only about two parts in a million. Thus the splitting arises because the electron and the proton spins change their direction from a parallel to an antiparallel configuration. The transition is forbidden and happens with a probability of 2.9×10^{-15} s^{-1}. This means that the time for a single isolated atom of neutral hydrogen to undergo this transition is around 10 million (10^7) years and so is unlikely to be seen in a laboratory on earth. However, since the total number of atoms of neutral hydrogen in interstellar medium is very large, these transitions could be detected. The hyperfine energy levels in $1S$ state is shown in Figure 5.21 and the hyperfine splittings in $n = 1$ and $n = 2$ are shown in Figure 5.22.

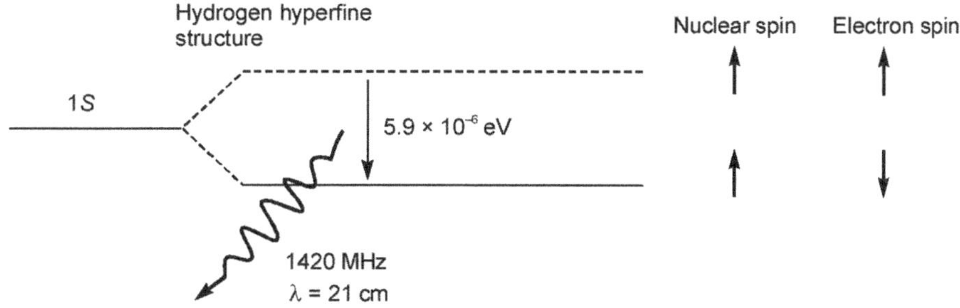

Figure 5.21   Hydrogen hyperfine structure

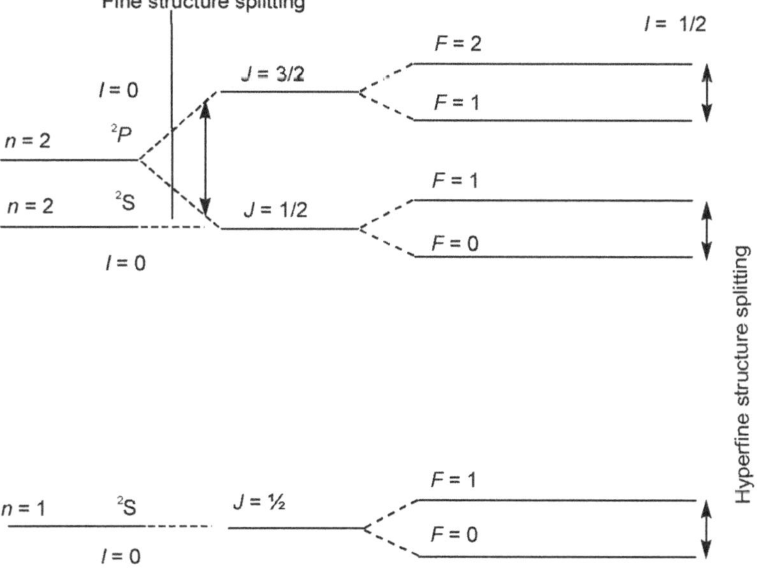

Figure 5.22   Hydrogen hyperfine structure splitting

In the coupling scheme, for $n = 1$, $l = 0$ (S state), $J = 1/2$

And for $n = 2$, $l = 0$ and 1. For $l = 1$, $J = 3/2$ or $1/2$ ($l + 1/2 ... l–1/2$)

whereas for $l = 0$, $J = 1/2$.

When J couple with the proton spin (or nuclear spin), the total spin F can take values from $J–I$ to $J+I$ if $J > I$ and hence for $J = \frac{1}{2}$, $F = 0$ or 1. For $J = 3/2$, $F = 1$ or 2. In this case F can take $2J + 1$ values.

The observation of the 21-cm line of hydrogen marked the birth of a new era in radio astronomy. The emission was first detected on March 25, 1951 by Harold Inving Ewen (b. March 5, 1922) and Edward Mills Purcell (Purcell August 30, 1912; Died March 7, 1997) at Harvard, followed soon afterward by observers Muller and Oort in Holland and Chritiansen and Hindman in Australia[16]. Dutch astronomer Hendrik van de Hulst[17] made the prediction that the 21-cm line should be observable in emission in 1944. The whole discovery originated from the detection of a radio "hiss" during 1930s, which appeared to be extraterrestrial in origin. It was then suggested that this might be coming from the centre of the Galaxy and then the emission line was observed.

The observation of the 21-cm line of the hydrogen atom was a great boon to astronomers. It is because that the electromagnetic radiations in the microwave and radio frequency regions can easily pass through the earth's atmosphere and can be observed from the earth with little hindrance. However interference due to television transmitters and ionosphere causes some difficulties in observing these faint transitions. The lines are found to be Doppler shifted (red shifted). Hence the line could be used to calculate the relative speed of each arm of our galaxy. The rotation curve of the galaxy is also calculated by using the 21-cm line. Thus the line is of great interest to the big bang theory. Most of the lines from the interstellar media were observed in emission. Hydrogen in its lower state absorbs 1420 MHz and the observation of 1420 MHz in emission implies a prior excitation to the upper state.

The 21-cm line is also known as HI transition or spin-flip transition.

The 21-cm hydrogen line is considered a favourable frequency to search for signals from another civilization, as part of the Search for Extra-Terrestrial Intelligence (SETI) program.

The hydrogen in our galaxy has been mapped by the observation of the 21-cm wavelength line of hydrogen gas.

Edward Mills Purcell shared the 1952 Nobel Prize with Felix Bloch for development of the nuclear magnetic resonance method of measuring magnetic fields in the nuclei of atoms.

5.13 AUGER EFFECT

The effect was discovered in 1925 by Pierre Victor Auger in a cloud chamber experiment with high energy X-rays and ionising the gas particles and observing the photoelectric electrons. It is a phenomenon in which the transition of an electron in an atom filling in an inner shell vacancy causes the emission of a second electron. Or in other words the filling of a vacancy in an inner electron energy level of an atom by an electron from an outer energy level of the same atom. The excess energy involved causes emission of another electron known as an Auger electron. When an electron is removed from an atom, the vacancy produced by the electron is occupied by another electron from a higher energy state releasing energy. In our usual knowledge the energy released by the second electron is in the form of a photon. However, the energy can also be transferred to another electron ejected from the atom. The second ejected electron is called Auger electron. The energy of the Auger electron corresponds to the difference between the initial state of the first electron and the ionization energy of the electron shell from which the Auger electron was ejected. Auger electron spectroscopy involves the emission of Auger electrons by bombarding the sample by X-rays or energetic electrons and measures the intensity with the Auger electron energy. The Auger combination is a similar effect as is observed in semiconductors. An electron and electron hole recombine giving up the energy to an electron in the conduction band (electron-hole pair).

Auger effect observations of electron tracks independent of frequency of the incident photon energy had suggested internal conversion of energy by radiationless transitions inside the atom. Thus Auger effect becomes a procedure by which one analyses the energetic electrons emitted from an excited atom after a series of internal relaxation events. The principle underlying the Auger effect is shown in Figure 5.23.

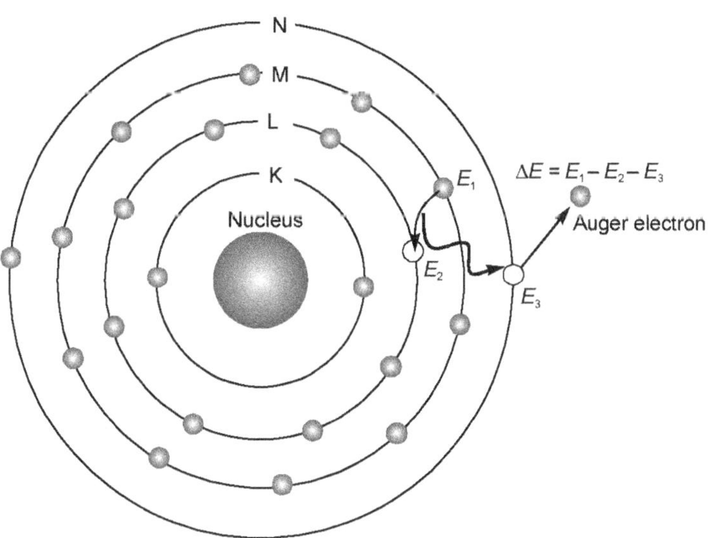

Figure 5.23   Auger effect

5.13.1 Electron Transitions and the Auger Effect

Thus the Auger effect is an electronic process resulting from the inter- and intra-state transitions of electrons in an excited atom. When an atom is probed by an external mechanism, such as a photon or a beam of electrons with energies in the range of 2 keV to 50 keV, an electron in the core shell can be removed leaving behind a hole. As this is an unstable state, the core hole is then filled by an outer shell electron, whereby the electron moving to the lower energy level loses an amount of energy equal to the difference in orbital energies. The transition energy can be coupled to a second outer shell electron which will be emitted from the atom if the transferred energy is greater than the orbital binding energy. An emitted electron will have a kinetic energy of

$$E_{\text{kin}} = E_{\text{core state}} - E_B - E_C'$$

where $E_{\text{core state}}, -E_B, -E_C'$ are the core level, first outer shell, and second outer shell electron energies respectively, measured from the vacuum level. An experimental set-up to study Auger effect is shown in Figure 5.24.

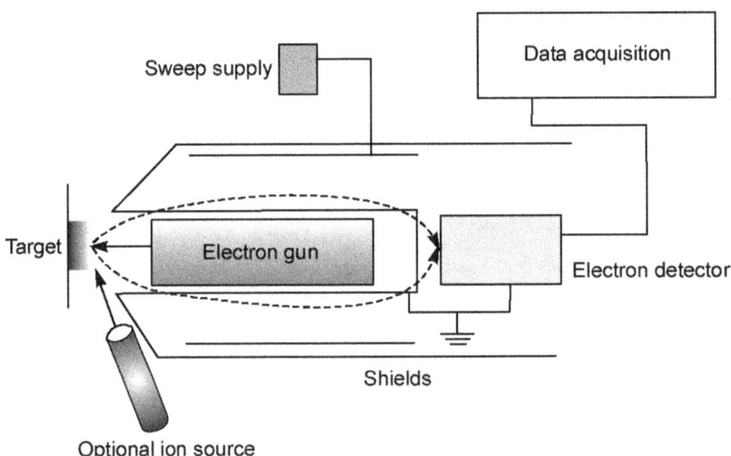

Figure 5.24   An experimental set-up to study Auger effect

Here an electron beam is focused into the specimen and the emitted electrons are deflected around the electron gun and then pass through an aperture to the analyser.

Since orbital energies are unique to an atom of a specific element, analysis of the ejected electrons gives information about the chemical composition of a surface.

The scanning Auger microscopes are used to obtain highly resolved chemical images.

Important facts to remember:

1. 2, 6, 10 and 14 are the number of electrons that can fill the s, p, d and f subshells (the $l = 0,1,2,3$ azimuthal quantum number)

2. The $1s$ subshell is the first s subshell, the $2p$ is the first p subshell ($n = 2$, $l = 1$, $3d$ is the first d subshell, and the $4f$ is the first f subshell.

SOLVED PROBLEMS

1. What is the electronic configuration for the element Niobium? (41)

   Solution

   $1s^2\ 2s^2\ 2p^6\ 3s^2\ 3p^6\ 4s^2\ 3d^{10}\ 4p^6\ 5s^2\ 4d^3$ is the expected.

   Niobium is actually:

   $1s^2\ 2s^2\ 2p^6\ 3s^2\ 3p^6\ 4s^2\ 3d^{10}\ 4p^6\ 5s^1\ 4d^4$

   The reason is that the $5s$ and $4d$ energy levels are quite close and certain electronic arrangements can result in the levels being slightly different than expected.

2. What is the electronic configuration of the element Nickel? (28)

   Solution

   $1s^2\ 2s^2\ 2p^6\ 3s^2\ 3p^6\ 4s^2\ 3d^8$

3. What is the electronic configuration for Nickel in terms of the nearest noble gas? How would the last valence orbital be filled?

   Solution

   $[\text{Ar}]\ 4s^2\ 3d^8$

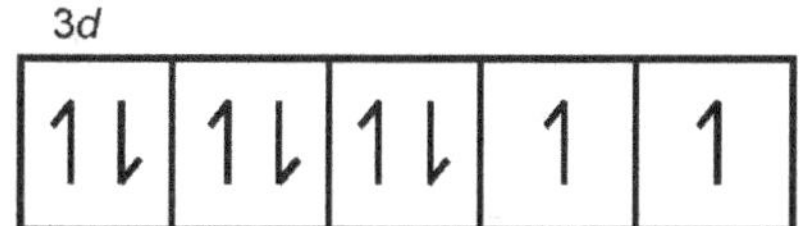

ENDNOTES

1. Wilhelm Conrad Röntgen (27 March 1845–10 February 1923) ,the first Nobel Prize in Physics in 1901.

2. Max Theodor Felix von Laue (9 October 1879–24 April 1960) was a German physicist, Nobel Prize in Physics in 1914.

3. Sir William Lawrence Bragg (31 March 1890–1 July 1971) shared the Nobel Prize in Physics in 1915 with his father Sir Willium Henry Bragg (2 July 1862–10 March 1942).

4. Henry Gwyn Jeffreys Moseley (23 November 1887–10 August 1915) was an English physicist.

5. Dmitri Ivanovich Mendeleev (8 February [Old Style-27 January] 1834–2 February [O.S. 20 January] 1907.

6. Wolfgang Ernst Pauli (25 April 1900–15 December 1958) Austrian theoretical physicist received Nobel Prize in 1945.

7. Kossel W, Verh. d. Deutsch. Physik. Gesselsch. 16 (1914) 398; 18(1916)339; *Z.Physik.* 1(1929)119.

8. Mathematische Hilfsmittel des Physikers (Mathematical help material for Physicists) Springer Verlag, Berlin. 1936.

9. George Eugene Uhlenbeck (December 6, 1900–October 31, 1988) was a Dutch American Samuel Abraham Goudsmit (July 11, 1902–December 4, 1978) was a Dutch American.

10. Otto Stern (17 February 1888–17 August 1969) was a German Physicist who won the Nobel Prize in Physics in 1943.

    Walther Gerlach (1 August 1889–10 August 1979) was a German Physicist.

11. Paul Adrien Maurice Dirac. (8 August 1902–20 October 1984) British theoretical physicist.

12. Otto Stern (17 February 1888–17 August 1969) was a German physicist and Nobel Laureate in Physics(1943).

    Walther Gerlach (1 August 1889–10 August 1979) was a German physicist.

13. Otto Stern, (1921). "Ein Weg zur experimentellen Pruefung der Richtungsquantelung im Magnetfeld." *Zeitschr. f. Physik.* 7 249–253

    Walther Gerlach & Otto Stern, (1922). "Das magnetische Moment des Silberatoms." *Zeitschrift für Physik* 9, 353–355.

14. George Eugene Uhlenbeck (6 Dec 1900–31 Oct 1988).

    Samuel Abraham Goudsmit (11 July 1902–4 December 1978).

15. Uhlenbeck, G.E. and Goudsmit, S. Naturwissenschaften 47 (1925) 953, *Nature.* 117 (1926) 264.

    Goudsmit, S. and Uhlenbeck, G.E. *Physica.* 6 (1926) 273.

16. Ewen, H. I. and Purcell, E.M. (1951). "Observation of a line in the galactic radio spectrum." *Nature.* 168 (4270): 35, 168, p.356,

    Muller, C.A.; J.H. Oort (September 1951). "The Interstellar hydrogen line at 1.420 Mc/sc and an Estimate of Galactic Rotation." *Nature.* 168 (4270): 357–358. Cosmology

    Christiansen and Hindman. (1952) Australian *J. Sci. Res.*, Vol. A5, p. 437.

    Edward Mills Purcell (August 30, 1912–March 7, 1997) Shared the 1952 Nobel Prize with Felix Bloch for development of the nuclear magnetic resonance Harold Irving Ewen (March 5, 1922–).

    P. Madau, A. Meiksin and M. J. Rees, "21-cm Tomography of the Intergalactic Medium at High Redshift." *Astrophysical Journal.* 475, 429 (199).

17. Hendrik Christoffel "Henk" van de Hulst (Nov. 19, 1918 – July 31, 2000).

REFERENCES

Briggs, David and Martin, P. Seah. (1983). *Practical Surface Analysis by Auger and X-ray Photoelectron Spectroscopy*. John Wiley and Sons, Chichester.

Grant, T. John (2003). S*urface Analysis by Auger and X-ray Photoelectron Spectroscopy*. IM Publications, Chichester.

Thomas, A. Carlson (1975). *Photoelectron and Auger Spectroscopy*. Plenum Press, New York.

Thompson, Michael, Baker, M.D., Christie, A. and Tyson, J.F. (1985). *Auger Electron Spectroscopy*. John Wiley and Sons, Chichester.

White, H.E. (1934). *Introduction to Atomic Spectra*. McGraw Hill, New York.

6

SPECTRA OF ALKALI METALS AND BORON GROUPS

6.1 OBSERVED SPECTRA

The absorption spectra of alkali metals Li, Na, K, Rb and Cs are in many ways similar to the hydrogen atom. And the spectra of these atoms were one of the first to have been systematically studied. The observation and the theoretical interpretations of these spectra were of much importance in the development of atomic spectroscopy. On the basis of the experimental observations the spectroscopists had arrived at certain conclusions and sets of rules in order to explain the salient features. It has been found that the observed spectral lines could be grouped into separate series. The electronic configuration in alkali metals is as follows.

Lithium	Li	$1s^2\, 2s$
Sodium	Na	$1s^2\, 2s^2\, 2p^6\, 3s$
Potassium	K	$1s^2\, 2s^2\, 2p^6\, 3s^2\, 3p^6\, 4s$
Rubidium	Rb	$1s^2\, 2s^2\, 2p^6\, 3s^2\, 3p^6\, 3d^{10}\, 4s^2\, 4p^6\, 5s$
Cesium	Cs	$1s^2\, 2s^2\, 2p^6\, 3s^2\, 3p^6\, 3d^{10}\, 4s^2\, 4p^6\, 4d^{10}\, 5s^2\, 5p^6\, 6s$

The characteristic feature of the spectra is due to the single valence electron.

Each of these elements provide a number of term series where each series is the difference between a fixed and a variable term and the differences give the wave number of the series of lines. The valence electron in the last s shell is responsible for the observed spectra in this group of atoms. The lowest level, for example in Lithium is an S state originating from the unbalanced s electron in the last shell giving rise to an S state. This is followed by a P state as the electron by excitation moves to the higher $2p$ shell, then an S state and a D state and then an F state. The observed spectral lines in Lithium could be grouped into separate series.

Principal series $\quad\quad v = 2S - nP \quad n = 2,3,4\dots$
Diffuse series $\quad\quad\; v = 2P - nD \quad n = 3,4,5\dots$
Sharp series $\quad\quad\;\; v = 2P - nS \quad n = 3,4,5\dots$
Fundamental series $\; v = 3D - nF \quad n = 4,5,6\dots$

Similar features are observed in other atoms of the group, excitation of $3s$ electron to different shells vis-à-vis states in sodium, $4s$ electron into higher shells in potassium and so on. In sodium for example

Principal series $\quad\quad v = \; 3S - nP \quad n = 3,4,5\dots$

Diffuse series $\quad\quad\; v = 3P - nD \quad n = 4,5,6\dots$

Sharp series $\quad\quad\;\; v = 3P - nS \quad n = 4,5,6\dots$

Fundamental series $\; v = 3D - nF \quad n = 4,5,6\dots$

As this group has a single valence electron, the spin of this electron only to be considered in order to derive the splitting of the levels. As the other shells are completely filled they do not take part in the transitions. We have already seen that the multiplicity of state depends on the total spin and the splitting is $2S + 1$ where S is the total spin. (Here again do not confuse with the state S and the total spin S, the nomenclature is the same, but they are two identities.) The multiplicity of levels in alkali metal atoms is $2S + 1 = 2 \cdot \frac{1}{2} + 1 = 2$. Thus we obtain doublets in the atom. That means all levels are doublets. The ground state of alkali atoms are $^2S_{1/2}$, but it is a single level as there can be only one J in the case of S state. [The subscript is the total quantum number $J - L + S$ and in this case $0 + \frac{1}{2} = \frac{1}{2}$. The superscript 2 on the left corresponds to the multiplicity and the subscript on the right is the total angular momentum j and S is the state which has L (or l) = 0.]. The electronic distribution in sodium atom is shown in Figure 6.1.

It was in order to explain these lines that Sommerfeld introduced the quantum number j called the inner quantum number or total orbital quantum number. We have already seen that n refers to the principal quantum number and the subshells have orbital quantum number l which have the values $l = 0, 1, 2, \dots(n-1)$. The total quantum number

$$j = l + s, \, l + s - 1, \dots | \, l - s \, |.$$

The selection rules for transitions are

$$\Delta l = \pm 1$$

$\Delta j = 0, \; \pm 1$ and Δn takes any integral value.

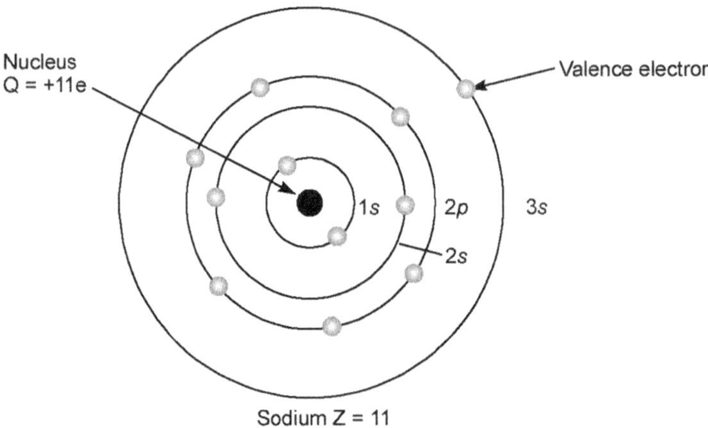

Figure 6.1   The electronic configuration of the sodium atom

Thus the total angular momentum is made of two parts, the orbital angular momentum due to the orbital motion of the electron and an additional angular momentum due to the spin of the electron about an axis through its centre of mass. A classical picture of spin and orbital motion of electron is shown in Figure 6.2.

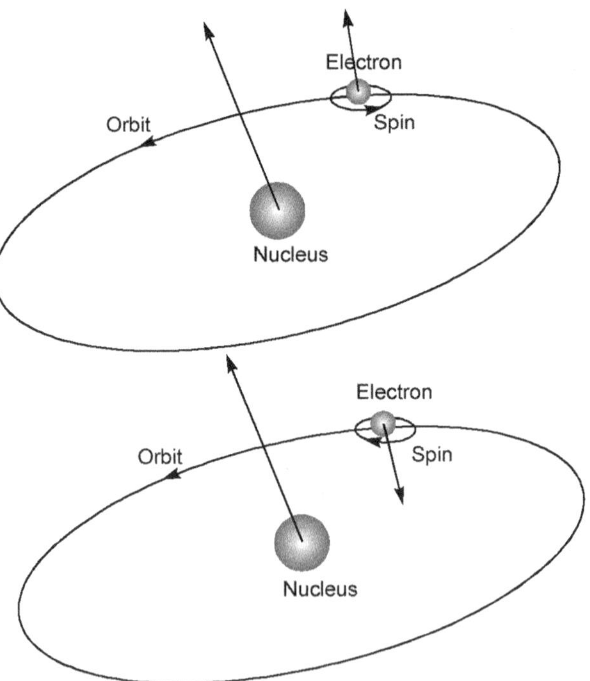

Figure 6.2   Spin and orbital motion of electrons

The value of the electron spin is half a quantum unit, i.e., $\frac{1}{2}\hbar$. It is also written as s where s is the spin $=\frac{1}{2}$. However, quantum mechanically the exact expression is $\sqrt{s(s+1)}\,\hbar$ which is sometimes written as s^*. In order to obtain the total angular momentum of the valence electron it is necessary to combine l and s vectorially. The procedure is that $\sqrt{s(s+1)}\hbar$ is vectorially added to $\sqrt{l(l+1)}\hbar$ and the resultant is $\sqrt{j(j+1)}\hbar$ where j under the square root is $l\pm s$.

Thus for a p electron $l=1$, $s=\frac{1}{2}$ and $j=|l\pm s|=\frac{3}{2}$ and $\frac{1}{2}$, the former giving

$$j^*\hbar=\sqrt{\frac{3}{2}\cdot\frac{5}{2}}\,\hbar=\frac{\sqrt{15}}{2}\,\hbar \text{ and the latter } j^*\hbar=\sqrt{\frac{1}{2}\cdot\frac{3}{2}}\,\hbar=\frac{\sqrt{3}}{2}\,\hbar.$$

For p electron there are two values of j, i.e., $j=l\pm s=\frac{3}{2}$ and $\frac{1}{2}$. Thus each value of l with the exception of $l=0$ is split into two levels with different energies. For $l=0$, there is only one level corresponding to $j=|l\pm s|=\frac{1}{2}$. Figure 6.3 gives the energy level diagram and the spectral series in lithium and sodium atoms. The first member of the principal series in sodium is the well known sodium D lines (*See* Figure 6.5).

To characterize an energy state arising from an electronic configuration, for example, in the case of sodium with a $3s$ valence electron we get a $^2S_{1/2}$ which is named as the ground state.

In the case of $^2P_{3/2}$, the superscript 2 on the left corresponds to the multiplicity and the subscript on the right is the total angular momentum J and P is the state which has L (or l) $=1$. In addition the principal quantum number can also be used in front of them. Thus the three states are

$3p\ ^2P_{1/2}$ and $3p\ ^2P_{3/2}$ and $3s\ ^2S_{1/2}$ or some times as

$3\ ^2P_{1/2}$ and $3\ ^2P_{3/2}$ and $3\ ^2S_{1/2}$

The two ^2P states are also written some times as $^2P_{1/2,3/2}$.

Here again the multiplicity is written on the left side as superscript, i.e., 2 × resultant electron spin quantum number +1.

Thus the nomenclature of any state is

$$^{2S+1}L_J$$

The Table 6.1 gives the value of j in different cases.

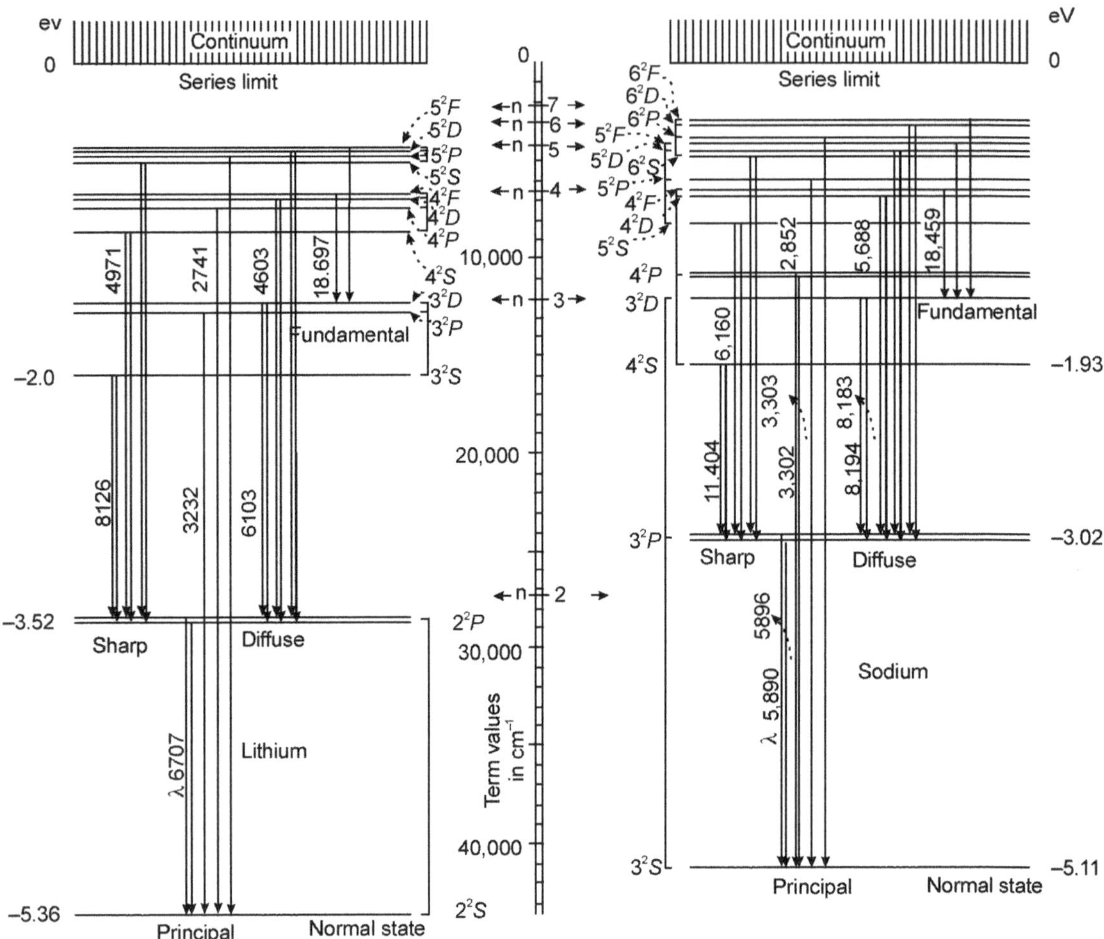

Figure 6.3 Energy level diagrams of neutral lithium and sodium atoms

Table 6.1 The value of the total quantum number in different cases

	l	j		
S	0	1/2		
P	1	1/2	3/2	
D	2		3/2	5/2
F	3			5/2	7/2

The well-known sodium D lines are from transitions in $n = 3$,

$$\left(\text{i.e., from } 3\,{}^2P_{\frac{1}{2}} - 3\,{}^2P_{\frac{1}{2}} \text{ and } 3\,{}^2P_{\frac{3}{2}} - 3\,{}^2S_{\frac{1}{2}}\right) \text{ as shown below.}$$

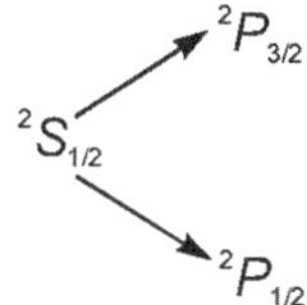

6.1.1 Intensity in Fine Structure Doublets

The observation has shown that the following intensity rules can be applied to the fine structure doublets.

The intensity pattern observed in the fine structure components can be used to arrive at certain conclusions in terms of quantum numbers of electron in the initial and final states. The rules are summarized as follows:

1. The strongest line in the doublet is the one in which the quantum numbers J and L change in the same way. It implies that a transition ${}^2S_{1/2} - {}^2P_{3/2}$ will be stronger than a transition ${}^2S_{1/2} - {}^2P_{1/2}$.

2. When there is more than one line satisfying the above rule, the line with the largest J value is the strongest.

3. The sum of the intensities of lines which come from a common initial level is proportional to $(2J + 1)$ of that level. $(2J + 1)$ is called quantum weight or weight factor.

4. The sum of intensities of line of a doublet, which end on a common level, is proportional to the quantum weight $(2J + 1)$ of that level.

Thus as shown above the line ${}^2S_{1/2} - {}^2P_{3/2}$ is stronger than ${}^2S_{1/2} - {}^2P_{1/2}$. The two lines start from the upper levels ${}^2P_{1/2}$ and ${}^2P_{3/2}$ ends in a common level ${}^2S_{1/2}$. In the former case L changes by 1 and also J changes by 1. In the second case though L changes by 1, J value does not change (in both states in this case $J = \frac{1}{2}$). Also the quantum weight of the common lower level is $2 \cdot \dfrac{1}{2} + 1 = 2$, and the total quantum weight of the upper two levels are $\left(2 \cdot \dfrac{3}{2} + 1\right) : \left(2 \cdot \dfrac{1}{2} + 1\right) = 4 : 2$, ${}^2P_{3/2}$ has more statistical weight and hence transition from or to that state is stronger.

Consider another example of the diffuse series in which there are two strong lines and a weak satellite. The transitions are ${}^2D_{5/2} - {}^2P_{3/2}$, ${}^2D_{3/2} - {}^2P_{1/2}$ and ${}^2D_{3/2} - {}^2P_{3/2}$. In the former two $\Delta L = 1$ and $\Delta J = 1$ and the line ${}^2D_{5/2} - {}^2P_{3/2}$ is stronger according to the rule 2 above.

The satellite line $^2D_{3/2} - {}^2P_{3/2}$ has $\Delta L = 1$ and $\Delta J = 1$ and hence is weak (*See* Figure 6.4).

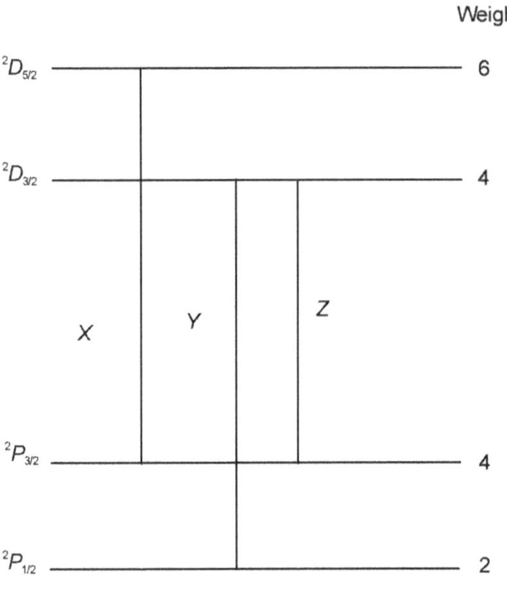

Figure 6.4   First member of the diffuse series

The relative intensities of the lines can be calculated by calculating the quantum weights of the levels. The quantum weights or statistical weights of the ^2D and ^2P states are given in Table 6.2.

Table 6.2   Statistical weights of ^2D amd ^2P states

Final state	Initial state	$^2D_{5/2}$	$^2D_{3/2}$
	$2J + 1$	6	4
$^2P_{3/2}$	4	X	Y
$^2P_{1/2}$	2	0	Z

X, Y and Z represent the unknown intensities of the three allowed transitions and zero for the forbidden transition. $^2P_{1/2} - {}^2D_{5/2}$ is forbidden.

From rule 3 we have $X : (Y + Z) = 6 : 4$

From rule 4 $(X + Y) : Z = 4 : 2$

$$\frac{X}{Y+Z} = \frac{6}{4} \text{ and } \frac{X+Y}{Z} = \frac{4}{2}$$

The above equations satisfy when $X = 9$, $Y = 1$ and $Z = 5$.

Generally the doublet D states are very close together and that they may not be resolved. The two lines observed may have then an intensity ratio $(9 + 1) : 5$ or $2 : 1$.

When the spectra of alkali metals are examined in detail, it is found that in the principal and sharp series of a particular element each line is really a doublet, i.e., consists of two lines. The well known sodium D line, which is a member of the principal series is actually two lines at $\lambda = 5896$ and $\lambda = 5890$ Å. When one goes from Li to Cs, it is found that the separations of lines in the doublet increase. The yellow lines of sodium are thus split due to spin–orbit coupling.

6.2 THE SODIUM D LINES

The two close prominent lines in the yellow region of the spectrum of neutral sodium atom at wavelengths of 588.9950 and 589.5924 nanometres (approx. 5890 Å and 5896 Å) are referred to as sodium D lines. They were first observed by Fraunhofer in the solar spectrum. And these doublets were conspicuous in stars similar in temperature of the sun. These lines are also called Fraunhofer lines. They also show up in the spectra of remote stars due to absorption by sodium atoms in the interstellar medium. We have already stated in our earlier discussions that the D lines are emitted by electron transitions from the $3p$ to the $3s$ levels of sodium. As discussed earlier the line at 589.0 nm has twice the intensity of the line at 589.6 nm. The transitions involved in sodium D lines are shown in Figure 6.5.

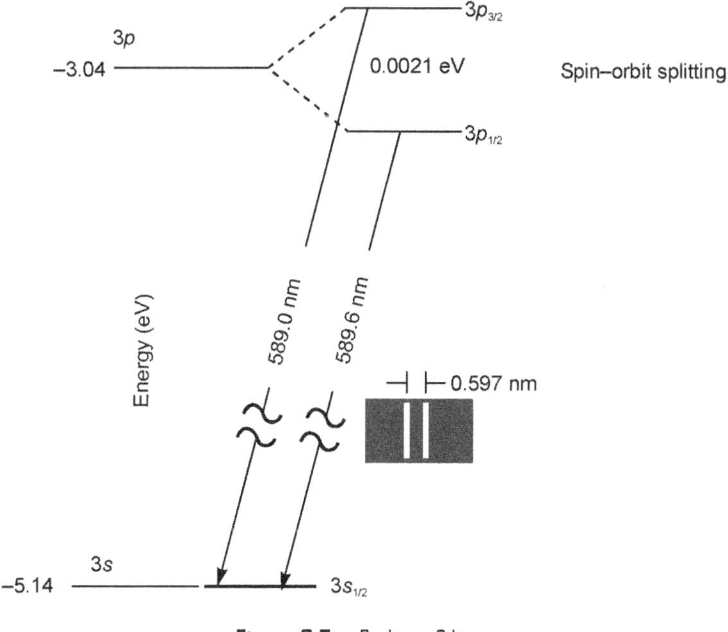

Figure 6.5   Sodium D lines

We have already seen that the sodium atom has 11 electrons in which ten of them constitute the core of the atom and the single valence electron is responsible for all spectral features. The spectral lines arise from the transition of this single electron, and the state of the atom is determined by the state of that electron. As sodium atom constitutes a single valence electron, the total spin $S = s = \frac{1}{2}$ and the splitting of the levels are $(2S+1) = 2$. Thus the state consists of doublets except the ground state. The ground state is with $L = 0$, $S = 1/2$ and $J = L+1/2 = \frac{1}{2}$ giving rise to $^2S_{1/2}$ state. The D line belongs to the principal series and is due to transition between a P state and the ground S state. For the upper P state $L = 1$, $S = 1/2$ and hence $J = \frac{1}{2}$ and $3/2$, thereby there are two P components, viz., $^2P_{1/2}$ and $^2P_{3/2}$. As derived above the ground S state is a single state $^2S_{1/2}$ and hence the transitions that can take place are with selection rules $\Delta L = \pm 1$ and $\Delta J = 0$ or ± 1. We obtain two transitions (doublet), viz.,

$$^2P_{1/2} \rightarrow {}^2S_{1/2} \text{ and } {}^2P_{3/2} \rightarrow {}^2S_{1/2}$$

This explains the doublet fine structure of sodium D lines in nitrogen. The D_1 component has a wavelength 5896 Å and is due to $3^2P_{1/2} \rightarrow 3^2S_{1/2}$ and the D_2 component has a wavelength 5890 Å and is due to $3^2P_{3/2} \rightarrow 3^2S_{1/2}$. The D_2 line is more intense since for D_2, $\Delta L = -1$ and $\Delta J = -1$ while for D_1, $\Delta L = -1$ and $\Delta J = 0$. The total intensity is $(2J+1)$ and hence the ratio of the intensities of the two lines is $2 : 1$.

6.3 FINE STRUCTURE IN THE OTHER SERIES

The sharp series arises due to transition between S and P orbitals and since S states are single, the transitions in the sharp series consist of doublets. However, in transitions in which the S states are not involved, the lines are triplets. For example the diffuse series arises due to transition between D and P states. The D states consist of $^2D_{5/2}$ and $^2D_{3/2}$ and the P states $^2P_{3/2}$ and $^2P_{1/2}$. Four transitions are therefore possible, viz.,

$$^2D_{5/2} \rightarrow {}^2P_{3/2}$$
$$^2D_{5/2} \rightarrow {}^2P_{1/2}$$
$$^2D_{3/2} \rightarrow {}^2P_{3/2}$$
$$^2D_{3/2} \rightarrow {}^2P_{1/2}$$

Out of these $^2D_{5/2} \rightarrow {}^2P_{1/2}$ is forbidden as it involves $\Delta J = 2$. $^2D_{5/2} \rightarrow {}^2P_{3/2}$ and $^2D_{3/2} \rightarrow {}^2P_{1/2}$ are intense as they involve $\Delta L = -1$ and $\Delta J = -1$. $^2D_{3/2} \rightarrow {}^2P_{3/2}$ is weak as it involves $\Delta L = -1$ and $\Delta J = 0$. Many times this line is not seen experimentally. Thus the spectrum of alkali metal consists of doublets and triplets. In most of the cases the doublet separation for the D state is smaller and it goes on decreasing as the L value increases.

6.4 SPECTRA OF BORON GROUP ELEMENTS

When the valence electron in the s orbit is responsible for the spectra in the alkali metal, the single unbalanced p electron in the outer shell contributes to the spectra in boron group of elements. The electron configuration in the boron group of elements is

B	5	$1s^2\, 2s^2\, 2p$
Al	13	$1s^2\, 2s^2\, 2p^6\, 3s^2\, 3p$
Ga	31	$1s^2\, 2s^2\, 2p^6\, 3s^2\, 3p^6\, 3d^{10}\, 4s^2\, 4p$
In	49	$1s^2\, 2s^2\, 2p^6\, 3s^2\, 3p^6\, 3d^{10}\, 4s^2\, 4p^6\, 4d^{10}\, 5s^2\, 5p$
Tl	81	$1s^2\, 2s^2\, 2p^6\, 3s^2\, 3p^6\, 3d^{10}\, 4s^2\, 4p^6\, 4d^{10}\, 4f^{14}\, 5s^2\, 5p^6\, 5d^{10}\, 6s^2\, 6p$

It is the single unbalanced p electron, which can be excited to various possible energy states. In alkali metals we have seen that the ^2S state is the ground state or the state which lies lowest in energy followed by ^2P and then other doublet states. In boron group, there is only one electron as in the alkali metal atoms, but this electron is a p electron, which gives rise to a ^2P state. Thus P state is the ground state of this group. As there is only one electron in the valence shell which is responsible for the excitation, the total spin is $1/2$ and hence the multiplicity of the states is $2 \cdot 1/2 + 1 = 2$. Thus the ground state is ^2P. In Boron, for example, when the electron is excited it goes to different states depending on the excitation energy. The first orbit to which the electron can be excited is the $3s$, which will give rise to a ^2S state. The single electron can then be excited to the $3d$ orbit giving rise to a ^2D level followed by ^2F level and so on. The ^2F levels as in alkali metals are hydrogen like indicating non-penetrating orbits. If we consider the first excited ^2S state, though it is doublet it gives rise to a single level, viz., $^2S_{1/2}$. The excited ^2P state will be a doublet, viz., $^2P_{1/2}$ and $^2P_{3/2}$. The doublet separation in the P state increases as we go to higher in Z, i.e., going from boron to thallium. This is shown schematically in Figure 6.6.

The same kind of spreading of the fine structure is observed in alkali atoms too when Z is increased, i.e., going from lithium to cesium. This is visible in the doublet splitting in the first member of the principal series. The fine structure separations in the first member for the alkali metal atoms are given in Tables 6.3 and 6.4 in cm^{-1} and in eV.

Table 6.3 The fine structure separations in lithium group

Li	Na	K	Rb	Cs
0.338	17.2	57.9	237.7	554.0 cm^{-1}
0.000042	0.00213	0.00717	0.0294	0.0686 eV

and in Boron group they are

Table 6.4   The fine structure separations in boron group

B	Al	Ga	In	Tl
5.20	112.07	526.10	2212.67	7792.42 cm^{-1}
0.000644	0.0139	0.0652	0.2741	0.9655 eV

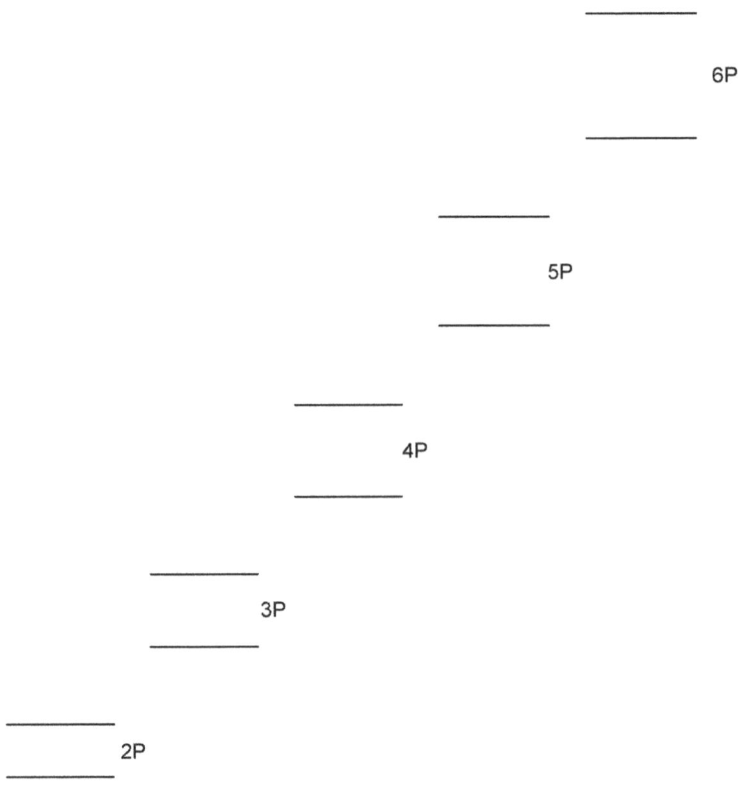

6P

5P

4P

3P

2P

Figure 6.6   Increase in doublet separation going from boron to thallium (not on scale)

The following points are important while considering the fine structure splitting.

1. It is observed that the fine structure intervals increase as Z increases.
2. It is observed that the separations in the ionized atom are larger than the corresponding splitting in the normal atom.
3. In each atom the doublet separation decreases in going to higher members.
4. Within each element the P doublets are wider than D doublets with the same n.

6.5 PENETRATING AND NON-PENETRATING ORBITS

The Bohr–Stoner scheme of electron configuration in atoms has shown that the alkali atoms consist of a core with completed shell and an s electron outside. The completed shell of electrons gives rise to spherically symmetrical potential. We have already seen that

1. Levels with the same n but different l have different energies.

2. Alkali atoms with levels having the same n have greatly different energies as compared to hydrogenic levels. It has been observed that when the value of l with the same value of n increases, the divergence from the hydrogenic levels decreases. This means that the largest divergence is when $l = 0$, i.e., for s orbits and decreases going to higher values of l with the same n.

For example in lithium the term value in the ^2S state, neglecting the fine structure splitting, is 43484.4 cm^{-1} whereas in the ^2P state it is 28581.4 cm^{-1}. The term value for the S state in hydrogen with the same principal quantum number is 27479.4 cm^{-1} which is obviously nearer to the ^2P state in Li. Rydberg suggested that the energy levels in alkali atoms could be characterized by the relation.

$$T_{n,l} = -\frac{R}{(n - \mu_1)^2} \quad \text{where } \mu_1 \text{ is known as quantum defect.} \tag{6.1}$$

Generally non-penetrating orbits are those for which the energies are very nearly equal to those corresponding to hydrogen like orbits. These orbits seldom penetrate into the atom core. The azimuthal quantum number l is large in these kinds of orbits. Generally orbits with l values nearing to n are non-penetrating orbits. In the above example $3d$ orbital is a non-penetrating orbit (*See* Figure 6.7).

It is difficult to find a sharp demarcation between penetrating and non-penetrating orbits. However, orbits for which the term values (energies) are approximately different from those of hydrogenic values with the same n can be penetrating orbits.

The spectra observed in alkali metals were first treated on the basis of Bohr theory. Even in the simplest alkali atom Li with the electron configurations $1s^2\,2s$, the valence electron $2s$ is not acted upon by the coulomb attraction of the nucleus, but has to be taken in the light of electron repulsion by the closed $1s^2$ shell. The interaction also destroys the spherical symmetry of the potential field making it angle dependent. It was Schrödinger[1] who suggested that an idealized model would be to look the atom in an atom core model in which the core electrons are thought of as being distributed uniformly over the surface of one or more concentric spheres. This along with the Sommerfeld theory of elliptical orbits could provide a reasonable picture of the spectra in alkali metals. In Sommerfeld theory the orbits with small value of l are more elliptical in shape whereas the orbits with larger l approach circular orbits. Thus the Sommerfeld theory is important as it had been suggested that orbits with small l and larger l for a given value of n could be classified into two groups, one elliptical and the other approaching circular. The orbits with smaller are those orbits which are nearer to the charge core closely, they may even penetrate into the closed shell while the orbits with larger l are outside the core and are quite away from it, thus came the classification of penetrating and non-penetrating orbits.

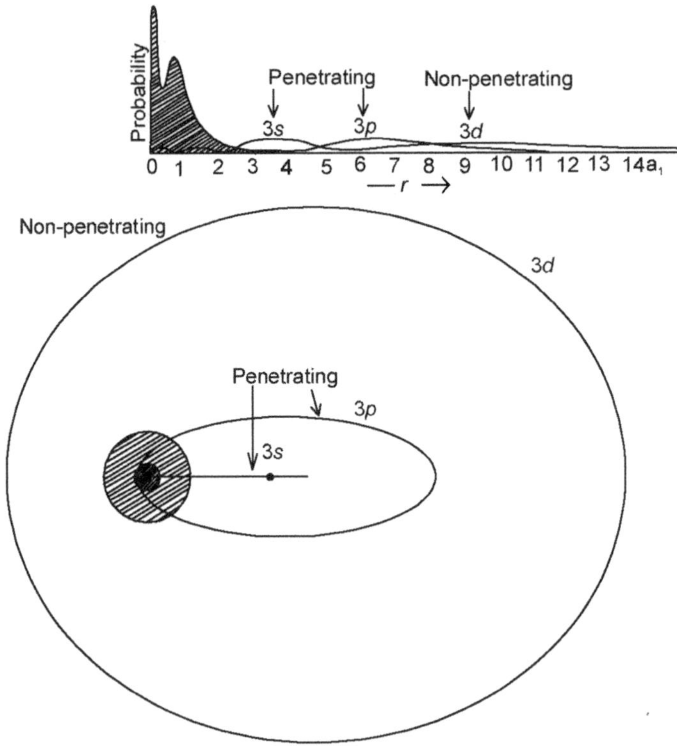

Figure 6.7   Three of the lowest possible states for the single valence electron (e.g., sodium atom)

By making a simple approach, we can consider the charge density produced by the completely filled inner shells as that of a uniformly charged sphere. Let us consider the effective charge experienced by the valence electron due to the nucleus and the closed shell is $Z_i\,e$ when it is in the penetrating region (for lower l) and Z_0 e be the electronic charge outside the shell. The potential energy of the valence electron when it is outside the inner shell is

$$V_0(r) = -\frac{Z_0 e^2}{r} \qquad (6.2)$$

While inside the shell it becomes

$$V_i(r) = -\left(\frac{Z_i e^2}{r}\right) + \frac{(Z_i - Z_0)e^2}{\rho} \qquad (6.3)$$

where $(Z_i - Z_0)\,e$ is the charge to be considered for the valence electron outside the spherical shell and the potential energy due to the charge contained in the spherical shell lying outside the orbit of the penetrating electron is

$$\frac{(Z_i - Z_0)e^2}{\rho} \quad r < \rho \tag{6.4}$$

The electron when it is completely outside the inner shells moves in an elliptical orbit appropriate to the potential $V_0(r)$. When it reaches the boundary of the inner shell, the effective potential changes to $V_i(r)$ and it changes to another orbit appropriate to the potential. The reverse is the case when the electron comes out of the inner region.

The kinetic energy is $\dfrac{1}{2}mv^2$ and in polar coordinates it is $m\,\dfrac{1}{2}\left(\dot{r}^2 + r^2\dot{\phi}^2\right)$

Thus the kinetic energy in polar coordinates $= \dfrac{1}{2}m\dot{r}^2 + \dfrac{1}{2}mr^2\dot{\phi}^2$

These involve two displacement coordinates, ϕ and r and two momentum coordinates

$$p_\phi = mr^2\dot{\phi} \quad \text{and} \quad p_r = m\dot{r}$$

For the two degrees of freedom, two phase space diagrams can be drawn as in Figures 2.22.

Applying the quantum conditions $\quad \oint p_\phi d\phi = kh \tag{6.5}$

And $\qquad\qquad\qquad\qquad\qquad \oint p_r dr = rh \tag{6.6}$

where p_r and p_ϕ are called radial and angular momenta, k and r can take integral values and are called azimuthal and radial quantum numbers.

According to Kepler's second law of planetary motion, the angular momentum p_ϕ remains a constant during the motion and

$$\oint p_\phi d\phi = \int_0^{2\pi} p_\phi d\phi = 2\pi\, p_\phi = 2\pi\, mr^2\dot{\phi} = kh \tag{6.7}$$

$$p_\phi = k\frac{h}{2\pi} \tag{6.8}$$

The total energy of the electron is

$$E = T + V = \frac{1}{2m}[p_r^2 + p_\phi^2] + V(r) \tag{6.9}$$

$V(r)$ has different values in the outer and inner part of the electron's orbit.

The quantum conditions for the radial moment p_r is

$$\oint p_r dr = \int_{inner} p_r dr + \int_{outer} p_r dr = rh \tag{6.10}$$

The total energy when the electron is outside the spherical shell is

$$E = T + V = \frac{1}{2m}[p_r^2 + p_\phi^2] - \frac{Z_0}{r^2} \tag{6.11}$$

The radial quantum conditions are obtained by evaluating the integrals over the whole circle

$$\oint \left[2m\left(E + \frac{Z_0 e^2}{r} \right) - \frac{p_\phi^2}{r} \right]^{\frac{1}{2}} dr = \oint R_0 dr = r_0 h \tag{6.12}$$

$$\oint \left[2m\left(E + \frac{Z_0 e^2}{r} - \frac{(Z_i - Z_0)e^2}{\rho} \right) - \frac{p_\phi^2}{r} \right]^{\frac{1}{2}} dr = \oint R_i dr = r_i h \tag{6.13}$$

The total quantum number or the principal quantum number is then

$$n_0 = k + r_0 \quad \text{and} \quad n_i = r_i + k \tag{6.14}$$

The total energy in the outside region considering the potential energy as zero when the electron is infinitely far removed from the nucleus

$$\text{P.E.} = V = -\frac{Z_0 e^2}{r} = -2T \text{ as kinetic energy is } \frac{1}{2}mv^2 = \frac{Z_0 e^2}{2r}.$$

[for equilibrium condition we have seen in Bohr's theory that

$$\frac{Z_0 e^2}{r^2} = \frac{mv^2}{r}, \text{ i.e., } v^2 = \frac{Z_0 e^2}{mr} \text{ and kinetic energy} = \frac{1}{2}mv^2 = \frac{Z_0 e^2}{2r}$$

$$E_0 = T + V = T - 2T = -T = \frac{V}{2} = -\frac{Z_0 e^2}{2r} = \frac{Z_0^2 e^2}{2a_1 n_0^2} \tag{6.15}$$

where, a_1 is the radius of the first Bohr circular orbit.

In the discussions of Bohr's theory we have seen that

$$\frac{e^2 Z}{2a} = \frac{2\pi^2 m e^4 Z^2}{n^2 h^2} \quad \text{from which} \quad a = \frac{h^2}{4\pi^2 m e^2} \cdot \frac{n^2}{Z} \quad \text{and the radius of the first Bohr orbit is}$$

$a_1 = \dfrac{h^2}{4\pi^2 m e^2}$. Note that for the correct expression, we have to take into account $\dfrac{1}{4\pi\varepsilon_0}$ and a_1 is

then $\dfrac{h^2 \varepsilon_0}{\pi m e^2}$. For the first orbit some authors use a_0 and some others use a_1.

The total energy inside is

$$E_i = -T = -\frac{V}{2} = -\frac{Z_i^2 e^2}{2 a_1 n_1^2} + \frac{(Z_i - Z_0)e^2}{\rho} \tag{6.16}$$

In the example if the outside orbit were a complete Keppler ellipse, the electron would never penetrate the shell. Similar is the case with inner orbit and if it were complete, the electron will remain

inside. And hence the integral $\displaystyle\oint_{outside} R_0 dr + \oint_{inside} R_i dr$ approach those of equations 6.12 and 6.13.

For stationary state $E_0 = E_i$

Expressing $r = r_0 + r_i$ and $n = r + k$ we get $n - n_0 = n_i - k$

The value of ρ is the smallest distance of the electron in the outer orbit from the nucleus and it should be the same as the longest distance of the electron from the nucleus when the electron is in the inner orbit. That means the perihelion distance of the outer orbit which is $a_0(1-\varepsilon_0)$ approaches the aphelion of the inner ellipse $a_i(1+\varepsilon_i)$, both approaching at the same time the radius of the spherical shell ρ.

Thus $a_0(1-\varepsilon_0) = a_i(1+\varepsilon_i) = \rho$ \hfill (6.17)

where a_0 and a_i are the semi-major axes of the outer and the inner ellipse and ε_0 and ε_i are the respective eccentricities.

$$\varepsilon_i = \frac{\rho}{a_1} - 1 \tag{6.18}$$

We have already seen that $k + r = n$, $n_0 = k + r_0$ and $n_i = r_i + k$

$n - n_0 = n_i - k = \mu$ which is the quantum defect.

The eccentricities $\varepsilon_i^2 = 1 - \dfrac{k^2}{n_i^2}$ and $a_i = \dfrac{a_0 n_i^2}{Z_i}$ \hfill (6.19)

Using these expressions in 6.18 we obtain $\mu = n - n_0 = n_i - k$ and

$$\varepsilon_i^2 = \left(\frac{\rho Z_i}{n_i^2 a_1} - 1\right)^2 = 1 - \frac{k^2}{n_I^2} \tag{6.20}$$

$$n_i = (\rho\, Z_i(2\rho a_1 Z_i - a_1^2 k^2)^{-1/2} \tag{6.21}$$

where a_1 is the first Bohr orbit.

We have already seen by comparing the old quantum theory with the quantum mechanics that the average value of r can be compared as

$$\frac{1}{r^2} = \frac{Z^3}{a_1^3 n^3 k^3} \quad \text{[old quantum theory (Bohr–Sommerfeld)]} \tag{6.22}$$

$$\frac{1}{r^2} = \frac{Z^3}{a_1^3 n^3 l(l+1)\left(l+\dfrac{1}{2}\right)} \quad \text{(quantum mechanics)} \tag{6.23}$$

We can replace now k^2 and k by their quantum mechanical equivalence $l(l+1)$ and $l+\dfrac{1}{2}$.

$$M = \rho Z_i [2\rho a_1 Z_i - a_1^2\, l(l+1)]^{-1/2} - \left(l + \frac{1}{2}\right) \tag{6.24}$$

6.6 QUANTUM MECHANICAL MODEL OF PENETRATING ORBIT

The Hamiltonian for an atom containing N number of electrons can be described as

$$H = \sum_i^N \left(-\frac{\hbar^2}{2m}\nabla_i^2 - \frac{Ze^2}{r_i}\right) + \sum_{j>i=1}^N \frac{e^2}{r_{ij}} \tag{6.25}$$

where the subscripts i and j corresponds to ith and jth electrons, r_i is the distance of the ith electron from the nucleus and r_{ij} is the distance between the ith and jth electrons. The second summation corresponds to the mutual repulsion between the ith and jth electrons. Because of this term in the Hamiltonian, a separation of variables is not possible and its effect in the quantum states is to be taken into account. As the number of electrons in the atom increases, these repulsive terms become important. The second repulsive term may tend to cancel the attractive term contained in the first term, the effect of the term is smaller than the attractive term. For very large value of r_i, i.e., when the ith electron is at a large distance compared to the other electrons from the nucleus (in the outermost orbit) we may consider $r_{ij} \approx r_i$ and in this case $\dfrac{e^2}{r_{ij}} \approx \dfrac{e^2}{r_i}$. If we consider in this way, and since for a particular value of i, j can have $N-1$ values $\sum_{i \Delta j} \frac{e^2}{r_{ij}}$, considering all other electrons will become $(N-1)\frac{e^2}{r_i}$. The repulsive potential acting on the ith electron is then

$$V(r_i) = -\frac{Ze^2}{r_i} + (N-1)\frac{e^2}{r_i} = -(Z-N+1)\frac{e^2}{r_i} \tag{6.26}$$

The effective charge is then $-(Z-N+1)\frac{e^2}{r_i}$ which shows that the charge of the electron is reduced and the effective charge is reduced by a factor of $(Z-N+1)$. When r_i large effective charge is small where when r_i is small it is large. However if we assume that r_i is very much smaller than all $r_j s$ ($i<<j$ and $i \neq j$) and if we assume a spherical charge distribution from the other $(N-1)$ electrons, the potential inside a shell of radius r_s having a charge e_s is the same at all points inside the shell and is equal to $\frac{e_s}{r_s}$. Thus the potential at the ith electron can be written as

$V(r_i) = -\frac{Ze^2}{r_s} + \frac{e_s^2}{r_s} = -\frac{Ze^2}{r_s} + (N-1)\frac{e^2}{r_s}$. The second term can be treated as a constant and thus we can consider that each electron moves under a potential

$$U(r) = -\frac{Ze^2}{r_s} + C \text{ when } r \text{ is small and} \tag{6.27}$$

$$U(r) = -(Z-N+1)\frac{e^2}{r_s} \text{ when } r \text{ is large} \tag{6.28}$$

We have already seen in our discussion on probability description that the orbital, we have found that the orbitals having a larger value of l for a particular n the probability goes on decreasing as we go from lower l to higher l. Thus we can consider that the electrons with lower l of a particular n remains in the neighbourhood of the nucleus than those electrons with higher l. The potential experienced by the higher l can be represented by the second potential whereas the first will be applicable for those electrons with lower l. Thus it is obvious that the potential of the first type will give a spectrum similar to hydrogenic atoms whereas the second potential will lead to deviations from the hydrogenic term values. Following the above approximations we can write the Schrödinger equation for the electron at the ith position as

$$\left[-\frac{\hbar^2}{2m}\nabla_i^2 + U(r_i) \right] \Psi(r_i) = E_i \Psi(r_i) \tag{6.29}$$

where E_i is the energy eigen value of the ith electron. In the case of alkali atoms where there is only one electron in the outer orbit, it can be thought that the excitation of the atoms occur due to the excitation of this single electron. Hence the spectrum would look like similar to the hydrogenic atoms. The total energy of the atom is the energy of the valence electron and the Schrödinger equation is similar to the one given in equation 6.29. It can be solved by many approximate methods. The solution gives the energy as

$$E_{n,l} = \frac{R}{(n - \mu_l)^2} \tag{6.30}$$

where μ_1 is called quantum defect. It is not possible to evaluate the potential and hence the experimental results are generally used to calculate the quantum defect in atoms. Some authors write the solution in terms of a shielding term by expressing the energy

$$E_{n,l} = \frac{R(Z - \sigma_{n,l})}{n^2} \tag{6.31}$$

The value of μ_l found in the case of sodium atom ($n = 3$, $l = 0$) is 1.37 cm^{-1} (Term value is 41444.9 cm^{-1}) whereas in Li ($n = 2$, $l = 0$) it is 0.41 cm^{-1} (Term value is 43484.4 cm^{-1}).

6.7 NORMAL AND INVERTED DOUBLET TERMS

In the case of the $^2P_{1/2}$ and $^2P_{3/2}$ states in sodium atom we have seen that the $^2P_{1/2}$ lies lower than $^2P_{3/2}$. Generally in doublet energy levels it is observed that the component with $J = L - 1/2$ lies lower in energy than the component with $J = L + 1/2$. The electron with its intrinsic spin may be regarded as a small magnet with a magnetic moment μ_s whose direction is always in opposite to the direction of the spin angular momentum because of the negative charge on the electron. The orbital motion of the electron also produces a magnetic field H at the position of the electron whose direction is the same as the orbital angular momentum. It can be easily understood if we imagine that the nucleus with its positive charge revolves around the electron and it will produce a magnetic field in the direction of the orbital angular momentum. The electron will have the minimum energy when μ_s is parallel to H. This is the case in the state where $J = L - 1/2$ and in the state where $J = L + 1/2$, μ_s is antiparallel to H. A normal term is the one in which the state with smallest J value lies deepest. In some elements it is observed that the states with largest J value lie deepest. This occurs in elements where the electrons are filled in the shells more than half its allowed number. The ground state of boron is a $^2P_{1/2}$ state as the boron has an electronic configuration.

$$B: 1s^2 \, 2s^2 \, 2p$$

which lead to $^2P_{1/2}$ and $^2P_{3/2}$ states. In the case of fluorine the electronic configurations is

$$F: 1s^2 \, 2s^2 \, 2p^5$$

In this case the outer shell needs one more electron to fill the shell. In classical argument this is like a one valence electron case where the doublets are inverted. We can consider this case in simple description as an electronic configuration of F: $1s^2 \, 2s^2 \, 2p^6$ with an electron with an imaginary positive charge equivalent in magnitude that of the electron. In this case the magnetic moment originating from the spin motion μ_s will have its direction in the same direction of the spin angular momentum as in this case we are considering a positive charge. Thus in this case $J = L + 1/2$ corresponds the case where μ_s and H are parallel. Thus a normal term arises when the electronic configuration is less than half of an incomplete subshell of electrons and an inverted state occurs when the electronic configuration is more than half the incomplete subshell of electrons.

ENDNOTE

1   Schrödinger, E. (1921). "Versuch zur modelmässigen Deutung des Terms der Scharfen Nebenserien." Z. f. *Physik*. 4, 347.

REFERENCES

Condon, E.U. and Shortley, G.H. (1935). *The Theory of Atomic Spectra.* Cambridge University Press, Cambridge.

Cowan, R.D. (1981). The Theory of Atomic Structure and Spectra. University of California Press, Berkeley.

Rai, D.K. and Thakur, S.N. (2005). *Atomic Structure and Modern Spectroscopy.* Vaivaswat Publications, Varanasi.

White, H.E. (1934). *Introduction to Atomic Spectrz.* Mc Graw-Hill Book Company, New York.

7

THE MULTIPLET STRUCTURE OF SPECTRAL LINES

7.1 VECTOR MODEL OF ATOM

We have already seen that the spectra of alkali atom consist of doublet fine structure. The atoms in the II A and II B groups of elements like beryllium, magnesium, calcium, strontium and barium, zinc, cadmium, mercury on the other hand give rise to singlet and triplet spectral lines. Here again the singlet and triplet series of lines may each be grouped into four groups of lines, viz., sharp, principal, diffuse and fundamental. It implies that there are two types of energy levels, singlets and triplets. The characteristics of spectra originating from the multiplet structure of energy levels are described in Section 8.2 of Chapter 8.

In order to explain the multiplet structure of spectral lines it was found that two quantum numbers we hitherto introduced, viz., n and l were not sufficient. A third quantum number m was also introduced by Sommerfeld, m having the values between $-l$ to $+l$ changing by one unit. It had become obvious that an atom, though could be oriented in a special direction in a magnetic field, it is not possible to single out a particular direction when there is no magnetic field. And hence its properties must be independent of m. It was possible to explain the normal Zeeman effect by this concept of orientation, however the anomalous Zeeman effect remained to be unexplained. As an another step to understand the multiplet structure Sommerfeld introduced a concept of inner quantum number. The fact that the alkali spectra were doublets and so also in other monovalent atoms like copper, silver and gold, it was concluded that the energy levels are doublets.

Similarly in the case of alkaline earth atoms there are two sets of energies, one singlets and the other triplets. Thus it has been found that in alkali atoms all levels are doublets and in the alkaline earth atoms all levels are either singlets or triplets except the *s* levels. The *s* levels are always observed as singles even when the series spectrum showed multiple character. This is particularly important in the diffuse series of lines in alkali atoms where the transitions are from different *d* levels to the lowest *p* level. Since the energy levels are doublets one would expect a quartet rather than doublets spectra. Even in the case of cesium atom where the doublet separation is large, there are only three lines and not four. Of these two are strong and the other weak usually called a satellite line. Similarly in the triplet diffusive series of alkaline earth atoms, the triplet nature of the two combining levels should give rise to nine lines, however even in the well resolved spectra there were only six lines of which two were intense and the rest four were weak. Thus Sommerfeld assigned additional quantum numbers arbitrarily, each component of the triplet *p* levels were termed as p_0, p_1, p_2 and the *d* levels as d_1, d_2, d_3. The suffixes were the inner quantum numbers named as *j* and the selection rules were $\Delta j = 0, \pm 1$. Similarly the doublet *p* levels were p_1 and p_2 and the doublet *d* levels as d_2 and d_3. The arbitrary values of the inner quantum numbers were later explained while describing Zeeman splitting by Lande. He had found that the number of levels into which a given multiplet level splits into Zeeman components in a magnetic field is $2j + 1$. The difference between the *j* introduced by Sommerfeld and that of Lande was by a value 1/2. Thus the correct value of *j* was related by the relation

$$J \text{ (correct)} = j \text{ (Sommerfeld)} - 1/2$$

The explanation for the number 1/2 was quite unintelligible till the discovery of the electron spin. The above developments have led to the vector atom model.

7.2 VECTOR MODEL FOR ORBITAL ANGULAR MOMENTUM

We have already seen that the electrons possess an orbital angular momentum. When an electron moves around in an orbit it has an orbital angular momentum and it can be visualized in terms of a vector.

The principle behind the vector atom model is that the atom itself can be visualized in terms of a vector in which the angular momentum vector is seen as precessing about a direction in space as shown in Figure 7.1(a). The angular momentum vector has the magnitude $|L| = \sqrt{l(l+1)}\ \hbar$. This is shown in Figure 7.1(b). However, only a maximum of *l* units can be measured along a given direction, where *l* is the orbital quantum number.

It should be remembered that it is a special kind of vector because it's projection along a direction in space is quantized to values one unit of angular momentum apart. That means it follows the quantum theory as well. We already know that as electron has a charge and therefore there is a magnetic moment associated with the orbital angular momentum. Hence the precession can be compared to the precession of a classical magnetic moment caused by the torque exerted by a magnetic field. This is depicted in Figure 7.2.

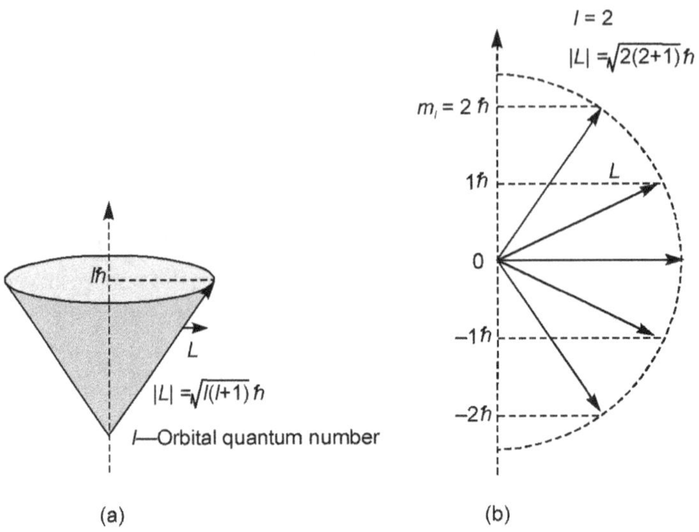

Figure 7.1 (a)   Vector model of atom (b)   Angular momentum orientation in vector model

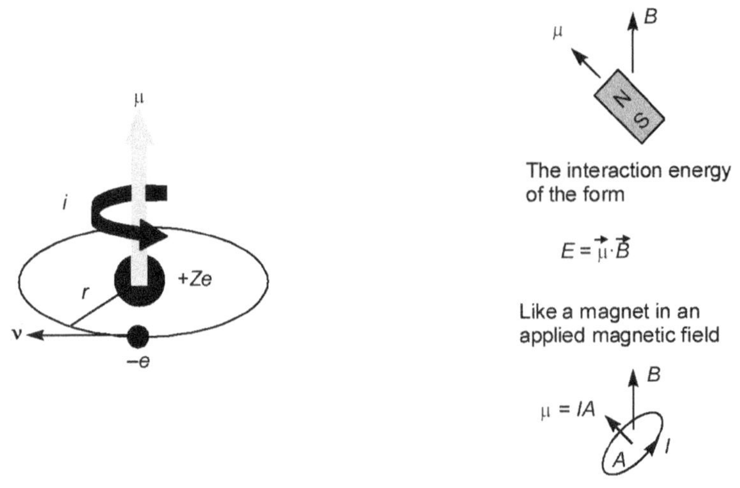

Figure 7.2   Classical magnetic moment caused by the torque

It can be considered as an "internal Zeeman" effect having magnetic quantum numbers. The Figure 7.1(b) above shows that the possible values for the "magnetic quantum number" m_l for $l = 2$ can take the values

$$m_l = -2, -1, 0, 1, 2$$

or, in general

$$m_l = -l, -l+1, \ldots, l-1, l \tag{7.1}$$

If we consider the total angular momentum j obtained by the vectorial addition of orbital angular momentum l and the electron spin angular momentum s, i.e., $j = l \pm s$ then the magnetic quantum numbers $m_j = -j, -j+1, \ldots j-1, j$.

7.3 ELECTRON ORBITAL MAGNETIC MOMENT

We know that the classical expression for magnetic moment is $\mu = IA$ where I is the current and A is the area of the loop. An electron in a circular orbit around a nucleus has a magnetic moment and an expression for this magnetic moment can easily be deduced from the above equation. We can see that it is proportional to the angular momentum of the electron. An electron in an orbit is shown in Figure 7.3.

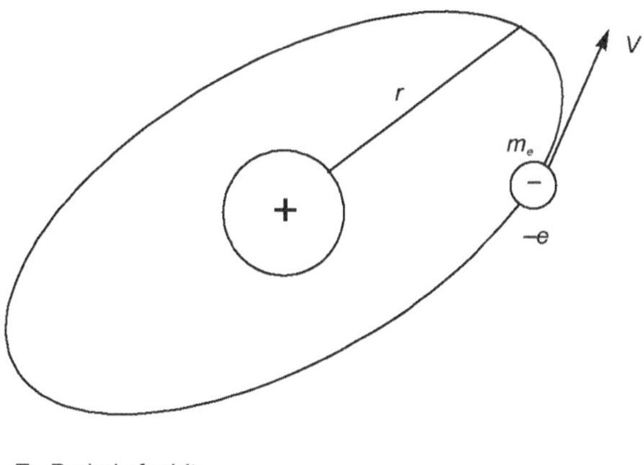

T—Period of orbit
L—Orbital angular momentum

Figure 7.3 Electron in an orbit

According to classical electromagnetic theory a current I conducted in a closed circuit of area A may be treated as a small magnet of magnetic moment μ.

As mentioned above the magnetic moment μ of a current in a single loop of circuit is given by the area of the circuit A times the current I in electrostatic units and divided by the velocity of light c.

$$\mu = \frac{area \cdot I}{c} = \frac{AI}{c} \tag{7.2}$$

IA is considered placed at the centre and standing perpendicular to the plane of the circuit. The rotation of the electron in it's orbit is equivalent to the rotation of a negative charge circumscribing the nucleus $\frac{\omega}{2\pi}$ times per second, or $\frac{1}{T} = \frac{\omega}{2\pi}$.

Hence current
$$I = \frac{-e}{T} = -\frac{\omega}{2\pi} e \tag{7.3}$$

For a circular orbit of radius r, the magnetic moment then becomes

$$\mu = \left(\frac{-\omega e}{2\pi c}\right)\pi r^2 = \frac{-\omega e r^2}{2c} \tag{7.4}$$

The angular momentum p due to the rotation of an electron of mass m is

$$p = m_e r^2 \omega$$

or

$$\omega r^2 = \frac{p}{m_e} \tag{7.5}$$

Hence $\mu = \dfrac{-e}{2m_e c} p$ or $\dfrac{\mu}{p} = -\dfrac{e}{2m_e c}$ $\tag{7.6}$

For an atom the angular momentum $p = \sqrt{l(l+1)}\hbar$ and hence

$$\mu = \frac{-e}{2m_e c}\frac{h}{2\pi}\sqrt{l(l+1)} = -\frac{eh}{4\pi m_e c}\sqrt{l(l+1)} = -\sqrt{l(l+1)}\,\mu_B \tag{7.7}$$

where $\mu_B = \dfrac{eh}{4\pi m_e c}$ is the unit of magnetic moment called the "Bohr magneton".

The Bohr magneton is named after Niels Bohr and is a physical constant of magnetic moment and its symbol is μ_B.

In SI units it is $\mu_B = \dfrac{eh}{4\pi m_e} = \dfrac{e\hbar}{2m_e}$.

In the SI system of units its value is

$$\mu_B = 9.274\ 009\ 49(80) \times 10^{-24}\ J.T^{-1}\ (T \text{ is tesla}).$$

In eV its value is

$$\mu_B = 5.7883\times10^{-5}\ eV \cdot T^{-1}$$

and in CGS units it is defined as

$$\mu_B = \frac{e\hbar}{2m_e c} = \frac{eh}{4\pi m_e c} \tag{7.8}$$

where

e is the elementary charge,

$\hbar$ is the reduced Planck's constant,

m_e is the electron rest mass,

c is the speed of light

In the CGS system of units its value is[1]

$$\mu_B = 0.927 \times 10^{-20} \ \text{erg.gauss}^{-1}$$

The Bohr magneton is the natural unit for expressing the electron magnetic dipole moment in the hydrogen atom. It was first calculated by Romanian physicist Stefan Procopiu around 1910, and in some Romanian literature it is called the Bohr–Procopiu magneton. An electron has an intrinsic magnetic dipole moment of approximately one Bohr magneton.[2]

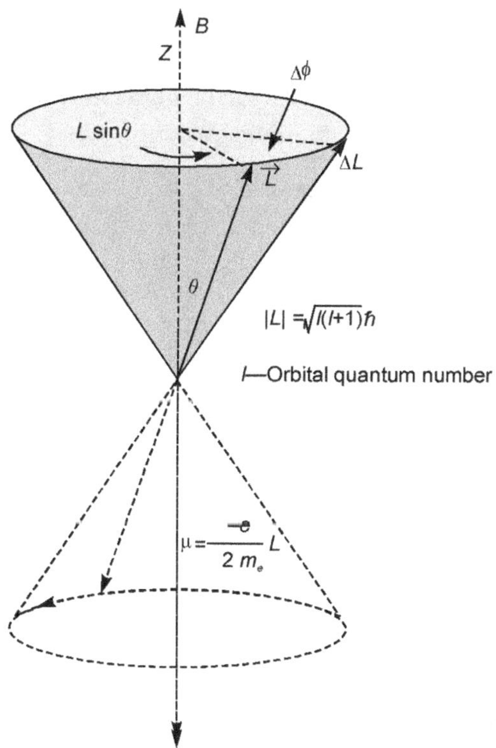

Figure 7.4 Larmor precession

The z-component of the magnetic moment is given by

$$\mu_z = -m_l \mu_B \tag{7.9}$$

Thus the quantization of angular momentum also implies quantization of magnetic moment.

As mentioned above, since there is a magnetic moment associated with the orbital angular momentum, the precession can be compared to the precession of a classical magnetic moment caused by the torque exerted by a magnetic field (*See* Figure 7.2). This kind of precession is called Larmor precession and the characteristic frequency originating from this precession is called the Larmor frequency. It is shown in Figure 7.4.

We already know that when a magnetic moment μ is placed in a magnetic field B, it experiences a torque which can be expressed as a vector product

$$\tau = \mu \times B \tag{7.10}$$

For a static magnetic moment or a classical current loop, the above torque tends to line up the magnetic moment with the magnetic field B, and this represents its lowest energy configuration. But if the magnetic moment arises from the motion of an electron in orbit around a nucleus, the magnetic moment is proportional to the angular momentum of the electron. The torque exerted then produces a change in angular momentum which is perpendicular to that angular momentum, causing the magnetic moment to precess around the direction of the magnetic field rather than settle down in the direction of the magnetic field. This is called the Larmor precession.

Thus when a torque is exerted perpendicular to the angular momentum L, it produces a change in angular momentum ΔL which is perpendicular to L, causing it to precess about the z axis. Labelling the precession angle as ϕ, we can describe the effect of the torque as follows:

$$\tau = \frac{\Delta L}{\Delta t} = \frac{L \sin \theta \Delta \phi}{\Delta t} = \left| \mu B \sin \theta \right| = \frac{e}{2m_e} LB \sin \theta \tag{7.11}$$

The precession angular velocity (Larmor frequency) can be obtained as $\dfrac{\Delta \phi}{\Delta t} \approx \dfrac{d\phi}{dt}$

and hence $$\omega_{\text{Larmor}} = \frac{d\phi}{dt} = \frac{e}{2m_e} B \tag{7.12}$$

These kinds of relationships for a finite current loop can be extended to the magnetic dipoles of electron orbits and to the intrinsic magnetic moment associated with electron spin.

We already know how a mechanical top precesses. The precession of the mechanical top is due to the gravitational field of the earth. The precession of electron in a magnetic field can be visualized in the same way a mechanical top in the presence of the gravitational field. We have called this precession as Larmor precession. In order to find the frequency of this precession of the electron in a magnetic field Larmor put forward a theory, which is called Larmor's theory. He has derived a relation between the magnetic moment, mechanical moment and the frequency of precession in terms of the specific charge e/m of the electron. Larmor's theory is very useful in deriving classically the splitting of energy levels in a magnetic field. Larmor has derived the theorem from classical electrodynamics and has shown that the ratio between the magnetic moment μ and the orbital angular moment p is given by the relation in electrostatic units

$$\frac{\mu}{p} = -\frac{e}{2mc} \tag{7.13}$$

p is also referred to as the mechanical moment of the orbiting electron.

It should be noted that in the case of spinning electron the above ratio

$$\frac{\mu}{p} = -2\frac{e}{2mc} \tag{7.14}$$

The change in the motion of an electron by a magnetic field of intensity B is a precession of the orbit about the field direction with a uniform angular velocity

$$\omega_L = -B \cdot \frac{\mu}{p} = B \cdot \frac{e}{2mc} \tag{7.15}$$

Thus the angular velocity of the Larmor precession is equal to the field strength B times the magnitude of the ratio between the magnetic and mechanical moments.

$$\omega_L = -B \cdot \frac{\mu}{p} = B \cdot \frac{e}{2mc} \tag{7.16}$$

and the precession frequency

$$v_L = \frac{B}{2\pi} \cdot \frac{\mu}{p} = B \cdot \frac{e}{4\pi mc} \tag{7.17}$$

The frequency of precession is independent of the orientation of the angle between the orbit's normal and the field direction.

7.4 VECTOR MODEL FOR TOTAL ANGULAR MOMENTUM

When orbital angular momentum L and electron spin angular momentum S are combined to produce the total angular momentum J in an atom, the combination process of these angular momenta can be visualized in terms of a vector model. Both the orbital and spin angular momenta precess around the direction of the total angular momentum J. Figure 7.5(a) can be seen as describing a single electron. However, multiple electrons for which the spin and orbital angular momenta have been combined to produce composite angular momenta S and L respectively can also be visualized in the same manner. In the case of two electrons $L = l_1 + l_2 ... |l_1 - l_2|$ and $S = s_1 + s_2 ... |s_1 - s_2|$. Here we have considered a coupling scheme which is known as LS coupling or Russel Saunder's (RS) coupling (*See* Chapter 8). This kind of coupling scheme is appropriate for light atoms with relatively small internal magnetic fields.

Figure 7.5(b) shows the vector diagram for a single electron case $l = 1$ and $s = 1/2$. The combination is a special kind of vector addition as is illustrated. As in the case of the orbital angular momentum, the projection of the total angular momentum along a direction in space is quantized to values differing by one unit of angular momentum. Here $J = 1 + 1/2$ and $1 - 1/2$, i.e., 3/2 and 1/2. In the former case where

$J = 3/2$, $m_j = 3/2, 1/2, -1/2, -3/2$ and in the latter case where $J = 1/2$, $m_j = 1/2$ and $-1/2$. The magnetic quantum number m_j are resolved only in an external magnetic field (Zeeman effect), otherwise they are all degenerate.

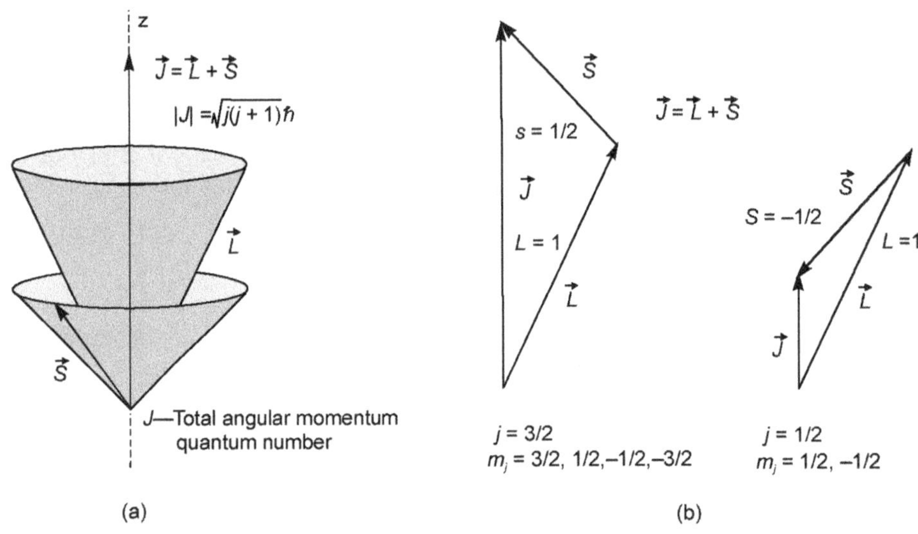

Figure 7.5   (a) A vector model for a single electron (b) The vector diagram for a single electron case $l = 1$ and $s = 1/2$.

Once we have combined orbital and spin angular momenta in the light of the vector atom model, the resulting total angular momentum can be visualized as precessing about any externally applied magnetic field. Note that in Figure 7.6(a), the L and S first form a resultant which is the total angular momentum which in turn precess around the magnetic field B.

This model has been first used to describe the Zeeman effect in sodium. The magnetic energy contribution is proportional to the component of total angular momentum along the direction of the magnetic field, which is usually defined as the z-direction.

The z-component of angular momentum is quantized in values one unit apart, so for the upper level of the sodium doublet $^2P_{3/2}$ with $j = 3/2$, the vector model gives the splitting as shown in Figure 7.6(b).

Even with the vector atom model, the determination of the magnitude of the Zeeman splitting is not trivial since the directions of S and L are constantly changing as they precess about J. This problem is handled with the Lande' g-factor.

As stated earlier this treatment of the angular momentum is appropriate for weak external magnetic fields where the coupling between the spin and orbital angular momenta can be presumed to be stronger than the coupling to the external field. This can be visualized with the help of a vector atom model of total angular momentum. If the external field is very strong, then it can decouple the spin and orbital angular momenta and the orbital angular momentum and spin angular momentum precess separately about the magnetic field B. This strong field case is called the Paschen–Back effect and leads to different patterns of splitting of the energy levels.

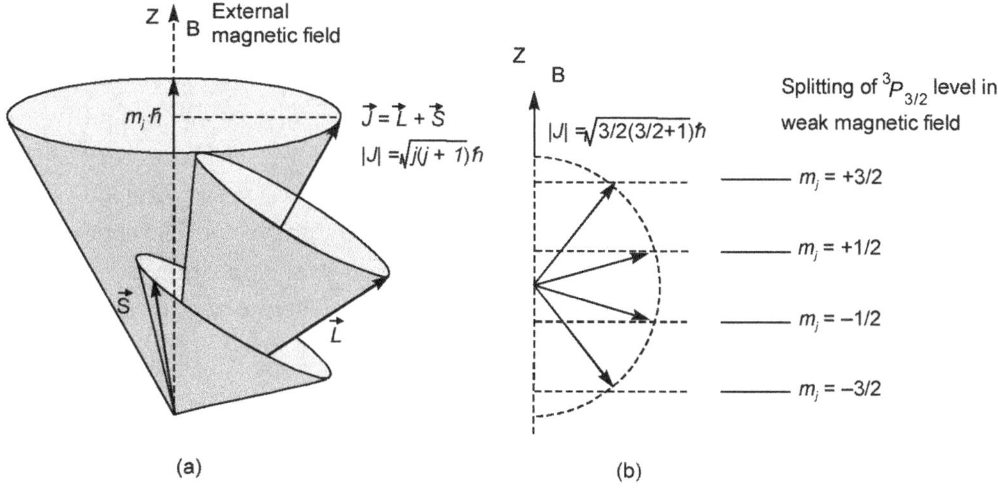

Figure 7.6 Angular momentum in a magnetic field

We have already seen from expression (7.12) that expressed in *SI* units $\omega_L = \dfrac{e}{2m_e}B$.

In the case of the electron spin precession, the angular frequency associated with the spin transition is usually written in the general form

$$\omega = \gamma B \tag{7.18}$$

where $\gamma = \dfrac{e}{2m_e}$ is called the gyromagnetic ratio (sometimes the magnetogyric ratio). This angular frequency is associated with the "spin flip" or spin transition, involving an energy change of $2\mu B$. An example for magnetic field 1 Tesla we can calculate

$$\omega_{\text{electron spin}} = \frac{2\mu_e B}{\hbar} = \frac{2\cdot2\cdot\frac{1}{2}(5.79\times10^{-5}\,\frac{eV}{T})(T)}{6.58\times10^{-16}\ \text{eV.s}} = 1.7608\times10^{11}\text{s}^{-1}$$

$$\nu = \frac{\omega}{2\pi} = 28.025\ \text{GHz}\quad\text{(Larmor frequency)}$$

$$\omega_{\text{proton spin}} = \frac{2\mu_p B}{\hbar} = \frac{2(2.79)(3.15\times10^{-8}\,\frac{eV}{T})(T)}{6.58\times10^{-16}\ \text{eV.s}} = 2.6753\times10^{8}\text{s}^{-1}$$

$$v = \frac{\omega}{2\pi} = 42.5781 \text{ MHz} \quad (\text{Larmor frequency}) \qquad (7.18a)$$

In electron spin resonance experiments the characteristic frequencies associated with electron spin are employed whereas those frequencies associated with the nuclear spin are used in nuclear magnetic resonance (NMR) experiments.

If the rotation of the total angular momentum vector about the magnetic field is as in the Figure 7.6, the electric current associated with the moment of negative charge may be regarded as flowing in the opposite direction in the electron orbit. The magnetic vector associated with the flow of current will act on the opposite direction to the angular momentum vector. The component of the magnetic moment in the direction of the field is $-\mu_J \cos \theta$ and the interaction energy is $-\mu_J \cdot B = -\mu_J B \cos \theta$. If E_0 is the energy of the system in the field free case, the energy when the field is applied to the potential energy of the system is

$$E_{mj} = E_0 + \mu_J B \cos \theta = E_0 - \mu_H B \qquad (7.19)$$

where μ_H is the magnetic moment component in the direction of the field. There will be $(2J + 1)$ orientation of J^*h and hence $(2J + 1)$ different energy values or splittings.

7.5 SPIN–ORBIT INTERACTIONS AND DOUBLET STRUCTURE

We have already seen that the alkali atom spectra consist of doublet fine structure. The doublet fine structure can be explained by considering the interaction of two types of angular momentum possessed by the valence electron. One of these angular momenta originates from the motion of the electron around the nucleus in an orbit, which is called the orbital angular momentum. Over and above the orbital angular momentum the electron possesses another angular momentum arising due to the spin of the electron, which is called spin angular momentum. The total angular momentum of the atom is that of the valence electron. The interaction energy is obtained by the quantum mechanical properties of the operator for the total angular momentum.

7.5.1 Spin–orbit Interaction

We have already seen that in order to explain the fine structure in atomic spectra and to explain the anomalous Zeeman effect it was necessary to introduce the concept of electron spin. Many experiments have later proved that the electron has an intrinsic spin as it orbits through the atom. Thus we consider two motions and the corresponding angular momenta, the orbital angular momentum and the spin angular momentum for the electron.

The energy levels of atomic electrons are affected by the interaction between the electron spin magnetic moment originating from the electron spin motion and the orbital angular momentum of the electron. It can be visualized as the interation of a magnetic field caused by the electron's orbital motion with the spin magnetic moment. The effective magnetic filed can be expressed in terms of the electron orbital angular momentum. The interaction energy is that of a magnetic dipole in a magnetic field and takes the form as in the Figure 7.7.

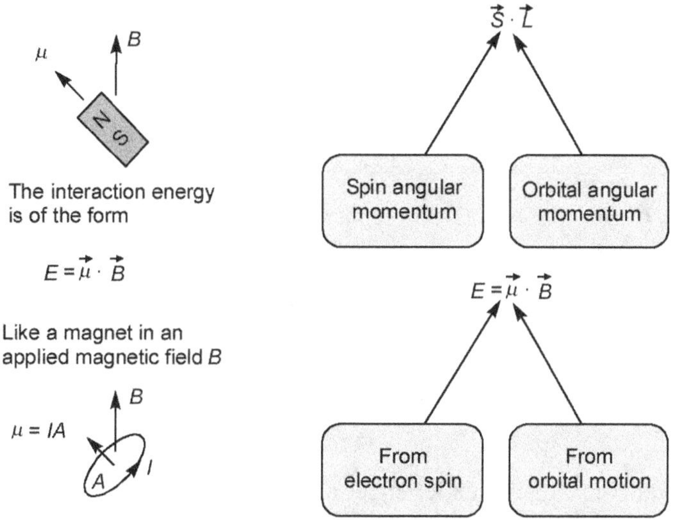

Figure 7.7 Spin–orbit interaction

The spin–orbit interaction results in a shift in the energy level or more exactly the splitting in an energy level of the electron due to the potential energy of the spin magnetic moment in a magnetic field it feels as it moves through the magnetic field of the nucleus. This is due to the internal magnetic field. However, when atomic spectral lines are split by the application of an external magnetic field, it is called the Zeeman effect.

Thus the spin–orbit interaction is a magnetic interaction, but with the magnetic field generated by the orbital motion of an electron within the atom itself. And hence it has been described as an "internal Zeeman effect". The splitting of the 3P state into 3 $^2P_{3/2}$ and 3 $^2P_{1/2}$ in sodium is due to the spin–orbit interaction.

The magnitude of the spin–orbit interaction has the form $\mu_z B = \mu_B S_z L_z$. The difference in energy for the 3 $^2P_{3/2}$ and 3 $^2P_{1/2}$ in the case of the sodium doublet, comes from a change of 1 unit in the spin orientation with the orbital part presumed to be the same. The change in energy is of the form

$$\Delta E = \mu_B gB = 0.0021 \text{ eV}$$

where μ_B is the Bohr magneton and g is the electron spin g-factor with value very close to 2. An estimate of the internal magnetic field needed to produce the observed splitting can be obtained as:

$$\mu_B gB = (5.79 \times 10^{-5} \text{ eV} / T)2B = 0.0021 \text{ eV}$$
$$B = 18 \text{ Tesla}$$

This is a very large magnetic field by laboratory standards.

In order to calculate the magnitude of the spin–orbit interaction a semi-classical approach can be made. When we consider the classical picture of hydrogen like atoms, the single electron moves in a central force in an orbit with an angular momentum $l*$ with magnitude $l* \dfrac{h}{2\pi}$.

$$\text{Hence} \qquad l* \frac{h}{2\pi} = m_e \, r \times v \qquad\qquad (7.20)$$

where m_e is the mass of the electron, v is the velocity, r is the radius vector and $l^* = \sqrt{l(l+1)}$.

According to classical electromagnetic theory, the electron moving in an orbit under the influence of the nuclear charge Ze experiences an electric field E. The electric field E at the electron by the charge Ze on the nucleus is given by

$$E = \frac{1}{4\pi\varepsilon_0} \frac{Ze}{r^3} \bar{r} \qquad\qquad (7.21)$$

Moving in this field the electron experiences a magnetic field which is given by the expression

$$B = \frac{-E \times v}{c} \qquad\qquad (7.22)$$

Substituting the value of E from the equation (7.21)

$$B = \frac{1}{4\pi\varepsilon_0} \frac{Ze}{cr^3} \bar{r} \times \bar{v} \qquad\qquad (7.23)$$

$$\text{From equation 7.20} \qquad 2\pi m_e \, r \times v = l*h \qquad\qquad (7.24)$$

$$\text{Hence} \qquad B = \frac{1}{4\pi\varepsilon_0} \, l* \frac{h}{2\pi} \frac{Ze}{m_e c} \frac{1}{r^3} \qquad\qquad (7.25)$$

When the electron moves under the influence of this magnetic field, it undergoes a Larmor precession around the field direction. The movement of the electron under the influence of this field is like the movement of a mechanical top. From the Larmor theorem, the product of the field strength B and the ratio of the magnetic and mechanical moment of the spinning electron gives the angular velocity of the Larmor precession. If we assume the ratio between the magnetic and mechanical moment of the spinning electron as $\dfrac{\mu}{p}$ the angular velocity is given by

$$\omega_L = -B\frac{\mu}{p} \qquad\qquad (7.26)$$

We have already seen from expression (7.13) that

$$\frac{\mu}{p} = -\frac{e}{2m_e c}$$

For a spinning electron $\dfrac{\mu}{p}$ is double of this ratio.

And hence for a spinning electron $\dfrac{\mu}{p} = -2\dfrac{e}{2m_e c}$.   (7.27)

Generally magnitude of $\dfrac{\mu}{p}$ can be written as

$$\frac{e}{2m_e c} = g \frac{e}{2m_e c}$$

for any motion, where g is called gyromagnetic ratio $g = 1$ for the orbital motion and $g = 2$ for spinning electron.

Therefore we obtain

$$\omega_L = B\frac{\mu}{p} = B2\frac{e}{2m_e c} = \frac{1}{4\pi\varepsilon_0} l * \frac{h}{2\pi} \frac{Ze^2}{m_e^2 c^2} \frac{1}{r^3}$$

$$v_L = \frac{\omega_L}{2\pi} = \frac{B}{2\pi}\frac{\mu}{p} = B\frac{e}{4\pi mc} = \frac{1}{4\pi\varepsilon_0} l * \frac{h}{4\pi^2} \frac{Ze^2}{m^2 c^2} \frac{1}{r^3}$$ (7.28)

The precession of L and S about J in vector model is shown in Figure 7.8. We have not, this far, taken into account the time dilation between the rest frames of the proton and electron; this effect is called Thomas precession and introduces a factor of 1/2. Thomas and Frenkel have shown that the relativistic transformation from the frame moving with the electron to a frame at rest introduces a factor 1/2. Or in other words A relativistic correction is necessary for this problem, as a relativistic precession also exists which found to be one half of the above in the opposite direction. Hence from equation 7.28,

$$\omega = \omega_L + \omega_r = \frac{1}{4\pi\varepsilon_0}\frac{1}{2}l * \frac{h}{2\pi}\frac{Ze^2}{m^2 c^2}\frac{1}{r^3}$$ (7.29)

The resultant precession of the spinning electron is therefore one half ω_L .

The interaction energy is the product of the angular velocity of the precession ω and the projection of the spin angular momentum of $l*$. Thus

$$\Delta W_{l,s} = \omega \cdot s * \frac{h}{2\pi}\cos(l*s*)$$

$$= \frac{1}{4\pi\varepsilon_0}\frac{Ze^2}{2m^2 c^2}\cdot\frac{h^2}{4\pi^2}\cdot\frac{1}{r^3}\cdot l * s * \cos(l*s*)$$ (7.30)

The electron–nuclear distance r in an atom is a function of Z, n and l and continuously changing and an average $\dfrac{1}{r^3}$ value is sufficient for our calculation as the interaction energy is small. The average value of $\dfrac{1}{r^3}$ has been calculated by quantum mechanics and perturbation theory to be

$$\frac{1}{r^3} = \frac{Z^3}{a_1^3 n^3 l \left(l + \dfrac{1}{2}\right)(l+1)} \tag{7.31}$$

where a_1 is the radius of the first orbit of the Bohr circular orbit.

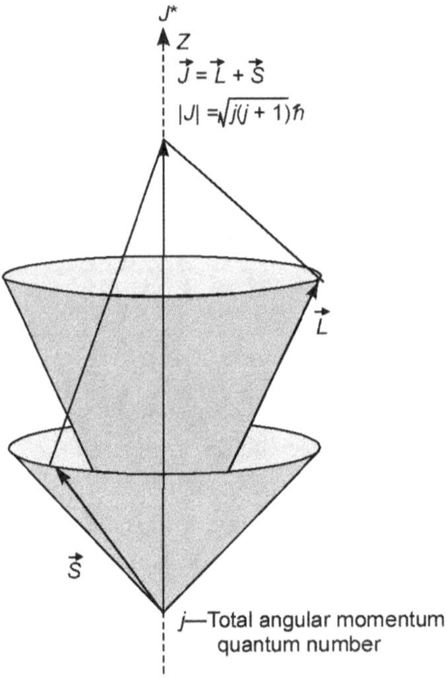

Figure 7.8   Precession of L and S about J

$$j^{*2} = l^{*2} + s^{*2} + 2l^* s^* \cos(l^* s^*) \tag{7.32}$$

$$\cos(l^* s^*) = \frac{1}{2}(j^{*2} - l^{*2} - s^{*2})$$

$$\Delta W_{l,s} = \frac{1}{4\pi\varepsilon_0} \frac{Ze^2}{2m^2 c^2} \cdot \frac{h^2}{4\pi^2} \cdot \frac{1}{r^3} \cdot \frac{1}{2}(j^{*2} - l^{*2} - s^{*2})$$

$$= \frac{1}{4\pi\varepsilon_0} \frac{Ze^2}{2m^2 c^2} \cdot \frac{h^2}{4\pi^2} \cdot \frac{Z^3}{a_1^3 n^3 l \left(l + \dfrac{1}{2}\right)(l+1)} \cdot \frac{1}{2}(j^{*2} - l^{*2} - s^{*2}) \tag{7.33}$$

By substituting the Rydberg constant $R = \left(\dfrac{1}{4\pi\varepsilon_0}\right)^2 \dfrac{2\pi^2 me^4}{ch^3} = \dfrac{me^4}{8\varepsilon_0^2 ch^3}$, the fine structure

constant $\alpha = \left(\dfrac{1}{4\pi\varepsilon_0}\right)\dfrac{e^2}{(h/2\pi)c} = \dfrac{e^2}{2\varepsilon_0 hc}$

(7.34)

and the first Bohr orbit $a_1 = \dfrac{h^2\varepsilon_0}{\pi me^2}$ we obtain

$$W_{l,s} = \frac{R\alpha^2 chZ^4}{n^3 l\left(l+\dfrac{1}{2}\right)(l+1)} \cdot \frac{1}{2}(j*^2 - l*^2 - s*^2) = -\Gamma$$

(7.35)

Thus spin–orbit interaction energy

$$\Gamma = a\frac{j*^2 - l*^2 - s*^2}{2} = a\, l*s*\cos(l*\, s*)$$

(7.36)

where

$$a = \frac{R\alpha^2 Z^4}{n^3 l\left(l+\dfrac{1}{2}\right)(l+1)}\ \text{cm}^{-1}$$

(7.37)

$$T = T_0 - \Gamma$$

A ^2P state has the levels $^2P_{1/2}$ and $^2P_{3/2}$ and the separation between these levels can be calculated as follows.

For P state $l = 1$

The multiplicity is 2 (as P is a doublet). Note that when a state is named as $^2P_{1/2}$, it means that the superscript is the multiplicity (in this case 2) and the subscript on the right is the total angular momentum j (in this case 1/2). The multiplicity of a state is calculated as $2S + 1$ where S is the total spin. Thus the multiplicity here $2 = 2S + 1$.

Therefore the spin $S = 1/2$.

Thus $\Gamma = a\dfrac{j*^2 - l*^2 - s*^2}{2}$

(7.38)

$$j*^2 = j(j+1),\ \ l*^2 = l(l+1),\ \ s*^2 = s(s+1)$$

(7.39)

Hence
$$\Gamma = a\,\frac{j(j+1)-l(l+1)-s(s+1)}{2} \tag{7.40}$$

For $^2P_{1/2}$ state $l = 1$, $s = 1/2$, $j = 1/2$

$$\Gamma = a\,\frac{\dfrac{1}{2}\left(\dfrac{1}{2}+1\right)-1.2-\dfrac{1}{2}\cdot\left(\dfrac{1}{2}+1\right)}{2} = a\,\frac{\dfrac{1}{2}\left(\dfrac{3}{2}\right)-2-\dfrac{1}{2}\left(\dfrac{3}{2}\right)}{2} = -a \tag{7.41}$$

For $^2P_{3/2}$ state $l = 1$, $s = 1/2$, $j = 3/2$

$$\Gamma = a\,\frac{\dfrac{3}{2}\left(\dfrac{3}{2}+1\right)-1.2-\dfrac{1}{2}\cdot\left(\dfrac{1}{2}+1\right)}{2} = a\,\frac{\dfrac{3}{2}\left(\dfrac{5}{2}\right)-2-\dfrac{1}{2}\left(\dfrac{3}{2}\right)}{2} = \frac{a}{2} \tag{7.42}$$

Thus $\Gamma = T_0 - a$ for $^2P_{1/2}$ state $\tag{7.43}$

And $\Gamma = T_0 + \dfrac{a}{2}$ for $^2P_{3/2}$ state $\tag{7.44}$

T_0 is the hypothetical un split position. The splitting is drawn schematically in Figure 7.9.

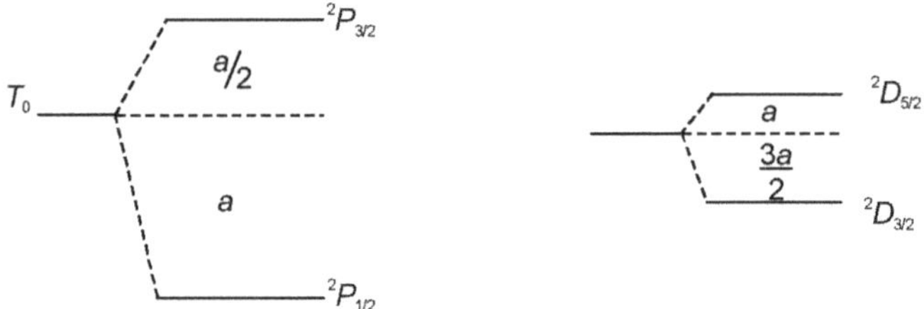

Figure 7.9   Spin–orbit interaction (fine structure) splitting in ^2P and ^2D states

7.6 QUANTUM MECHANICAL DERIVATION

We have seen that the charge Ze on the nucleus gives rise to an electric field (equation 7.21)

$$E = \frac{1}{4\pi\varepsilon_\circ}\frac{Ze}{r^3}\bar{r}$$

which in turn produces magnetic field (expression 7.23)

$$B = \frac{1}{4\pi\varepsilon_\circ}\frac{Ze}{cr^3}\bar{r}\times\bar{v} = \frac{1}{4\pi\varepsilon_\circ}l*\frac{h}{2\pi}\frac{Ze}{m_e c}\frac{1}{r^3}$$

The interaction energy is

$$\Delta H = -\mu \cdot B \tag{7.45}$$

$$\mu = -\frac{e}{m_e c} s^*$$

$$\Delta H = \frac{1}{4\pi\varepsilon_o} \frac{Ze^2}{m_e^2 c^2} \frac{1}{r^3} L^* \cdot S^* \tag{7.46}$$

The above expression is true if the electron was at rest and the nucleus was revolving around it. As mentioned earlier Thomas and Frenkel have shown that there need to have a relativistic correction and introduced a factor 1/2 in the above energy expression.

$$\text{Hence } \Delta H = \frac{1}{4\pi\varepsilon_o} \frac{Ze^2}{2m_e^2 c^2} \frac{1}{r^3} L^* \cdot S^* = \xi(r) L^* \cdot S^* \tag{7.47}$$

$$\text{where } \xi(r) = \frac{1}{4\pi\varepsilon_o} \frac{Ze^2}{2m_e^2 c^2} \frac{1}{r^3}$$

In the vector atom model $l^* \cdot s^*$ is a constant of motion but not $\frac{1}{r^3}$. Hence average value of $\frac{1}{r^3}$ must be determined.

Thomas precession is a correction to the spin–orbit interaction in Quantum Mechanics, which takes into account the relativistic time dilation between the electron and the nucleus in hydrogenic atoms.

7.6.1   Evaluating the Energy Shift: Quantum Mechanical

We can consider $\xi(r) \, L.S$ as a perturbation and in quantum mechanics the energy is evaluated by using $\xi(r) \, L.S$ as an operator to be added as a perturbing potential $V(r)$ in the ordinary form of the Schrödinger equation

$$H\Psi = \left[\frac{P_r^2}{2m} + \frac{L^2}{2mr^2} + V(r) + \xi(r)L.S \right]\Psi = E\Psi \tag{7.48}$$

where ψ stands for a two component Pauli's wavefunction.

In (7.48) $x\,(r)\,L \cdot S$ is the Hamiltonian corresponding to spin–orbit interaction.

$$\Delta H = \hat{H}_{so} = \xi L \cdot S = \frac{1}{2m_e c^2} = \left(\frac{1}{r}\right)\left(\frac{\partial v}{\partial r}\right)\hat{L} \cdot \hat{S}$$

Where V is the coulomb potential of the electron in the field of atom. A proper derivation requires a relativistic treatment of the electron which is beyond the scope of the book.

We solve it by degenerate perturbation theory. Basically

$$\Delta E = \langle \Psi \mid \Delta H \mid \Psi \rangle = \langle \Psi \mid \xi L \cdot S \mid \Psi \rangle \tag{7.49}$$

From this we will be able to get two expectation values which must be evaluated;

The first one is the evaluation of $L.S$

The total angular moment $J^2 = (L + S)^2 = L^2 + S^2 + 2\,L.S$ $\tag{7.50}$

$$L.S = \frac{1}{2}\left(J^2 - L^2 - S^2\right) \tag{7.51}$$

Since H, L^2, J^2 and J_z are commuting set of operators the above equation shows that $L.S$ commutes with all of them. We replace the operator with its eigen value

$$\langle \psi \mid L.S \mid \psi \rangle = \left\langle \psi \left| \frac{1}{2}\left(J^2 - L^2 - S^2\right) \right| \psi \right\rangle = \frac{1}{2}\hbar[J(J+1) - L(L+1) - S(S+1)] \tag{7.52}$$

The Schrödinger equation is written separately for $J = l + \dfrac{1}{2}$ and $J = l - \dfrac{1}{2}$ and ψ is one or the other of the two space operators $\psi_1 \psi_2$. The first order perturbation gives

$$\Delta E = \langle \psi \mid \Delta H \mid \psi \rangle$$

$$= \frac{1}{2}[J(J+1) - L(L+1) - S(S+1)]\hbar^2 \int \psi^*_{n,l,m}\xi(r)\psi_{n,l,m}\,\partial\tau \tag{7.53}$$

Substituting the value of $\xi(r) = \dfrac{1}{4\pi\varepsilon_\circ}\dfrac{Ze^2}{2m^2c^2r^3}$

$$\Delta E = \frac{1}{4\pi\varepsilon_\circ}\frac{Ze^2}{2c^2m^2}\left\langle \frac{1}{r^3} \right\rangle \frac{\hbar^2}{2}\,[J(J+1) - L(L+1) - S(S+1)] \tag{7.54}$$

where $\left\langle \dfrac{1}{r^3} \right\rangle = \int \psi^*_{n,l,m}\dfrac{1}{r^3}\psi_{n,l,m}\,\partial\tau = \dfrac{1}{a_0^3}\dfrac{Z^3}{n^3 L\left(L + \dfrac{1}{2}\right)(L+1)}$ $\tag{7.55}$

$$\Delta E = \frac{1}{4\pi\varepsilon_\circ}\frac{Ze^2}{2c^2m^2}\frac{h^2}{2}\frac{1}{a_0^3}\frac{Z^3}{n^3 L\left(L + \dfrac{1}{2}\right)(L+1)}\,[J(J+1) - L(L+1) - S(S+1)]$$

Substituting the Rydberg constant $R = \left(\dfrac{1}{4\pi\varepsilon_0}\right)^2 \dfrac{2\pi^2 me^4}{ch^3}$ and the fine structure

constant $\alpha = \dfrac{Ze^2}{2\varepsilon_0 hc}$ and $a_0 = \dfrac{h^2\varepsilon_0}{\pi me^2}$.

$$\Delta E = \frac{Rch\alpha^2 Z^4}{n^3}\frac{J(J+1)-L(L+1)-S(S+1)}{L(L+1)(2L+1)} \tag{7.56}$$

For $J = L + \dfrac{1}{2}$ $\qquad E = \dfrac{Rch\alpha^2 Z^4}{n^3}\dfrac{1}{(L+1)(2L+1)}$

For $J = L - \dfrac{1}{2}$ $\qquad E = \dfrac{Rch\alpha^2 Z^4}{n^3}\dfrac{1}{L(2L+1)}$

The difference between the two levels is

$$\Delta E = \frac{R\alpha^2 Z^4}{n^3 L(L+1)}\,\text{cm}^{-1}$$

$$\Delta E = \frac{\beta}{2}(J(J+1)-L(L+1)-S(S+1)) \tag{7.57}$$

For hydrogen, we can write the explicit result

$$\beta(n,l) = \frac{\mu_0}{4\pi}Z^2 g_s \mu_B^2 \frac{1}{n^3 a_0^3 L\left(L+\dfrac{1}{2}\right)(L+1)} \tag{7.58}$$

It should be noted that J, L, S are used frequently as capital letters or small letters. Capital letters are used when the total angular momenta of all valence electrons are concerned. If there is only one valence electron, the capital and small letters do not make any difference.

7.7 NORMAL AND INVERTED DOUBLET TERMS

We have seen above that in the doublet levels the $J = L - 1/2$ lies lower than that of the component $J = L + 1/2$. This is the normal state. In some cases where the outer shell is filled with more than half its allowed number of electrons one obtains inverted states.

Inverted doublets for example arise from electron shells lacking one electron to complete the shell. Example is a p^5 configuration. By inverted means that the term $j = l + 1/2$ lie deepest in an energy level diagram. However, in the same atom some levels are found to be normal and some inverted. This is explained as perturbation due to other energy levels. Normal and inverted states are explained in more detail in Chapter 11 Section 11.2.2.

7.8 SPECTRA OF IONIZED ATOMS

Atoms in the IIA and IIB groups have electronic configurations

Be [4]	$1s^2\ 2s^2$
Mg [12]	$1s^2\ 2s^2\ 2p^6\ 3s^2$
Ca [20]	$1s^2\ 2s^2\ 2p^6\ 3s^2\ 3p^6\ 4s^2$
Sr [38]	$1s^2\ 2s^2\ 2p^6\ 3s^2\ 3p^6\ 3d^{10}\ 4s^2\ 4p^6\ 5s^2$
Ba [56]	$1s^2\ 2s^2\ 2p^6\ 3s^2\ 3p^6\ 3d^{10}\ 4s^2\ 4p^6\ 4d^{10}\ 5s^2\ 5p^6\ 6s^2$
Ra [88]	$1s^2\ 2s^2\ 2p^6\ 3s^2\ 3p^6\ 3d^{10}\ 4s^2\ 4p^6\ 4d^{10}\ 5s^2\ 5p^6\ 6s^2\ 6p^6\ 6d^{10}\ 6s^2$
Zn [30]	$1s^2\ 2s^2\ 2p^6\ 3s^2\ 3p^6\ 3d^{10}\ 4s^2$
Cd [48]	$1s^2\ 2s^2\ 2p^6\ 3s^2\ 3p^6\ 3d^{10}\ 4s^2\ 4p^6\ 4d^{10}\ 5s^2$
Hg [80]	$1s^2\ 2s^2\ 2p^6\ 3s^2\ 3p^6\ 3d^{10}\ 4s^2\ 4p^6\ 4d^{10}\ 4f^{14}\ 5s^2\ 5p^6\ 5d^{10}\ 6s^2$

The last two electrons in the s orbit are responsible for the chemical valence and also for the spectrum. We may consider that the valence electrons may be moving in orbits which may be penetrating the core and coming closer to the nucleus or in nonpenetrating orbits depending on their orbital angular momentum. In the calcium atom for example we have two electrons in the valence shell $4s$. This gives rise to a $^1S_0$ ground state to the atom. If the atom is excited by some means, one or both of these two valence electrons are being excited. We would now consider the case where only one of these electrons is excited resulting in atleast eight groups of spectral lines. The first excited state of atom should be either D or P states corresponding to a $3d$ or a $4p$ orbital the electron could take. Experimental evidence shows that the first excited state is a P state ^1P and ^3P lying close together which means that only one of the electrons is excited and thus one electron remains in $4s$ and the other is excited to $4p$. The next excited state is a D state (^1D and ^3D close together) arising from one electron being in $4s$ orbit and the other in the $3d$ orbit. The next P state originates from a $4s$ $5p$ combination of the valence electron. Thus the successive members of the P series have the combinations $4s\ 4p$, $4s\ 5p$, $4s\ 6p$...$4s\ \alpha p$ or the electron configuration is $4s\ np$ where $n = 4,5,6,...\alpha$. This constitutes the Principal series. The diffuse series is due to the excitation $4s\ nd$ giving rise to ^1D and ^3D states, fundamental series $4s\ nf$ giving rise to ^1F and ^3F states. When the electron is in the $n = \alpha$ orbit, regardless of the series of orbits by which the electron reached this limit, the electron gets removed from the atom which is ionization. In the ionized atom only one electron is left in the $4s$ orbit. All eight series approach a common limit which is the ionizing limit. To understand further the nomenclature of the state we shall consider a $4s\ 5p$ configuration. In this case $l_1 = 0$, $l_2 = 1$ and hence $L = 0 + 1...|0{-}1| = 1$ and the spins $s_1 = 1/2$ and $s_2 = 1/2$ and hence $S = 0$ or 1. One gets singlet and triplet states and in this case the states are ^1P and ^3P. The J values are $J = L + S...|L–S|$ and for triplets $J = 1+1...1–1 = 2,1,0$. The states are thus $^3P_{2,1,0}$. For singlets $J = 1 + 0 = 1$ and thus a $^1P_1$.

Thus by some excitation if one of the electrons in the valence shell gets away from the shell, the atom is said to be ionized. The electron configuration then becomes identical to the alkali metal atoms. The neutral atoms are generally written as Be I, Mg I, Ca I, Sr I, Ba I and so on. The singly ionized atoms are Be II, Mg II, Ca II, Sr II, Ba II and so on.

The valence electron of the ionized atoms are

Be II,	Mg II,	Ca II,	Sr II,	Ba II
$2s$	$3s$	$4s$	$5s$	$6s$

and those of neutral alkali atoms are

Li I	Na I	K I	Rb I	Cs I
$2s$	$3s$	$4s$	$5s$	$6s$

Obviously the resulting spectrum will be similar in both cases, the only differences being in the energies and thereby in the frequency of the resulting spectrum. There are many methods in ionizing neutral atoms like photon ionization, however, if the ionization is due to a fast moving electron, it should have sufficient energy to remove the electron or to bring it to the ionization limit and to excite the other valence electron to an excited state of the ionized atom left. The energy levels and the transitions in the case of ionized calcium is shown in Figure 7.10.

The ground state of the ionized calcium is the state of the single valence s electron. It gives rise to a $^2S_{1/2}$ state. In the case of alkali metal atoms the first excited state is a ^2P state, however, in the case of ionized calcium the first excited state is a ^2D state. This has been concluded from experimental observation of the difference series of the spectra. The relative positions of the groups of diffusive and fundamental series are interchanged in this case. Thus the first excited state in ionized calcium is 3^2D. Owing to the fact that a transition is not possible from this level to the ground state because of the selection rule that $\Delta l = \pm 1$ is the only allowed transition, the excited electron to the 3^2D state remains there for a long period. Such states are said to have long life time and are called metastable state. Once the electron is in this state it may return to the ground state only by nonradiative transition which can be caused by collision by other electrons or other particles. Following the metastable state the next state is the 4^2P state and the first member of the principal series originates from this state. The next higher energy state is 5^2S from which the first member of the sharp series originates. The strong lines in the alkali atoms lie in the near infrared and the visible region while the corresponding lines in the ionized alkali earths are in the visible and near ultraviolet region. It should be remembered that the arrangements of electrons are thus made through experimental observation of the energy levels. The atomic picture of the ionized atom is now to be viewed in such a way that the nuclear charge Ze surrounded by a cloud of $(Z - 2)$ electrons and outside this a single electron. The net charge $2e$ makes the spectra similar to ionized helium. The term values in that case is

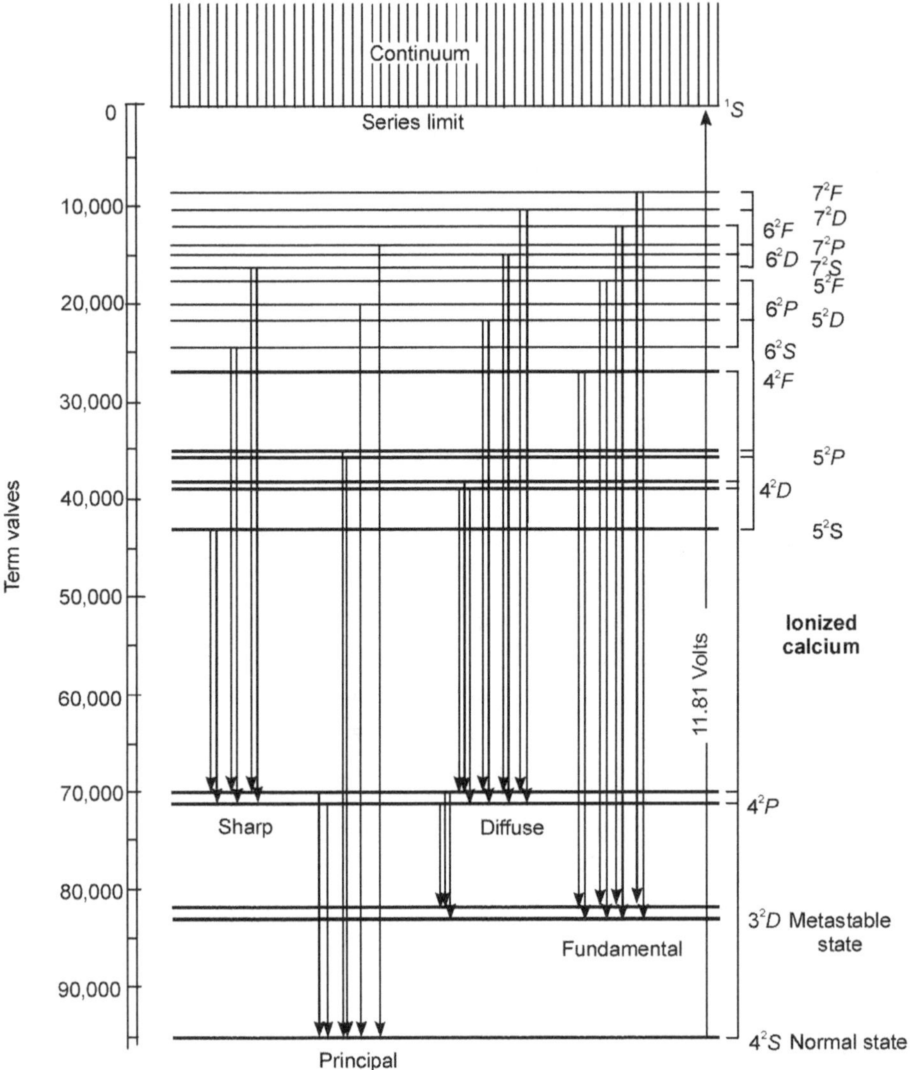

Figure 7.10   Energy level diagram of singly ionized calcium

$$T = \frac{Z^2 R}{n^2} = \frac{4R}{n^2}$$

In a similar manner the term values of other atoms may be written as

$T = 4R / n_{\text{eff}}^2 = 4R / (n - \mu)^2$ where μ is the quantum defect.

The transition frequency can be expressed as $\nu = 4R / (n_1 - \mu_1)^2 - 4R / (n^2 - \mu_2)^2$.

7.9 BRANCHING RULE

Branching rule is a convenient procedure to construct the energy states of the atoms from the ionized atom to the normal atom. As an example we consider the energy state of the ionized calcium atom. It has $4s$ as the valence electron. By excitation of this electron we assume that the electron has gone to the $3d$ state thereby giving the atomic state as ^2D. If we imagine that the second electrons already knocked out from the neutral calcium atom returns and consider three possibilities: to the $4s$ or to $4p$ or to $4d$.

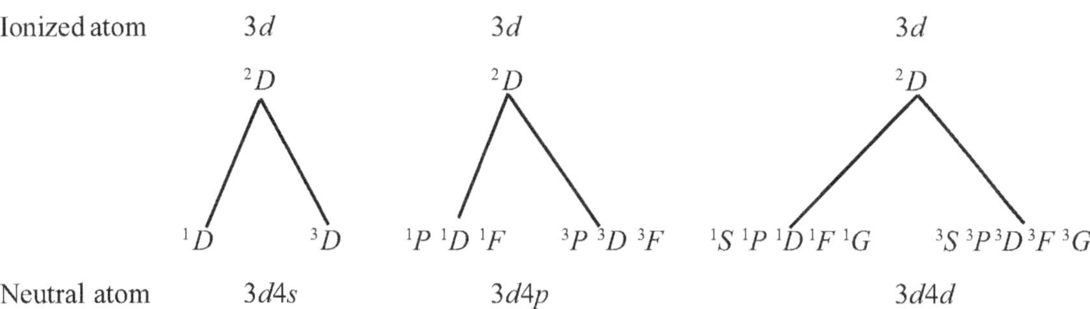

If the L values of the parent term is equal to or greater than the l of the added electron then the possible terms are $L - l$ to $L + l$. If l is greater than L then the possible terms are $l - L$ to $l + L$. In above the third case $3d4d$, L and l are equal and hence $L - l = 0$ and $L + l = 4$ and hence 0, 1, 2, 3, 4, i.e., S, P, D, F, G are possible.

SOLVED PROBLEMS

1. Calculate the doublet splitting in the case of (1) a ^2D state and (2) for a ^2F states originating from spin–orbit interaction.

Solution

(1) for a ^2D state the two states are designated by $^2D_{3/2}$ and $^2D_{5/2}$.

Here the multiplicity, i.e., $2S + 1 = 2$ where S is the total spin

Hence $S = \tfrac{1}{2}$

$l = 2$ and hence $j = l + 1/2$ to $l - 1/2 = 5/2$ and $3/2$ respectively.

Thus we have the above two states with $j = 5/2$ and $3/2$

For $^2D_{5/2}$ state

$$\Gamma = a\frac{j(j+1)-l(l+1)-s(s+1)}{2} = a\frac{\dfrac{5}{2}\left(\dfrac{5}{2}+1\right)-2.3-\dfrac{1}{2}\cdot\left(\dfrac{1}{2}+1\right)}{2}$$

$$= a\frac{\dfrac{5}{2}\left(\dfrac{7}{2}\right)-6-\dfrac{1}{2}\cdot\left(\dfrac{1}{2}+1\right)}{2} = a$$

For $^2D_{3/2}$ state

$$\Gamma = a\frac{j(j+1)-l(l+1)-s(s+1)}{2} = a\frac{\dfrac{3}{2}\left(\dfrac{3}{2}+1\right)-2.3-\dfrac{1}{2}\cdot\left(\dfrac{1}{2}+1\right)}{2}$$

$$= a\frac{\dfrac{3}{2}\left(\dfrac{5}{2}\right)-6-\dfrac{1}{2}\cdot\left(\dfrac{3}{2}\right)}{2} = -\frac{3a}{2}$$

Thus the shift is $\Gamma = T_0 + a$ for the $^2D_{5/2}$ state

And $\Gamma = T_0 - \dfrac{3a}{2}$ for the $^2D_{3/2}$ state.

The splitting is $\dfrac{5a}{2}$.

(2) ^2F state

The two doublet states are obtained as $^2F_{5/2}$ state and $^2F_{7/2}$ state.

The splitting $2S+1 = 2$, $S = 1/2$,

For an F state $l = 3$ and hence $j = l + s$ and $l - s = 7/2$ and $5/2$.

For $^2F_{5/2}$

$$\Gamma = a\frac{j(j+1)-l(l+1)-s(s+1)}{2} = a\frac{\dfrac{5}{2}\left(\dfrac{5}{2}+1\right)-3.4-\dfrac{1}{2}\cdot\left(\dfrac{1}{2}+1\right)}{2}$$

$$= a\frac{\dfrac{5}{2}\left(\dfrac{7}{2}\right)-12-\dfrac{1}{2}\cdot\left(\dfrac{3}{2}\right)}{2} = -2a$$

For $^2F_{7/2}$

$$\Gamma = a\frac{j(j+1)-l(l+1)-s(s+1)}{2} = a\frac{\dfrac{7}{2}\left(\dfrac{7}{2}+1\right)-3.4-\dfrac{1}{2}\cdot\left(\dfrac{1}{2}+1\right)}{2}$$

$$= a\frac{\dfrac{7}{2}\left(\dfrac{9}{2}\right)-12-\dfrac{1}{2}\cdot\left(\dfrac{3}{2}\right)}{2} = \frac{3a}{2}$$

Total splitting is $\dfrac{7a}{2}$.

ENDNOTES

1   Robert, C. O'Handley (2000). *Modern Magnetic Materials: Principles and Applications.* John Wiley and Sons, New York.

2   Mahajan, A. and Rangwala, A. (1989). *Electricity and Magnetism.* Tata McGraw-Hill Publishing Company Ltd., New Delhi.

REFERENCES

Aruldhas, G. (2001). *Molecular Structure and Spectroscopy.* (2001). Prentice Hall of India, New Delhi.

Bernath, P.F. (1995). *Spectra of Atoms and Molecules.* Oxford University Press, Oxford.

Condon, U. and Shortley, G.H. (1935, 1999). *The Theory of Atomic Spectra.* Cambridge University Press, Cambridge, U.K.

Griffiths, D.J. (2005). *Introduction to Quantum Mechanics*, 2nd (edn.). Prentice Hall, NJ.

Landau, L. and Lifshitz, L.M. (2003). *Quantum Mechanics: Non-Relativistic Theory.* Volume 3. Elsevier Science, Oxford.

Rai, D.K. and Thakur, S.N. (2005). *Atomic Spectra, Structure and Modern Spectroscopy.* Vaivaswat Publications, Varanasi.

Straughan, B.P. and Walker, S. (1976) Chapman and Hall, London.

White, H.E. (1934). *Introduction to Atomic Spectra.* McGraw-Hill, New York.

8

THE CHARACTERISTICS OF SPECTRA

8.1 ATOMS WITH TWO VALENCE ELECTRONS

The energy levels of the atoms are classified on the basis of their multiplicity. It was observed that the atoms containing two valence electrons give rise to energy levels, which are singlets or triplets. The II A and II B groups in the periodic table containing the atoms like beryllium, manganese, calcium, strontium, barium, zinc, cadmium and mercury, give rise to singlet and triplet energy levels. Russel and Saunders in 1925 analysed these spectra on the basis of angular momentum coupling which was an important development in the atomic spectroscopy.

8.2 OBSERVED SPECTRA

The singlet and triplet series of spectral lines are grouped mainly into four types, viz., sharp, principal, diffuse and fundamental series. In analysing these spectra the following observations were made. In these atoms there are four chief series of singlets and four chief series of triplets.

1. Sharp and diffuse series of triplets have a single common limit and the singlets series have another limit.

2. The difference in frequency in cm^{-1} of the above series limit with the series limit of the triplet principal series is the same as the frequency of the first member of the triplet principal series.

3. The frequency difference in cm^{-1} of the above common limit with the series limit of the triplet fundamental series is the same as the first member of the triplet diffuse series.

4. The differences as described above are applicable to singlet series also.

5. In most of the spectra the fine structure is very small, however the above relations hold good when the fine structure is taken into account.

6. All members of the sharp triplet series consist of three components having constant separation and approaches a triple limit.

7. The triplet principal series comprise of three lines each with decreasing separations and approach a common single limit.

8. Triplet diffuse and fundamental series contain six components—three strong and three weak lines and they approach a triple limit.

9. All observed triplets and singlets series can be represented by a general series formula

$$V_n = \frac{R}{(n_1 - \mu_1)^2} - \frac{R}{(n_2 - \mu_2)^2} \tag{8.1}$$

where μ_1 and μ_2 are quantum defects. n_1 is a fixed number and n_2 takes successive integral values and $n = n_2 - n_1$ where $n_2 > n_1$.

10. There are also some anomalous series lines which do not fit into the above formula

We have already seen that an electron is identified with an orbital angular momentum $l * \dfrac{h}{2\pi}$ where $l* = \sqrt{l(l+1)}$ and a spin angular momentum $s * \dfrac{h}{2\pi}$ where $s* = \sqrt{s(s+1)}\left(s = \frac{1}{2}\right)$ and their magnetic quantum numbers. The angular momentum l takes values $0,1,2,\ldots (n-1)$ and are termed s, p, d, f, g, h, i, etc., electrons respectively. All electrons making up completed shells are quantized with respect to each other and their resultant angular momentum vanishes. Thus only the valence electrons (the outer electrons in the incomplete subshells) contribute in the resultant angular momenta of the atom.

In order to explain the observed features of the spectra of atom with two valence electrons Russel and Saunders proposed approximate schemes in which the angular momentum vectors of both electrons are considered and a coupling scheme had been proposed.

8.3 *LS* COUPLING SCHEME OR RUSSEL SAUNDERS COUPLING SCHEME (RS COUPLING SCHEME)

For atoms having many electrons, the so called multi-electron atoms where the spin–orbit coupling is weak, it can be presumed that the orbital angular momenta of the individual electrons add to form a resultant orbital angular momentum L.

$$L = (l_1 + l_2),\ (l_1 + l_2 - 1),\ (l_1 + l_2 - 2),\ \ldots |(l_1 - l_2)| \tag{8.2}$$

Likewise, the individual spin angular momenta are assumed to couple to produce a resultant spin angular momentum S.

$$S = (s_1 + s_2),\ (s_1 + s_2 - 1),\ \ldots \left| (s_1 - s_2) \right| \tag{8.3}$$

Thus in this scheme it is assumed that the orbital motions of the two electrons are first coupled together to form a resultant L so also the spin motions are coupled together to form a resultant S. L and S then combine to form the resultant J (*See* Figures 8.1 and 8.2).

$$J = L + S \tag{8.4}$$

If we consider a two valence electron arrangement in an atom and if l_1 is the orbital quantum number of the first electron and l_2 the respective orbital quantum number of the second electron, then l_1^* and l_2^* are their respective angular momenta with $l_1^* \dfrac{h}{2\pi}$ and $l_2^* \dfrac{h}{2\pi}$. Then l_1 and l_2 are coupled with respect to each other forming a resultant L^* where the total orbital angular momentum is $L^* = \sqrt{L(L+1)}$ with the value $L^* \dfrac{h}{2\pi}$. The two spins s_1^* and s_2^* couple to form a resultant $S^* = \sqrt{S(S+1)}$ with the value $S^* \dfrac{h}{2\pi}$. In the case of the electrons $s = \dfrac{1}{2}$ and hence

$$s_1^* = \sqrt{\left(\frac{1}{2}\right)\cdot\left(\frac{3}{2}\right)} = \frac{1}{2}\sqrt{3} \quad \text{and} \quad s_2^* = \sqrt{\left(\frac{1}{2}\right)\cdot\left(\frac{3}{2}\right)} = \frac{1}{2}\sqrt{3} \quad \text{and the resultant } S = 0 \text{ or}$$

$$\left[S = s_1 + s_2 \ldots \left| s_1 - s_2 \right| = \frac{1}{2} + \frac{1}{2} \text{ and } \frac{1}{2} - \frac{1}{2} \right].$$ This gives $S^* = \sqrt{0 \times 1} = 0$ for $S = 0$ and $S^* = \sqrt{1 \times 2} = \sqrt{2}$ for $S = 1$.

Let us now consider a configuration of the electronic structure with two valence electrons having p electrons $l_1 = 1$ and $l_2 = 1$. In this case the resultant $L = 0, 1, 2$ (L can take values from $l_1 + l_2,\ (l_1 + l_2 - 1), \ldots \left| l_1 - l_2 \right|$. Thus the angular momenta are $l_1^* = \sqrt{l_1(l_1 + 1)} = \sqrt{2}$ and $l_2^* = \sqrt{l_2(l_2 + 1)} = \sqrt{2}$ and $L^* = \sqrt{6}, \sqrt{2}, \sqrt{0}$ for $L = 2, 1, 0$. The state with $L = 0$ is named as an S state, with $L = 1$ as a P state and with $L = 2$ as a D state. The multiplicity of a state is defined as $2S + 1$ where S is the total electron spin. (Total electron spin nomenclature should not be confused with the state named S). A state is usually written as

$$^{2S+1}L_J$$

where left superscript is the multiplicity, L is the total orbital quantum number and the right subscript is the total quantum number of the atom which is J. J can take values from $L + S$ to $\left| L - S \right|$. In the above examples of two p valence electrons we have S, P, D states.

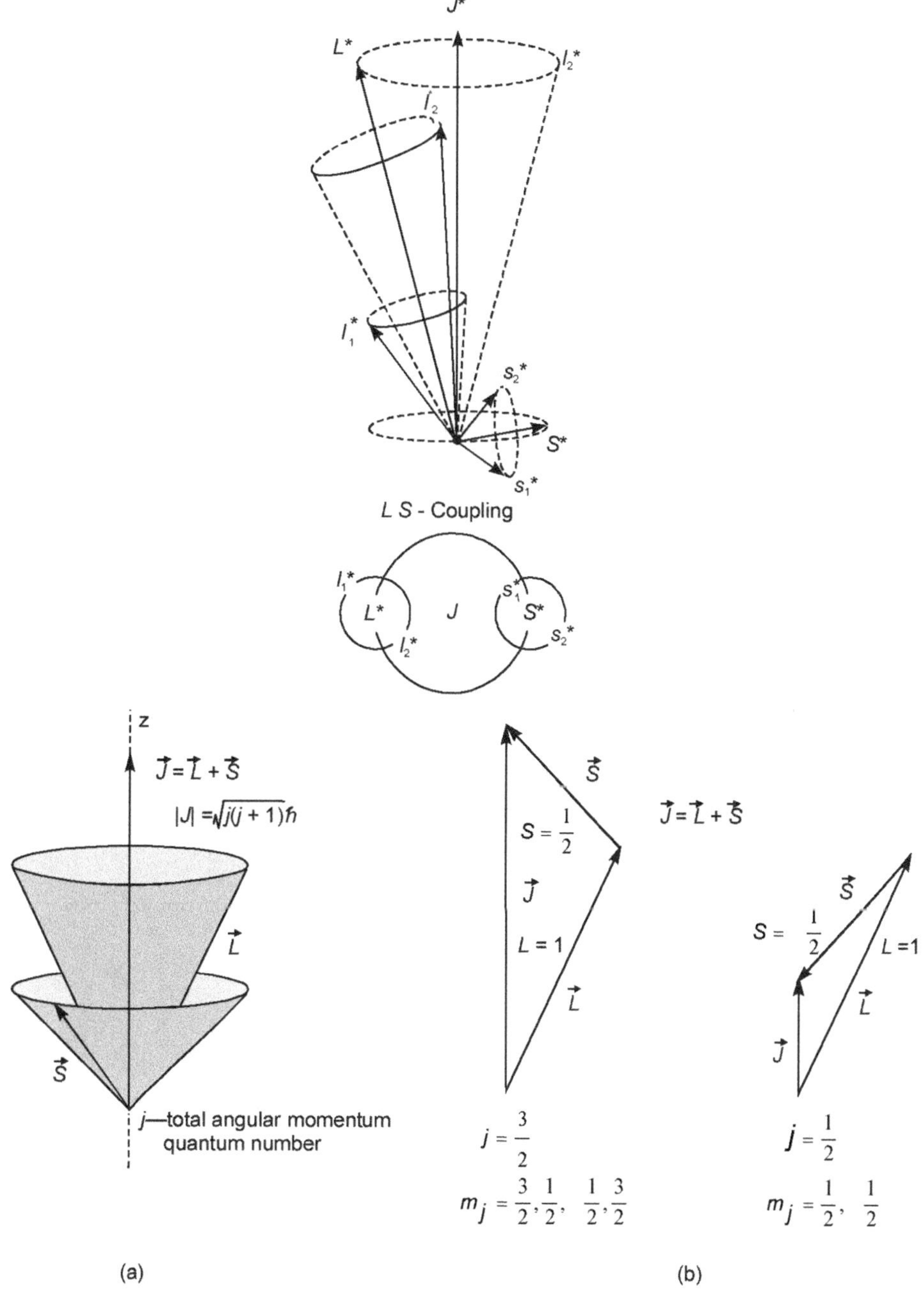

Figure 8.1 *LS* coupling

As the total spins = 0 or 1, two types of states arise, viz., singlets and triplets. $2S + 1 = 1$ in the first case with $S = 0$ (singlet) and $2S + 1 = 3$ in the second case (triplet). So the possible states are ^3S, ^1P, ^3P and ^1D, ^3D.

It should be noted that the various types of levels described by the capital letter L are universally accepted as

$$L = 0, 1, 2, 3, 4, 5, \ldots$$

as $S, P, D, F, G, H, \ldots$

Again the nomenclature of the S state (i.e., $L = 0$ state) should not be confused with the total electron spin S. Ironically they have the same nomenclature.

Thus the orbital motions of the electrons couple to give a total orbital angular momentum L^* and the spins coupled together to give a total spin angular momentum S^*. The total angular momentum arising out of the coupling of L and S is then the total angular momentum $J^* = \sqrt{J(J+1)}$ with total angular quantum number having non-negative values of $J = L \pm S$. For the above example of a ^1S state, $J = 0$ and hence the state is termed as $^1S_0$. The state originating from $L = 0$ and S(spin) $= 1$, gives a ^3S state and the J is $J = L \pm S = 1$, thereby we get a $^3S_1$. For a ^3P state, the possible J values are $J = L \pm S = l +1 \ldots |l - 1| = 2, 1, 0$. Thus the states are written either as $^3P_{0,1,2}$ or as $^3P_0, ^3P_1, ^3P_2$.

The states arising from the two equivalent p electrons are

^1S, ^3P and ^1D and with the J numbers they are $^1S_0$, $^3P_{0,1,2}$ and $^1D_2$.

For a pd electron configuration with one electron each in the subshells, the total spin becomes $s_1 \pm s_2 = 1$ or 0 and hence there exists triplets or singlets. The total orbital angular momentum can have values $L = l_1 + l_2, l_1 + l_2 - 1, |l_1 - l_2| = 3, 2, 1$ and thus we obtain S, P, D states with multiplicities singlets or triplets.

Thus with total spin $= 0$ we obtain singlets

With total spin $= \dfrac{1}{2}$ one obtains doublets (case of single valence electrons)

With total spin $= 1$ one obtains triplets

It should be noticed that with $L = 0$ and $S = 1$, though it is expected to have a triplet, it gives only one J and hence all S levels are single in agreement with experiments. But in a magnetic field it reveals its characteristics, that means a ^3S level will split into three components with $m_J = -1, 0, +1$.

A ^2S state will split in a magnetic field into $m_J = -\dfrac{1}{2}$ or $+\dfrac{1}{2}$ and a ^1S level in a magnetic field will remain as singlet with $m_J = 0$, but with a possible shift.

In case of identical valence electrons, the proper combination should produce an antisymmetric wave function upon exchange of identical electrons. For example consider the case of d^2 electrons as in the case of the titanium atom. The electron configuration of titanium is [Ar] $3d^2\, 4s^2$. We could see here that the two s electrons make a complete closed shell and the $3d^2$ electrons contribute to the net angular momentum. For d electrons $l_1 = 2$ and $l_2 = 2$. And hence L can have values $4, 3, 2, 1, 0$. and S (spin) has values 1 or 0. Thus we should have triplets and singlets.

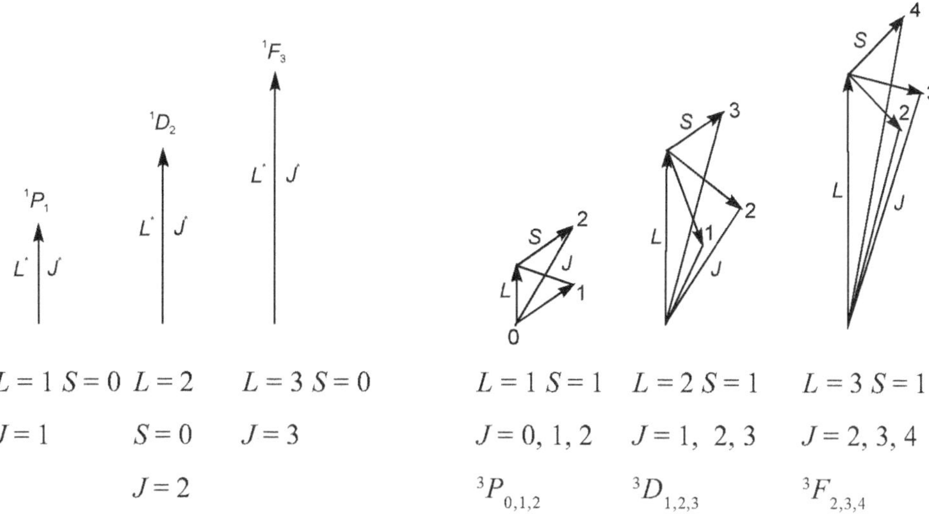

Figure 8.2   Vector diagram for two valence electrons in *LS* couplings

$J = 5, 4, 3$ for the combination $L = 4$ and spin $S = 1$ or 0 G state

$J = 4, 3, 2$ for the combination $L = 3$ and spin $S = 1$ or 0 F state

$J = 3, 2, 1$ for the combination $L = 2$ and spin $S = 1$ or 0 D state

$J = 2, 1, 0$ for the combination $L = 1$ and spin $S = 1$ or 0 P state

$J = 1$ for the combination $L = 0$ and spin $S = 1$ or 0 S state

The possibility of $J = 5$ is excluded by symmetry considerations for the identical electrons. The vector model can visualize the other states.

The coupling of two equivalent d electrons ($L = 2$) thus gives the following states. The identical electrons must have an antisymmetric wave function $\Psi = \Psi_{spin}\ \Psi_{orbital}$.

$$\vec{L}_1 + \vec{L}_2 = \vec{L}$$
$$\vec{2} + \vec{2} = 0, 1, 2, 3, 4$$
$$\vec{S}_1 + \vec{S}_2 = \vec{S}$$
$$\frac{\vec{1}}{2} + \frac{\vec{1}}{2} = 0, 1$$
$$\vec{J} = \vec{L} + \vec{S} = 0, 1, 2, 3, 4$$

Since $S = 1$ represents a symmetric spin state, it must couple with antisymmetric orbital states $L = 1$ and $L = 3$.

The coupling scheme mentioned above is called *L–S* coupling or Russell–Saunders coupling. The *LS* coupling scheme gives good agreement with the experimental spectral details for many light atoms. In dealing with heavier atoms it is found that a coupling scheme called *j–j* coupling yields better agreement with experimental spectrum.

8.4 *J–J* COUPLING

In light atoms, we have seen that the interactions between the orbital angular momenta of individual electrons is stronger than the interaction between the electron spin and the orbital angular momentum, the so called, spin–orbit coupling between the spin and orbital angular momenta of the electrons. In cases where spin–orbit interaction is small and the interaction between the orbital motions are strong, the *L–S* coupling scheme can be applied. However, in atoms where the spin–orbit interaction is large, the spin and the orbital motion couple first for the individual electron and the resultant angular momenta then couple each other. This is the basic of the *j–j* coupling scheme.

$$j_1 = l_1 + s_1$$
$$j_2 = l_2 + s_2 \quad j = \sum_i j_i$$

As mentioned above, *j–j* coupling scheme is mainly observed in heavy atoms with larger nuclear charge. In those cases spin–orbit interactions are strong. *j–j* coupling scheme is shown in Figure 8.3.

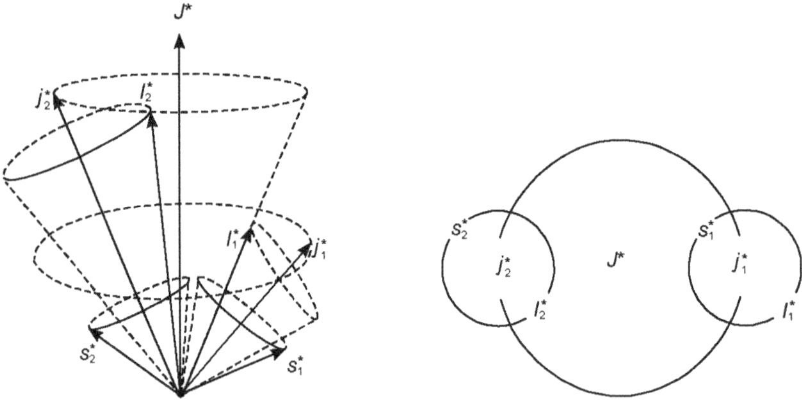

Figure 8.3   *j–j* coupling

8.5 INTERACTION ENERGIES

The fine structure splititings were first developed in terms of the vector atom model. It gives qualitatively the energy separations, though later quantum mechanics had provided exact calculation of these energies. However, in many cases it had been found that the energies obtained from vector model approach gives quantitative agreement as well. Here we discuss the vector model approach.

We have already seen in Chapter 7 the interaction energy due to the spin–orbit interaction of a single electron, i.e., the shift of each fine structure level from the hypothetical unsplit position is given by

$$-\Delta T_{l,s} = a \cdot l * s * \overline{\cos(l * s*)} = a \frac{j*^2 - l*^2 - s*^2}{2} = \Gamma \qquad (8.5)$$

where $a = \dfrac{R\alpha^2 Z^4}{n^3 l \left(1 + \dfrac{1}{2}\right)(l+1)}$ cm^{-1}

In the case of an atom with electron configuration with two valence electron we should consider six such interactions as there are four angular momenta, viz., l_1, l_2, s_1 and s_2.

These interactions are

    (1) s_1* with s_2*
    (2) l_1* with l_2*,
    (3) l_1* with s_1*,
    (4) l_2* with s_2*,
    (5) l_1* with s_2*,
    (6) l_2* with s_1*

We can obtain similar interaction terms as in equation 8.5. Thus

$$\begin{aligned}
\Gamma_1 &= a_1\, s_1 * s_2 * \cos(s_1 * s_2*) \\
\Gamma_2 &= a_2\, l_1 * l_2 * \cos(l_1 * l_2*) \\
\Gamma_3 &= a_3\, l_1 * s_1 * \cos(l_1 * s_2*) \\
\Gamma_4 &= a_4\, l_2 * s_2 * \cos(l_2 * s_2*) \\
\Gamma_5 &= a_5\, l_1 * s_2 * \cos(l_1 * s_2*) \\
\Gamma_6 &= a_6\, l_2 * s_1 * \cos(l_2 * s_1*)
\end{aligned} \qquad (8.6)$$

In the vector model it is assumed that the predominating interaction between any two vectors precess more rapidly around their mutual field than around any other. And they undergo Larmor precession.

In the case of LS coupling it is assumed that l_1 and l_2 precess more rapidly in the field of L and s_1 and s_2 precess more rapidly in the field of S. L and S then precess around J. And in this case the interaction energies Γ_1 and Γ_2 predominate over Γ_3 and Γ_4. The interaction energies Γ_5 and Γ_6 are assumed to be negligible and neglected. The LS coupling scheme is shown in Figure 8.1.

The Γ factors in LS coupling

$$\Gamma_1 = a_1\, s_1 * s_2 * \cos(s_1 * s_2*)$$

with s_1^* and s_2^* precessing around their resultant S^* with fixed angles of inclination

Applying cosine law for triangles $S^{*2} = s_1^{*2} + s_2^{*2} + 2s_1^* \, s_2^* \, \cos \, (s_1^* \, s_2^*)$

$$\Gamma_1 = \frac{1}{2} a_1 \, (S^{*2} - s_1^{*2} - s_2^{*2})$$

$$\text{i.e., } \Gamma_1 = \frac{1}{2} \, a_1[S(S+1) - s_1(s_1+1) - s_2(s_2+1)] \tag{8.7}$$

Similarly

$$\Gamma_2 = \frac{1}{2} \, a_2 (L^{*2} - l_1^{*2} - l_2^{*2}) \tag{8.8}$$

$$\text{i.e., } \Gamma_2 = a_2[L(L+1) - l_1(l_1+1) - l_2(l_2+1)] \tag{8.9}$$

The precession of L^* and S^* around J^* can be considered in the same as that of the precession of a single electron with l^* and s^* around j^*. In calculating the interaction energy of the precession of L^* and S^*, the average values of the cosines $\overline{\cos(l_1^* s_1^*)}$ are to be taken since the angles between the vectors are continuously changing. Trigonometrically it can be shown that

$$\overline{\cos(l_1^* s_1^*)} = \cos(l_1 * L) \cdot \cos(L * S^*) \cdot \cos(S * s_1^*)$$

$$\overline{\cos(l_2^* s_2^*)} = \cos(l_2 * L) \cdot \cos(L * S^*) \cdot \cos(S * s_2^*) \tag{8.10}$$

In this case it is assumed that the components normal to the respective axes cancel.

Thus

$$\Gamma_3 = a_3 \, l_1^* \, s_1^* \cos \, (l_1 * s_1^*) = a_3 \, l_1^* \, s_1^* \cos \, (l_1 * L) \cdot \cos \, (L * S^*) \cdot \cos(S * s_1^*)$$

$$= a_3 \, l_1^* \, s_1^* \frac{1}{2}(L^{*2} - l_1^{*2} - l_2^{*2}) \cdot \frac{1}{2}(S^{*2} - s_1^{*2} - s_2^{*2})\cos(L * S^*)$$

$$\Gamma_4 = a_4 \, l_2^* \, s_2^* \cos(l_2 * s_2^*) = a_4 \, l_2^* \, s_2^* \cos \, (l_2 * L) \cdot \cos(L * S^*) \cdot \cos(S * s_2^*)$$

$$= a_4 \, l_2^* \, s_2^* \frac{1}{2}(L^{*2} - l_2^{*2} - l_1^{*2}) \cdot \frac{1}{2}(S^{*2} - s_2^{*2} - s_1^{*2})\cos(L * S^*) \tag{8.11}$$

$$\Gamma_3 + \Gamma_4 = (a_3\alpha_3 + a_4\alpha_4)L * S * \cos(L * S^*) = \frac{1}{2}(a_3\alpha_3 + a_4\alpha_4)(J^{*2} - L^{*2} - S^{*2}) \tag{8.12}$$

where,

$$\alpha_3 = \frac{s_1^{*2} - s_2^{*2} + S^{*2}}{2S^{*2}} \cdot \frac{l_1^{*2} - l_2^{*2} + L^{*2}}{2L^{*2}} \tag{8.13}$$

$$\alpha_4 = \frac{s_1^{*2} - s_2^{*2} + S^{*2}}{2S^{*2}} \cdot \frac{l_2^{*2} - l_1^{*2} + L^{*2}}{2L^{*2}}$$

Considering $A = a_3\alpha_3 + a_4\alpha_4$

$$\Gamma_3 + \Gamma_4 = A L^* S^* \cos(L^* S^*) = \frac{1}{2}A(J^{*2} - L^{*2} - S^{*2}) \tag{8.14}$$

Thus under the *LS* coupling scheme, the energies of the various atomic levels originating from a given electronic configuration with definite values of n_1, l_1 and n_2, l_2 is

$$\Gamma = \Gamma_0 + \Gamma_1 + \Gamma_2 + (\Gamma_3 + \Gamma_4) \tag{8.15}$$

For a given value of L and S, $\Gamma_0 + \Gamma_1 + \Gamma_2$ is a constant, but $(\Gamma_3 + \Gamma_4)$ is different according to the value of J.

Comparing the experimental and calculated spin–orbit interaction splittings it is presumed that the interaction is magnetic in nature whereas the orbit–orbit or spin–spin interactions are attributed to electrostatic effects. It has been seen that the singlet levels lie higher than the triplet levels. Thus the coefficient a_1 in expression (8.7) is considered to be negative. Let us now calculate the energy levels of a *sp* electron configuration. We have now $l_1 = 0$ and $l_2 = 1$; $s_1 = \frac{1}{2}$ and $s_2 = \frac{1}{2}$.

Thus for $S = 0$ (singlet state)

$$\Gamma_1 = \frac{1}{2}a_1(S^{*2} - s_1^{*2} - s_2^{*2}) = \frac{1}{2}a_1\left[0 - \left(\frac{1}{2}\right)\cdot\left(\frac{3}{2}\right) - \left(\frac{1}{2}\right)\cdot\left(\frac{3}{2}\right)\right]$$

$$= -\frac{1}{2}a_1\left(\frac{3}{4} + \frac{3}{4}\right) = -\frac{3}{4}a_1 \tag{8.16}$$

For $S = 1$

$$\Gamma_1 = \frac{1}{2}a_1(S^{*2} - s_1^{*2} - s_2^{*2}) = \frac{1}{2}a_1\left[2 - \left(\frac{1}{2}\right)\cdot\left(\frac{3}{2}\right) - \left(\frac{1}{2}\right)\cdot\left(\frac{3}{2}\right)\right]$$

$$= -\frac{1}{2}a_1\left(2 - \frac{3}{4} - \frac{3}{4}\right) = \frac{1}{4}a_1 \tag{8.17}$$

As a_1 is negative Γ_1 becomes positive and Γ_2 negative. Γ_2 originates from $l_1 l_2$ interaction giving rise to various S, P, D type of states. The most stable state is the one in which the two magnetic moments and hence the mechanical moments are most nearly parallel to each other. This means that the state with the maximum L (originating from $l_1 + l_2$) will lie deepest. Hund has proposed the following rules for the states. Out of all terms arising the term with the highest L will lie deepest and out of all terms with the same L arising from any electron configuration, the term with the highest multiplicity, i.e., with the highest S value will lie deepest. This is known as Hund's rule.

$${}^{3}P_{0}$$

$$\Gamma_{3}+\Gamma_{4}=A\,L*S*\cos(L*S*)=\frac{1}{2}A(J*^{2}-L*^{2}-S*^{2})=\frac{1}{2}A(0-1\times2-1\times2)=\frac{1}{2}A(-4)=-2A \quad (8.18)$$

$${}^{3}P_{1}$$

$$\Gamma_{3}+\Gamma_{4}=A\,L*S*\cos(L*S*)=\frac{1}{2}A(J*^{2}-L*^{2}-S*^{2})=\frac{1}{2}A(1\times2-1\times2-1\times2)$$

$$=\frac{1}{2}(-2)=-A \quad (8.19)$$

$${}^{3}P_{2}$$

$$\Gamma_{3}+\Gamma_{4}=A\,L*S*\cos(L*S*)=\frac{1}{2}A(J*^{2}-L*^{2}-S*^{2})=\frac{1}{2}A(2\times3-1\times2-1\times2)=\frac{1}{2}A(2)=A \quad (8.20)$$

The splittings for the *sp* configuration is shown in Figure 8.4.

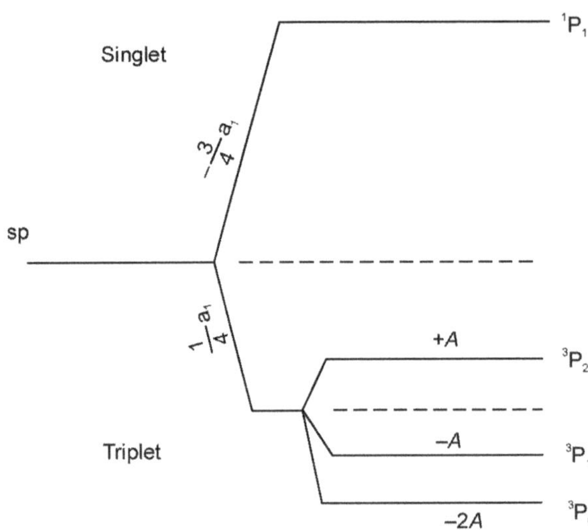

Figure 8.4   The splittings for the *sp* configuration

For an electronic configuration *pd*, $l_{1}=1$ and $l_{2}=2$ and the states originating are $L=l_{1}+l_{2}\ldots|l_{1}-l_{2}|=3,2,1$ giving rise to *F, D, P* states. And since there are two electrons involved the total spin will be either 1 or 0 $\left(S=s_{1}+s_{2}\text{ or }|s_{1}-s_{2}|\right)$. Hence singlets and triplets are possible. Remember multiplicity is 2*S*+1, which is 3 or 1 in either case. This means that the states ${}^{3}F$, ${}^{3}D$, ${}^{3}P$ and ${}^{1}F$, ${}^{1}D$, ${}^{1}P$ are possible. The splitting in *pd* configuration is shown in Figure 8.5.

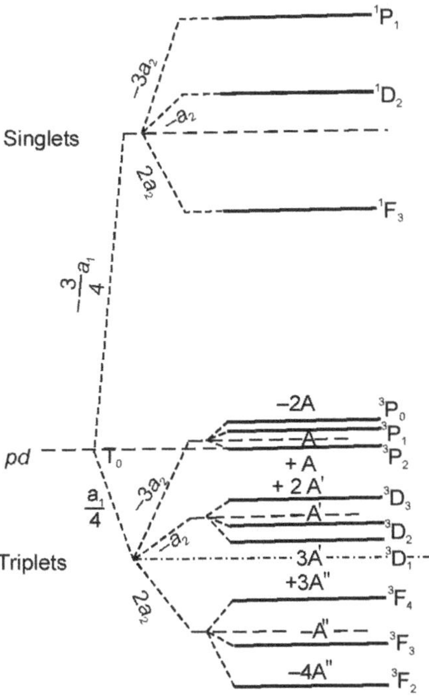

Figure 8.5   Schematic diagram of splitting in *pd* configuration

For ^1P state, $J = L + S = 1 + 0 = 1$ and hence the state is designated as $^1P_1$ state and

$$\Gamma_1 = \frac{1}{2}a_1(S^{*2} - s_1^{*2} - s_2^{*2}) = \frac{1}{2}a_1\left(0 - \frac{3}{4} - \frac{3}{4}\right) = -\frac{3}{4}a_1 \tag{8.21}$$

$$\Gamma_2 = \frac{1}{2}a_2(L^{*2} - l_1^{*2} - l_2^{*2}) = \frac{1}{2}a_2[2 - 2 - 6] = -3\,a_2 \tag{8.22}$$

For ^1D state $J = 2 + 0 = 2$ and hence the state is designated as $^1D_2$ state.

Here

$$\Gamma_1 = \frac{1}{2}a_1(S^{*2} - s_1^{*2} - s_2^{*2}) = \frac{1}{2}a_1\left(0 - \frac{3}{4} - \frac{3}{4}\right) = \frac{3}{4}a_1 \tag{8.23}$$

$$\Gamma_2 = \frac{1}{2}a_2(L^{*2} - l_1^{*2} - l_2^{*2}) = \frac{1}{2}a_2[6 - 2 - 6] = -a_2 \tag{8.24}$$

For ^1F, $J = 3$ and hence the state is $^1F_3$. For this state

$$\Gamma_1 = \frac{1}{2}a_1(S*^2 - s_1*^2 - s_2*^2) = \frac{1}{2}a_1\left(0 - \frac{3}{4} - \frac{3}{4}\right) = \frac{3}{4}a_1 \tag{8.25}$$

$$\Gamma_2 = \frac{1}{2}a_2(L*^2 - l_1*^2 - l_2*^2) = \frac{1}{2}a_2[12 - 2 - 6] = 2a_2 \tag{8.26}$$

The triplet states are ${}^3P_{0,1,2}$ ${}^3D_{1,2,3}$ and ${}^3F_{2,3,4}$.

For ${}^3P_0$

$$\Gamma_1 = \frac{1}{2}a_1(S*^2 - s_1*^2 - s_2*^2) = \frac{1}{2}a_1\left(1 \times 2 - \frac{3}{4} - \frac{3}{4}\right) = \frac{a_1}{4} \tag{8.27}$$

$$\Gamma_2 = \frac{1}{2}a_2(L*^2 - l_1*^2 - l_2*^2) = \frac{1}{2}a_2(2 - 2 - 6) = -3a_2 \tag{8.28}$$

$$\Gamma_3 + \Gamma_4 = A\,L*S*\,\cos(L*S*) = \frac{1}{2}A\,(J*^2 - L*^2 - S*^2)$$

$$= A(0 - 2 - 2) = -2A \tag{8.29}$$

For ${}^3P_1$

$$\Gamma_1 = \frac{a_1}{4}, \qquad \Gamma_2 = -3a_2 \tag{8.30}$$

$$\Gamma_3 + \Gamma_4 = \frac{1}{2}A(1 \times 2 - 1 \times 2 - 1 \times 2) = -A \tag{8.31}$$

For ${}^3P_2$

$$\Gamma_1 = a_1/4, \qquad \Gamma_2 = -3a_2 \tag{8.32}$$

$$\Gamma_3 + \Gamma_4 = \frac{1}{2}A(6 - 2 - 2) = A \tag{8.33}$$

For ${}^3D_{1,2,3}$

${}^3D_1$

$$\Gamma_1 = \frac{1}{2}a_1(S*^2 - s_1*^2 - s_2*^2) = \frac{1}{2}a_1\left(1 \times 2 - \frac{3}{4} - \frac{3}{4}\right) = \frac{a_1}{4} \tag{8.34}$$

$$\Gamma_2 = \frac{1}{2}a_2(L*^2 - l_1*^2 - l_2*^2) = \frac{1}{2}a_2(2 \times 3 - 1 \times 2 - 2 \times 3) = \frac{1}{2}a_2(6 - 2 - 6) = -a_2 \tag{8.35}$$

$$\Gamma_3 + \Gamma_4 = \frac{1}{2}A(J*^2 - L*^2 - S*^2) = \frac{1}{2}A(1 \times 2 - 2 \times 3 - 1 \times 2) = -3A \tag{8.36}$$

$^3D_2$

$$\Gamma_1 = a_1(S*^2 - s_1*^2 - s_2*^2) = \frac{1}{2}a_1\left(1 \times 2 - \frac{3}{4} - \frac{3}{4}\right) = \frac{a_1}{4} \tag{8.37}$$

$$\Gamma_2 = \frac{1}{2}a_2(L*^2 - l_1*^2 - l_2*^2) = \frac{1}{2}a_2(2 \times 3 - 1 \times 2 - 2 \times 3) = -a_2 \tag{8.38}$$

$$\Gamma_3 + \Gamma_4 = \frac{1}{2}A(J*^2 - L*^2 - S*^2) = \frac{1}{2}A(2 \times 3 - 2 \times 3 - 1 \times 2) = -A \tag{8.39}$$

$^3D_3$

$$\Gamma_1 = \frac{1}{2}a_1(S*^2 - s_1*^2 - s_2*^2) = \frac{1}{2}a_1\left(1 \times 2 - \frac{3}{4} - \frac{3}{4}\right) = \frac{a_1}{4} \tag{8.40}$$

$$\Gamma_2 = \frac{1}{2}a_2(L*^2 - l_1*^2 - l_2*^2) = \frac{1}{2}a_2(2 \times 3 - 1 \times 2 - 2 \times 3) = -a_2 \tag{8.41}$$

$$\Gamma_3 + \Gamma_4 = \frac{1}{2}A(J*^2 - L*^2 - S*^2) = \frac{1}{2}A(3 \times 4 - 2 \times 3 - 1 \times 2) = 2A \tag{8.42}$$

8.5.1 *J–J* Coupling Scheme

In this coupling scheme, Γ_3 and Γ_4 are larger as l_1 and s_1 first form j_1, and l_2 and s_2 first form j_2, and then j_1 and j_2 couple to form J.

$$\Gamma_3 = a_3\, l_1 * s_1 * \cos(l_1 * s_2*) = a_3\, l_1 * s_1 * \cos(l_1 * L) \cdot \cos(L * S*) \cdot \cos(S * s_1*)$$

$$= a_3 \frac{j_1^{*2} - l_1^{*2} - s_1^{*2}}{2} = a_3 \frac{j_1(j_1+1) - l_1(l_1+1) - s_1(s_1-1)}{2} \tag{8.43}$$

$$\Gamma_4 = a_3\, l_1 * s_1 * \cos(l_1 * s_2*) = a_3 l_1 * s_1 * \cos(l_1 * L) \cdot \cos(L * S*) \cdot \cos(S * s_1*)$$

$$= a_3 \frac{j_2^{*2} - l_2^{*2} - s_2^{*2}}{2} = a_3 \frac{j_2(j_2+1) - l_2(l_2+1) - s_2(s_2-1)}{2} \tag{8.44}$$

Since the angle between l_1 and l_2 and that between s_1 and s_2 are no longer constant of motion we cannot use the simple formula as earlier. Average values of the cosine of these angles can be written as

$$\overline{\cos(l_1, l_2)} = \cos\left(\overline{l_1}, \overline{j_1}\right)\cos\left(\overline{j_1}, \overline{j_2}\right)\cos\left(\overline{l_2}, \overline{j_2}\right)$$

$$\text{and} \qquad \overline{\cos(s_1, s_2)} = \cos\left(\overline{s_1}, \overline{j_1}\right)\cos\left(\overline{j_1}, \overline{j_2}\right)\cos\left(\overline{s_2}, \overline{j_2}\right) \tag{8.45}$$

Thus $\Gamma_1 + \Gamma_2 = \dfrac{1}{2}(a_1\beta_1 + a_2\beta_2)\,[(J(J+1) - j_1(j_1+1) - j_2(j_2+1)]$

$$= \frac{1}{2}A\,[J(J+1) - j_1(j_1+1) - j_2(j_2+1)] \tag{8.46}$$

$$\beta_1 = \frac{s_1(s_1+1) - j_1(j_1+1) - l_1(l_1+1)}{2} \cdot \frac{s_2(s_2+1) - j_2(j_2+1) - l_2(l_2+1)}{2} \tag{8.47}$$

$$\beta_2 = \frac{l_1(l_1+1) - j_1(j_1+1) - s_1(s_1+1)}{2} \cdot \frac{l_2(l_2+1) - j_2(j_2+1) - l_2(l_2+1)}{2} \tag{8.48}$$

$A = a_1\beta_1 + a_2\beta_2$

Thus $T(j_1, j_2, J) = \Gamma_0 + \Gamma_3 + \Gamma_4 + \Gamma_1 + \Gamma_2$ $\hspace{2cm}$ (8.49)

Another case where the overall L and S are decoupled is the case where there is a very strong external magnetic field is applied. This is called the Paschen–Back effect.

8.6 ALKALINE EARTH SPECTRA

The electronic configurations of alkaline earth atoms are

Be : $1s^2\,2s^2$

Mg : $1s^2\,2s^22p^63s^2$

Ca : $1s^2\,2s^22p^63s^23p^64s^2$

Sr : $1s^2\,2s^22p^63s^23p^64s^24p^65s^2$

Ba : $1s^2\,2s^22p^63s^23p^6\;\;3d^{10}\;\;4s^24p^6\;4d^{10}\;5s^2\;6s^2$

They represent the ground state electronic configuration. If given sufficient energy to the atom to excite an electron, one of the s electrons in the valence shell is excited to the higher orbitals depending on the energy of excitation the other s electron remaining in the valence shell itself. Taking an example of calcium we can have the following configurations and the resulting terms

(1) $1s^2\,2s^22p^63s^23p^64s^1\;\;np^1$ giving rise to ^1P and ^3P terms. In this case $n \geq 4$

(2) $1s^2\,2s^22p^63s^23p^64s^1\;\;nd^1$ giving rise to ^1D and ^3D terms. In this case $n \geq 3$

(3) $1s^2\,2s^22p^63s^23p^64s^1\;\;nf^1$ giving rise to ^1F and ^3F terms. In this case $n \geq 3$

(4) $1s^2\,2s^22p^63s^23p^64s^1\;\;ns^1$ giving rise to ^1S and ^3S terms. In this case $n \geq 5$

(5)   In addition there are terms involving excitation to other orbitals with $n \geq 3$

According to Hund's rule the triplet states lie lower than the singlet states arising from the same electron configuration. This explains the different limits approached by the triplet and singlet series in sharp and diffuse series of the spectra.

If we consider the triplet S state, viz., ^3S, it behaves as a singlet though it is a triplet from the argument given below. We have seen the multiplicity which is $2s + 1$ in this case is 3 (s is electron spin), and hence $s = 1$ and J can have values from $L + S$ to $|L - S|$ and it gives only one value in this case which is $J = 1$. And hence we have a $^3S_1$ state which is single. Only in a magnetic field it will split. Therefore the sharp series consist of triplet members with constant splitting which correspond to the lowest $^3P_{0,\,1,2}$ terms.

From equations 8.12–8.15 we can see that the term $\Gamma_3 + \Gamma_4$ decreases in magnitude and L increases. Because of this the splittings in P state will be larger than in D or F states which confirms the experimental observation in triplet states of alkaline earth atoms. As Γ_3 and Γ_4 arise from magnetic interactions their magnitude increases with their atomic number. Hence the triplet splittings increase as one goes from Be to Ba. If an electron is in the $3d$ orbital is excited to varying states with n greater than 4 we get the following terms:

(7) $3d^9\, ns^1$ gives ^1D and ^3D terms with $n \geq 3$

(8) $3d^9\, np^1$ gives $^1P^\mathrm{o}$, $^1D^\mathrm{o}$, $^1F^\mathrm{o}$ and $^3P^\mathrm{o}$, $^3D^\mathrm{o}$ and $^3F^\mathrm{o}$ terms with $n \geq 4$

And an excitation to nd orbitals gives rise to singlet and triplet S, P, D, F, G terms.

The superscript 'o' means odd parity. Transitions between singlets and triplets are generally spin forbidden.

8.7 THE LANDE INTERVAL RULE

The expression

$$\Gamma_3 + \Gamma_4 = \frac{1}{2}A(J^{*2} - L^{*2} - S^{*2}) = \frac{1}{2}A\big(J(J+1) - L(L+1) - S(S+1)\big) \tag{8.50}$$

is significant as it represents the Lande interval rule. Let us consider triplets ^3P, ^3D or ^3F.

In all cases $S = 1$. And the three states can be written as

$$^3P_0 \quad ^3P_1 \quad ^3P_2 \qquad\qquad ^3D_1 \quad ^3D_2 \quad ^3D_3 \qquad\qquad ^3F_2 \quad ^3F_3 \quad ^3F_4$$

The ratios in separation in these cases are

$$1:2 \qquad\qquad 2:3 \qquad\qquad 3:4$$

Thus for a given S and L each fine structure term difference is proportional to the larger of the two J values associated with them. This can also be verified by equation (8.50) by taking the differences between the term J and the adjacent term $J + 1$.

REFERENCES

Bernath, P.F. (1995). *Spectra of Atoms and Molecules.* Oxford University Press, New York.

Condon, E.U. and Shortley, G.H. (1935, 1999). *The Theory of Atomic Spectra.* Cambridge University Press, U.K.

Foot, C.J. (2005). *Atomic Physics.* Oxford University Press, Oxford, U.K.

Herzberg, G. (1945). *Atomic Spectra and Atomic Structure.* Dover, New York.

Hollas, J.M. (2004). *Modern Spectroscopy*, 4th (edn.). John Wiley & Sons, West Sussex, England.

Johnson, W.R. (2007, 2010). *Atomic Structure and Theory.* Springer Verlag, Berlin, New York.

White, H.E. (1934). *Introduction to Atomic Spectra.* McGraw Hill., New York.

9

THE EFFECT OF MAGNETIC AND ELECTRIC FIELDS ON THE SPECTRAL TRANSITIONS

9.1 ZEEMAN EFFECT

The Zeeman effect is the splitting of spectral lines into several components in the presence of a magnetic field while Stark effect is the splitting of spectral lines in the presence of an electric field. The Zeeman effect was first observed in the yellow lines of sodium flames by the Dutch physicist Pieter Zeeman (1865–1943)[1] in 1896 and the effect was subsequently named as the Zeeman effect. It should be noted that the Bohr theory was not known at that time. It is remarkable that so much detailed spectroscopy had been carried out long before the atomic structure proposed by Niels Bohr and it is still more notable that Zeeman's first study of the sodium Zeeman splitting was done even before J. J.Thomson's discovery of the electron in 1897.

After Thomson's discovery of the electron, Zeeman and Lorentz did extensive study on the influence of magnetic fields on the spectral emissions from the atoms.

When the sodium flame is placed in a static magnetic field the first principal doublets are found to be broadened. The spectral lines showed 3 components when viewed along the transverse direction of the magnetic field, the middle one being unshifted from the original position. Later it was found that it contains as many as 15 components.

The Zeeman Effect and the Stark effect were two analogous effects which result in the splitting of a spectral line into several components in the presence of magnetic or electric field. Stark observed the Stark effect in 1913 and the Zeeman effect in 1896. The Zeeman Effect is very important in applications such as electron spin resonance spectroscopy and magnetic resonance imaging (MRI).

Thus when a source of light is placed under a static magnetic or electric field, the spectrum observed is found to shift from its original position observed (or split) where there were no fields. Typical example of energy level splitting and the splitting in the spectral lines are shown in Figure 9.1 and 9.2.

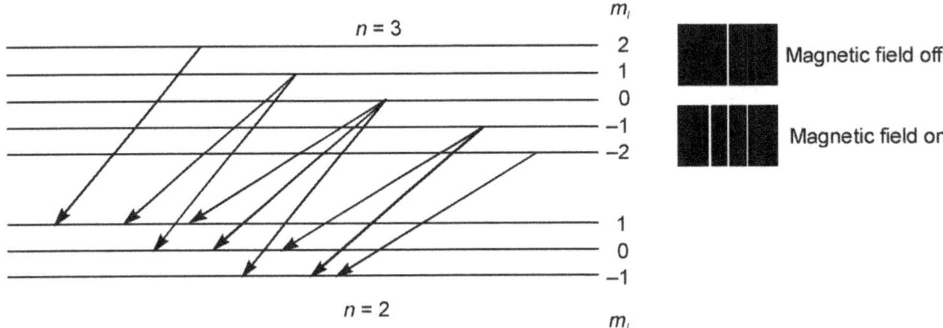

Figure 9.1   The sodium Zeeman splitting

The Zeeman effect for the hydrogen atom gave experimental support for the quantization of angular momentum which arose from the solution of the Schrödinger equation. Normal Zeeman effect is observed for spin 0 states. In this case the spin does not contribute to the angular momentum.

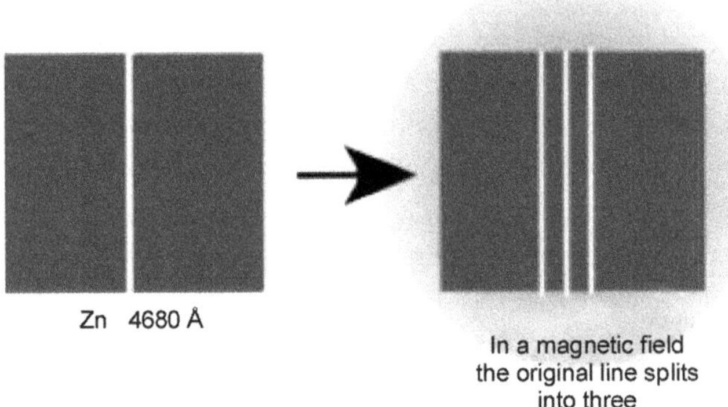

Figure 9.2   Zeeman effect

In the above diagram the selection rules are

$$\Delta l = \pm 1$$

$$\Delta m_l = 0, \pm 1$$

The example is of Normal Zeeman effect. This type of splitting is observed in zinc where the states are singlets (meaning electron spin = 0). As stated above the spin does not contribute to the angular momentum in the above case.

The central component was linear polarized with its electric vector along the magnetic field direction whereas the other two components are linearly polarized with their electric vectors perpendicular to the magnetic field. When viewed along the longitudinal direction, i.e., along the direction of the magnetic field, the central component was found to be absent while the other two components show circular polarization in opposite directions. One line is on the higher frequency side and the other on the lower frequency side.

Lorentz, from consideration of classical theory, pointed out that when the light source is placed in a magnet or electric field it forces the electrons to modify their motion in such a way as to change their period of motion. In the case of the atoms in a magnetic field which is normal to the plane of the electron orbit, the magnetic field makes the electron to speed up or to slow down by an amount, depending upon the strength of the magnetic field. It can be interpreted that the electron either gets a "push" or it is "pulled" down in their motion in the orbit. This push and pull in fact depend upon the strength of the field, mass and charge of the electron and the velocity of light.

Thus when an atom is placed in a magnetic field there is shift or splitting in the frequencies of spectral lines observed. This is due to the fact that the energy levels of the atom are either shifted or split due to the interaction of the applied magnetic field with the magnetic moment of the atom.

The splitting of the energy levels is due to two types of interactions:

1.   the interaction of the applied magnetic field with the orbital motion of the electron
2.   the interaction of the magnetic field with the electron spin.

And two types of Zeeman effect is known.

1.   Normal Zeeman effect
2.   Anomalous Zeeman effect

Example of normal Zeeman effect is the type of splittings observed in zinc singlets. Normal Zeeman effect occurs for spin = 0 states, since it does not contribute to the angular momentum.

Anomalous Zeeman effect appears on transitions where the net spin of the electron is not zero, the number of Zeeman sublevels being even instead of odd if there is an even number of electrons involved. It was called anomalous because the electron spin had not yet been discovered at that time, and there were no good explanations to it when Zeeman observed the effect.

9.2   THE VECTOR MODEL OF A ONE ELECTRON SYSTEM IN A WEAK MAGNETIC FIELD

Lande's vector model of the atom and the calculation of the famous Lande g factor helped very much in explaining semiclassically the Zeeman effect. Vector model atom in a magnetic field is given pictorially again in Figure 9.3.

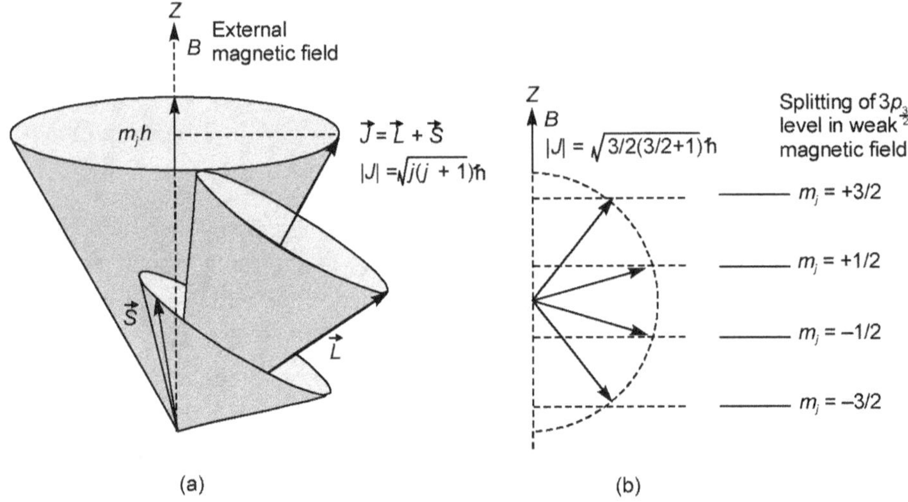

Figure 9.3   Vector model atom in a magnetic field

In a weak magnetic field, the magnetic moment associated with the total mechanical moment $p_j = j^*$ causes the atom to precess like a top along the field direction B. The quantum condition involved is that the projection of the angular momentum j^* in the direction of the magnetic field takes up half integral values from $-j$ to $+j$ and the values are given by m where

$$m = \pm\frac{1}{2}, \pm\frac{3}{2}, \ \pm\frac{5}{2}, \ \ldots \pm j.$$

We have already seen that the ratio between the magnetic and mechanical moment of an electron in an orbit according to classical theory is (7.13)

$$\frac{\mu_l}{p_l} = \frac{e}{2mc} \tag{9.1}$$

However the ratio between the magnetic and mechanical moment for a spinning electron is found to be empirically (7.14)

$$\frac{\mu_s}{p_s} = 2\frac{e}{2mc} \tag{9.2}$$

A schematic vector diagram is shown in the Figure 9.3. It is seen that the resultant magnetic moment μ_{ls} is not in line with the resultant mechanical moment j^*. This is shown in Figure 9.4.

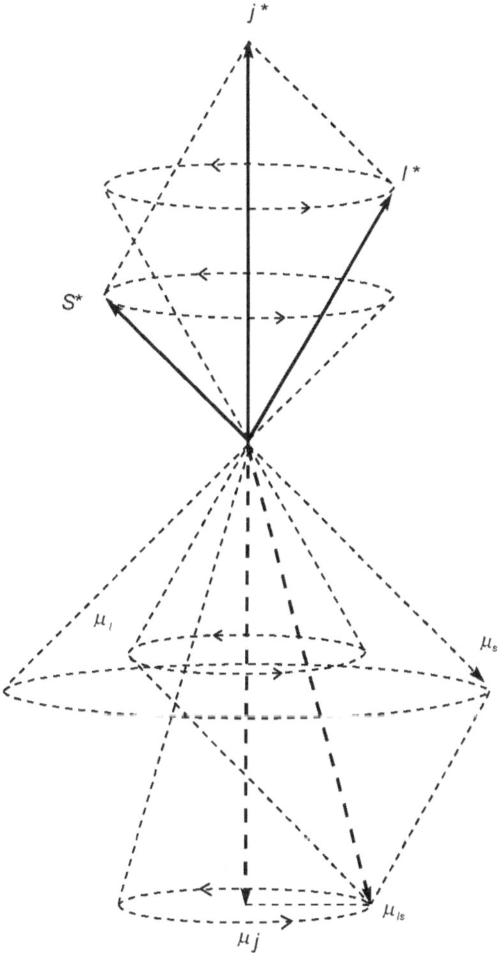

Figure 9.4   Vector model showing the magnetic and mechanical moments

9.2.1 Normal Zeeman Effect

We know from classical electrodynamics that the current I conducted in a closed loop of area A gives the dipole moment $\mu = I\,A$.

The rotation of the electron with a negative charge around the nucleus makes $\dfrac{\omega}{2\pi}$ revolution per second and hence $I = -\dfrac{\omega}{2\pi}e$.

$$\mu = \mathrm{I} \cdot \text{Area} = -\frac{\omega}{2\pi}\, e \cdot \pi r^2 = -\frac{\omega\, e\, r^2}{2}$$

$$p = m\, r^2 \omega \quad \text{and hence} \quad r^2 = \frac{p}{m\,\omega}$$

hence
$$\mu = -\frac{\omega e}{2}\cdot\frac{p}{m\omega} = -\frac{e}{2m}\,p = -\frac{e}{2m}\frac{h}{2\pi}\sqrt{J(J+1)}$$

$$= -\frac{eh}{4\pi m}J^* = \mu_B\, J^* \tag{9.3}$$

Where $J^* = \sqrt{J(J+1)}$ and in the above we have substituted the quantum mechanical equivalent of angular momentum P as $\sqrt{J(J+1)}\dfrac{h}{2\pi}$

We have already seen in Chapter 7 that in SI unit

$\mu_B = \dfrac{eh}{4\pi m}$ and is called Bohr magneton and in CGS unit $\mu_B = \dfrac{eh}{4\pi mc}$. And in SI unit $\dfrac{\mu}{p} = -\dfrac{e}{2m}$.

In CGs unit $\dfrac{\mu}{p} = -\dfrac{e}{2mc}$

Larmor frequency from the Larmor precession of the electron is

$$\omega_L = -B\cdot\frac{\mu}{p} = B\frac{e}{2m}$$

$$v_L = B\cdot\frac{e}{2m}\frac{1}{2\pi} = B\frac{e}{4\pi m} \tag{9.4}$$

Thus the magnetic moment of an atom $\mu = -\dfrac{e}{2m}\dfrac{h}{2\pi}\sqrt{J(J+1)} = -\dfrac{e}{2m}\dfrac{h}{2\pi}J^*$

$$E_{M_J} = E_0 - \mu_J B \cos\theta$$

$$E_{M_J} - E_0 = -\mu_J B \cos\theta = -\mu_J B\frac{M_J}{J}$$

For atom with $s = 0$　$\mu_J = \mu$ and hence

$$E_{M_J} - E_0 = -\mu_J B\frac{M_J}{J} = \frac{e}{2m}\frac{h}{\pi}B\frac{M_J}{J} = h\, v_L M_J \tag{9.5}$$

The selection rule

✧　$\Delta L = 0, \pm 1$

✧　$\Delta M_L = 0, \pm 1$

❖ $\Delta S = 0$

❖ $\Delta M_S = 0$

For transverse observation $\Delta M_J = 0$ and hence

$$\Delta E = E_0' + h\nu_L M_J' - (E_0'' + h\nu_L M_J'')$$
$$= (E_0' - E_0'') + h\nu_L (M_J' - M_J'') \tag{9.6}$$

$$\nu = \nu_0 + \nu_L \Delta M_J \qquad \nu_0 = \frac{E_0' - E_0}{h} \tag{9.7}$$

for longitudinal observation $\Delta M_J = 0, \pm 1$

thus we have the Zeeman spectrum consisting of $\nu_1 = \nu_0 + \nu_L$

$$\nu_2 = \nu_0$$
$$\nu_3 = \nu_0 - \nu_L \tag{9.8}$$

Normal Zeeman effect is observed in Cd atom. And since the observed spectrum is simple it is termed as normal Zeeman effect.

To get a quantum mechanical derivation of Zeeman effect we start with the energy of a magnetic moment $\vec{\mu}$ in a magnetic field $\vec{B}$ and is given by $-\vec{\mu} \cdot \vec{B}$. The Quantum mechanical approach is given in Section 9.4.

9.2.2 Anomalous Zeeman Effect: The Sodium Zeeman Effect

The splitting due to a magnetic field observed in sodium D lines is an example of anomalous Zeeman effect. For all atoms for which the total spin S is greater than zero anomalous Zeeman effect occurs when a magnetic field is applied. D_1 lines split into four components and D_2 lines into six components. This is due to the fact that the upper and lower levels split in different magnitudes. The Sodium D doublets and their splittings in a magnetic field are shown in Figures 9.5 and 9.6.

When electron spin is included, there is a greater variety of splitting patterns.

The transition which gives rise to the doublet is from the $3p$ to the $3s$ level, levels which would be the same in the hydrogen atom. The $3p$ level is split into states with total angular momentum $j = 3/2$ and $j = 1/2$ by the magnetic energy of the electron spin in the presence of the internal magnetic field caused by the orbital motion. This is the earlier described spin–orbit interaction. In the presence of an additional externally applied magnetic field, these levels are further split by the magnetic interaction, showing dependence of the energies on the z-component of the total angular momentum. This splitting gives the Zeeman effect in the sodium D lines (Figure 9.7).

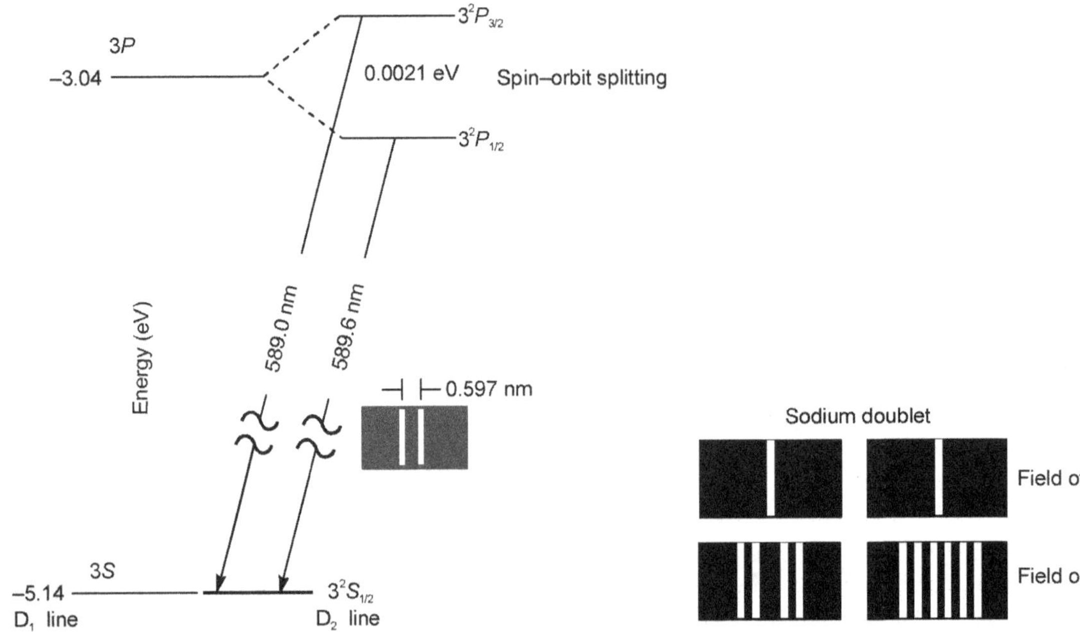

Figure 9.5   The Sodium D doublets

Figure 9.6   "Anomalous" Zeeman effect

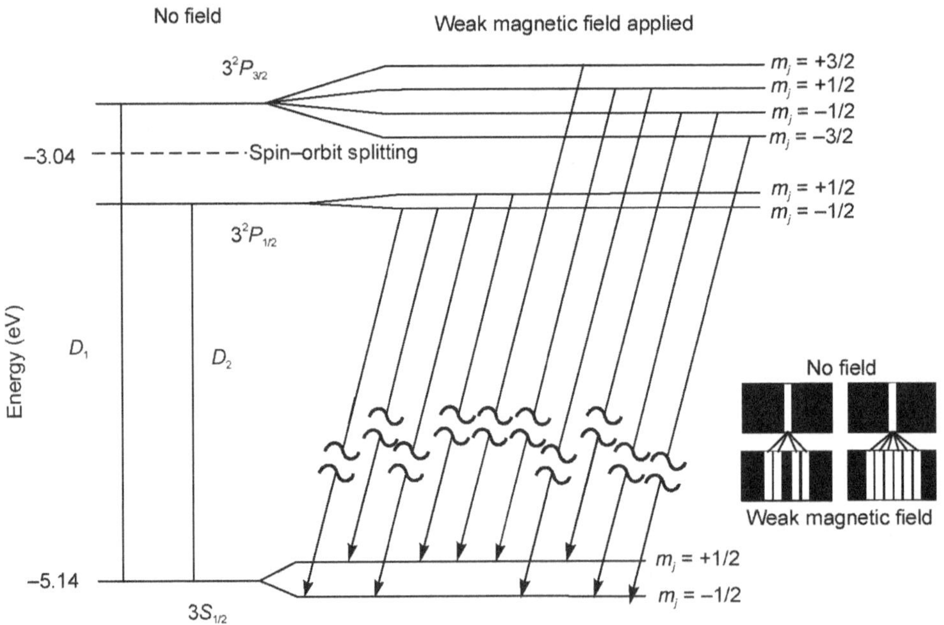

Figure 9.7   The sodium Zeeman effect

9.3 INTERACTION ENERGY IN A MAGNETIC FIELD

Figure 9.3 as well as Figure 9.4 depicts a schematic vector diagram for the magnetic and mechanical moments.

It can be seen from the Figure 9.4 that the resultant magnetic moment μ_{ls} is not in line with the resultant mechanical moment $j * h / 2\pi$. The mechanical moment $j * h / 2\pi$ is invariant, the other angular momenta due to $l*$, $s*$, μ_l, μ_s, μ_{ls} precess around $j*$. If we resolve μ_{ls} into two components one parallel to $j*$ and other perpendicular, only the component parallel to μ_{ls} contribute to the magnetic moment of the atom. As $j*$ is a constant of motion the components perpendicular to $j*$ average out to zero owing to the continual change in direction.

We have already seen that the ratio of the magnetic moment to the mechanical moment in CGS unit

$$\frac{\mu_L}{p_L} = -\frac{e}{2mc} = \frac{\mu_L}{L^* \dfrac{h}{2\pi}} \tag{9.9}$$

$$\frac{\mu_S}{p_S} = -2\frac{e}{2mc} = \frac{\mu_S}{S^* \dfrac{h}{2\pi}} \tag{9.10}$$

$$\mu_L = -L^* \frac{h}{2\pi}\frac{e}{2mc}\frac{erg}{gauss} \tag{9.11}$$

$$\mu_s = -2S^* \frac{h}{2\pi}\frac{e}{2mc}\frac{erg}{gauss} \tag{9.12}$$

Note that for spinning electron the ratio is twice. In general

$$\frac{\mu}{p} = -g\frac{e}{2mc} \tag{9.13}$$

The magnitude of the component of μ_L and μ_s along $j*$

$$\text{component of } \mu_L \text{ along } j* = L^* \frac{h}{2\pi}\frac{e}{2mc}\cos(L * j*) \tag{9.14}$$

$$\text{component of } \mu_s \text{ along } j* = 2S^* \frac{h}{2\pi}\frac{e}{2mc}\cos(S * j*) \tag{9.15}$$

$$\mu_L + \mu_S \text{ along } j^* = [L^*\cos(L^*j^*) + 2s^*\cos(S^*j^*]\frac{h}{2\pi}\frac{e}{2mc} \tag{9.16}$$

Also $\mu_L + \mu_S = \mu_J = [L^*\cos(L^*j^*) + 2s^*\cos(S^*j^*]\dfrac{h}{2\pi}\dfrac{e}{2mc}$

$$= [L^*\cos(L^*j^*) + 2s^*\cos(S^*j^*]\text{Bohr Magneton} \tag{9.17}$$

$$1 \text{ Bohr magneton} = \frac{eh}{4\pi mc}$$

Now $\quad \dfrac{\mu_J}{p_J} = -g\dfrac{e}{2mc}$

$$\text{i.e.,} \quad \frac{\mu_J}{p_J} = -g\frac{e}{2mc} = -\frac{\mu_J}{J^*\dfrac{h}{2\pi}} \tag{9.18}$$

$$\mu_J = -g\frac{e}{2mc}\frac{h}{2\pi}J^* = [L^*\cos(L^*j^*) + 2s^*\cos(S^*j^*]\frac{h}{2\pi}\frac{e}{2mc} \tag{9.19}$$

Thus $J^*g_J = L^*\cos(L^*j^*) + 2s^*\cos(S^*j^*)$

Application of cosine law gives

$$S^{*2} = L^{*2} + J^{*2} - 2L^*J^*\cos(L^*J^*)$$
$$L^{*2} = S^{*2} + J^{*2} - 2S^*J^*\cos(S^*J^*)$$

$$\cos(L^*J^*) = \frac{L^{*2} + J^{*2} - S^{*2}}{2L^*J^*} \tag{9.20}$$

$$\cos(S^*J^*) = \frac{S^{*2} + J^{*2} - L^{*2}}{2S^*J^*} \tag{9.21}$$

On substitution

$$g_J = \frac{L^*}{J^*}\left\{\frac{L^{*2} + J^{*2} - S^{*2}}{2L^*J^*}\right\} + \frac{2S^*}{J^*}\frac{S^{*2} + J^{*2} - L^{*2}}{2S^*J^*}$$

$$= \left\{\frac{L^{*2} + J^{*2} - S^{*2}}{2J^{*2}}\right\} + 2\left\{\frac{S^{*2} + J^{*2} - L^{*2}}{2J^{*2}}\right\} \tag{9.22}$$

$$g_J' = \frac{3J^{*2} + S^{*2} - L^{*2}}{2J^{*2}} = 1 + \frac{J^{*2} + S^{*2} - L^{*2}}{2J^{*2}} \tag{9.23}$$

We already know that $J^* = \sqrt{J(J+1)}$, $L^* = \sqrt{L(L+1)}$, $S^* = \sqrt{S(S+1)}$

Hence $$g_J = 1 + \frac{J(J+1) + S(S+1) - L(L+1)}{2J(J+1)} \tag{9.24}$$

g_J is called Lande g-factor. It is very important in the calculation of Zeeman levels. For a single valence electron $L = l$, $S = s$ and $J = j$

$$g_J = 1 + \frac{j(j+1) + s(s+1) - l(l+1)}{2j(j+1)} \tag{9.25}$$

In a magnetic field the orbital magnetic moment and the spin magnetic moment precess around the applied field. Due to the anomalous electron spin the precession of S^* around the magnetic field is twice that of the orbital angular momentum L^*. If the applied field is not too strong the coupling between l^* and s^* are strong and hence j^* which is the resultant of l^* and s^* precess around the magnetic field. The angular velocity of this precession is given by Larmor's theory and it is the magnetic field multiplied by the ratio of the magnetic moment and the mechanical moment.

$$\omega_l = -B \cdot \frac{\mu_J}{P_J} = B\, g \frac{e}{2mc} \tag{9.26}$$

The energy of interaction is given by the precessional angular velocity ω_L multiplied by the component of the resultant mechanical moment j^* on the axis of rotation B.

Thus $$\Delta W = \omega_L\, j^* \frac{h}{2\pi} \cos(j^* B) \tag{9.27}$$

Substituting for ω_L from equation 9.26

$$\Delta W = B g \frac{e}{2mc}\, j^* \frac{h}{2\pi} \cos(j^* B) \tag{9.28}$$

$j^* \dfrac{h}{2\pi} \cos(j^* B)$ is the projection of j^* on the magnetic field B and in terms of magnetic quantum number it is $m\dfrac{h}{2\pi}$.

Thus $$\Delta W = B\, g \frac{e}{2mc}\, m \frac{h}{2\pi} = m\, g\, B \frac{eh}{4\pi mc} \tag{9.29}$$

Interaction energy is obtained by dividing by hc

$$\frac{\Delta W}{hc} = -\Delta T = m\ g\ B\frac{e}{4\pi mc^2}\,cm^{-1}$$

(9.30)

$\dfrac{\Delta W}{hc}$ is also sometimes called the term value of the energy state.

The applied magnetic field B is the same for all levels of a given atom and we can write the above expression as

$$\frac{\Delta W}{hc} = -\Delta T = m\ g\ \frac{eB}{4\pi mc^2} = m\ g\ L\ cm^{-1}$$

(9.31)

where $L = \dfrac{eB}{4\pi mc^2}$ called the Lorentz unit.

We can see form the expression 9.31 that the magnetic interaction energy is proportional to the magnetic field. We can also notice that the g factor is the primary deciding factor of the shift or splitting in a given magnetic field as also the magnetic quantum number m. We can calculate the g-factors for different l and s.

The expression 9.31 greatly simplifies the calculation of energy separation of magnetic sublevels. It can be seen that the relative separations of the magnetic sublevels are determined by the g factor. Figure 9.8 shows the Zeeman pattern in case of $^1D_2 - {}^1P_1$ transition.

Table 9.1   The Lande g factor for some levels

l	Term	g	m	mg
0	$^2S_{1/2}$	2	$\pm\dfrac{1}{2}$	± 1
1	$^2P_{1/2}$	$\dfrac{2}{3}$	$\pm\dfrac{1}{2}$	$\pm\dfrac{1}{3}$
	$^2P_{3/2}$	$\dfrac{4}{3}$	$\pm\dfrac{3}{2},\pm\dfrac{1}{2}$	$\pm\dfrac{6}{3},\pm\dfrac{2}{3}$
2	$^2D_{3/2}$	$\dfrac{4}{5}$	$\pm\dfrac{3}{2},\pm\dfrac{1}{2}$	$\pm\dfrac{6}{5},\pm\dfrac{2}{5}$
	$^2D_{5/2}$	$\dfrac{6}{5}$	$\pm\dfrac{5}{2},\pm\dfrac{3}{2},\pm\dfrac{1}{2}$	$\pm\dfrac{15}{5},\pm\dfrac{9}{5},\pm\dfrac{3}{5}$

Note that there are 5 sublevels (m-components)in the upper state $[(2j + 1)$ sublevels] and 3 m components in the lower level.

The Lande g-factor for some levels are given in Table 9.1 and the splitting of a $^2P_{3/2}$ is as shown in Figure 9.7.

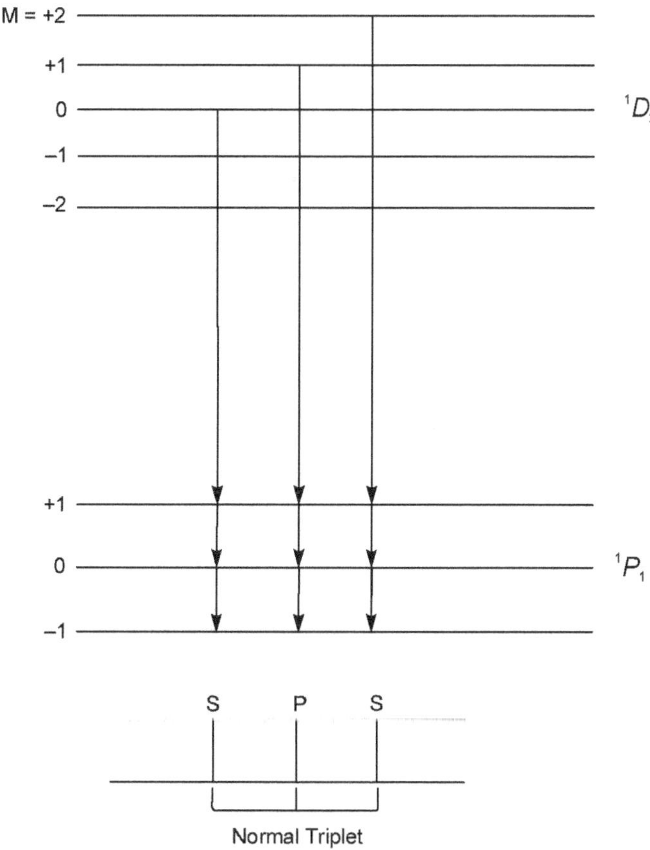

Figure 9.8   The Zeeman pattern in case of $^1D_2 - {}^1P_1$ transition

$$g \text{ for } {}^2S_{1/2},\ g_J = 1 + \frac{j(j+1)+s(s+1)-l(l+1)}{2j(j+1)} = 1 + \frac{\frac{1}{2}\left(\frac{1}{2}+1\right) + \frac{1}{2}\left(\frac{1}{2}+1\right) - 0(0+1)}{2\frac{1}{2}\left(\frac{1}{2}+1\right)}$$

$$= 1 + \frac{\frac{1}{2}\left(\frac{3}{2}\right) + \frac{1}{2}\left(\frac{3}{2}\right) - 0}{2 \times \frac{1}{2}\left(\frac{3}{2}\right)} = 1 + \frac{\frac{3}{4} + \frac{3}{4} - 0}{2 \times \frac{3}{4}} 1 + \left(\frac{6}{4} \times \frac{4}{6}\right) = 2$$

Similarly for $^2P_{1/2} \cdot g_J = 1 + \dfrac{\frac{1}{2}\left(\frac{1}{2}+1\right)+\frac{1}{2}\left(\frac{1}{2}+1\right)-1(1+1)}{2\times\frac{1}{2}\left(\frac{1}{2}+1\right)} = 1 + \dfrac{\frac{3}{4}+\frac{3}{4}-2}{2\times\frac{3}{4}}$

$$= 1 + -\frac{2}{4}\times\frac{4}{6} = \frac{2}{3}$$

and so on.

9.3.1 Selection Rules

The selection rule for anomalous Zeeman effect

✦ $\Delta J = 0, \pm 1$

✦ $\Delta M_J = 0, \pm 1$ (9.32)

The selection rules for the transition is that for any transition the magnetic quantum number m changes by $+1$, 0 or -1. Thus $\Delta m = 0, \pm 1$.

For the D_1 line there are six allowed transitions and for D_2 line there are four allowed transitions. The polarization rules are

Viewed perpendicular to the field: $\Delta m = \pm 1$ plane polarized $\perp$ to the magnetic field (s components)

$\Delta m = 0$, plane polarized $\parallel$ to the magnetic field (p components)

Viewed $\parallel$ to the field : $\Delta m = \pm 1$ circularly polarized (s components)

$\Delta m = 0$, forbidden (p components)

p stands for parallel components and s for perpendicular components. s is derived from the German word Senkrecht meaning perpendicular.

9.3.2 Intensities

The sum of all the transitions originating from any initial Zeeman level is equal to the sum of all transitions leaving any other level having the same n and l values. Similarly the sum of all transitions arriving at any Zeeman level is equal to the sum of all transitions arriving at any other level having the same n and l values.

It may be stated as follows:

Transitions $j \to j$ $m \to m \pm 1,$ $I = A(j \pm m + 1)(j \mp m)$

$m \to m,$ $I = 4A\,m^2$ (9.33)

$$\text{Transition } j \to j+1 \quad m \to m \pm 1, \quad I = B(j \pm m + 1)(j \pm m + 2)$$

$$m \to m, \quad I = 4B(j + m + 1)(j - m + 1) \tag{9.34}$$

where A and B are constants. If one determines the relative intensities, it is not necessary to know A and B.

For a rapid calculation of Zeeman pattern we follow the method given below. The separation pattern mg for the initial and final states are first written down in two rows with equal values of m directly below or above each other. For example for a transition of type $^2D_{5/2} - ^2P_{3/2}$ we write

Initial $^2D_{5/2}$ state m 5/2   3/2   1/2   −1/2   −3/2   −5/2

mg 15/5   9/5   3/5   −3/5   −9/5   −15/5

Final $^2P_{3/2}$ state mg 6/3   2/3   −2/3   −6/3

The vertical arrows indicates p components $(\Delta m = 0)$ and the diagonal ones the s components $(\Delta m = \pm 1)$.

$$g_J = 1 + \frac{J(J+1) + S(S+1) - L(L+1)}{2J(J+1)}$$

For $^2D_{5/2} \; j = 5/2, \; s = \tfrac{1}{2}, \; l = 2$

$$g_J = 1 + \frac{\frac{5}{2}\left(\frac{5}{2}+1\right) + \frac{1}{2}\left(\frac{1}{2}+1\right) - 2(2+1)}{2\frac{5}{2}\left(\frac{5}{2}+1\right)} = 1 + \frac{\left[\frac{5}{2}\left(\frac{7}{2}\right) + \frac{1}{2}\left(\frac{3}{2}\right) - 6\right]}{2\frac{5}{2}\left(\frac{7}{2}\right)} = 1 + \frac{\left(\frac{35}{4}\right) + \left(\frac{3}{4}\right) - 6}{2\left(\frac{35}{4}\right)}$$

$$= 1 + \frac{\frac{14}{4}}{\frac{70}{4}} = 1 + \frac{14}{4} \times \frac{4}{70} = 1 + \frac{2}{10} = \frac{6}{5}$$

mg values originating from the g_j value is shown in the second row of the above table.

9.3.3 The Sodium Zeeman Effect

We have already seen that the sodium spectrum is dominated by the bright doublet known as the Sodium D-lines at 588.9950 and 589.5924 nm. From the energy level diagram it can be seen that these lines occur from a transition from the $3p$ to the $3s$ levels. The amount of splitting depends on the magnetic field and the Lande g-factor.

Term	J	L	S	g_l
$3^2P_{3/2}$	3/2	1	1/2	4/3
$3^2P_{1/2}$	1/2	1	1/2	2/3
$3^2S_{1/2}$	1/2	0	1/2	2

Examination of the size of the Lande g-factor g_L for the three levels will show why the splittings of the different levels are different in magnitude. The number of transitions are governed by the selection rule and this explains why the transitions shown are allowed and others not. Zeeman effect in the principal series doublet in sodium is shown in Figure 9.7.

The intensity rules say that the sum of all transitions originating from any initial Zeeman level is equal to the sum of all transitions leaving any other level having the same n and l values. Similarly sum of all transitions arriving at any initial Zeeman level is equal to the sum of all transitions arriving at any other level having the same n and l values. Weak and strong field energies in the case of principal series doublets in sodium is given in Table 9.2.

Table 9.2   Weak and strong field energies for the principal series doublet

No field		Weak field (Zeeman effect)			Strong field (Paschen–Back effect)			
State		m	g	mg	m_l	m_s	$m_l m_s$	m_l+2m_s	$am_l m_s$
$^2P_{3/2}$ $+a/2$		$+\dfrac{3}{2}$		$+\dfrac{6}{3}$	$+1$	$+\dfrac{1}{2}$	$+\dfrac{1}{2}$	$+2$	$+\dfrac{a}{2}$
		$+\dfrac{1}{2}$		$+\dfrac{2}{3}$	0	$+\dfrac{1}{2}$	0	$+1$	0
			$\dfrac{4}{3}$						
		$-\dfrac{1}{2}$		$-\dfrac{2}{3}$	-1	$+\dfrac{1}{2}$	$-\dfrac{1}{2}$	0	$-\dfrac{a}{2}$
		$-\dfrac{3}{2}$		$-\dfrac{6}{3}$	$+1$	$-\dfrac{1}{2}$	$-\dfrac{1}{2}$	0	$-\dfrac{a}{2}$

(Contd.)

Table 9.2   (Continued)

No field		Weak field (Zeeman effect)			Strong field (Paschen–Back effect)				
State		m	g	mg	m_l	m_s	$m_l m_s$	$m_l + 2m_s$	$am_l m_s$
$^2P_{1/2}$	$-a$	$+\dfrac{1}{2}$	$\dfrac{2}{3}$	$+\dfrac{1}{3}$	0	$-\dfrac{1}{2}$	0	-1	0
		$-\dfrac{1}{2}$		$-\dfrac{1}{3}$	-1	$-\dfrac{1}{2}$	$\dfrac{1}{2}$	-2	$+\dfrac{a}{2}$
$^2S_{1/2}$	0	$+\dfrac{1}{2}$	2	$+1$	0	$+\dfrac{1}{2}$	0	$+1$	0
		$-\dfrac{1}{2}$		-1	0	$-\dfrac{1}{2}$	0	-1	0

The explanations for the different patterns of splitting gave additional insight into the effects of electron spin.

9.4 QUANTUM MECHANICAL APPROACH

It has been stated earlier that in order to get a quantum mechanical picture of Zeeman effect, we should start with the energy of a magnetic moment $\vec{\mu}$ in a magnetic field $\vec{B}$ which is given by $-\vec{\mu} \cdot \vec{B}$. The magnetic interaction energy between the electron and the external magnetic field can be written as

$$W_l = \frac{e}{2\mu c} \vec{B} \cdot \vec{L} \text{ (interaction between external}$$
$$\text{magnetic field and orbital motion)} \tag{9.35}$$

$$W_s = \frac{e}{2\mu c} \vec{B} \cdot 2\vec{S} \text{ (interaction between}$$
$$\text{electron spin and magnetic field)} \tag{9.36}$$

The interaction energy due to the orbital motion of the electron in a magnetic field can be obtained by considering the magnetic field

$$\vec{B} = \text{Curl } \vec{A} \tag{9.37}$$

where $\vec{A}$ is the vector potential. In this case $\nabla \cdot \vec{A} = 0$.

The Hamiltonian for the electron of charge $-e$ in the field of the vector potential $\vec{A}$ is obtained by considering the momentum as $\left(\vec{p} + \dfrac{e}{c}\vec{A}\right)$ instead of $\vec{p}$. Thus the kinetic energy part of the Hamiltonian becomes $\dfrac{1}{2\mu}\left(\vec{p} + \dfrac{e}{c}\vec{A}\right)^2$ instead of the usual $\dfrac{p^2}{2m}$.

$$\frac{1}{2\mu}\left(\vec{p} + \frac{e}{c}A\right)^2 = \frac{1}{2\mu}\left(\vec{p}^2\right) + \frac{e}{2\mu c}\left[\vec{p}\cdot\vec{A} + \vec{A}\cdot\vec{p}\right] + \frac{e^2}{2\mu c^2}\vec{A}^2 \tag{9.38}$$

The first term corresponds to field free Hamiltonian and the second and third terms correspond to the Hamiltonian due to the applied magnetic field.

The usual method is to convert the classical kinetic energy into quantum mechanical operators. This is achieved by converting

$$\vec{p} = -i\hbar\nabla$$

thus $\vec{p}\cdot\vec{A}\psi = -i\hbar\nabla\cdot(\vec{A}\psi) = -i\hbar(\Psi\nabla\cdot\vec{A} + \vec{A}\cdot\nabla\Psi)$ \hfill (9.39)

Since $\nabla\cdot\vec{A} = 0$ we obtain $\vec{p}\cdot\vec{A}\psi = \vec{A}\cdot\vec{p}\psi$

The interaction Hamiltonian thus becomes

$$\frac{e}{2\mu c}[\vec{p}\cdot\vec{A} + \vec{A}\cdot\vec{p}] + \frac{e^2}{2\mu c^2}\vec{A}^2 = \frac{e}{\mu c}[\vec{A}\cdot\vec{p}] + \frac{e^2}{2\mu c^2}\vec{A}^2 \tag{9.40}$$

In the case of weak magnetic fields the second term is small and hence can be neglected. Thus the interaction Hamiltonian due to the orbital motion is

$$W_l = \frac{e}{\mu c}[\vec{A}\cdot\vec{p}] \tag{9.41}$$

For a uniform magnetic field $\vec{A} = \dfrac{1}{2}\left(\vec{B}\times\vec{r}\right)$ \hfill (9.42)

From (9.41) and (9.42), $W_l = \dfrac{e}{2\mu c}\vec{B}\times\vec{r}\cdot\vec{p}$

$$= \frac{e}{2\mu c}\vec{B}\cdot\vec{r}\times\vec{p} \tag{9.43}$$

We have already seen that $\bar{r} \times \bar{p} = \bar{L}$ the orbital angular momentum

Hence $W_l = \dfrac{e}{2\mu c} \bar{B} \cdot \bar{L}$ (9.44)

Thus from (9.35), (9.36) and (9.44) the total interaction due to the spin and orbital motion of the electron is

$$H = \frac{e}{2\mu c} \bar{B} \cdot [\bar{L} + 2\bar{S}]$$ (9.45)

Considering the magnetic field in Z direction

$$H = \frac{e}{2\mu c} \bar{B} \cdot \left[\bar{L}_z + 2\bar{S}_z \right]$$ (9.46)

$$L_Z + S_Z = J_Z$$

Thus $H = \dfrac{e\bar{B}}{2\mu c} \cdot \left[\bar{J}_z + \bar{S}_z \right]$ (9.47)

We consider a vector model where $\bar{L}$ and $\bar{S}$ precess around $\bar{J}$ and $\bar{J}$ in turn precesses around the magnetic field $\bar{B}$.

The interaction energy

$$E = \int \Psi^* H \Psi \, d\tau = \frac{e\bar{B}}{2\mu c} \int \Psi^* \left(\bar{L}_z + 2\bar{S}_z \right) \psi \, d\tau$$

$$= \frac{c\bar{B}}{2\mu c} < (LSJM \mid \bar{L}_z + 2S_z \mid LSJM >$$ (9.48)

From the theory of angular momentum we can write

$$E = \frac{e\bar{B}}{2\mu c} M \hbar < (LSJ \mid \bar{L}_z + 2\bar{S}_z \mid LSJ >$$

And $< (LSJ \mid \bar{L}_z \mid LSJ > = \dfrac{L(L+1) - S(S+1) + J(J+1)}{2J(J+1)}$ (9.49)

$< (LSJ \mid \bar{S}_z \mid LSJ > = \dfrac{S(S+1) - L(L+1) + J(J+1)}{2J(J+1)}$ (9.50)

Thus $E = \dfrac{e\bar{B}}{2\mu c} \, g \, M \hbar$

where g is known as Lande g-factor and

$$g = 1 + \frac{J(J+1) + S(S+1) - L(L+1)}{2J(J+1)} \qquad (9.51)$$

The interaction energy in wave number unit is

$$-\Delta T = g\,\mu_B\,MB = gML \qquad (9.52)$$

where $\mu_B = \dfrac{e\hbar}{2\mu c}$ is the Bohr magneton. The magnitude of the splitting is shown in Figure 9.9.

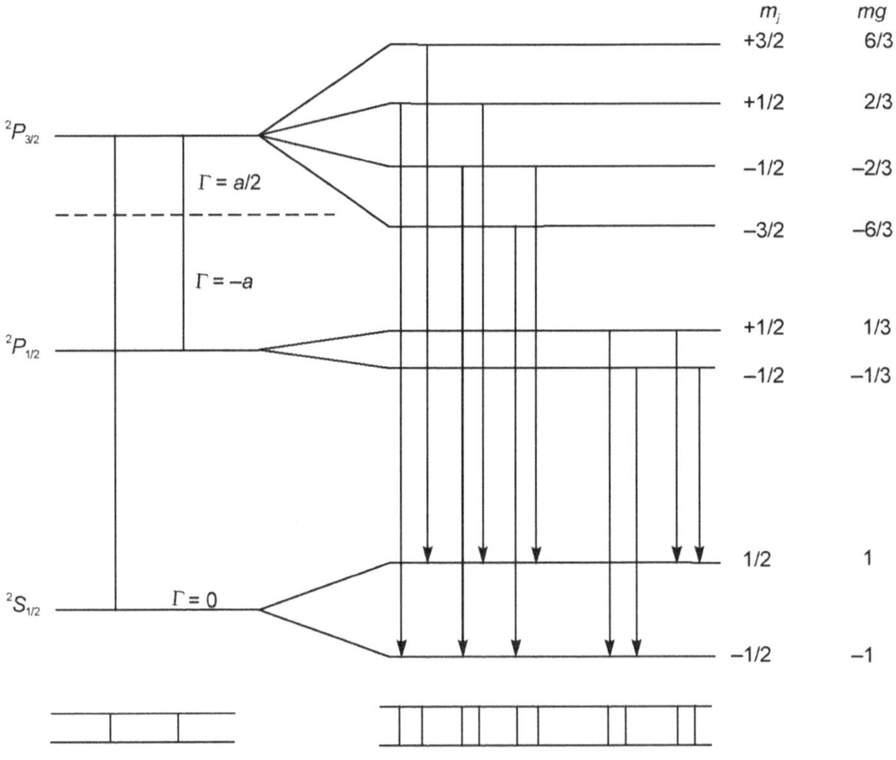

Figure 9.9   Zeeman effect in a weak magnetic field

The selection rule is that the magnetic quantum number m changes by $+1, 0$ or -1, i.e., $\Delta m = 0, \pm 1$. As mentioned earlier, the D_1 line has six allowed transitions and for the D_2 line there are four allowed transitions.

Viewed $\perp$ to the field: $\Delta m = \pm 1$ is plane polarized $\perp$ to the magnetic field (s components)

$\Delta m = 0$ plane polarized $\parallel$ to the field (p components)

Viewed $\parallel$ to the field: $\Delta m = \pm 1$ is circularly polarized (s components)

$$\Delta m = 0 \quad \text{forbidden } (p \text{ components})$$

It is shown that the sum of all transitions starting from any intial Zeeman level is equal to the sum of all transitions leaving any other level having the same n and l values. Similarly the sum of all transitions arriving at any Zeeman level is equal to the sum of all transitions arriving at any other level having the same n and l values. In short as explained in equations 9.33 and 9.34 for transitions

$$j \to j \qquad m \to m \pm 1, \ \text{Intensity } I \ = A(j \pm m + 1) \ (j \mp m)$$
$$m \to m, \qquad \text{Intensity } I \ = 4Am^2 \tag{9.53}$$

$$j \to j+1 \quad m \to m \pm 1, \ \text{Intensity } I \ = B(j \pm m + 1) \ (j \pm m + 2)$$
$$m \to m, \qquad \text{Intensity } I \ = 4B(j + m + 1)(j - m + 1) \tag{9.54}$$

where A and B are constants and relative intensities can be obtained.

For atoms having α valence electrons the total Hamiltonian of an atom in a magnetic field is:

$$H = H_0 + H_1 = H_0 + \sum_{\alpha} \xi(\vec{r}_\alpha)\vec{L} \cdot \vec{S} - \sum_{\alpha} \vec{\mu}_\alpha \cdot \vec{B} \tag{9.55}$$

where H_0 is the unperturbed Hamiltonian of the atom, and the sums over α are sums over the electrons in the atom. The term

$$\xi(\vec{r}_\alpha)\vec{L} \cdot \vec{S} \tag{9.56}$$

is the LS-coupling for each electron (indexed by α) in the atom. The sum vanishes if there is only one electron. The magnetic coupling

$$\vec{\mu}_\alpha \cdot \vec{B} = \frac{\mu_B}{\hbar}\left(g_L \vec{L} + g_S \vec{S}\right) \cdot \vec{B} \tag{9.57}$$

is the energy due to the magnetic moment μ of the α-th electron. It can be written as sum of contributions of the orbital angular momentum and of spin angular momentum, with each multiplied by the gyroscopic or Lande g-factor. By projecting the vector quantities on to the z-axis, the Hamiltonian may be written as

$$H = H_0 + \xi(r)\vec{L} \cdot \vec{S} + \mu_B(g_L L_z + g_S S_z)B_z \approx H_{at} + \frac{\mu_B}{\hbar}(J_z + S_z)B_z \tag{9.58}$$

where the g-factors are $g_L = 1$ and $g_s \approx 2$. The summation over the electrons was omitted for readability. Here, $J_z = L_z + S_z$ is the total angular momentum, and the LS-coupling term has been folded into H_0.

The size of the interaction term H' is not always small, and can induce large effects on the system. In the Paschen–Back effect, described below, H' cannot be treated as a perturbation, as its magnitude is comparable to or larger than the unperturbed system H_{at}. The H' term does not commute with H_{at}. In particular, Sz does not commute with the spin–orbit interaction H_{at}.

Examples

1.   Calculate the frequency shift associated with Zeeman effect in a 10 T field.

$\mu_B = 5.788 \times 10^{-5}$ eV/T basic unit of magnetic moment

$\Delta E = \mu_B \, B = h \, \Delta v$

$\Delta v = \mu_B \, B / h = 5.788 \times 10^{-5}$ eV/T $\times 10T / (4.14 \times 10^{-15}$ eV s$) = 1.39 \times 10^{11}$ Hz $= 0.14$ GHz

2.   What is the photon energy associated with the two satellites in the above case?

$\Delta E = \mu_B \, B = 5.788 \times 10^{-5}$ eV/T $\times 10 \ T = 5.788 \times 10^{-4}$ eV $= 0.6$ meV

3.   An electron absorbs a photon to flip from spin down to spin up in a magnetic field of 10 Tesla. What is the energy of the photon and the frequency?

$\Delta E = \mu_B \, B = 2 \times 5.788 \times 10^{-15}$ eV/T $\times 10 = 1.15 \times 10^{-3}$ eV $= 1.15$ meV

$hv = \Delta E$

$v = 1.15 \times 10^{-3}$ eV $/ 4.14 \times 10^{-15}$ eVs $= 2.80 \times 10^{11}$ Hz $= 0.28$ GHz

9.5 PASCHEN–BACK EFFECT

The splitting of energy levels in an external magnetic field had been described as Zeeman effect. However, when the magnetic field is strong, the splitting observed is different than those in a weak magnetic field. In Zeeman effect (i.e., for sufficiently weak magnetic fields) the splitting is small compared to the energy difference between the unperturbed levels,. We have seen that we could describe the splitting in a weak magnetic field with the aid of a vector atom model. However, if the magnetic field is large enough, the coupling becomes different. The strong magnetic field disrupts the coupling between the orbital and spin angular momenta, resulting in a different pattern of splitting. This effect is called the Paschen–Back effect. Or in other words the Paschen–Back effect is the splitting of atomic energy levels in the presence of a strong magnetic field.

In the strong-field case (Figure 9.10b), S and L couple more strongly to the external magnetic field than to each other, and can be visualized as independently precessing about the external field direction. Figure 9.10a is given as a comparison which is the case in Zeeman effect.

We have already seen the splitting of sodium D lines in weak magnetic fields in Figure 9.6.

The following is a model of the changes in the pattern if the magnetic field were strong enough to decouple L and S. The resulting spectrum would be a triplet with the centre line twice the intensity of the outer lines. The splitting of sodium D lines in a strong magnetic field is shown in Figure 9.11.

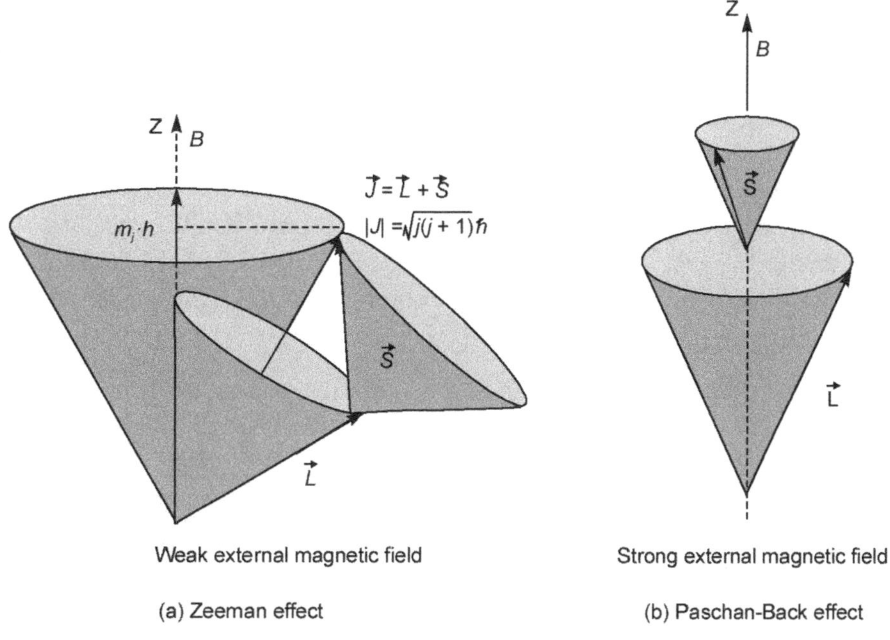

Figure 9.10   The splitting of atomic energy levels in a weak or strong magnetic field

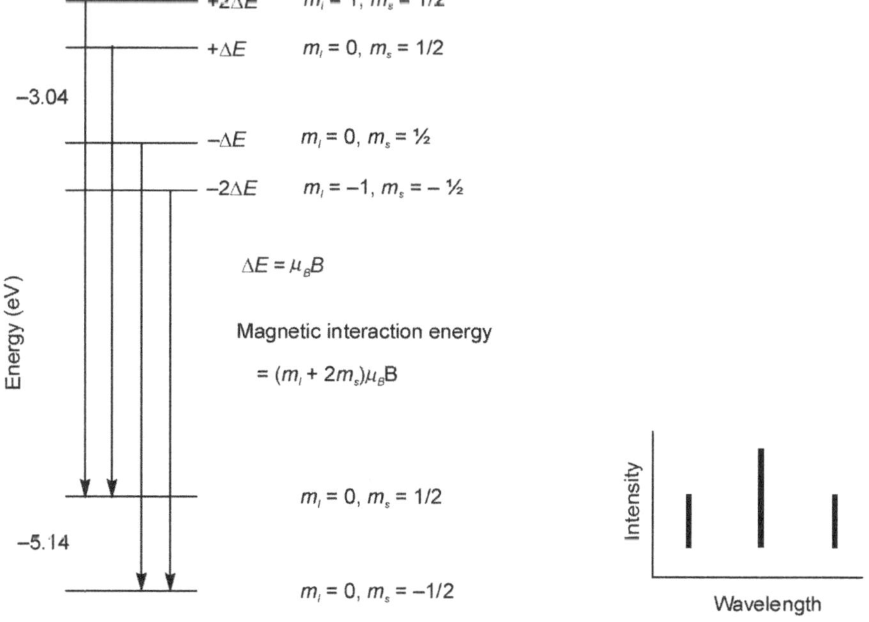

Figure 9.11   The splitting of sodium D lines in a strong magnetic field

To obtain the above pattern, the projections of L and S in the z-direction have been treated independently and the m_s multiplied by the spin g-factor. The energy shift is expressed in terms of Bohr magneton μ_B. The selection rules explain why the transitions shown are allowed and others not.

Sodium atom spectrum is used as the basis of the model for convenience, but the fields required to create Paschen–Back effect in sodium are unrealistically high. Lithium, on the other hand, has a spin–orbit splitting of only 0.00004 eV compared to 0.0021 eV for sodium. Such small energy values are sometimes expressed in "wave numbers", or $1/\lambda$ in cm^{-1}. In the units of wave numbers the lithium separation is about 0.3 cm^{-1}and the sodium separation is about 17 cm^{-1}. The Paschen–Back conditions are met in some lithium spectra observed on the Sun, so this effect does have astronomical significance.

In short when the external magnetic field is larger than the internal field due to the spin and the orbital motions of the electron, the $l*$ and $s*$ precess separately around B instead of the resultant $J*$ revolving around B as considered earlier in the case of Zeeman effect. Zeeman effect is the splitting of the energy levels in a weak magnetic field where we have considered that $l*$ and $s*$ couple together to form $J*$ and the resultant $J*$ precess around the magnetic field. In the large field case— Paschen–Back effect—$J*$ is no longer fixed in magnitude.

The quantum conditions imposed are m_l takes values from $-l$ to $+l$ thus having $(2l+1)$ levels and m_s takes values $+\dfrac{1}{2}$ and $-\dfrac{1}{2}$. Thus there are $2(2l+1)$ levels. The orientation of l and s on B are shown in Figure 9.9 b.

The interaction energy now consists of three parts.

1. energy due to the precession of $l*$ around B
2. energy due to the precession of $s*$ around B
3. interaction energy between $l*$ and $s*$

We have already seen that by Larmor's theorem that angular velocity is given by

$$\omega_{l*} = -B\frac{\mu_l}{p_l} = B\frac{e}{2mc} \tag{9.59}$$

$$\omega_{s*} = -B\frac{\mu_s}{p_s} = B2\frac{e}{2mc} \tag{9.60}$$

The interaction energy is then the above expressions multiplied by the projection of these angular momenta (due to the orbital motion and the spin on the magnetic field B)

$$\Delta W_{l,B} = B\frac{e}{2mc}l*\frac{h}{2\pi}\cos(l*B) = B\frac{e}{2mc}m_l\frac{h}{2\pi} \tag{9.61}$$

$$\Delta W_{s,B} = 2B\frac{e}{2mc}s*\frac{h}{2\pi}\cos(s*B) = 2B\frac{e}{2mc}m_s\frac{h}{2\pi} \tag{9.62}$$

The total energy shift from the above interaction is

$$\Delta W_B = (m_l + 2m_s)\, B \frac{eh}{4\pi mc}$$

then dividing by hc, we get $\dfrac{\Delta W_B}{hc} = (m_l + 2m_s)\, B \dfrac{e}{4\pi mc^2}\, cm^{-1} = (m_l + 2m_s)\, L\ cm^{-1}$

where $L = \dfrac{eB}{4\pi mc^2}$ is the Lorentz unit $\hfill$ (9.63)

The third contribution of the interaction energy is from the interaction of l^* and s^*. This is the usual spin–orbit interaction we already derived in section 7.5.

$$\Gamma = -T_{ls} = a\, l^* s^* \cos(l^* s^*)$$

where $a = \dfrac{R\alpha^2 Z^4}{n^3 l \left(l + \dfrac{1}{2}\right)(l+1)}\, cm^{-1} \hfill$ (9.64)

Since the angle between l^* and s^* is constantly changing, the average value of $\cos(l^* s^*)$ is to be calculated. From a well-known theory in trigonometry, with l^* and s^* independently precessing around B, it can be shown that

$$\overline{\cos l^* s^*} = \cos(l^* B)\cos(s^* B)$$

and hence the Γ factor

$$\Gamma = -\Delta T_{ls} = al^* \cos(l^* B)s^* \cos(s^* B) = a\, m_l\, m_s \hfill (9.65)$$

thus the total shift in energy is

$$-\Delta T_{ls} = (m_l + 2m_s)\, L + a\, m_l\, m_s\ cm^{-1} \hfill (9.66)$$

$$T = T_0 - (m_l + 2m_s)\, L - a\, m_l\, m_s\ cm^{-1} \hfill (9.67)$$

We can now calculate the Paschen–Back effect in sodium D lines

The selection rules are $\Delta m_L = 0, \pm 1$ and $\Delta m_S = 0$

m_L takes values from $+l$ to $-l$, i.e., $l, l-1, l-2, \ldots -l$

And m_S takes values between $s, s-1, \ldots -s$

For the sodium D lines the transition is $3^2 P_{3/2} - 3^2 S_{1/2}$ and $3^2 P_{1/2} - 3^2 S_{1/2}$

The transitions are shown in Figure 9.12.

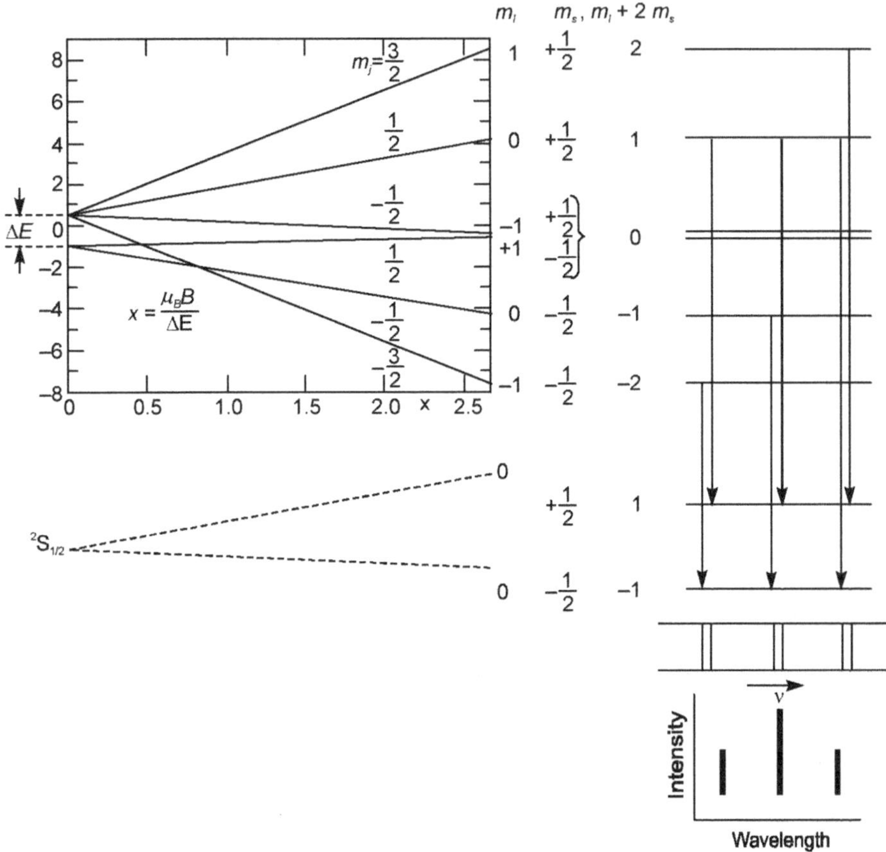

Figure 9.12   Paschen–Back effect

When the external magnetic field is so strong that the Zeeman splitting is greater than the spin–orbit splitting, effectively decoupling L and S, the level splitting is uniform for all atoms and only three spectral lines are seen, as the normal Zeeman effect. Each of the three lines is actually a closely spaced doublet, as illustrated by the transitions shown at right of Figure 9.12.

The D_1 line is the transition $3^2P_{1/2} \rightarrow 3^2S_{1/2}$. From equations (9.24) we can compute the Lande g-factors and use in computing the ΔE values from equation (9.29) is follows:

$$\text{For the } 3^2P_{1/2} \text{ level} \quad g = 1 + \frac{\frac{1}{2}\left(\frac{1}{2}+1\right) + \frac{1}{2}\left(\frac{1}{2}+1\right) - 1(1+1)}{2\left(\frac{1}{2}\right)\left(\frac{1}{2}+1\right)} = 2/3$$

For the $3^2S_{1/2}$ level $g = 1 + \dfrac{\dfrac{1}{2}\left(\dfrac{1}{2}+1\right) + \dfrac{1}{2}\left(\dfrac{1}{2}+1\right) - 0}{2\left(\dfrac{1}{2}\right)\left(\dfrac{1}{2}+1\right)} = 2$

and from equation 9.29

For the $3^2S_{1/2}$ level $\Delta E = \left(\dfrac{2}{3}\right)\left(\pm\dfrac{1}{2}\right)(5.79\times10^{-9} \text{ eV/gauss})B$

For the $3^2S_{1/2}$ level $\Delta E = 2\left(\pm\dfrac{1}{2}\right)(5.79\times10^{-9} \text{ eV/gauss})B$

The longest wavelength line $\left(m_l = -\dfrac{1}{2} \rightarrow m_l = +\dfrac{1}{2}\right)$ will have undergone a net energy shift of

$-1.93\times10^{-9}B - 5.79\times10^{-9}B = -7.72\times10^{-9}B$ eV.

The shortest wavelength line $\left(m_l = +\dfrac{1}{2} \rightarrow m_l = -\dfrac{1}{2}\right)$ will have undergone a net energy shift of

$1.93\times10^{-9}B + 5.79\times10^{-9}B = 7.72\times10^{-9}B$ eV.

The total energy difference between these two photons is

$$\Delta E = -1.54\times10^{-8}B\text{eV}$$

Since $\lambda = 1/v = hc/E,$ then $\Delta\lambda = -(hc/E^2)\Delta E = 0.022$ nm. We then have

$$\Delta E = -0.022 \text{ nm}(E^2/hc) = -1.54\times10^{-8}B$$

where $E = hc/\lambda = hc/(589.9 \text{ nm})$.

Finally we have

$$B = \frac{(0.022\times10^{-9})hc}{(589.8\times10^{-9})^2(1.54\times10^{-8})(1.60\times10^{-19})}$$
$$B = 0.51\,T = 5100 \text{ gauss}$$

For comparision, earth's magnetic field averages about 0.5 gauss.

Zeeman and Paschen–Back effect for sodium D lines are shown in Figure 9.13 starting with no field case and ending in strong field case

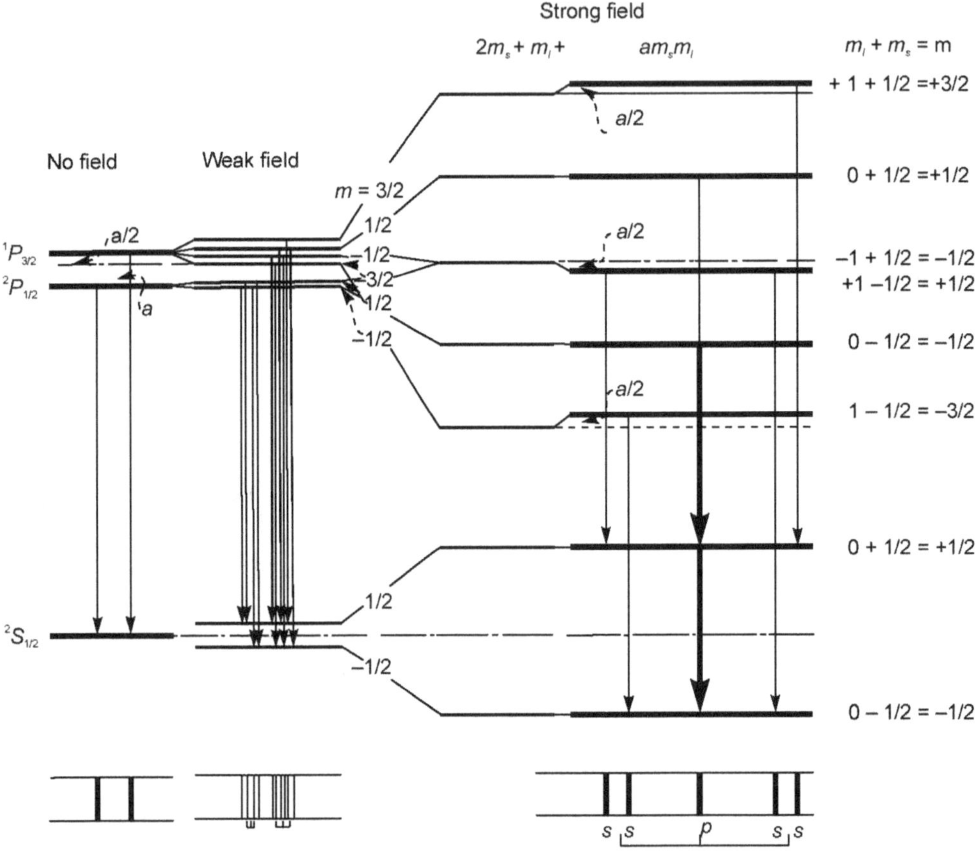

Figure 9.13   Energy levels for a principal-series doublet starting with no field at the left and ending with a strong field (Paschen–Back effect) at the right

SUMMARY

Paschen–Back effect is the splitting of energy levels in strong external magnetic field. In a weak external field, the energy levels in the magnetic field are small compared to the spin–orbit interaction and hence spin and orbit couple together first and then couple to the field. In a strong magnetic field the spin and orbit couple separately to the external magnetic field. Since spin–orbit coupling increases rapidly with increasing charge Z, the condition of a strong field is accomplished in light atoms than heavy.

The total interaction energy is

1.   the energy interaction of the orbital magnetic moment μ_L

$$E_L = -\mu_L B \cos(\mu_L \ B)$$

2. the energy interaction of the orbital magnetic moment μ_S

$$E_S = -\mu_S B \cos(\mu_S \ B)$$

3. the energy of interaction between μ_L and μ_S

$$E_{L,S} = hc \ A \ M_L \ M_S \text{ where } A \text{ is the coupling constant}$$

The total energy $E = \ E_0 + E_L + E_S + E_{L,S}$

$E_0 + E_{L,S}$ gives the energy of the atom in the absence of an external field.

$$E = E_0 - \mu_L \ B \cos(\mu_L B) - \ \mu_S B \cos(\mu_S B) + hc \ A \ M_L \ M_S$$

$$\mu_L = -\frac{e}{2m} L * \frac{h}{2\pi}$$

$$\mu_S = -2\frac{h}{2\pi} S * \frac{e}{2m}$$

$$\cos(\mu_L B) = \frac{M_L}{L*}, \ \cos(\mu_S B) = \frac{M_S}{S*}$$

Hence $E = E_0 - \mu_L \ B\dfrac{M_L}{L*} - \mu_S \ B\dfrac{M_S}{S*} + hc \ A \ M_L \ M_S$

For single valence electron small m, l and s can be used.

9.6 RELATION BETWEEN WEAK AND STRONG FIELD QUANTUM NUMBERS

A method of obtaining a relation between weak and strong field quantum numbers were given by Breit[2]. The scheme starts with strong magnetic field quantum numbers for each electron and then develops weak field quantum numbers and then to field free quantum numbers.

We construct an array of quantum numbers for P state as shown below. The elements of the array are filled with all possible sums of m_l and m_s. The sums are the weak field quantum numbers. These elements are divided into two parts by dotted lines. We can correlate now each weak field level m (the elements in the array) with the strong field level and they are the value given by the value of m_l directly above and the value of m_s directly to the right of the m value. Thus for example $m = 3/2$ of $^2P_{3/2}$ corresponds to the state $m_l = 1$ and $m_s = 1/2$. The $m = 1/2$ state correspond to $m_l = 0$ and $m_s = 1/2$, the $m = -3/2$ correspond to $m_l = -1$ and $m_s = -1/2$ and so on.

The selection rules are $\Delta m_l = 0, \pm 1$ and $\Delta m_s = 0$.

We consider the case of an *sp* configuration. In this case we have $l_1 = 0$ and $l_2 = 1$. In LS coupling $L = l_1 + l_2 \ldots |l_1 - l_2|$ and thus $L = 1$ and hence a P state. The j values are obtained first by obtaining the total spin. The total spin $S = \frac{1}{2} + \frac{1}{2} \ldots |\frac{1}{2} - \frac{1}{2}| = 1$ and 0. We thus obtain triplets and singlets. We have already seen the multiplicity is $2S + 1$. Thus the states are ^3P and ^1P. The J values for ^3P are obtained from $J = L + S \ldots |L - S|$ and thus $J = 1 + 1 \ldots |1 - 1| = 2,1,0$. The triplet P states are thus $^3P_{0,1,2}$. Similarly for ^1P we obtain $^1P_1$.

The electrons are generally designated by four quantum numbers n, l, m_l, m_s. However we need not consider n as it does not take part in the type of terms which we are discussing. Hence we have to take into consideration only $l_1 = 0$ (s electron) $m_l = 0$, $m_s = \pm 1/2$ and $l_2 = 1$ (p electron) $m_l = 1, 0, -1$ and $m_s = \pm 1/2$. In LS coupling case we write down the values of m_{l_1} and m_{l_2} as shown in Table 9.3.

Table 9.3 Magnetic quantum numbers for *sp* electron in two coupling cases

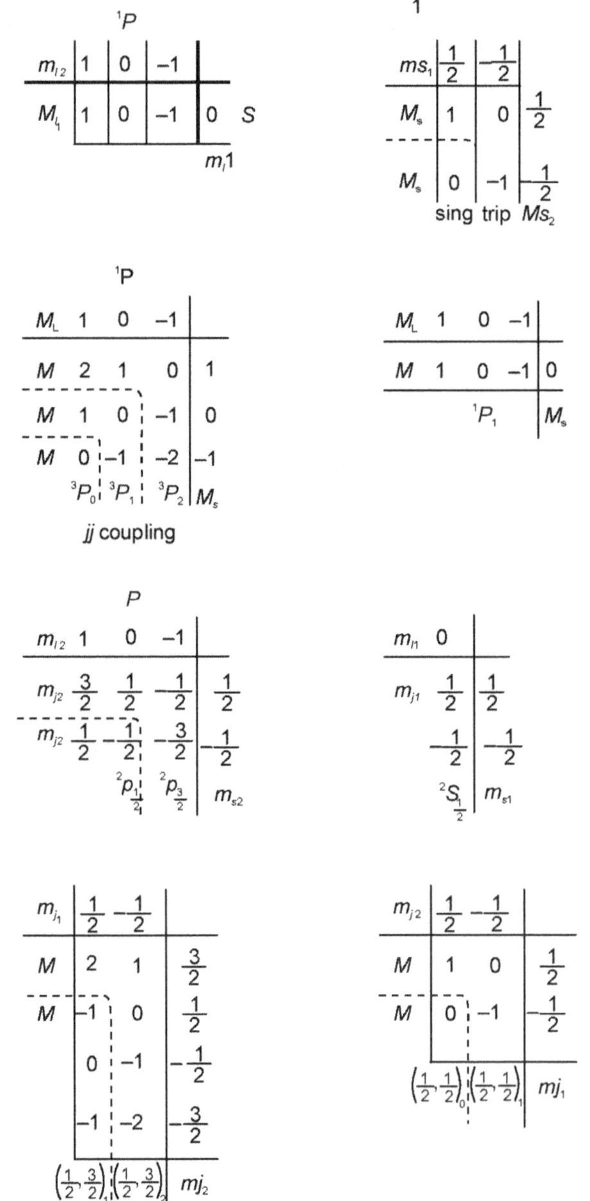

The above arrangement can be understood in the following way. In the first part of the table at the top left $m_{l1} = 0$ and $m_{l2} = 1$ give $M_L = 1$. $m_{l1} = 0$ and $m_{l2} = 0$ give $M_L = 0$ and $m_{l1} = 0$ and $m_{l2} = -1$ give $M_L = -1$. On the right hand at the top $M_s = 0$, $M_s = 0$ and ± 1 separated by the L-shaped marking give the M values for singlets and triplets. Similarly the quantum sums divided by the dotted lines are just the weak field quantum numbers for the $^3P_{0,1,2}$ and $^1P_1$. The case of jj coupling is shown at the lower portion of the table.

Table 9.4   Correlation of weak- and strong-field quantum numbers *Breit Scheme*

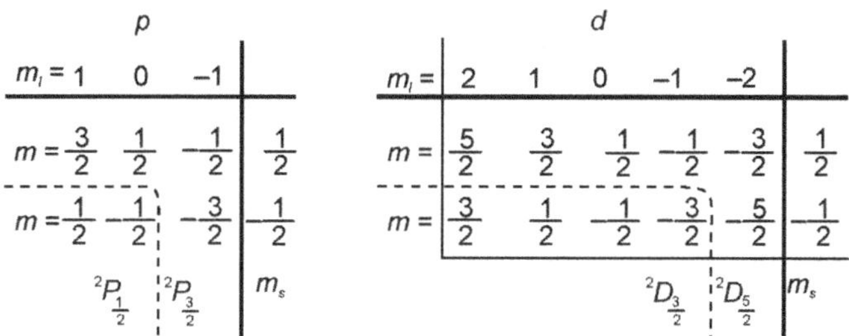

It should be noted that the sum of the projections of the various angular momenta on the magnetic field B should remain constant according to the law of conservation of momenta. It means that the magnetic quantum number m obtained in the weak field case should be the same in the strong field case as well. Hence we can correlate $m = m_l + m_s$. However this alone is not sufficient for a correlation since in many cases there will be more than one level with the same. Considering quantum mechanical rules the levels with the same m never cross. An easy method to correlate the weak field and strong field case as proposed by Breit is shown as an example in Table 9.4 for p and d.

We write down the values of m_l in horizontal row and m_s in vertical column. We fill the array then with all possible sums of m_l and m_s. These are weak field m quantum numbers. They are shown inside the dotted lines in Table 9.4. These weak field m can then be correlated with the strong field m_l and m_s. The value of m_l directly above a particular m and m_s directly right give the m_l and m_s values by which the m is obtained. For example in Table $m = \dfrac{3}{2}$ belonging to $^2P_{\frac{3}{2}}$ goes to states $m_l = 1$ and $m_s = \dfrac{1}{2}$ in strong field case. Though there are two ways of drawing the L-shaped dotted curves, only one set will give the correct correlation.

9.7   PAULI'S EXCLUSION PRINCIPLE

The Pauli's principle in its simplest form can be understood in the following way. No two electrons can have all their quantum numbers n, l, m_l and m_s the same.

This is of particular interest in the case of equivalent electrons. Equivalent electrons are any two electrons having the same n and l values. Let us now consider two equivalent p electrons, the electron configuration ending p^2 shell. We start with the strong field case and write down the possible m_s and m_l values.

$$m_s = \quad 1/2 \quad 1/2 \quad 1/2 \quad -1/2 \quad -1/2 \quad -1/2$$

$$m_l = \quad 1 \quad 0 \quad -1 \quad 1 \quad 0 \quad -1$$

$$\quad\quad\quad a \quad\quad b \quad\quad c \quad\quad d \quad\quad e \quad\quad f$$

These are the possible states in which a single p electron may exist in the atom. When we consider two equivalent electrons the Pauli's exclusion principle insists that one of the quantum values m_s or m_l must be different. We can now write the combinations in which this condition can apply. They are

$$ab$$

$$ac \quad\quad bc$$

$$ad \quad\quad bd \quad\quad cd$$

$$ae \quad\quad be \quad\quad ce \quad\quad de$$

$$af \quad\quad bf \quad\quad cf \quad\quad df \quad\quad ef$$

We can now write the combination in the following way (taking first in row wise)

$$M_s = \quad 1 \quad 1 \quad 0 \quad 0 \quad 0 \quad 1 \quad 0 \quad 0 \quad 0 \quad 0 \quad 0 \quad 0 \quad -1 \quad -1 \quad -1$$

$$M_L = \quad 1 \quad 0 \quad 2 \quad 1 \quad 0 \quad -1 \quad 1 \quad 0 \quad -1 \quad 0 \quad -1 \quad -2 \quad 1 \quad 0 \quad -1$$

We now start with the highest value of M_L

$$M_L = \quad 2 \quad 1 \quad 0 \quad -1 \quad -2$$

$$M_s = \quad 0 \quad 0 \quad 0 \quad 0 \quad 0$$

These are the different M_s components of $L = 2$ and $S = 0$ and thus it gives a ^1D state.

Out of the remaining we can collect

$$M_L = \quad 1 \quad 0 \quad -1 \quad 1 \quad 0 \quad -1 \quad 1 \quad 0 \quad -1$$

$$M_s = \quad 1 \quad 1 \quad 1 \quad 0 \quad 0 \quad 0 \quad -1 \quad -1 \quad -1$$

giving ^3P. The remaining $M_L = 0$ and $M_s = 0$ gives a ^1S state. The two equivalent p electrons thus give rise to ^1S, ^3P and ^1D terms. If they are nonequivalent the terms obtained are ^1S, ^1P ^1D, ^3S, ^3P and ^3D. One can derive the terms for other equivalent electrons in the same way.

The Breit's scheme when applied to equivalent d^2 electrons, we start with all possible m_{s1} and m_{s2} combinations in one array and all possible m_{l1} and m_{l2} in another array.

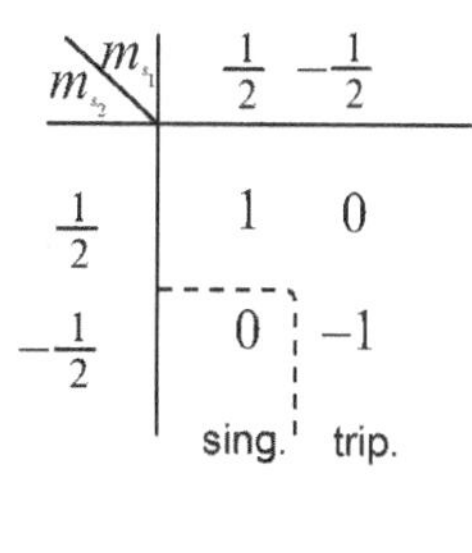

The M_L and M_s corresponding to a pair of m_l and m_s appear on the body of the table. If the total quantum numbers are different all possible combinations magnetic quantum numbers are allowed. This is the case of nonequivalent electrons. However, if the electrons are equivalent the M_L values lying on the diagonal cannot be taken with M_s values lying on the diagonal since it would mean that $m_{l_1} = m_{l_2}$ and $m_{s_1} = m_{s_2}$. Since the values on the lower left of the diagonal are the same on the upper right we will have to consider only one part above or below the diagonal. Now we are left with

$$M_s = 1 \qquad M_L \quad 3 \quad 2 \quad 1 \quad 0 \quad -1 \quad -2 \quad -3$$

$$M_s = -1 \qquad M_L \quad 3 \quad 2 \quad 1 \quad 0 \quad -1 \quad -2 \quad -3$$

$$M_s = 1 \qquad M_L \qquad\qquad\quad 1 \quad 0 \quad -1$$

$$M_s = -1 \qquad M_L \qquad\qquad\quad 1 \quad 0 \quad -1$$

If the M_L values are alike, values $M_s = \pm 1$ are forbidden. And since the two $M_s = 0$ values are identical one of them should be excluded. We have left with now the following combinations

$$M_s = 0 \qquad M_L \quad 4 \quad 3 \quad 2 \quad 1 \quad 0 \quad -1 \quad -2 \quad -3 \quad -4$$

$$M_s = 0 \qquad M_L \quad 3 \quad 2 \quad 1 \quad 0 \quad -1 \quad -2 \quad -3$$

$$M_s = 0 \qquad M_L \qquad\quad 2 \quad 1 \quad 0 \quad -1 \quad -2$$

$$M_s = 0 \qquad M_L \qquad\qquad\quad 1 \quad 0 \quad -1$$

$$M_s = 0 \qquad M_L \qquad\qquad\qquad 0$$

The second and the fourth rows go with the preceding tabulation and give rise to ^3F and ^3P terms and the remaining gives rise to ^1G, ^1D and ^1S terms.

9.8   STRONG FIELD CASE (PASCHEN–BACK EFFECT)

To simplify the solution, it is useful to assume that $[H_{at}, S_z] = 0$, so that L_z and S_z have a set of common eigen functions with respect to H_{at}. This allows the expectation values of L_z and S_z to be easily evaluated on a general state $|A\rangle$:

$$\left(H_{at} + \frac{B_z \mu_B}{\hbar}(L_z + 2S_z) \right)|A\rangle = \left(E_{at} + B_z \mu_B(m_l + 2m_s) \right)|A\rangle \tag{9.68}$$

The above may be read as implying that the *LS*-coupling is completely broken by the external field. The system re-arranges substantially according to the B_z field. The m_l and m_s are still "good" quantum numbers. This means that the selection rules obtained from $\Delta S = 0$, $\Delta L = \pm1$ are still very likely for the system. In particular, apart from the line splittings one might normally expect, only three spectral lines will be visible, corresponding to the $\Delta m = \pm1$ transition rule. The splitting depends upon the *l* level being considered. The spectral lines depend on the transition frequencies, that is, on the difference of energy.

9.9 THE ZEEMAN EFFECT AND PASCHEN–BACK EFFECT IN TWO VALENCE ELECTRONS

The Lande's vector model has given reasonable explanations on the spin–orbit interactions and the shifting or splittings of lines (energy levels) in magnetic and electric fields. The rules governing to the one electron valence system can be very successfully applied to the two valence system. Instead of dealing with only two quantum numbers, viz. *l* and *s* for a single electron, now these quantum numbers for both the electrons should be considered. With two electrons there are two individual quantum numbers $l_1{}^*$, $l_2{}^*$, $s_1{}^*$ and $s_2{}^*$. Depending upon the coupling of these angular moment we can arrive at the resultant magnetic moments in the *LS* coupling or in *jj* coupling schemes. We have already seen that the magnetic moments from the orbital motion and the spin motion are

$$\mu_l = -l * \frac{h}{2\pi} \frac{e}{2mc} \frac{ergs}{gauss} \tag{9.69}$$

$$\mu_s = -2\,s * \frac{h}{2\pi} \frac{e}{2mc} \frac{ergs}{gauss} \tag{9.70}$$

If we consider the *LS* coupling the two *l*'s are coupled together to form a resultant L^* and the two *s*'s are coupled together to form a resultant S^*

$$\text{Thus } \mu_L = -[l_1 * \cos(l_1 * L^*) + l_2 * \cos\,(l_2 * L^*)]\frac{h}{2\pi}\frac{e}{2mc} = L * \frac{h}{2\pi} \cdot \frac{e}{2mc} \frac{ergs}{gauss} \tag{9.71}$$

$$\mu_S = -[s_1 * \cos(s_1 * S^*) + s_2 * \cos(s_2 * S^*)]\frac{h}{2\pi}\frac{2e}{2mc} = S * \frac{h}{2\pi}\frac{2e}{2mc}\frac{ergs}{gauss} \tag{9.72}$$

L^* and S^* are coupled to form a resultant J^* which in turn precess around the applied magnetic field. Thus projecting μ_L and μ_S on J^* we get the total magnetic moment of the atom as

$$\mu_J = -[L*\cos(L*J*)+2S*\cos(S*J*)]\frac{h}{2\pi}\frac{e}{2mc}\frac{ergs}{gauss}$$

$$= -[L*\cos(L*J*) + 2S*\cos(S*J*)]\frac{eh}{4\pi mc}\frac{ergs}{gauss} \tag{9.73}$$

where $\dfrac{eh}{4\pi mc}$ is one Bohr magneton.

We may write $L*\cos(L*J*) + 2S*\cos(S*J*) = g\,J*$ and hence from equation 9.73

we get $\mu_J = -gJ*\dfrac{eh}{4\pi mc}\dfrac{ergs}{gauss}$

In LS coupling the angles between $L*$, $S*$ and $J*$ are all fixed and hence the cosine terms can be readily evaluated. From the geometrical consideration

$$L*\cos(L*J*) = \frac{J*^2 + L*^2 - S*^2}{2J*} \quad \text{and} \tag{9.74}$$

$$S*\cos(S*J*) = \frac{J*^2 + S*^2 - L*^2}{2J*} \tag{9.75}$$

Substituting in equation

$$g_{J*} = \frac{J*^2 + L*^2 - S*^2}{2J*} + 2\frac{\left(J*^2 + S*^2 - L*^2\right)}{2J*} \tag{9.76}$$

$$g = 1 + \frac{J*^2 + S*^2 - L*^2}{2J*^2} \tag{9.77}$$

This is almost the same expression we derived in the case of a single valence electron. The main difference is that instead of the individual $l*$ and $s*$ and $j*$, the resultant $L*$, $S*$ and $J*$ are used. Thus the Lande g-factor can be calculated easily in case of different configurations.

In jj coupling we have already seen that the $l*$ and $s*$ of each electron precess around $j*$ and the different $j*$ s make a resultant J which in turn precess around the applied magnetic field. We have already seen that magnetic moment of a single electron as

$$\mu_J = -[l*\cos(l*j*)+2s*\cos(S*j*)]\frac{h}{2\pi}\frac{e}{2mc}$$

$$= -j*g_j\frac{h}{2\pi}\frac{e}{2mc} \tag{9.78}$$

In the case of a two electron system the magnetic moments are

$$\mu_1 = -j_1 * g_1 \frac{h}{2\pi}\frac{e}{2mc}$$

$$\mu_2 = -j_2 * g_2 \frac{h}{2\pi}\frac{e}{2mc} \qquad (9.79)$$

The j_1* and j_2* are coupled to form a resultant $J*$ in the jj coupling scheme and the $J*$ in turn couple in the direction of the magnetic field giving rise to various m components.

$$\mu_J = -\left[j_1 * g_1 \frac{h}{2\pi}\frac{e}{2mc}\cos(j_1 * J*) + j_2 * g_2 \frac{h}{2\pi}\frac{e}{2mc}\cos(j_2 * J*) \right]$$

$$= -[j_1 * g_1 \cos (j_1 * J*) + j_2 * g_2 \cos(j_2 * J*)]\frac{h}{2\pi}\frac{e}{2mc}\frac{ergs}{gauss}$$

$$= -J * g \frac{h}{2\pi}\frac{e}{2mc}\frac{ergs}{gauss} \qquad (9.80)$$

where $J * g = [j_1 * g_1 \cos(j_1 * J*) + j_2 * g_2 \cos(j_2 * J*)]$

The angles between j_1*, j_2* and $J*$ are constants, the cosine formula gives

$$j_1 * \cos(j_1 * J*) = \frac{J*^2 + j_1^{*2} - j_2^{*2}}{2J*}$$

$$j_2 * \cos (j_2 * J*) = \frac{J*^2 + j_2^{*2} - j_1^{*2}}{2J*}$$

$$g = g_1 \frac{J*^2 + j_1^{*2} - j_2^{*2}}{2J*^2} + g_2 \frac{J*^2 + j_1^{*2} - j_2^{*2}}{2J*^2} \qquad (9.81)$$

In an sp configuration $j_1 = 1/2$ and $j_2 = 1/2$, J can have values 0 or 1.

The vector models are given in Figure 9.14.

In the weak field case the total angular momentum $J * \dfrac{h}{2\pi}$ precesses around the applied magnetic field B. The projection of $J * \dfrac{h}{2\pi}$ is $J * \dfrac{h}{2\pi}\cos(J * B) = M\dfrac{h}{2\pi}$. M takes values from $+J$ to $-J$.

As in the case of a single valence electron where $j*$ makes a Larmor precession around B, in this case the resultant $J*$ makes a Larmor precession around B and the angular velocity is

$$\omega_L = B. g.\frac{e}{2mc} \qquad (9.82)$$

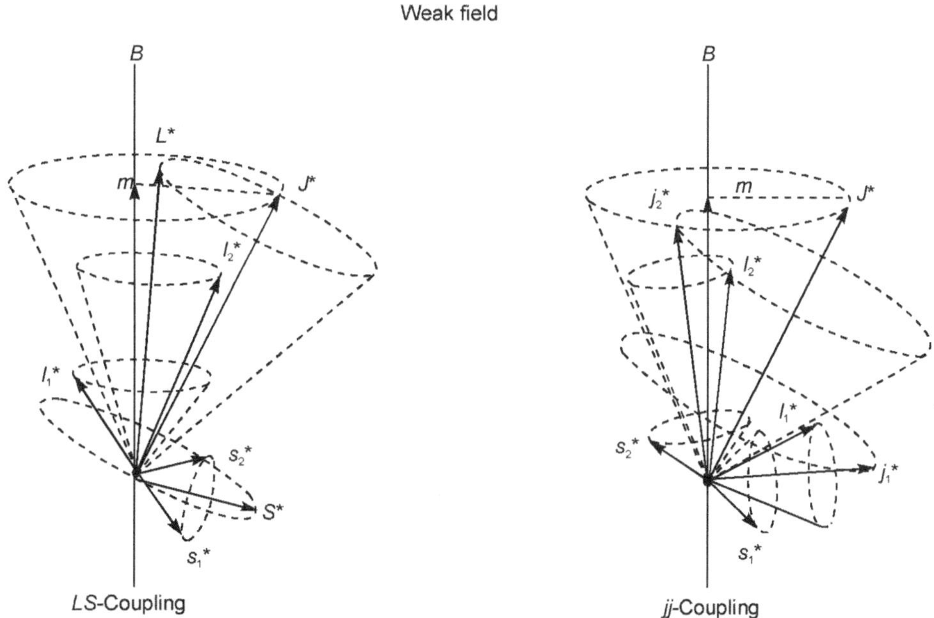

Figure 9.14   The vector model for *LS*-and *jj*-coupling in a weak magnetic field (Zeeman effect).

The interaction energy is

$$\Delta W = B.\, g.\, \frac{e}{2mc} J * \frac{h}{2\pi} \cos(J * B) = B.\, g.\, \frac{e}{2mc} M \frac{h}{2\pi} = M.g.\, B \frac{eh}{4\pi mc} \tag{9.83}$$

Energy in wave numbers $-\Delta E_M = \dfrac{\Delta W}{hc} = M.g.\dfrac{Be}{4\pi mc^2}\,\text{cm}^{-1}$ 　$\tag{9.84}$

If E_0 is the term value when there is no field, the term value in a weak magnetic field is

$$E_M = E_0 - Mg\frac{Be}{4\pi mc^2}\,\text{cm}^{-1} = E_0 - M.g.L\ \text{cm}^{-1} \tag{9.85}$$

In the above discussions we have assumed that the $l_1{}^*$ and $l_2{}^*$ precess rapidly and form the resultant L^*, similar is the case with $s_1{}^*$ and $s_2{}^*$ where $s_1{}^*$ and $s_2{}^*$ precess rapidly around S^*. L^* and S^* precess more slowly around J^* and J^* precesses around B still slower. In such kind of precessions the components perpendicular to the axes cancel out. Similar arguments apply in *jj*-coupling as well.

An example of splittings in the magnetic field in $^3D_3$ is shown in Figure 9.15. Zeeman pattern in a $^1D_2 - {}^1P_1$ transition has already been shown in Figure 9.8.

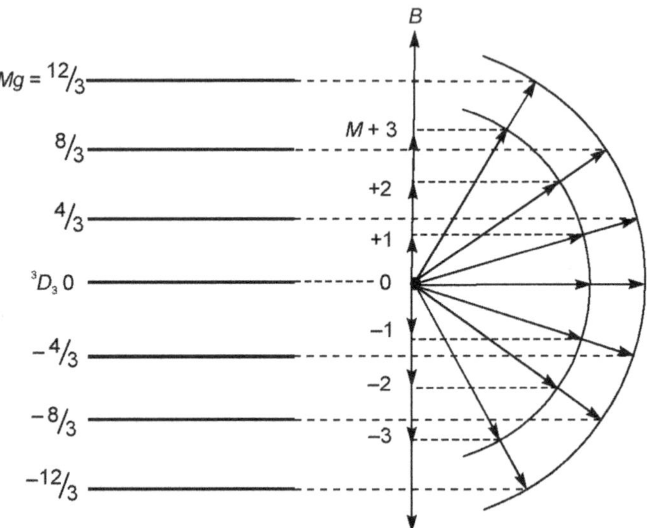

Figure 9.15   Graphical representation of the splitting of a $^3D_3$ level in a weak magnetic field

The selection rules are

When viewed perpendicular $\Delta M = \pm 1$; plane polarized $\perp B$ (s components)

$\Delta M = 0$; plane polarized $\parallel B$ (p components)

viewed parallel to the field $\Delta M = \pm 1$; circularly polarized $\perp B$ (s components)

$\Delta M = 0$; forbidden (p components)

For a $^1D_2 - {}^1P_1$ transition there are nine allowed transitions, however as the levels are equally spaced the Zeeman pattern is a normal triplet. The g-factor for all singlet levels are unity, transitions in singlet levels will show normal triplets.

The Zeeman transitions from a ^3P to ^3S is shown in Figure 9.16.

For a ^3P level there are three sub-levels $^3P_{0,1,2}$ and for ^3S we have $^3S_1$. The g-factors in the case of $^3P_0$, $^3P_1$ and $^3P_2$ are $\dfrac{0}{0}, \dfrac{3}{2}, \dfrac{3}{2}$ and for $^3S_1$ the g value is 2.

The intensity rules for Zeeman patterns in the case of atoms with two valence electrons are that the sum of the intensities of all transitions starting from any initial Zeeman level is equal to the sum starting from any other level. Similarly, the sum of the intensities of all the transitions arriving at any final level is equal to the sum arriving at any other level.

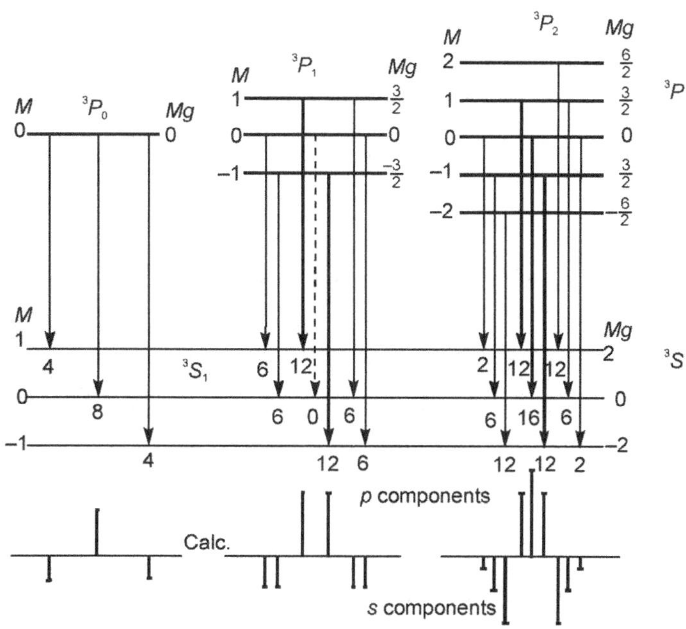

Figure 9.16   Zeeman Pattern in ^3p–^3s transition

For transitions $J \to J$

$$M \to M \pm 1, \quad I = A(J \pm M + 1)(J \text{ G } M)$$
$$M \to M, \qquad I = 4\,AM^2$$

Transition $J \to J + 1$

$$M \to M \pm 1, \quad I = (J \pm M + 1)(J \pm M + 2)$$
$$M \to M, \qquad I = 4\,B(J + M + 1)(J - M + 1) \tag{9.86}$$

The g sum rule

When we start analysing the Zeeman patterns of a new spectrum, it is difficult to assign the lines unless we know the type of coupling in the atom under consideration. Since the g values are different in the LS-coupling and jj-coupling it is difficult to calculate the expected splitting. The Pauli's g sum rule can be utilized for the analysis. It states that all the states arising from an electron configuration, the sum of the g factors for levels with the same J value is constant independent of the coupling scheme. If we consider, as an example, the sp configuration we arrive at a ^3P and a ^1P state.

Here $l_1 = 0$ and $l_2 = 1$ and hence $L = l_1 + l_2, \ldots| l_1 - l_2 | = 1$

$L = 1$ is a P state.

As there are 2 electrons we arrive at the total spin S as

$$S = s_1 + s_2 \ldots |s_1 - s_2| = 1 \text{ or } 0$$

From $S = 1$ we get a triplet state ($2S + 1 = 3$) and from $S = 0$ we get a singlet state ($2S + 1 = 1$)

We can now calculate the three J values for the ^3P state.

$$J = L + S, \ldots |L - S| = 1 + 1, \ldots |1 - 1| = 2, 1, 0$$

Thus we obtain $^3P_0$, $^3P_1$, $^3P_2$ and the three states are sometimes written as $^3P_{0,1,2}$

For ^1P state $J = L + S \ldots |L - S| = 1 + 0 = 1$ and hence we obtain a $^1P_1$ state.

g value for a $^3P_2$ state is $g = 1 + \dfrac{J^{*2} + S^{*2} - L^{*2}}{2J^{*2}} = 1 + \dfrac{J(J+1) + S(S+1) - L(L+1)}{2J(J+1)}$

$$g = 1 + \frac{2.3 + 1.2 - 1.2}{2(2.3)} = \frac{3}{2}$$

for $^3P_1$ state $\qquad g = 1 + \dfrac{1.2 + 1.2 - 1.2}{2(1.2)} = \dfrac{1}{2}$

for $^3P_0$ state $\qquad g = 1 + \dfrac{0 + 1.2 - 1.2}{2(2.3)} = 1$

from jj-coupling

$$g = g_1 \frac{J^{*2} + j_1^{*2} - j_2^{*2}}{2J^{*2}} + g_2 \frac{J^{*2} + j_2^{*2} - j_1^{*2}}{2J^{*2}}$$

9.10 PASCHEN–BACK EFFECT IN TWO VALENCE ELECTRONS

As explained earlier in Pachen–Back effect applied magnetic field is large enough to break the spin–orbit coupling. In this case the interaction energy between J^* and B becomes so strong that the coupling between L^* and S^* in the case of LS-coupling and j_1^* and j_2^* in the case of jj-coupling brake. The vector model in the case of Paschen–Back effect is shown in Figure 9.17.

The derivation of energy levels is almost the same as in single valence electron. In the case of LS-coupling, the L^* and S^* precess around B and they are independently quantized. The projection of $L^* \dfrac{h}{2\pi}$ on B gives the magnetic angular momentum $M_L \dfrac{h}{2\pi}$ where M_L takes values from $-L$ to $+L$. The projection of $S^* \dfrac{h}{2\pi}$ on B is then $M_S \dfrac{h}{2\pi}$ and M_S takes values from $-S$ to $+S$.

Thus M_L has values $L, L - 1, \ldots -L$ and

M_S has values from $S, S - 1, \ldots -S$

For singlets $M_S = 0$ because total spin for singlets is zero.

And for triplet $M_S = +1, 0, -1$ as the total spin $S = 1$

(remember the splitting $= 2S+1$ where S is the spin).

The Larmor precession frequencies due to the precessions of L^* and S^* on B are

$$\omega_L = B \cdot \frac{e}{2mc} \quad \text{and} \quad \omega_S = B \cdot 2\frac{e}{2mc} \tag{9.87}$$

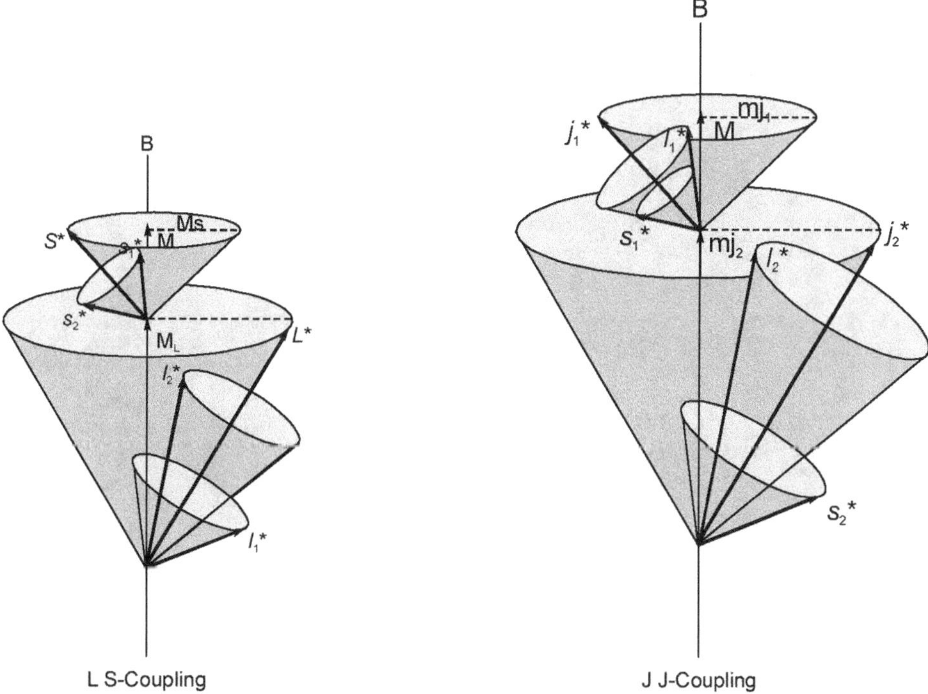

Figure 9.17   Classical vector model for LS and JJ-coupling schemes in a strong magentic field

The interaction energy due to the angular momentum $L^* \frac{h}{2\pi}$ is obtained by the product of the component of angular momentum L^* on B and the angular velocity ω_L. Similar is the case due to the angular momentum $S^* \frac{h}{2\pi}$.

Thus we have

$$\Delta W_{L,B} = B \cdot \frac{e}{2mc} L^* \frac{h}{2\pi} \cos(L^* B) = B \cdot \frac{e}{2mc} M_L \frac{h}{2\pi} \tag{9.88}$$

370   *Atomic Spectroscopy*

$$\Delta W_{S,B} = B \cdot 2 \frac{e}{2mc} S * \frac{h}{2\pi} \cos(S*B) = B \cdot 2 \frac{e}{2mc} M_S \frac{h}{2\pi} \qquad (9.89)$$

The energy shift is the sum of these two interaction energies

$$\Delta W_B = (M_L + 2M_S) \, B \frac{eh}{4\pi mc} \qquad (9.90)$$

The energy in wave numbers $-\Delta T_B = -\dfrac{\Delta W_B}{hc} = (M_L + 2M_S)\dfrac{Be}{4\pi mc^2}$

$$= (M_L + 2M_S) \, L \ \mathrm{cm}^{-1} \qquad (9.91)$$

Table 9.5 Magnetic energies in weak and strong fields assuming *LS*-coupling

| | Weak field | | | Strong field | | | | | |
Term	Γ_{weak}	M	Mg	M	M_S	M_L	$2M_S$	$2M_S+M_L$	$\Gamma_{\text{strong}}=A.M_L M_S$
		+2	5/2	+2	+1	+1	+2	+3	+A
		+1	3/2	+1	+1	0	+2	+2	0
$^3P_2$	+A	0	0	0	+1	−1	+2	+1	−A
		−1	−3/2	+1	0	+1	0	+1	0
		−2	−5/2	0	0	0	0	0	0
				−1	0	−1	0	−1	0
		+1	+3/2						
$^3P_1$	−A	0	0						
		−1	−3/2						
				0	−1	+1	−2	−1	−A
$^3P_0$	−2A	0	0	−1	−1	0	−2	−2	0
				−2	−1	−1	−2	−3	+A

We have already assumed that the spin–orbit interaction energy is small compared to the energy due to the applied magnetic field. By adding this small energy to the above we obtain the total shift. We have already seen that the spin–orbit interaction energy is derived as

$$-T_{L,S} = A.M_L\, M_S \quad \text{where} \quad A = a_3\alpha_3 + a_4\alpha_4 \tag{9.92}$$

Thus the total energy shift is

$$-T_{L,S} = (M_L + 2M_S)L + A \cdot M_L M_S \tag{9.93}$$

Splitting of levels in $^3S_1$ and $^3P_{0,1,2}$ are shown in Figure 9.15 and the table gives the energies in a weak and strong magnetic fields. In the case of $^3P_{0,1,2}$ the possible values of M_L are 1, 0, −1 and the possible values of M_S are 1, 0, −1.

In going from weak field case to strong field case the value of $M = M_L + M_S$ has the same values in both cases. And two levels with the same M never cross. The selection rules for the strong field case is

$$\Delta M_S = 0$$

$$\Delta M_L = 0 \ (p \text{ components})$$

$$= \pm 1 (s \text{ components}) \tag{9.94}$$

9.10.1 Paschen–Back Effect in Atoms with *jj* -Coupling

The theoretical treatment is almost similar to that in the case of *LS* coupling. The conditions in *jj* couplings are

$j_1{}^*$ and $j_2{}^*$ precess around the magnetic field and are quantized independently with the field direction B. The projection of $j_1{}^*\dfrac{h}{2\pi}$, i.e., $j_1{}^* \cos\,(j_1{}^*B)\dfrac{h}{2\pi}$ is $m_{j1}\dfrac{h}{2\pi}$ where m_{j1} takes values $-j_1$ to $+j_1$. The projection of $j_2{}^*\dfrac{h}{2\pi}$, i.e., $j_2{}^*\cos(j_2{}^*B)\dfrac{h}{2\pi}$ is $m_{j2}\dfrac{h}{2\pi}$ where takes values $-j_2$ to $+j_2$.

The total magnetic interaction energy is the sum of the interaction of

1. $j_1{}^*$ with the magnetic field B
2. $j_2{}^*$ with the magnetic field B
3. interaction between j_1 and j_2

The angular frequencies due to the Larmor precession of $j_1{}^*$ around the magnetic field B and due to the Larmor precession of $j_2{}^*$ around the magnetic field B are

$$\omega_{J1} = B \cdot g_1 \frac{e}{2mc} \quad \text{and} \quad \omega_{j2} = B \cdot g_2 \frac{e}{2mc} \tag{9.95}$$

where g_1 and g_2 are g factors for the electrons.

The interaction energy is obtained as earlier by multiplying the angular velocities by the projection of the corresponding angular momenta on the magnetic field B.

Thus

$$\Delta W_{j_1}, B = B \cdot g_1 \frac{e}{2mc} j_1{}^* \frac{h}{2\pi} \cos(j_1 {}^* B) = B \cdot g_1 \frac{e}{2mc} m_{j_1} \frac{h}{2\pi} \tag{9.96}$$

$$\Delta W_{j_2}, B = B \cdot g_2 \frac{e}{2mc} j_2{}^* \frac{h}{2\pi} \cos(j_2 {}^* B) = B \cdot g_2 \frac{e}{2mc} m_{j_2} \frac{h}{2\pi} \tag{9.97}$$

and the coupling between $j_1{}^*$ and $j_2{}^*$ in analogy with the derivation of the spin–orbit interaction in LS-coupling is given by

$$\Delta T_{ij} = A \cdot m_{j1} m_{j2} \ \text{cm}^{-1} \tag{9.98}$$

Adding the first two we obtain

$$\Delta W_B = B \cdot g_1 \frac{e}{2mc} m_{j_1} \frac{h}{2\pi} + B \cdot g_2 \frac{e}{2mc} m_{j2} \frac{h}{2\pi}$$

$$= (g_1 m_{j_1} + g_2 m_{j2}) B \cdot \frac{eh}{4\pi mc} \tag{9.99}$$

By dividing by hc

$$\frac{\Delta W_B}{hc} = (g_1 m_{j_1} + g_2 m_{j2}) \frac{eB}{4\pi mc^2} = (g_1 m_{j_1} + g_2 m_{j2}) L \ \text{cm}^{-1} \tag{9.100}$$

Total shift is now

$$-\Delta T_M = (g_1 m_{j_1} + g_2 m_{j2}) L + A \cdot m_{j_1} m_{j_2} \ \text{cm}^{-1} \tag{9.101}$$

the shifted line is $T_M = T_1 - \Delta T_M$ where T_M is the unshifted position.

The selection rules for jj-coupling are

$$\Delta m_{j_1} = 0$$

$$\Delta m_{j_2} = 0 \ (p \text{ components})$$

$$\Delta m_{j_2} = \pm 1 \ (s \text{ components}) \tag{9.102}$$

The energy levels for np^2-configuration is shown in Figure 9.18.

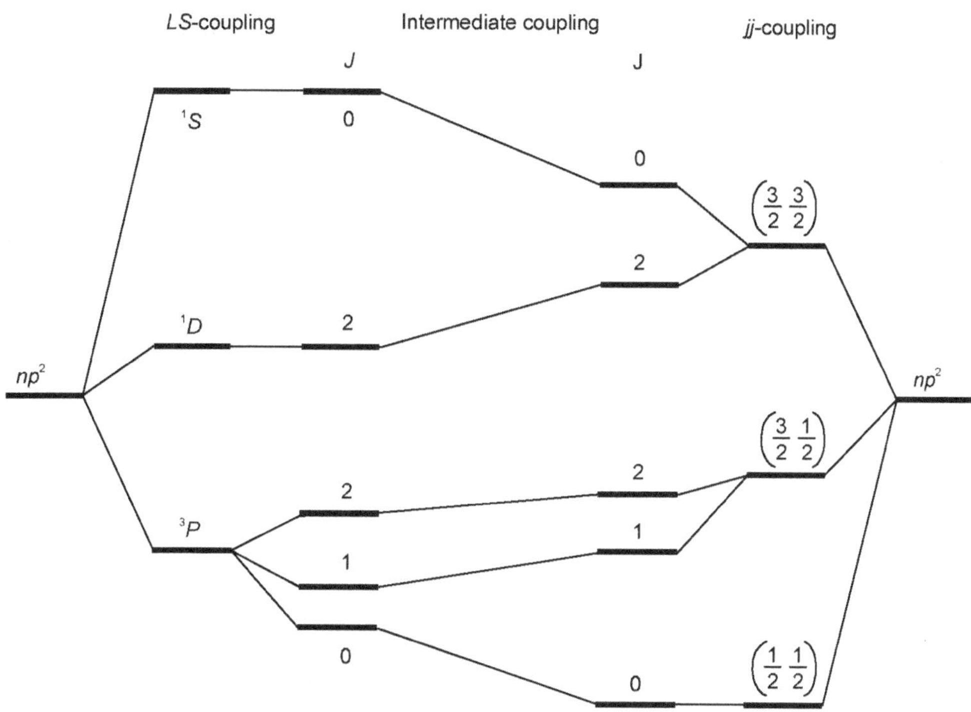

Figure 9.18   Energy levels for np^2-configuration

9.11 STARK EFFECT

The splitting of atomic spectral lines as a result of an externally applied electric field is called the Stark effect. Zeeman demonstrated the splitting of spectral lines in a magnetic field in 1897, which was later, called Zeeman effect and it took another 16 years to demonstrate a similar effect in an applied electric field. It was in the year 1913 that Johannes Stark[3] demonstrated the splitting of spectral lines in the Balmer series of hydrogen atom in an electric field. This was in the same year the Bohr theory of hydrogen atom was proposed by Neils Bohr. It was independently discovered in the same year by the Italian physicist Antonino Lo Surdo, and in Italy it is thus sometimes called the Stark-Lo Surdo effect. The satisfactory theoretical treatment by Epstein and Schwarzschild[4] gave additional success to the quantized orbits proposed in the Bohr's theory. Later Kramers made attempts to explain Stark effect neglecting the electronic spin[5]. The difficult task of producing high electric field in a discharge tube was believed to be the reason in hampering the early detection of Stark effect in the time of the detection of Zeeman effect.

In 1913 Stark observed that every line in the Balmer series was split into a number of components when an electric field of more or around 100 KV per centimetre was applied. It was then observed that linear Stark effect occurs in hydrogen like spectra and that too with electric field large enough for the fine structure to be neglected. When viewed perpendicular to the field, each line pattern is observed to be plane polarized with electric vector parallel to the electric field (*p*-components) and the others polarized with electric vector normal to the field (*s*-components). The main features are the following.

1. All lines in the hydrogen atom show a symmetrical pattern in an electric field. The number of lines and the total width of the pattern increase with n.

2. The wave number difference is proportional to field strength E and is valid for all lines in the hydrogen atom.

3. Some components appear only in transverse observation and are polarized with electric vector parallel to the field and the other components appear polarized with electric vector perpendicular to the field in transverse observation and unpolarized in longitudinal observation.

In a simple description of energy difference in an electric field, the interaction energy can be written as

$$\Delta T = AF + BF^2 + CF^3 + \ldots \tag{9.103}$$

where ΔT represents the change in term values of the atom in wave numbers and F is the field strength. That means ΔT is the shift in the energy levels from the field free position. Only first order Stark effect is found in hydrogen like atoms. Thus first term in equation (1) is the only contributing term. And it is called a first order effect. In general, one distinguishes first- and second-order Stark effects. The first-order effect is linear in the applied electric field, while the second-order effect is quadratic in the field. The second term involving F to the second power is called second-order effect and so on. Epstein, Wentzel, Waller, Van Vleck, Doi, and Schrödinger and others have calculated the coefficients A, B and C using parabolic coordinates and have values

$$A = \frac{3h}{8\pi^2 mec} n(n_2 - n_1)$$

$$B = \frac{h^5}{2^{10}\pi^6 m^3 e^6 c} n^4 \left\{ 17n^2 - 3(n_2 - n_1) - 9m_i^2 + 19 \right\}$$

$$C = \frac{3h^9}{2^{15}\pi^{10} m^5 e^{11} c} n^7 \left\{ 23n^2 - (n_2 - n_1)^2 + 11m_i^2 + 39 \right\} \tag{9.104}$$

where n is the total quantum number, n_1, n_2 and m_1 are electric quantum numbers with

$$m_1 = n - n_2 - n_1 - 1$$

n takes the values 1, 2, 3, ... α

n_1 takes values   0, 1, 2, 3, ...$(n - 1)$

n_2 takes values   0, 1, 2, 3, ... $(n-1)$

$m_1 = 0, \pm 1, \pm 2 \ldots \pm(n-1)$

When the field is expressed in volt/cm, the constants A, B and C have the numerical values

$A = 6.42 \times 10^{-5}$

$B = 5.22 \times 10^{-16}$

$C = 1.53 \times 10^{-25}$

When the electric field is small, say less than that 100 KV/cm the lower energy states of the hydrogen atom show a linear Stark effect and the result is symmetrical splitting of every level. In the case of higher fields and higher states the second order effect will also be appreciable which results in unidirectional displacement of each line.

The splitting of the energy levels by an electric field first requires that the field polarize the atom and then interact with the resulting electric dipole moment. That dipole moment depends upon the magnitude of M_j, but not its sign, so that the energy levels are split into $J+1$ or $J+1/2$ levels, for integer and half-integer spins respectively. m_l gives the projection of orbital angular momentum $\left(m_1 \dfrac{h}{2\pi} \right)$ along the electric field axis and is similar to the magnetic quantum number in a magnetic field. The only difference is that in an electric field the orbital angular momentum is not necessarily centred as in the case of magnetic field. The quantum numbers n_1 and n_2 are called parabolic quantum numbers and are analogous to the radial and azimuthal quantum numbers in field free case. The essential difference between the Zeeman effect and the Stark effect is that in each pair of levels $+m_j$ and $-m_j$ have the same energy. Thus for example in an electric field $m_j = \dfrac{3}{2}$ has the same energy as that of $m_j = -\dfrac{3}{2}$. This means $m_j = \pm \dfrac{3}{2}$ are degenerate in an electric field where as it is not so in a magnetic field.

For hydrogen $Z = 1$

$$T = -\frac{E - E_0}{hc} = \frac{3h}{8\pi^2 mec} n(n_2 - n_1) \, F = 6.42 \times 10^{-5} \, F \, n(n_2 - n_1)\text{cm}^{-1} \qquad (9.105)$$

If the electric field applied is in the Z-direction, then the potential energy to which the electron is subjected to is given by

$$V(r) = \frac{Ze^2}{r} + (-e \, F \, z) \qquad (9.106)$$

where $V' = e \, F z$ is a small perturbation on the potential $\dfrac{Ze^2}{r}$. The action of the external electric field on an orbital electron is quite different from the effect due to a magnetic field. The classical equation is no longer separable in polar coordinates and the required solutions can be found using parabolic coordinates.

In this case the orbits are not closed and in very long time they will completely fill the space between two pairs of paraboloids. A complete calculation gives the solutions as given in equation (1).

In quantum mechanics $V' = e\,F\,z$ can be considered as a perturbation. The quantum mechanical perturbation operator is

$$V' = -e\,F\,z = V' = -e\,F\,r\,\cos\,\theta = F\mu \tag{9.107}$$

The matrix elements of the perturbation operator between two sets of degenerate wavefunction is

$$H_{nl'm';n,l,m} = \int \Psi_{nl'm'}\langle eFr\cos\theta\rangle \Psi_{nlm}\,d\tau$$

$$\Psi_{n,l,m} = R(r)\Theta\,(\theta)\,\Phi\,(\phi)$$

$$d\tau = r_2\sin\theta\,dr\,d\theta\,d\phi$$

Hence

$$H_{nl'm';n,l,m} = e\,F\left[\int_0^\infty R_{nl'}\,R_{nl}r^3\,dr\right] \times \left[\int_0^\pi P_{l'}^{m'}(\cos\theta)P_l^m(\cos\theta)\cos\theta\sin\theta\,d\theta\right] \times \left[\int_0^{2\pi}\exp i(m'-m)\phi\,d\phi\right]$$

The term containing ϕ is zero unless $m' = m$ and the term containing θ is zero unless $l' = l\pm1$.

9.11.1 Weak Field Stark Effect in Hydrogen Atom

We consider a hydrogen atom in a weak electric field. In this case the interaction between the resultant angular momentum j and the electric field is comparatively less than the interaction between the $l*$ and $s*$ (the spin–orbit interaction). In other words the Stark splitting is small compared with fine structure splitting originating from the interaction of the electron spin and the orbital motion.

In this case the electron mechanical moment $j*\dfrac{h}{2\pi}$ precesses around the direction of the applied electric field F. The projection of $j*$ on F is m_j which can take values from $+j$ to $-j$ differing from each other by unity. As mentioned earlier the important difference between Zeeman and Stark effect is that in Stark effect each pair of levels $+m_j$ and $-m_j$ arising from a given level j have the same energy unlike in Zeeman effect where these two pair of levels have different energies. Thus a level with $j=\dfrac{3}{2}$ splits into $m_j=\pm\dfrac{3}{2}$ and $\pm\dfrac{3}{2}$. That means the levels having $m_j=+\dfrac{3}{2}$ and $-\dfrac{3}{2}$ have the same energy so also the levels with $m_j=+\dfrac{1}{2}$ and $-\dfrac{1}{2}$. In a magnetic field a given level j splits into $(2j+1)$ levels whereas in an electric field it splits into only $j+1$ components. The reason for this can be understood in terms of classical model or quantum mechanical model. We know that the nature of forces which act on the electron are purely electrostatic and hence the energy of interaction on n and l depends only on the inclination of the orbital plane or on the charge of the electron and does not depend on the direction of motion of the electron in the orbit. Thus the pair of levels having $+m_j$ and $-m_j$ correspond to the same inclination of the orbital plane. In magnetic field the energies depend on the direction of the orbital motion and hence the energies change sign when m_j changes sign. Figure 9.19 shows the energy level diagram for the $n=2$ and $n=3$ levels of the hydrogen atom.

The quantum numbers for weak field Stark effect are n, j and m_j. The selection rules are $\Delta m_j = 0$ gives p components and $\Delta m_j = \pm 1$ gives s components.

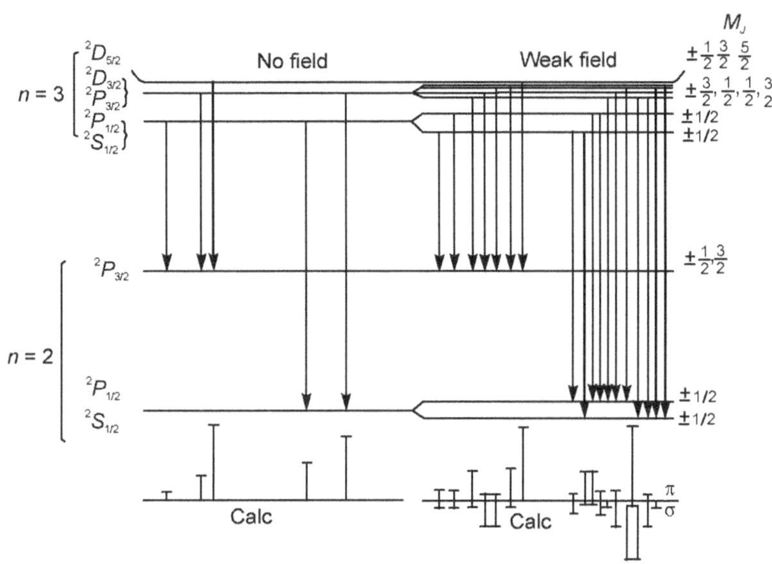

Figure 9.19   Fine structure and Stark pattern for Balmer α-line of hydrogen atom

9.11.2 Strong Field Stark Effect in Hydrogen

A strong electric field is considered when the Stark splitting is higher than the spin–orbit interaction splitting (fine-structure splitting). In such a coupling the interaction between l^* and s^* are broken and they quantize separately around the field direction. Resolving l^* along the field F and normal to F, the component normal to F average out to zero. m_l takes values from $-l$ to $+l$ and m_s from $-\dfrac{1}{2}$ to $+\dfrac{1}{2}$. Epstein and Schwarzschild have derived the first order Stark effect in such a coupling as

$$\Delta T = AF = 6.42 \times 10^{-5}(n_2 - n_1)F \tag{9.108}$$

Although the coupling between l^* and s^* is broken down, there still exists a small interaction and the energy of interaction is as in Zeeman effect which gives

$$\Gamma = a\, l^* s^* \cos(l^* s^*)\ \text{cm}^{-1} \tag{9.109}$$

Though l is not a good quantum number in strong field case, averaging the cosine gives

$$\Gamma = a\, m_l m_s\ \text{cm}^{-1} \tag{9.110}$$

The total energy shift is then

$$\Delta T = 6.42 \times 10^{-5}\, n\, (n_2 - n_1)F + a\, m_l\, m_s \tag{9.111}$$

Here a is the spin–orbit coupling constant determined from fine structure and F is the field measured in volts/cm.

Because the amount of Zeeman splitting and of polarization depend on the magnetic field strength, this effect provides a powerful tool for investigating cosmic magnetic fields. Since the distance between the Zeeman sub-levels is proportional with the magnetic field, this effect was used by astronomers to measure the magnetic field of the sun and other stars.

ENDNOTES

1　Pieter Zeeman (25 May 1865–9 October 1943)shared the 1902 Nobel Prize in physics with Hendrik Lorentz for his discovery of the Zeeman effect.

2　G. Breit, *Phys.Rev.* 28, 334 (1926).

3　Johannes Stark (15 April 1874–21 June 1957) was a German physicist. In 1919 Stark was awarded the Nobel Prize for Physics for his "discovery of the Doppler effect in canal rays and the splitting of spectral lines in electric fields".

4　Stark, J. Berlin Akad. (1913). *Wissenschaft* 40, 932. Epstein, P.S. *Ann. Der Physik,* 50, 489, 1916. *Phys. Zeits.* 17, 148, 1916. Schwarzchild; *Sitz. Ber. Berlin. Akadem.Wiss.* p 548, 1916.

5　Kramer, H.A. *Zeitscrift f. Physik.* 3, 199, 1920.

REFERENCES

Condon, E.U. and Shortley, G.H. (1935). *The Theory of Atomic Spectra.* Cambridge University Press, Cambridge.

Forman, Paul. (1970). "Alfred Lande and the anomalous Zeeman Effect." 1919–1921. *Historical Studies in the Physical Sciences.* 2: 153–261.

Griffiths, J. David. (2004). *Introduction to Quantum Mechanics,* 2nd (edn.). Prentice Hall, New Jersey.

Liboff, L. Richard. (2002). *Introductory Quantum Mechanics.* Addison-Wesley. Sanfrancisco, CA.

Rai, D.K. and Thakur, S.N. (2005). *Atomic Spectra, Structure and Modern Spectroscopy.* Vaivaswat Publications, Varanasi.

White, H.E. (1934). *Introduction to Atomic Spectra,* McGraw-Hill, NY.

Wolf, H.C. and Brewer, W.D. (1983, 2004) Atom and Quantumphysik, Springer Verlag, Berlin, Heidelberg.

Zeeman, P. (1897). "On the influence of magnelism on the nature of the light emitted by a substance." *Phil.Mag.* 43: 226. (1897).

Zeeman, P. (1897). "The effect of magnetisation on the nature of light emitted by a substance." *Nature.* 55: 347.

10

CORRELATION OF THE ATOMIC STRUCTURE WITH MORE THAN ONE ELECTRON

10.1 INTRODUCTION

Though Bohr's theory could explain to a certain extent the characteristic spectra of hydrogen atom, it was unsuccessful in treating the features of the atomic structure with more than one electron. It was due to the fact that there was no method of correlating the motions of one electron with the others. The Schrödinger equation was able to overcome these difficulties by taking into account the electronic repulsion in atoms with many electrons, but the mathematical solutions were found to be tedious.

There are only few one electron systems. Most of the atoms and molecules contain many electrons where Schrödinger equation cannot be solved exactly. However, approximation techniques in solving the Schrödinger equation were possible.

Before going into the elaborate treatment of helium atom, which involve two electrons, let us consider some aspects of systems containing two or more identical particles. By identical particles we mean that the two particles are indistinguishable by any physical measurements of their properties. Two electrons are one example. In classical mechanics there are no differences between identical particles or nonidentical particles. Classically, for example, the laws of classical mechanics holds good in a hand ball game whether we use two hand balls in the game or use one hand ball and a tennis ball. The point is that one of the identical balls can be identified by some means for example its use in a previous game and so on. This is however, not possible in quantum mechanics. Here we can utmost get the information from the probability of finding a hand ball in a particular region and we do not get any information about which hand ball it is. Thus in quantum mechanics the information we get about the particles is without labelling them. Let us consider two noninteracting identical particles in one dimension. The Hamilton operator for that system is given by

$$\hat{H} = -\frac{\hbar^2}{2m}[\nabla^2(1) + \nabla^2(2)] \tag{10.1}$$

If we interchange the labels (1) and (2), the equation is identical. That means the Hamiltonian is unchanged and hence the Hamiltonian is symmetric in the two particles. In the above eigen value equation the values of particles (1) and (2) are separable and the solutions are

$$\psi_1 = A\sin\frac{n_1\pi x}{a}$$

$$\psi_2 = A\sin\frac{n_2\pi x}{a} \tag{10.2}$$

Suppose we now consider two electrons, one in the lowest orbital and the second one in the next orbital. An acceptable solution for the eigen value equation is then

$$\psi = \psi_1(1)\ \psi_2(2) \tag{10.3}$$

The subscripts 1 and 2 refer to the values of quantum numbers n_1 and n_2. The above solution implies that we can specifically identify the particles (1) and (2) and such a function is hence physically unacceptable. Such a conclusion is impossible to verify experimentally in the case of two equivalent particles. The only thing we can ascertain is that one particle is in state ψ_1 and the second particle is in state ψ_2, however we do not know which one in state (1) and (2).

We shall now construct a wavefunction which takes into account the indistinguishability. We have said that the equation (10.1) is symmetric in the two electrons, an equally good solution to the eigen value equation can be

$$\psi = \psi_2(1)\ \psi_1(2) \tag{10.4}$$

A linear combination of solutions is also a good solution. Two linear combinations are

$$\psi_+ = \frac{1}{\sqrt{2}}[\psi_1(1)\ \psi_2(2) + \psi_2(1)\psi_1(2)] \tag{10.5a}$$

$$\psi_- = \frac{1}{\sqrt{2}}[\psi_1(1)\ \psi_2(2) - \psi_2(1)\psi_1(2)] \tag{10.5b}$$

The first equation (10.5a) is a symmetrical linear equation and the second equation (10.5b) is an antisymmetrical linear equation. It means the first equation does not change by interchanging the particle (1) and (2) whereas the second equation changes by an interchange of particles. The importance of the above solutions is that we can definitely say that there is a particle in state (1) and another in state (2), but we cannot definitely say which one is in which.

Thus it is clear that it is the general property of wave functions for systems containing similar particles that they must be properly symmetrized to take into account the indistinguishability of the particles. This can be considered as an additional postulate. Pauli's exclusion principle is one form of this postulate and could be allied to identical particles of half integral spin.

Thus all acceptable wave functions for half integral spin particles must be antisymmetric upon permutation of the coordinates of any two particles. Such particles are called Femi particles or fermions. For particles with integral spin, the wave function should be symmetric with respect to permutation of the coordinates of two particles. Such particles are called Bose–Einstein particles or bosons. Antisymmetric wave functions are discussed in more detail in Section 10.3.

In the following we consider the case of two electron system He atom. We can consider that the electrons move around the nucleus independently. This is called the method of independent electrons. The other method called the method of self-consistent field is more sophisticated and the two methods are complimentary to each other.

10.2 INDEPENDENT ELECTRON APPROXIMATION

10.2.1 Helium Atom

The helium atom has two electrons and a nucleus of charge $+2e$. We shall consider a coordinate system with nucleus as origin and the coordinates of electron 1 as (x_1, y_1, z_1) and electron 2 as (x_2, y_2, z_2). We consider the nucleus to be at rest. As a general consideration we can take the nuclear charge as Ze where Z is the atomic number. We can treat the helium like ions such as Li^+, Be^{++}, etc. in this manner. The Hamiltonian operator is then

$$H = -\frac{\hbar^2}{2m}\nabla_1^2 - \frac{\hbar^2}{2m}\nabla_2^2 - \frac{Ze^2}{r_1} - \frac{Ze^2}{r_2} + \frac{e^2}{r_{12}} \qquad (10.6)$$

where m is the mass of the electron, r_1 and r_2 are the distances of electrons 1 and 2 from the nucleus, ∇_1 and ∇_2 are the Laplacian operators for electrons 1 and 2 respectively; r_{12} is the distance between the two electrons and Z is the nuclear charge. The first two terms in the expression are the operators for the electron kinetic energy, the third and the fourth terms are the potential energies of attraction between the electrons and the nucleus and the last term is the electron repulsion term.

The coordinate system of the two electrons is shown in Figure 10.1.

The Schrödinger equation is

$$H\psi = E\psi \qquad (10.7)$$

We can also write with a different nomenclature, i.e., $H(1,2)\,\psi(1,2) = E(1,2)\,\psi(1,2)$

where $\psi(1,2)$ or ψ is the eigen function and $E(1,2)$ is the eigen value of the two electrons taken together in the atom.

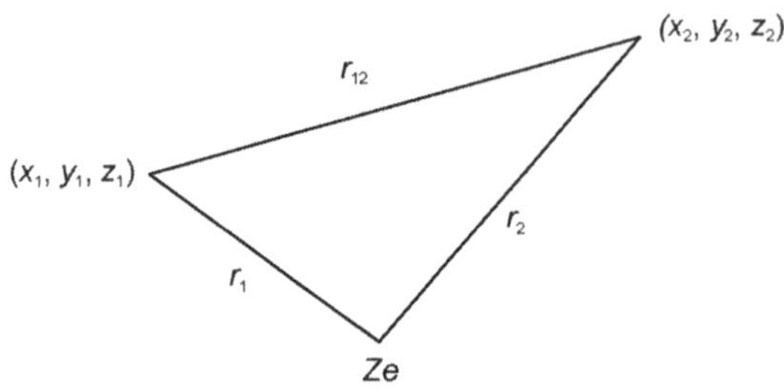

Figure 10.1   The coordinate system of the two electrons

The function $\psi(1,2) = \psi$ has six coordinates, three for each electron, which can be written as and $\psi(x_1, y_1, z_1, x_2, y_2, z_2)$ and

$$r_{12} = \left[(x_1 - x_2)^2 + (y_1 - y_2)^2 - (z_1 - z_2)^2\right]^{1/2}$$

$$\Psi^2 = \left[\psi(x_1, y_1, z_1, x_2, y_2, z_2)\right]^2 \quad \text{and}$$

$$\Psi^2 d\tau = \left[\psi(x_1, y_1, z_1, x_2, y_2, z_2)\right]^2 dx_1\, dy_1\, dz_1 \cdot dx_2\, dy_2\, dz_2 \tag{10.8}$$

where ψ^2 gives the probability per unit volume that the electron 1 is at the point (x_1, y_1, z_1) and the electron 2 is at the point (x_2, y_2, z_2) at any given instant. $\Psi^2 d\tau$ gives the probability that at any given instant the electron 1 is in the volume element $d\tau_1$ $(dx_1\, dy_1\, dz_1)$ and the electron 2 is in the volume element $d\tau_2$ $(dx_2\, dy_2\, dz_2)$.

In spherical polar coordinate

$$\psi(1,2) = \psi(r_1,\ \theta_1,\ \phi_1;\ r_2, \theta_2, \phi_2) \tag{10.9}$$

The normalization condition is $\int[\psi(x_1, y_1, z_1, x_2, y_2, z_2)]^2\, d\tau_1\, d\tau_2 = 1$

In spherical polar coordinates $d\tau_1 = r_1^2\, dr_1\, \sin\, \theta_1\, d\theta_1\, d\phi_1$

$$d\tau_2 = r_2^2\, dr_2\, \sin\, \theta_2\, d\theta_2\, d\phi_2 \tag{10.10}$$

In independent electron approximation, the helium atom Hamiltonian can be rearranged in the form

$$H = \left(-\frac{\hbar^2}{2m}\nabla_1^2 - \frac{Ze^2}{r_1}\right) + \left(-\frac{\hbar^2}{2m}\nabla_2^2 - \frac{Ze^2}{r_2}\right) + \frac{e^2}{r_{12}}$$

$$= H_1^0 + H_2^0 + \frac{e^2}{r_{12}} \tag{10.11}$$

where $H_1^0 = -\frac{\hbar^2}{2m}\nabla_1^2 - \frac{Ze^2}{r_1}$ and

$$H_2^0 = -\frac{\hbar^2}{2m}\nabla_2^2 - \frac{Ze^2}{r_2}$$

are considered to be the unperturbed Hamiltonian, say H^0 of the helium atom. The unperturbed system is thus a helium atom in which the two electrons exert no forces on each other. Although there is no physical reality in considering so, it becomes easy to apply perturbation theory to such a system to get approximation to the actual helium atom. The unperturbed Hamiltonian thus is the sum of the Hamiltonians of two independent hydrogen like particles. Hence H_1^0 and H_2^0 have hydrogen like eigen functions (orbital) say $\phi(1)$ and $\phi(2)$.

Thus $H_1^0\ \phi(1) = E(1)\ \phi(1)$

$$H_2^0\ \phi(2) = E(2)\ \phi(2) \tag{10.12}$$

where $E(1)$ and $E(2)$ are the eigen values.

We can write $H^0\psi^0 = [H(1) + H(2)]\ [\phi(1)\ \phi(2)]$

$$\begin{aligned}
&= H(1)\ [\phi(1)\phi(2)] + H(2)\ [\phi(1)\ \phi(2)]\\
&= [H(1)\ \phi(1)\ ]\ \phi(2) + [H(2)\ \phi(2)]\ \phi(1)\\
&= E^0(1)\ \phi(1)\ \phi(2) + E^0(2)\ \phi(1)\ \phi(2)\\
&= [E^0(1) + E^0(2)]\ [\phi(1)\ \phi(2)] \tag{10.13}
\end{aligned}$$

Thus the unperturbed eigen values and eigen functions can be approximated as

$$E^0 = E^0(1) + E^0(2)$$
$$\psi^0 = \phi\ (1)\ \phi\ (2) \tag{10.14}$$

$H(1)$ operates on $\phi(1)$ and $H(2)$ on $\phi(2)$ only and hence $\phi(1)\ \phi(2) = \phi(2)\ \phi(1)$

In independent electron approximation the function $\psi^0 = \phi(1)\ \phi(2)$

We can approach the problem in the following way.

We have seen above that for an atom of two electrons

$$H = -\frac{\hbar^2}{8\pi^2 m}\left(\nabla_1^2 + \nabla_2^2\right) - \frac{Ze^2}{r_1} - \frac{Ze^2}{r_2} + \frac{e^2}{r_{12}}$$

$$H = -\frac{\hbar^2}{8\pi^2 m}\nabla_1^2 - \frac{Ze^2}{r_1}\nabla_2^2 - \frac{Ze^2}{r_2} + \frac{e^2}{r_{12}} \tag{10.15}$$

where ∇_1 and ∇_2 are the Laplacian operators for electrons 1 and 2; r_1 and r_2 are the distances of these electrons from the nucleus; r_{12} is the distance between the two electrons and Z is the nuclear charge.

The first two terms represent the kinetic energy operators of the two electrons and the last three terms represent the potential energy operators of which the last term represent the coulomb repulsion and the other two terms coulomb attraction. The subscripts 1 and 2 refer to the two electrons. For the purpose of simplifying the calculations of the integrals involved it is found to be convenient to represent the quantities in atomic units. The Bohr's first radius is taken as unit of length.

Expressing the distances in terms of Bohr's first radius makes the transformation to atomic units.

$$a_0 = \frac{h^2}{4\pi^2 m e^4} = 0.52917 \times 10^{-8} \text{ cm}$$

We write $r_1 = a_0 R_1$, $r_2 = a_0 R_2$, $r_{12} = a_0 R_{12}$

$$\frac{\partial^2}{\partial x_1^2} = \frac{1}{a_0^2} \frac{\partial^2}{\partial X_1^2}, \text{ etc.} \qquad (10.16)$$

The electron mass $m_e = 9.1091 \times 10^{-23}$ g is taken as the unit of mass.

The time taken to travel 1 a.u. of length in Bohr's first orbit a_0 $h/2\pi e^2 = 2.42 \times 10^{-17}$ sec is taken as the unit of time. The electron velocity in the first Bohr orbit is $(2\pi e^2/h) \cdot (e^2/a_0) = 27.210$ eV is taken as the unit of energy.

With this transformation the Hamiltonian operator is

$$H = -\frac{h^2}{8\pi^2 m} \frac{1}{a_0^2} \left(\nabla_1^2 + \nabla_2^2 \right) - \frac{Ze^2}{a_0 R_1} - \frac{Ze^2}{a_0 R_2} + \frac{e^2}{a_0 R_{12}} \qquad (10.17)$$

In units of $\dfrac{e^2}{a_0}$ the Hamiltonian is thus

$$H = -\frac{1}{2} \left(\nabla_1^2 + \nabla_2^2 \right) - Z \left(\frac{1}{R_1} + \frac{1}{R_2} \right) + \frac{1}{R_{12}} \qquad (10.18)$$

The Hamiltonian for an atom of N electrons is

$$H = -\frac{1}{2} \sum_{i=1}^{n} \nabla_i^2 - \sum_{j=1}^{n} \frac{Z}{R_j} + \sum_{i<j}^{n} \frac{1}{R_{ij}} \qquad (10.19)$$

or

$$H = -\frac{1}{2} \sum_{i=1}^{n} \nabla_i^2 - \sum_{j=1}^{n} \frac{Z}{R_j} + \frac{1}{2} \sum_{i\neq j}^{n} \frac{1}{R_{ij}} \qquad (10.20)$$

the index $i < j$ is necessary in order to prevent counting the electronic repulsions twice.

This is called nonrelativistic Hamiltonian.

Thus
$$H = -\frac{1}{2}\left(\nabla_1^2 + \nabla_2^2 + \cdots + \nabla_n^2\right) - \left(\frac{Z}{R_1} + \frac{Z}{R_2} + \cdots + \frac{Z}{R_n}\right) + \frac{1}{2}\left(\frac{1}{R_{12}} + \frac{1}{R_{23}} + \cdots + \frac{1}{R_{mn}}\right) \qquad (10.21)$$

where $\nabla_i^2 = \dfrac{\partial^2}{\partial x_i^2} + \dfrac{\partial^2}{\partial y_i^2} + \dfrac{\partial^2}{\partial z_i^2}$

It can be solved by approximate methods.

The subscripts i and j can have any value from 1 to n except that $i \neq j$. It should be noted that if $i = j$ it would mean that the term $\dfrac{1}{R_{ii}} = \dfrac{1}{R_{jj}}$ corresponds to a nonexistent self repulsion. Also $\dfrac{1}{2}$ in the last term of equation is necessary to avoid terms like $\dfrac{1}{R_{ij}}$ and $\dfrac{1}{R_{ji}}$ entering twice in the summation.

For helium atom $Z = 2$ and using the atomic units we can write

$$H = -\frac{1}{2}\left(\nabla_1^2 + \nabla_2^2\right) - \frac{2}{R_1} - \frac{2}{R_2} + \frac{1}{R_{12}} \qquad (10.22)$$

The Schrödinger equation is

$$H(1,2)\psi(1,2) = E(1,2)\psi(1,2) \qquad (10.23)$$

or simply $H\psi = E\psi$

where ψ of the eigen function of the two electrons taken together and E is the total energy of the atom. ψ has six coordinates, three for each electron

Thus $\psi = \psi(x_1, y_1, z_1; x_2, y_2, z_2)$

Probability per unit volume

$$\psi^2 = [\psi(x_1, y_1, z_1; x_2, y_2, z_2)]^2$$

and $\psi^2 d\tau = [\psi(x_1, y_1, z_1; x_2, y_2, z_2)]^2 dx_1\, dy_1\, dz_1 \cdot dx_2\, dy_2\, dz_2$

In the above equations $x_1 = a_0 x_1$, $y_1 = a_0 y_1$ and so on.

where ψ^2 is the probability per unit volume that the first electron is at the point with coordinates x_1, y_1, z_1 and the second electron at the point with coordinates x_2, y_2, z_2. $\psi^2 d\tau$ is the probability that at any given instant the first electron is in the volume element $d\tau_1 = dx_1\, dy_1\, dz_1$ and the second electron is in the volume element $d\tau_2 = dx_2\, dy_2\, dz_2$. The normalization condition implies that

$$\int[\psi(x_1, y_1, z_1, x_2, y_2, z_2)]^2 dx_1\, dy_1\, dz_1 \cdot dx_2\, dy_2\, dz_2 = 1 \qquad (10.24)$$

In spherical polar coordinates

$$d\tau_1 = dx_1\,dy_1\,dz_1 = r_1^2\,dr_1\,\sin\,\theta_1\,d\theta_1\,d\phi_1 \qquad \text{and}$$

$$d\tau_2 = dx_2\,dy_2\,dz_2 = r_2^2\,dr_2\,\sin\,\theta_2\,d\theta_2\,d\phi_2$$

The Hamiltonian for the helium atom can be written in the form

$$H = \left[-\frac{1}{2}\nabla_1^2 - \frac{2}{R_1}\right] + \left[-\frac{1}{2}\nabla_2^2 - \frac{2}{R_2}\right] + \frac{1}{R_{12}} = H(1) + H(2) + \frac{1}{R_{12}} \tag{10.25}$$

where $H(1) = \left[-\dfrac{1}{2}\nabla_1^2 - \dfrac{2}{R_1}\right]$ and

$$H(2) = \left[-\frac{1}{2}\nabla_2^2 - \frac{2}{R_2}\right]$$

The Hamiltonian is the sum of the two one electron Hamiltonian plus the term $1/R_{12}$.

$H(1) + H(2)$ can be considered as the unperturbed Hamiltonian say H^0 and the term $1/R_{12}$ as a small perturbation. $H(1)$ and $H(2)$ has hydrogen like eigen values $\phi\,(1)$ and $\phi\,(2)$.

Hence

$$H(1)\,\phi\,(1) \; = E(1)\,\phi\,(1) \;\; \text{and}$$

$$H(2)\,\phi\,(2) = \; E(2)\,\phi\,(2) \tag{10.26}$$

$E(1)$ and $E(2)$ are the corresponding eigen values. We have seen in equation 10.13 that the unperturbed eigen functions and eigen values could be given by

$$\psi^0 = \phi\,(1)\,\phi\,(2) \;\; \text{and} \;\; E^0 = E(1) + E(2)$$
$$H^0\,\psi^0 = [H(1) + H(2)][\phi\,(1)\,\phi(2)]$$
$$= H(1)\,\phi\,(1)\,\phi\,(2)] + H(2)\,\phi\,(1)\,\phi\,(2)]$$
$$= [H(1)\,\phi\,(1)]\,\phi\,(2)] + [H(2)\,\phi\,(2)]\,\phi(1) \tag{10.27}$$
$$= E(1)\,\phi\,(1)\,\phi\,(2)\; + E(2)\,\phi\,(2)\,\phi\,(1)$$
$$= [E(1) + E(2)][\phi\,(1)\,\phi\,(2)]$$

$H(1)$ operates on $\phi(1)$ only and $H(2)$ on $\phi(2)$ only and hence $\phi\,(1)\,\phi\,(2) = \phi\,(2)\,\phi\,(1)$.

Thus the function $\psi^0 = \phi\,(1)\,\phi\,(2)$ which is the true eigen function of H^0 can be used as an eigen function of the Hamiltonian $H = H(1) + H(2) + \dfrac{e^2}{r_{12}}$ can be considered as an approximate eigen function of H^0 and thus the eigen functions of the one electron Hamiltonians $H(1)$ and $H(2)$ may serve as a basis wave function for two electron system. The two electron problem can thus be considered as that of two one electron hydrogen problems.

The energies are then

$$E^0(1) = -\frac{Z^2}{n_1^2}\frac{e^2}{2a_0}$$

$$E^0(2) = -\frac{Z^2}{n_2^2}\frac{e^2}{2a_0} \quad \text{and} \tag{10.28}$$

$$E^0 = -Z^2\left(\frac{1}{n_1^2}+\frac{1}{n_2^2}\right)\frac{e^2}{2a_0} \quad n_1 = 1, 2, 3, \text{ etc. and } n_2 = 1, 2, 3, \text{ etc.} \tag{10.29}$$

where a_0 is the Bohr radius.

For helium atom the two electrons are in the $1s$ orbital and thus both of them have the same orbital function, which can be written as

$$1s(1) = \sqrt{\frac{Z^3}{\pi a_0^3}}\; e^{\left(-\frac{Zr_1}{a_0}\right)} \quad \text{and}$$

$$1s(2) = \sqrt{\frac{Z^3}{\pi a_0^3}}\; e^{\left(-\frac{Zr_2}{a_0}\right)} \tag{10.30}$$

Thus $\psi = 1s(1)\,1s(2) = \sqrt{\dfrac{Z^3}{\pi a_0^3}}\; e^{\left(-\frac{Z(r_1+r_2)}{a_0}\right)} \tag{10.31}$

For hydrogen like atom $E = -\dfrac{1}{2}\dfrac{Z^2 e^2}{n^2 a_0} \tag{10.32}$

And for He$^+$ $Z = 2$ and electron in $1s$ state and hence $n = 1$.

And $E^{He} = E^{He^+}(1) + E^{He^+}(2) = -Z^2\left[\dfrac{1}{1^2}+\dfrac{1}{1^2}\right]\dfrac{e^2}{2a_0} = -2Z^2\cdot\dfrac{e^2}{2a_0} \tag{10.33}$

The expression $-\dfrac{e^2}{2a_0}$ is the ground state energy of the hydrogen atom which is equal to -13.606 eV.

For helium $Z = 2$ and hence $E = -2\times 4\ \times\dfrac{e^2}{2a_0} = -8\times 13.606 = -108.8$ eV $= -1.74\times 10^{-17}$ J

which is equal to $-\dfrac{1.74\times 10^{-17}}{1.6^{-19}} = 108$ eV. $\tag{10.34}$

However the experimental value of the helium atom is $-$ 78.4 eV (-1.25×10^{-17} J). Thus the energy calculated is much below the observed value and error is almost 38%. This clearly indicates that we cannot neglect the $\dfrac{1}{r_{12}}$ term in the Hamiltonian.

10.2.2 Perturbation Approach

In applying the methods of perturbation theory we set

$H = H^0 + H^{(1)}$ where

$$H^0 = -\frac{1}{2}\left(\nabla_1^2 + \nabla_2^2\right) - Z\left(\frac{1}{R_1} + \frac{1}{R_2}\right) \quad \text{and} \tag{10.35}$$

$$H^{(1)} = \frac{1}{R_{12}}$$

The zeroth order solution is

$$H^0 \psi^0 = E^0 \psi^0$$

If we now set $\psi^0 = \phi^0(1)\,\phi^0(2)$; $E^0 = E^0(1) + E^0(2)$ the above equation becomes

$$-\frac{1}{2}\nabla_1^2\,\phi^0(1) + \left(E^0(1) + \frac{Z}{R_1}\right)\phi^0(1) = 0$$

$$-\frac{1}{2}\nabla_2^2\,\phi^0(2) + \left(E^0(2) + \frac{Z}{R_2}\right)\phi^0(2) = 0 \tag{10.36}$$

These equations are of the hydrogen like equations and for the ground state of helium with two $1s$ electrons

$$\phi^0(1) = \frac{1}{\sqrt{\pi}} Z^{3/2} \exp\ (-ZR_1)$$

$$\phi^0(2) = \frac{1}{\sqrt{\pi}} Z^{3/2} \exp\ (-ZR_2) \tag{10.37}$$

$$\psi^0 = \phi^0(1)\,\phi^0(2) = \frac{Z^3}{\pi}\exp(-Z(R_1 + R_2)) \tag{10.38}$$

$$E^0 = E^0(1) + E^0(1) = 2Z^2 E_{1s}(H) \tag{10.39}$$

where $E_{1s}(\text{H})$ is the energy of the ground state of hydrogen atom and in ordinary units $\left(-\dfrac{1}{2}\dfrac{e^2}{a_0}\right)$.

The first order correction is

$$E(1) = \int\int \psi^{0*} H^{(1)} \psi^0 \, d\tau_1 \, d\tau_2 = \int\int \phi^0(1) \, \phi^0(2) \frac{1}{R_{12}} \phi^0(1) \, \phi^0(2) \, d\tau_1 \, d\tau_2$$

$$= \left(\frac{e^2}{a_0}\right)\left(\frac{Z^6}{\pi^2}\right) \int\int [\exp(-Z(R_1 + R_2)) \frac{1}{R_{12}} \exp(Z(R_1 + R_2))] \, d\tau_1 \, d\tau_2 \qquad (10.40)$$

$$d\tau_1 = R_1^2 \sin \theta_1 \, dR_1 \, d\theta_1 \, d\phi_1$$
$$d\tau_2 = R_2^2 \sin \theta_2 \, dR_2 \, d\theta_2 \, d\phi_2$$

The integral represents the average coulomb repulsion energy. It is generally expanded in terms of the associated Legendre polynomials.

$$\int\int \phi^0(1)\phi^0(2)\frac{1}{R_{12}} \phi^0(1) \, \phi^0(2) d\tau_1 \, d\tau_2 = J$$

$$= \left(\frac{Z^3}{\pi}\right)^2 \left(\frac{e^2}{a_0}\right) \int\int \exp[-Z(R_1 + R_2)] \frac{1}{R_{12}} \exp[-Z(R_1 + R_2)] \, d\tau_1 d\tau_2$$

The solution is tedious and the result after evaluation gives $J = \dfrac{5Z}{8}\left(\dfrac{e^2}{a_0}\right)$ $\qquad (10.41)$

In units of $\dfrac{e^2}{a_0}$, $J = \dfrac{5Z}{8}$ $\qquad\qquad (10.42)$

$$E = E^0 + E^{(1)} = \left(2Z^2 - \frac{5Z}{4}\right)\left(-\frac{1}{2}\frac{e^2}{a_0}\right) = \left(-Z^2 + \frac{5Z}{8}\right)\frac{e^2}{a_0} \qquad (10.43)$$

For helium atom $Z = 2$ and hence

$$E = E^0 + E^{(1)} = \left(-4 + \frac{10}{8}\right) \text{a.u} = -\frac{11}{4} = -2.75 \text{ a.u} = -74.8 \text{ eV} \qquad (10.44)$$

This value is fairly in agreement with the experimental value.

Ionization potential

He($1s^2$) = He$^+$ ($1s$) + e($1s$)

Ionization potential I.P. $= E_{1s}^{\text{He+}} - E_{1s^2}^{\text{He}} = -2 - (-2.75) = 0.75 \text{ a.u} = 20.4 \text{ eV} = 0.326 \text{ J}$

The energy of the ground state of He$^+$ is $Z^2 E_{1s}(\text{H})$

$$\left(Z^2 + \frac{5Z}{8}\right) E_{1s}(\text{H}) = \frac{3}{2} E_{1s}(\text{H}) = \left(\frac{3}{2}\right) 13.60 = 20.40 \text{ electron volt.} \qquad (10.45)$$

10.2.3 Variation Method

In variation method the solution of the problem starts with a trial function. In order to have an approximate function in the case of helium atom we consider that out of the two electrons, one electron shield the nucleus and the other electron feels that the nuclear charge as effectively reduced. Therefore the value of Z in the function

$$\Psi(1,2) \text{ or simply } \Psi = 1s(1)\ 1s(2) = \frac{Z^3}{\pi}\exp\ [-Z(r_1 + r_2)] \tag{10.46}$$

is less than 2. Thus Z may be used as a variable parameter and its value is determined by minimizing the energy. In doing so we set

$$\frac{dE_{1s^2}}{dZ} = 0 \tag{10.47}$$

Using the variation integral as per variation principle we can write

$$E_{1s^2} = \frac{\int \psi H \psi d\tau}{\int \psi^2 d\tau} = \int \psi H \psi d\tau \tag{10.48}$$

$\int \psi^2 d\tau = 1$ as Ψ is real and is normalized.

$$H = H(1) + H(2) + \frac{1}{r_{12}} \tag{10.49}$$

$$E_{1s^2} = \int\int 1s(1)1s(2)\left[H(1) + H(2) + \frac{1}{r_{12}}\right]1s(1)1s(2)d\tau_1 d\tau_2$$

$$E_{1s^2} = \int\int 1s(1)1s(2)H(1)1s(1)1s(2)d\tau_1 d\tau_2 + \int\int 1s(1)1s(2)H(2)1s(1)1s(2)d\tau_1 d\tau_2$$

$$+ \int\ \int 1s(1)1s(2)\left(\frac{1}{r_{12}}\right)1s(1)1s(2)d\tau_1 d\tau_2 \tag{10.50}$$

where $H(1)$ and $H(2)$ are Hamiltonian operators which operate on electron (1) and electron (2) respectively.

Hence $H(1)\phi(1) = E(1)\phi(1)$

$H(2)\phi(2) = E(2)\phi(2)$

$$E_{1s^2} = \int\int 1s(1)1s(2)E_{1s}^{\text{He}^+}(1)1s(1)1s(2)d\tau_1 d\tau_2 + \int\int 1s(1)1s(2)E_{1s}^{\text{He}^+}(2)1s(1)1s(2)d\tau_1 d\tau_2 + J$$

where
$$J = \int\int 1s(1)1s(2)\left(\frac{1}{r_{12}}\right)1s(1)1s(2)d\tau_1 d\tau_2 \tag{10.51}$$

$$E_{1s}^{He^+}(1) = E_{1s}^{He^+}(2) = E_{1s} \quad (\text{say}) \tag{10.52}$$

Therefore $E_{1s^2} = 2E_{1s}\int\int [1s(1)1s(2)]^2 d\tau_1 d\tau_2 + J = 2E_{1s} + J$ $\qquad$ (10.53)

The first part is normalized.

E_{1s} is the energy of the He$^+$ ion in $1s$ state

$$E_{1s} = \int 1s\left(-\frac{1}{2}\nabla^2 - \frac{2}{r}\right)1s\, d\tau \tag{10.54}$$

$1s$ is the eigen function of $-\dfrac{1}{2}\nabla^2 - \dfrac{2}{r}$ with $Z \le 2$ and the eigen value is $-\dfrac{Z^2}{2}$

$$E_{1s} = \int 1s\left(-\frac{1}{2}\nabla^2 - \frac{Z}{r} - \frac{2-Z}{r}\right)1s d\tau = \int 1s\left(-\frac{1}{2}\nabla^2 - \frac{Z}{r}\right)1s d\tau - \int 1s\left(\frac{2-Z}{r}\right)1s d\tau$$

$$= -\frac{Z^2}{2}\int (1s)^2 d\tau - (2-Z)\int 1s\left(\frac{1}{r}\right)1s d\tau = -\frac{Z^2}{2} - (2-Z)\int 1s\left(\frac{1}{r}\right)1s d\tau \tag{10.55}$$

$$1s = \sqrt{\frac{Z^3}{\pi}}\exp(-Zr) \text{ and } d\tau = r^2 dr \sin\theta\, d\theta\, d\phi \tag{10.56}$$

$$\int 1s\left(\frac{1}{r}\right)1s d\tau = \frac{Z^3}{\pi}\int_0^\pi \frac{1}{r}\exp(-2Zr)\, r^2 dr \int_0^\pi \sin\theta d\theta \int_0^{2\pi} d\phi = \frac{Z^3}{\pi}\cdot 4\pi \int_0^\alpha re^{-2Zr} dr$$

By evaluating the integral we get

$$\int 1s\left(\frac{1}{r}\right)1s d\tau = 4Z^3 \cdot \frac{1}{(2Z)^2} = Z$$

Hence from equation (10.54)

$$E_{1s} = -\frac{Z^2}{2} - (2-Z)Z \tag{10.57}$$

$$E_{1s^2} = 2[-\frac{Z^2}{2} - (2-Z)Z + J]$$

$$= -Z^2 - 2Z(2-Z) + \frac{5}{8}Z \text{ using equation (10.42) } J = \frac{5}{8}Z$$

$$= Z^2 - \frac{27}{8}Z \tag{10.58}$$

$$\frac{dE_{1s^2}}{dZ} = 2Z - \frac{27}{8} = 0, \quad Z = \frac{27}{16}$$

$$E_{1s^2} = \left(\frac{27}{16}\right)^2 - \frac{27}{8} \times \frac{27}{16} = -2.84 \text{ a.u} = -77.48 \text{ eV} = -1.24 \times 10^{-17} \text{ J} \tag{10.59}$$

The value $Z = \dfrac{27}{16}$ represents the effective nuclear charge. The difference $2 - \dfrac{27}{16} - \dfrac{5}{16}$ represent the screening effect.

10.3 SYMMETRIC AND ANTISYMMETRIC WAVE FUNCTIONS

In a two electron atom we have seen that the Hamiltonian can be represented by

$$H = \left[-\frac{1}{2}\nabla_1^2 - \frac{2}{R_1}\right] + \left[-\frac{1}{2}\nabla_2^2 - \frac{2}{R_2}\right] + \frac{1}{R_{12}} = H(1) + H(2) + \frac{1}{R_{12}} \tag{10.60}$$

where $\quad H(1) = \left[-\dfrac{1}{2}\nabla_1^2 - \dfrac{2}{R_1}\right]$ and

$$H(2) = \left[-\frac{1}{2}\nabla_2^2 - \frac{2}{R_2}\right] \tag{10.61}$$

The Hamiltonian is the sum of the two one electron Hamiltonian plus the term $1/R_{12}$.

$H(1) + H(2)$ can be considered as the unperturbed Hamiltonian say H^0 and the term $1/R_{12}$ as a small perturbation. $H(1)$ and $H(2)$ has hydrogen like eigen values $\varphi(1)$ and $\varphi(2)$.

Hence $\quad H(1)\,\phi(1) = E(1)\,\phi(1)$ and

$$H(2)\,\phi(2) = E(2)\,\phi(2)$$

$$\psi^0 = \phi(1)\,\phi(2)$$

We have seen that $H^0\,\psi^0 = [E(1) + E(2)][\phi(1)\,\phi(2)]$.

In a two electron atom ϕ (1) and ϕ (2) may be same or different. For example in helium atom both electrons are in the $1s$ state and the combined wave function is then

$$\Psi\,(1,2) = \phi\,(1)\,\phi\,(2) = 1s(1)\,1s\,(2) \tag{10.62}$$

where 1 and 2 refer to the two electrons which are equivalent and indistinguishable. In the excited state one of the electrons is excited and the two electrons may have the orbitals $1s$ and $2s$. And since the electrons are indistinguishable the complete wave function ψ may be either

$$\Psi\,(1,2) = 1s(1)\,2s\,(2)$$
$$\Psi\,(2,1) = 1s(2)\,2s\,(1) \tag{10.63}$$

Both of them are equally good eigen functions of the Hamiltonian with the same eigen value. Such states are called degenerate states.

The two functions Ψ $(1, 2)$ and Ψ $(2, 1)$ differ in an interchange of the electrons in the two orbitals.

Though this is a good approximation there occurs a serious difficulty in explaining the probability distribution of the electrons. The probability distribution is governed by the squares of the wave functions and as Ψ $(1, 2)$ and Ψ $(2, 1)$ differ, their squares are also different and lead to a different probability distribution for the same two electrons in the same state. Since the electrons are indistinguishable the probability must not change by simply interchanging the electron positions, the above result makes no physical sense.

In other words

$$[\Psi(1,2)]^2 = [1s(1)2s(2)]^2 = [\Psi\,(2,1)]^2 = [1s(2)2s(1)]^2$$

i.e., $\quad [\Psi(1,2)]^2 = [\Psi(2,1)]^2$

$$\Psi(1,2) = \pm\,\Psi(2,1) \tag{10.64}$$

This means that the two wave functions $\Psi(1,2) = 1s(1)2s(2)$ and $\Psi(2,1) = 1s(2)\,2s\,(1)$ ought to be either the same or simply the negative of the other in order to obtain the same probability distribution. If a wavefunction remains unchanged by an interchange of electrons it is called a symmetric function and if it changes sign it is called an antisymmetric. Thus the function should be a linear combination (either sum or difference) in order to have the same probability distribution. Thus the two wave functions should be

$$\Psi_+ = N\,[1s(1)\,2s(2) + 1s(2)2s\,(1)]$$
$$\Psi_- = N\,[1s(1)\,2s(2) - 1s(2)2s\,(1)] \tag{10.65}$$

where N is the normalization function. N can be obtained by taking the probability as 1.

Thus

$$\int \psi^2 d\tau = N^2[1s(1)\ 2s(2) \pm 1s(2)\ 2s(1)]^2\ d\tau_1\ d\tau_2 = 1$$
$$N^2\ \int 1s(1)^2\ d\tau_1 \int 2s(2)^2\ d\tau_2 + \int 2s(1)^2\ d\tau_1 \int 1s(2)^2\ d\tau_2$$
$$\pm 2\ \int 1s(1)\ 2s(1)\ d\tau_1\ \int 1s(2)\ 2s(2)\ d\tau_2 = 1 \qquad (10.66)$$

Since $1s$ and $2s$ functions are orthogonal we obtain

$$N^2\ [1 + 1 + 0] = 1 \text{ and hence } N = \frac{1}{\sqrt{2}}$$

Thus $\Psi_+ = \dfrac{1}{\sqrt{2}}[1s(1)\ 2s\ (2) + 1s(2)\ 2s\ (1)]$ symmetric

$$\Psi_- = \frac{1}{\sqrt{2}}[1s(1)\ 2s(2) - 1s(2)\ 2s\ (1)] \text{ antisymmetric} \qquad (10.67)$$

These two wave functions have the same probability distribution as ψ_+^2 and ψ_-^2 remain unchanged.

10.3.1 Spin

We have already seen that the spin angular momentum has the value

$$S = \sqrt{s(s+1)}\ h/2\pi \qquad (10.68)$$

where s is the spin quantum number. The spin angular momentum vector has three components S_x, S_y, S_z, however only one among them can have specific value and usually S_z is chosen. We have already seen that two values of S_z are possible having the value $m_s h/2\pi$ where $m_s = 1/2$ or $-1/2$. Generally the state corresponding to the two quantum numbers $\pm 1/2$ are degenerate and the degeneracy is removed only in magnetic or electric field. The eigen value equation corresponding to the spin vector can be written as

$$S_z \alpha(s) = \frac{1}{2}\frac{h}{2\pi}\alpha(s)$$
$$S_z \beta(s) = -\frac{1}{2}\frac{h}{2\pi}\beta(s) \qquad (10.69)$$

where $\alpha(s)$ and $\beta(s)$ are eigen functions corresponding to the eigen values $+1/2$ and $-1/2$. The spin coordinate is independent of the orbital coordinate which means that the operator depending only on the space coordinate does not affect the spin function. Thus

$$\nabla^2\left[\Psi(x,y,z)\alpha(s)\right] = \alpha(s)\nabla^2\left[\Psi(x,y,z)\right] \qquad (10.70)$$

Similarly the spin operator does not affect the orbital part of wave function. The spin functions are orthonormal, i.e.,

$$\int \alpha^2 ds = \int \beta^2 ds = 1$$

$$\int \alpha\beta ds = \int \beta\alpha ds = 0 \tag{10.71}$$

where *ds* is the element of space coordinate. In order to completely describe the electron we have to now consider along with the probability of finding the electron at any instant at a given point in space, the probability that it is in one of the spin states $\alpha(s)$ or $\beta(s)$. Thus the actual state of the electron is obtained by the product of the orbital and spin functions. For two electron systems the following four spin functions are possible (*See* Table 10.1).

Table 10.1   Spin product functions for two electron systems

Spin product functions	Resultant spin quantum number $S_z = \Sigma s_z$	Resultant angular momentum $M_s \dfrac{h}{2\pi}$
$\alpha(1)\,\alpha(2)$	$+1$	$+1\dfrac{h}{2\pi}$
$\beta(1)\,\beta(2)$	-1	$-1\dfrac{h}{2\pi}$
$\alpha(1)\,\beta(2)$	0	0
$\alpha(2)\,\beta(1)$	0	0

The above functions are symmetric, however, Pauli's exclusion principle implies that for a system of two or more electrons the complete wave function including spin must be antisymmetric with respect to interchange of electron positions. As in the case of orbital wave functions we can consider linear combinations of the spin functions. They are then

$$\alpha(1)\ \alpha(2)$$

$$\beta(1)\ \beta(2)$$

$$\frac{1}{\sqrt{2}}[\alpha(1)\ \beta(2) + \alpha(2)\ \beta(1)]$$

$$\frac{1}{\sqrt{2}}[\alpha(1)\ \beta(2) - \alpha(2)\ \beta(1)] \tag{10.72}$$

Out of the above the last combination is the only one which is antisymmetric. If we take under consideration the orbital wave functions also, then the combination of spin orbital functions will be

$$
\begin{aligned}
&[1s(1)\ 1s(2)]\ [\alpha(1)\ \alpha(2)] \\
&[1s(1)\ 1s(2)]\ [\beta(1)\ \beta(2)] \\
&[1s(1)\ 1s(2)]\ \left[\frac{1}{\sqrt{2}}[\alpha(1)\ \beta(2)+\alpha(2)\ \beta(1)]\right] \\
&[1s(1)\ 1s(2)]\ \left[\frac{1}{\sqrt{2}}[\alpha(1)\ \beta(2)-\alpha(2)\ \beta(1)]\right]
\end{aligned}
\tag{10.73}
$$

The excited electron configuration in Helium is $1s\,2s$ and the spin–orbital wave function will be

$$
\frac{1}{\sqrt{2}}[1s(1)\ 2s\ (2)+1s(2)\ 2s\ (1)]\ \alpha(1)\ \alpha(2)
$$

$$
\frac{1}{\sqrt{2}}[1s(1)\ 2s(2)+1s(2)\ 2s\ (1)]\ \beta(1)\ \beta(2)
$$

$$
\frac{1}{\sqrt{2}}[1s(1)\ 2s(2)+1s(2)\ 2s(1)]\frac{1}{\sqrt{2}}[\alpha(1)\ \beta(2)+\alpha(2)\ \beta(1)]
$$

$$
[1s(1)\ 2s(2)+1s(2)\ 2s(1)]\ \frac{1}{\sqrt{2}}[\alpha\ (1)\ \beta\ (2)-\alpha\ (2)\ \beta\ (1)]
$$

$$
\frac{1}{\sqrt{2}}[1s(1)\ 2s(2)-1s(2)\ 2s(1)]\ \alpha\ (1)\ \alpha\ (2)
$$

$$
\frac{1}{\sqrt{2}}[1s(1)\ 2s(2)-1s(2)\ 2s(1)]\ \beta\ (1)\ \beta\ (2)
$$

$$
\frac{1}{\sqrt{2}}[1s(1)\ 2s(2)-1s(2)\ 2s(1)]\frac{1}{\sqrt{2}}[\alpha\ (1)\ \beta\ (2)+\alpha\ (2)\ \beta\ (1)]
$$

$$
\frac{1}{\sqrt{2}}[1s(1)\ 2s(2)-1s(2)\ 2s(1)]\frac{1}{\sqrt{2}}[\alpha(1)\ \beta(2)-\alpha\ (2)\ \beta\ (1)]
\tag{10.74}
$$

Out of these the following combinations are antisymmetric

$$
\left[1s(1)\ 2s(2)\ +\ 1s(2)\ 2s\ (1)\right]\frac{1}{\sqrt{2}}\left[\alpha(1)\ \beta(2)-\alpha(2)\ \beta(1)\right]
\tag{10.75(a)}
$$

$$
\left[1s(1)\ 2s(2)-1s(2)\ 2s(1)\right]\alpha(1)\ \alpha(2)
\tag{10.75(b)}
$$

$$
\left[1s(1)\ 2s(2)-1s(2)\ 2s(1)\right]\beta(1)\ \beta(2)
\tag{10.75(c)}
$$

$$\left[1s(1)\,2s(2) - 1s(2)\,2s(1)\right]\frac{1}{2}\left[\alpha(1)\beta(2) + \alpha(2)\beta(1)\right] \qquad (10.75)(d)$$

It should be remembered in selecting the above functions

$$\text{symmetric} \times \text{symmetric} = \text{symmetric}$$

$$\text{antisymmetric} \times \text{antisymmetric} = \text{symmetric}$$

$$\text{symmetric} \times \text{antisymmetric} = \text{antisymmetric}$$

An electron orbital is characterized by quantum numbers n, l, m_l and m_s. When two electrons are in the same orbital, for example $1s^2$ electrons in helium, $n = 1$, $l = 0$, $m_l = 0$ and hence m_s cannot be the same for the two electrons. Hence for one electron $m_s = +1/2$ and the other $m_s = -1/2$. On the other hand if they are in different orbitals, for example the excited state electron configuration in helium $1s\,2s$, the n is different and hence m_s can also have the same values. Thus no two electrons in an atom should have all the four quantum numbers identical.

10.4 EXCITED STATE OF HELIUM

The energy
$$E_{1s,2s} = \frac{\Psi * H\Psi d\tau}{\Psi * \Psi d\tau} \qquad (10.76)$$

where Ψ is an antisymmetric wave function. It is real and normalized and hence

$$E_{1s,2s} = \int \Psi H\Psi d\tau \qquad (10.77)$$

If we assume the orbital wave function as $\psi_\pm$ we can write down the energy from the above equation as

$$E_{1s,2s} = \int \Psi_\pm \frac{1}{\sqrt{2}}\left[\alpha(1)\beta(2) \mp \alpha(2)\beta(1)\right] H \Psi_\pm \frac{1}{\sqrt{2}}\left[\alpha(1)\beta(2) \mp \alpha(2)\beta(1)\right] d\tau \qquad (10.78)$$

where $d\tau$ is used for integration over space and spin coordinates of both electrons 1 and 2. However there is no spin coordinate in the Hamiltonian to operate on the spin functions α and β.

Hence $E_{1s,2s} = \dfrac{1}{2}\left[\int \Psi_\pm H\Psi_\pm d\tau\right]\left[\int\int \alpha(1)\beta(2)\mp\alpha(2)\beta(1)\right]^2 ds_1\,ds_2 \qquad (10.79)$

where $d\tau$ corresponds to space coordinates and ds_1 and ds_2 corresponds to spin coordinates of electrons 1 and 2.

Double integral can be written as

$$\int \alpha(1)^2 \, ds_1 \int \beta(2)^2 \, ds_2 + \int \alpha(2)^2 \, ds_2 \int \beta(1)^2 \, ds_1 \mp 2\int \alpha(1)\beta(1) ds_1 \int \alpha(2)\beta(2) ds_2$$

$$\quad\quad 1 \quad\quad\quad\quad 1 \quad\quad\quad\quad 1 \quad\quad\quad\quad 1 \quad\mp\quad 2\times 0 \quad\quad 0$$

$$=\quad 1 + 1 + 0 = 2 \tag{10.80}$$

Therefore $\quad E_{1s,2s} = \dfrac{1}{2}\left[\int \Psi_{\pm} H \Psi_{\pm} d\tau\right]\times 2 = \int \Psi_{\pm} H \Psi_{\pm} d\tau$ $\tag{10.81}$

Thus it is independent of the spin part.

We have already seen that the Hamiltonian for the two electron system is

$$H = H(1) + H(2) + \frac{1}{r_{12}} \tag{10.82}$$

And hence $\quad E_{1s,2s} = \dfrac{1}{2}\left[\int \Psi_{\pm} H(1)\Psi_{\pm} d\tau + \int \Psi_{\pm} H(2)\Psi_{\pm} d\tau + \int \Psi_{\pm}\left(\dfrac{1}{r_{12}}\right)\Psi_{\pm} d\tau\right]$ $\tag{10.83}$

$$\int \Psi_{\pm} H(1)\Psi_{\pm} d\tau = \frac{1}{2}\int\int \left[1s(1)2s(2)\pm 1s(2)2s(1)\right]H(1)[1s(1)2s(2)\pm 1s(2)2s(1)]d\tau_1 d\tau_2$$

$$= \frac{1}{2}\int\int 1s(1)2s(2)H(1)1s(1)2s(2)d\tau_1 d\tau_2 \pm \frac{1}{2}\int\int 1s(2)2s(1)H(1)1s(2)2s(1)d\tau_1 d\tau_2$$

$$\pm \frac{1}{2}\int\int 1s(1)2s(2)H(1)1s(2)2s(1)d\tau_1 d\tau_2 \pm \frac{1}{2}\int\int 1s(2)2s(1)H(1)1s(1)2s(2)d\tau_1 d\tau_2$$

$$\tag{10.84}$$

This may be written as

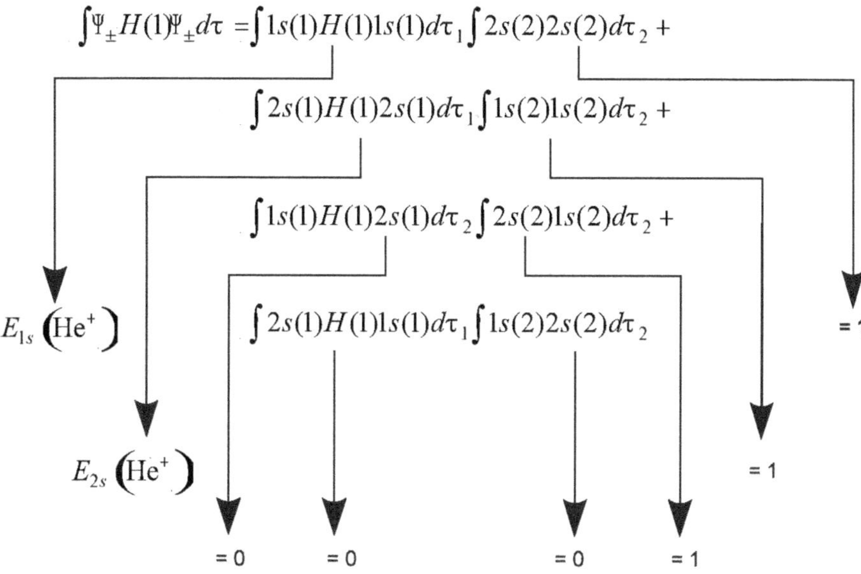

Note that $1s$ and $2s$ functions are orthogonal and disappears.

$$\int \Psi_\perp H(1)\Psi_\perp d\tau = E_{1s}(\text{He}^+) + E_{2s}(\text{He}^+) \tag{10.85}$$

Similarly
$$\int \Psi_\pm H(2)\Psi_\pm d\tau = E_{1s}(\text{He}^+) + E_{2s}(\text{He}^+) \tag{10.86}$$

The third integral $\displaystyle \int \Psi_\pm \frac{1}{r_{12}}\Psi d\tau = \int\int [1s(1)2s(2) \pm 1s(2)2s(1)]^2 \frac{1}{r_{12}} d\tau_1 d\tau_2$

$$= \int\int [1s(1)2s(2)]^2 \frac{1}{r_{12}} d\tau_1 d\tau_2 + \int\int [1s(2)2s(1)]^2 \frac{1}{r_{12}} d\tau_1 d\tau_2 \pm$$

$$2 \int\int 1s(1)2s(2)1s(2)2s(1)\frac{1}{r_{12}} d\tau_1 d\tau_2 \tag{10.87}$$

It can be seen that the first and the second integrals in expression are the same as it represents the expectation value of $\dfrac{1}{r_{12}}$ in $1s\,2s$ state. This is the electronic repulsion energy. This will be the same whether the configuration is $1s(1)\,2s(2)$ or $1s(2)2s(1)$. Hence

$$\int \Psi_{\pm} \frac{1}{r_{12}} \Psi_{\pm} d\tau = 2[\int (1s(1)2s(2))^2 \frac{1}{r_{12}} d\tau_1 \ d\tau_2 \pm 2 \int \int 1s(1)2s(2)1s(2)2s(1) \frac{1}{r_{12}} d\tau_1 d\tau_2 = 2(J \pm K)$$

$$(10.88)$$

From equation (10.83) the total energy of the helium atom in the $1s\,2s$ state is then

$$E_{1s,2s} = \frac{1}{2}[2(E_{1s}(He^+) + 2E_{2s}(He^+) + J \pm K)] = E_{1s}(He^+) + E_{2s}(He^+) + J \pm K \qquad (10.89)$$

$(J \pm K)$ is the first order perturbation correction to the unperturbed energy $E_{1s}(He^+) + E_{2s}(He^+)$. The J is the coulomb repulsion energy between the $1s$ and $2s$ orbitals and K is the average energy of the interchange of the electrons. K is also called exchange energy and has no classical counter part and is said to be quantum mechanical correction. Thus there are two energy levels for the $1s\,2s$ configuration in the helium atom corresponding to the $J + K$ and $J - K$ contributions. $J + K$ corresponds to the wave function (10.75) (a) and the $J - K$ corresponds to the three antisymmetric wave functions (10.75) (b), (c), (d). $J - K$ is thus threefold degenerate or is a triplet while the level corresponding to $J + K$ is a singlet.

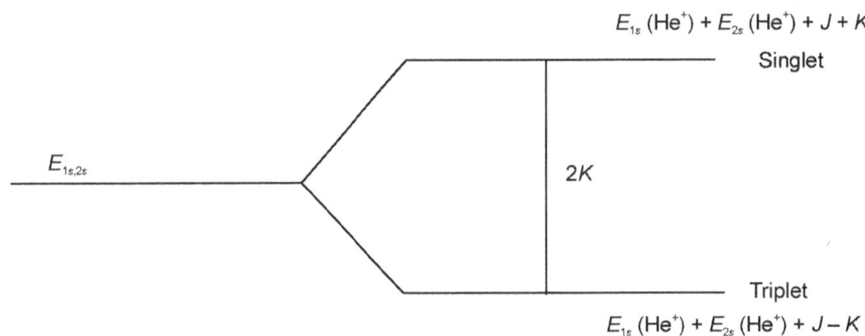

Figure 10.2   The energy level for helium atom

The use of hydrogen like wave function as the basis set does not bring the correct results and hence Hartree used the wave function of the atom as a single product of atomic orbitals $\psi = \phi_1(1) \ \phi_2(2) \ \phi_3(3) \ \dots \ \phi_n(n)$.

$$H(i)\phi_i = E_i \phi_i \qquad (10.90)$$

The Hamiltonian in atomic units is

$$H(i) = -\nabla_i^2 - \frac{Z}{r_i} + V_i \qquad (10.91)$$

V_i is the total potential energy of the electron due to the repulsion of all other electrons.

$$V_i = \sum_{j \neq i} \int \phi_i \frac{1}{r_{ij}} \phi_j \, d\tau \tag{10.92}$$

where the potential energy of repulsion of electron ϕ_i by ϕ_j $V_{ij} = \int \phi_i \frac{1}{r_{ij}} \phi_j d\tau$

We can consider this as a small perturbation and the Hamiltonian is

$$H = H^0 + \sum_{j \neq i} \int \phi_i \frac{1}{r_{ij}} \phi_j \, d\tau \tag{10.93}$$

In order to solve we assume a reasonable set of wave functions $\phi_1 \; \phi_2 \; \phi_3 \ldots \phi_n$, usually Slater type orbitals are used. The first step is to solve the Hamiltonian equation by the assumed wave functions numerically and then obtain a new set of $\phi_1' \; \phi_2' \; \phi_3' \ldots \phi_n'$ wave functions which is then used to solve the wave equation. At the end we arrive at a set of wave functions which are not different from the previous set. This method is called self-consistents field approximation (SCF method) and the wave functions are called Self-consistent field wave functions. The energy can be obtained by the first order perturbation theory as $\Delta E = \int \psi H \psi d\tau$.

Thus $$E = \int \phi_i(i) H^0 \phi_i d\tau + \int \phi_i \left[\sum_{j=1} \phi_j(j) \frac{1}{r_{ij}} \phi_j(j) \right] d\tau \tag{10.94}$$

We can show that the energy of n electron atom is

$$E = \sum_{i}^{n} E_i^0 + \frac{1}{2} \sum_{i=1}^{n} \sum_{j \approx i}^{n} J_{ij} \tag{10.95}$$

E is minimized with respect to the variable parameter ϕ_i's. This method was first introduced by Hatree. But Hatrees SCF method lacks from the fact that the orbital product wave function was not antisymmetric. The method was modified by Fock who introduced antisymmetric function which was conveniently expressed as a determinant. In this case all atomic orbitals are doubly occupied with paired electrons.

$$\psi = \frac{1}{\sqrt{n!}} \left| \phi_1(1) \phi_1(2) \phi_2(2) \phi_2(3) \ldots \right| \tag{10.96}$$

Fock introduced a Fock operator similar to the Hamilton operator in order to solve the eigen value equations. Hartree–Fock method is beyond the scope of this book and the readers are advised to refer books on quantum chemistry.

10.5 SLATER-TYPE ORBITALS (STO)

In the self-consistent field theory for many electron atoms, by calculating SCF orbital of each electron, we consider the shielding effects of all the other electrons which effectively reduce the nuclear charge. Generally hydrogen like orbital is used for simple calculations. However, we have seen that even for the calculation of helium atomic energy levels, the assumption is not sufficient. Hence a better form of atomic orbitals are needed to start with. J.C. Slater put forward a basis set of orbital, which are now terms as the Slater-type orbitals.

We have seen that the potentials due to the nuclear attraction of electrons and the electronic repulsions are spherically symmetric (dependent on r only). Hence the atomic orbitals, as in the case of hydrogen like systems, should be characterized by the angular part of the wave function rather than the radial part. If we consider the area close to the nucleus, i.e., r small, the nuclear charge is Z as there is no screening. The radial function in such parts is of the hydrogen like form r^l. If we consider the parts at larger distances from the nucleus, the effective charge is reduced and when r is very large, the effective charge is one as all the electrons except the one whose wave function is being considered screen the nucleus. The radial function behaves as $e^{-\alpha r}$. The angular part will be the same in hydrogen like atoms. Slater gave the approximate expression

$$\psi_{nlm}(r,\theta,\phi) = [Nr^{n-1}(\exp(-Z'r/n)][Y_{l,m}(\theta,\phi)]$$

where N is the normalization factor, n' is the effective principal quantum number, Z' is the effective nuclear charge and $Y_{l,m}(\theta,\phi)$ the angular part (spherical harmonics). Functions of this type are called Slater type Orbitals.

REFERENCES

Goodishman, J. *Contemporary Quantum Chemistry*. Plenum Press, New York.

Levin, I.N. (1991). *Quantum Chemistry*. Prentice Hall, New York.

Lowe, J.P. and Peterson, K. (2006). *Quantum Chemistry*. Elsevier, Amsterdam.

McQuarrie, D.A. (1983). *Quantum Chemistry*. University Science Books, Sausalite, California.

Prasad, R.K. (1992). *Quantum Chemistry*. Wiley Eastern Ltd, New Delhi.

11

SPECTRAL TERMS FOR NEUTRAL AND IONIZED ATOMS

11.1 COMPLEX SPECTRA

Till now we have seen characteristic spectra as singlets, doublets and triplets. There are in general multiplet lines composed of four, five, six, seven or even eight regularly spaced lines. These originate evidently from transitions among multiplet energy levels. These multiplets are usually called quartet, quintet, sextet, septet and octet lines or levels. Multiplicities of lines in an element are either even or all are odd. These generally depend on the valence electrons. In the atomic spectrum of cobalt all lines are even, doublets, quartets or sextets. On the other hand the spectrum of iron is of odd multiplicity, triplets, quintets and septets. In order to find whether a level is quartet, quintet or sextet, the atom should be placed in a weak magnetic field and counting the Zeeman levels one could clearly say the multiplicity of the level. For example a ^4P, ^5P or a ^6P will each have only three fine structure levels and the multiplicity of the levels are observed only when the atom is placed in a weak magnetic field and observing its magnetic levels.

In 1919 Kossel and Sommerfeld[1] proposed the displacement law for the spectra of neutral and ionized atoms. The displacement law is as follows: the spectrum and energy levels of any neutral atom with atomic number Z closely resemble the spectrum and energy levels of the ionized atom $Z+1$ succeeding it in the periodic table. It means if we consider the ionized atoms like Be^+, Mg^+, Ca^+, Sr^+, Ba^+, they have the same electronic configuration as those of Li, Na, K, Rb and Cs and the spectra are doublets. The other law concerning the multiplicities of the levels in atoms is alternation law of multiplicities. It states that spectral terms arising from successive elements in the periodic table alternate between even and odd multiplicities. The alternation of multiplicities in the first long period is as shown in Table 11.1.

Table 11.1 The alternation of multiplications in the first period

K	Ca	Sc	Ti	V	Cr	Mn	Fe	Co	Ni	Cu	Zn	Ga	Ge
	Sing		Sing		Sing		Sing		Sing		Sing		Sing
Doub.		Doub		Doub		Doub		Doub		Doub		Doub	
	Trip		Trip		Trip		Trip		Trip		Trip		Trip
		Quart		Quart		Quart		Quart		Quart			
			Quint		Quint		Quint		Quint				
				Sext		Sext		Sext					
					Sept		Sept						
						Oct							

11.2 THE BRANCHING RULE

Let us consider the case of the doubly ionized germanium. Germanium has atomic number 32 and belongs to the carbon group of elements. The common type of coupling in this group is *jj* coupling. The electronic configuration of neutral germanium is

Ge $1s^2\,2s^2\,2p^6\,3s^2\,3p^6\,3d^{10}\,4s^2\,4p^2$

All electrons except the last two similar electrons ($4p^2$) form completed subshells. These two electrons are two equivalent electrons. When two electrons are equivalent unlike in nonequivalent electrons, certain terms are forbidden. Two equivalent p electrons give rise to $^1S_0$, $^3P_{0,1,2}$, $^1D_2$.

We now think the above description of an atom in the reverse order. Consider that the germanium is doubly ionized. A doubly ionized germanium has two electrons knocked off from the neutral germanium and has an electron configuration Ge^{++} $1s^2\,2s^2\,2p^6\,3s^2\,3p^6\,3d^{10}\,4s^2$. It has then $^1S_0$ as its ground state. The electron configuration $1s^2\,2s^2\,2p^6\,3s^2\,3p^6\,3d^{10}\,4s^2$ of doubly ionized Ge, i.e., Ge^{++} can also be written as

Ge^{++} or ($1s^2\,2s^2\,2p^6\,3s^2\,3p^6\,3d^{10}\,4s^2\,4p^0$)

If we allow one electron to return it will occupy the $4p$ orbit.

The state originating from the configuration $1s^2\,2s^2\,2p^6\,3s^2\,3p^6\,3d^{10}\,4s^2\,4p^1$ gives rise to a ^{2}P state which is shown in the middle of the Figure 11.1. This is also the doublet limit of the various series of the neutral atom. If the second electron is now added to the orbit in a $5s$, $5p$ or a $4d$ orbit the various possible states originating from these configurations can be depicted in the following manner.

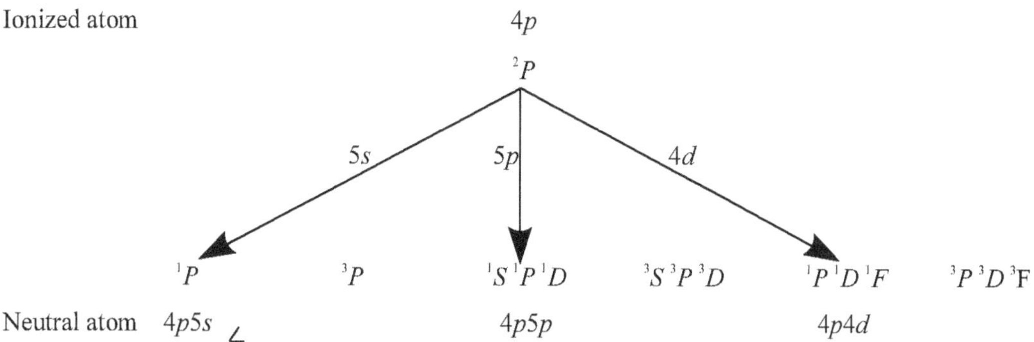

Figure 11.1   Depiction of the various possible states on adding second electron

By adding electrons we see that the multiplicities change. If the electron added is to an s orbit the L value is not changed as $l = 0$ for an s state and thus by adding $5s$, P states are obtained. The electron added is to $5p$ ($l = 1$), the L value goes one up and one down. By adding the electron to the $4d$ orbit, still there are two valence electrons and hence the multiplicity is 1 or 3 as the total spin of the electrons is 0 or 1. And one can derive the same P, D and F states. It is clear from above that by spectroscopic observation we can conclusively say which electron configuration atom has.

Generalizing we start with an atom stripped of all but one valence electron and return electrons to it one by one until the atom becomes again neutral atom. If l_1, l_2, l_3, … are the orbital quantum numbers and s_1, s_2, s_3… are the spin quantum numbers, and j_1, j_2, j_3… are the total quantum numbers and we add spins to form the resultant we get a following scheme as shown in Figure 11.2.

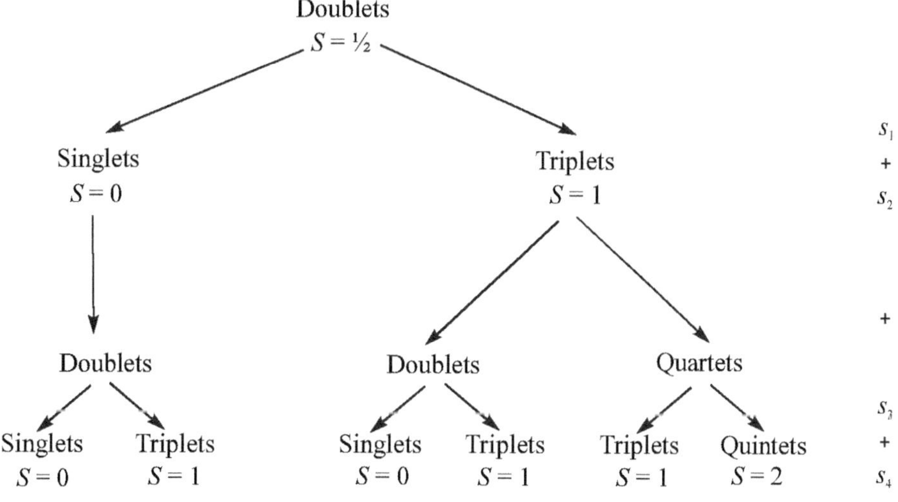

Figure 11.2   The scheme representing a change of an ionized atom to a neutral atom

Thus by adding, an electron raises and lowers the multiplicity by one. This can be explained in terms of the vector model as well. Similarly by considering the l's in the same manner, we start with $l = 0$, and adding $l_2 = 1, 2$, etc. we can make a branching picture as in Figure 11.3.

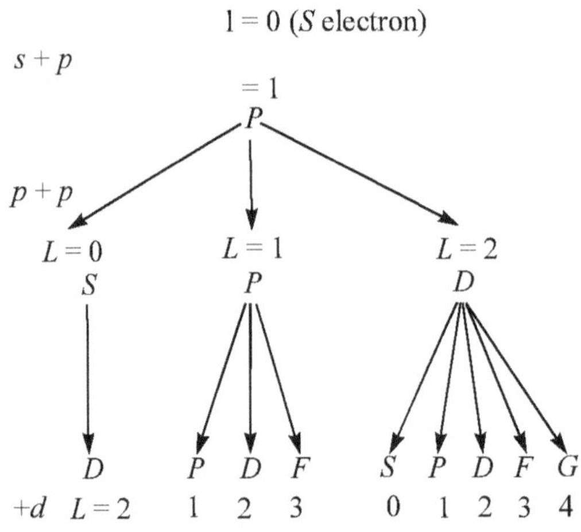

Figure 11.3   Branching rule

the two p electrons are assumed to have different total quantum numbers. Combining L and S will give the resultant J values.

11.2.1 Hund's Rule

Hund has proposed a rule in determining the lowest state arising out of an electron configuration of atom. It states that out of all terms with the same L value arising from a given electron configuration: (1) the term with the highest multiplicity, i.e., highest S value (remember that the multiplicity is $2S + 1$ where S is the total electron spin), will in general lie deepest; and (2) of these the term with the highest L value will lie deepest[2]. And (3) the state with the lowest J lies lowest. Out of the above states arising from the electron configuration in germanium according to Hund's rule the $^3P_{0,1,2}$ terms should lie deepest followed by $^1D_2$ and then $^1S_0$. This is also as observed. If we excite the atom, one electron in the $4p^2$ state will be raised to the lowest possible available state $5s$, thus changing the configuration to $4p5s$. A ps configuration will give rise to (they are no longer equivalent after excitation) $^1P_1$ and $^3P_{0,1,2}$ terms as follows: $l_1 = 1$ and $l_0 = 0$ and hence $L = 1$, thus giving rise to P state. Now $s_1 = 1/2$ and $s_2 = 1/2$ and hence $S = 0$ and 1 giving rise to singlet and triplet states. Thus one arrives at ^1P and ^3P. For the state ^1P, $J = L \pm S = 1 + 0 = 1$ (singlet arises from $S = 0$), thereby giving rise to $^1P_1$ state. For the state ^3P, $J = 1 + 1 \ldots |1 - 1| = 2,1,0$ thus giving rise to $^3P_{0,1,2}$ terms. It is observed that in germanium the coupling is very close to jj coupling. If one p electron in the $4p^2$ configuration is raised to different $4pns$ configurations like $4p5s$, $4p6s$, $4p7s$, etc. various $^1P_1$ and $^3P_{0,1,2}$ states will be observed as shown in Figure 11.1.

When the limit of this series, i.e., $4p \propto s$ is reached, the atom will be ionized by loosing one p electron from the $4p$ orbit, thus remaining only one electron in $4p$. The ionized atom is then Ge^+ having the electronic configuration Ge^+ $1s^2\ 2s^2\ 2p^6\ 3s^2\ 3p^6\ 3d^{10}\ 4s^2\ 4p$. There is only one electron in the p shell in this case thereby giving $L = 1$ and $s = 1/2$ and thus a doublet state $(2S + 1 = (2 \times 1/2) + 1 = 2)$. J in this case is $J = L + S...L - S = 1 + 1/2$ and $1 - 1/2 = 3/2$ or $1/2$ and thus $^2P_{1/2,3/2}$. Thus the normal state of the singly ionized germanium is a ^2P state, out of which the state with the lowest J lies deepest, i.e., $^2P_{1/2}$.

If we go on exciting an electron further from the ionized germanium Ge^+, the electron may go to $5p$ or $4d$ state as shown in the Figure 11.1 until one more electron can be ejected from the atom and the germanium atom thus becoming doubly excited. The doubly excited germanium atom Ge^{++} has the ground state $^1S_0$. In this case it has only 30 electrons remaining in the closed shells.

Of the spectroscopic terms arising from equivalent electrons, those with the largest multiplicity have the lowest energy and of these the largest L value has the lowest energy.

Multiplets formed from equivalent electrons are regular when less than half of the subshells are filled with electrons and they are inverted when more than half the subshells are filled with electrons.

11.2.2 Inverted Terms

A brief discussion on the normal and inverted states have already been given in Chapter 7. A more detailed discussion is given in this section. In some elements, especially in atoms in which the terms arise from an electron configuration where more than half the uncompleted shell is filled, it is observed that nearly all of the terms have inverted fine structure. This is opposite to normal terms. A normal term is the state in which the fine structure level with the smallest J value lies deepest on the energy level diagram. Inverted level is the one, in which the level with the largest J value lies lowest. A normal term arises from electron configuration involving less than half an incomplete subshell.

For example in boron, the electronic configuration is $1s^2\ 2s^2\ 2p$ and leads to $^2P_{3/2}$ and $^2P_{1/2}$ states. The $1s^2\ 2s^2$ does not contribute to the angular momentum of the atom and hence only angular momentum of $2p$ electron is to be considered to derive the states of the atom. And it is less than half filled and hence we get a normal state. From Hund's rule $^2P_{1/2}$ state lie lower.

In some elements, especially in atoms in which the terms arise from an electron configuration where more than half the shell is filled, for example a p^5 electron configuration in fluorine, it is observed that nearly all of the terms have inverted fine structure. This is opposite to normal terms. We have stated earlier that a normal term arises from electron configuration involving less than half an incomplete subshell.

In a doublet energy level of atomic system it is generally observed that the level in which $J = |L - S|$ lies lower in energy than the state with $J = L + S$. It means that in ^2P states $^2P_{1/2}$ lies lower than $^2P_{3/2}$. Qualitatively we can give an explanation in the following manner. The electron with its intrinsic motion, i.e., the electron spin can be regarded as a small magnet with a magnetic angular momentum μ_s. The direction of this magnetic moment is opposite to the spin angular mechanical momentum vector because of the negative charge of the electron. The orbital motion of electron around the nucleus gives rise to another magnetic field at the position of the electron. This magnetic field is in the same direction of the orbital angular momentum. This can be easily visualized if we imagine that the electron is at rest and the positively charged nucleus revolves around it. In a magnetic field the spinning electron will have minimum energy when μ_s is parallel to the magnetic field. μ_s is parallel to the magnetic field when $J = |L-1/2|$ and is antiparallel when $J = L + 1/2$ and hence the former lies lower. The two motions of the electron orbital and spin are shown in Figure 11.4. In the case of boron the ground state is $1s^2\,2s^2\,2p$ and hence it has a single valence electron, viz. $2p$. The $1s^2$ and $2s^2$ do not contribute to the angular momenta of the atom. This configuration gives rise to two states $^2P_{1/2}$ and $^2P_{3/2}$. From the above discussion the $^2P_{1/2}$ lies below $^2P_{3/2}$.

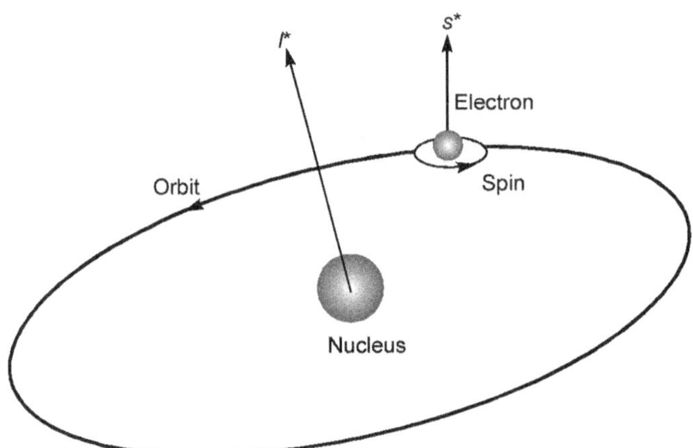

Figure 11.4   The orbital motion of an electron around the nucleus

If we now consider the electronic configuration of the fluorine atom we have

F : $1s^2\,2s^2\,2p^5$

Thus it lacks one electron to be completely filled in the p shell.

In this configuration we can imagine a hypothetical situation where one more electron is added to make the configuration $1s^2\,2s^2\,2p^6$ and an electron with hypothetical electron with "positive" charge. Thus the system becomes like a single valence system. Following the earlier argument the magnetic moment μ_s will be this time in the direction of the spin angular momentum of the single electron outside the orbit because of its "positive" charge. This means that $J = L + 1/2$ will be more stable und lies lower than $J = L - 1/2$.

This configuration also gives rise to $^2P_{3/2}$ and $^2P_{1/2}$ states, however in such an electronic configuration where more than half the shell is filled with electrons the $^2P_{3/2}$ state lies lower than the

$^2P_{1/2}$ state. We can qualitatively explain as follows: The multiplet is said to be regular if the level with the smallest J value lies lowest and they are said to be inverted if the level with the largest value of J lie deeper. Thus in the case of boron we have a regular state whereas in fluorine it is inverted. As a general rule it is found that atoms with electronic configuration nl^p and $nl^{[2(2l+1)-p]}$ have the same multiplet ground state, but the state is regular in the former and inverted in the latter. Here p and $[2(2l+1)-p]$ represent the number of electrons in the outermost valence shell of the atom characterized by the quantum numbers n and l. Some examples of regular and inverted states are given in Table 11.2.

Table 11.2   Examples of regular and inverted states

Atom	Z	Valence shell electron configuration	Lowest state	Type of state
B	5	$2p^1$	$^2P_{1/2}$	Regular
F	9	$2p^5$	$^2P_{3/2}$	Inverted
C	6	$2p^2$	$^3P_0$	Regular
O	8	$2p^4$	$^3P_2$	Inverted
Al	13	$3p$	$^2P_{1/2}$	Regular
Cl	17	$3P^5$	$^2P_{3/2}$	Inverted
Ti	22	$3d^2\,4s^2$	$^3F_2$	Regular
V	23	$3d^3 4s^2$	$^4F_{3,2}$	Regular
Co	27	$3d^7 4s^2$	$^4F_{9,2}$	Inverted
Ni	28	$3d^8 4s^2$	$^3F_4$	Inverted

11.2.3 Selection Rules

For LS coupling, the selection rules are

$$\Delta S = 0, \ \Delta L = 0, \pm 1, \ \Delta J = 0, \pm 1$$

In *jj* coupling further restrictions are

$$\Delta j_1 = 0, \pm 1$$
$$\Delta j_2 = 0, \pm 1$$
$$\Delta J = 0, \pm 1$$

11.2.4 Intensity Relations

The ratio between the intensities of two lines in emission may be written as

$$I_{m \to n} / I_{p \to q} = [\exp(-h v_m / kT) / \exp(-h v_p / kT)] \cdot [v^4_{m \to n} / v^4_{p \to q}] \cdot [P_{m \to n} / P_{p \to q}]$$

The first term arises from the Boltzmann factor having the temperature T of the initial states m and p and the final states n and q. The second term is Einstein's v^4 correction. The third term comes from the sum rule which states that (1) the sum of the intensities of all lines of a multiplet which start from a common initial level is proportional to the quantum weight $(2J + 1)$ of the initial level and (2) the sum of the intensities of all lines of a multiplet which end on a common final level is proportional to the quantum weight $(2J + 1)$ of the final level. Of course these are used for approximate calculations and to a correct prediction of intensities one should go for quantum mechanical derivations.

11.2.5 Breits's Scheme

This helps in the derivation of spectral terms from magnetic quantum numbers. It starts with magnetic quantum numbers in very high magnetic fields and then to weak fields and finally to field free levels. We shall consider the example of an electron configuration sp. The states are then obtained from the total quantum number L and the total spin quantum number S.

$$L = l_1 + l_2 \; ... |l_1 - l_2| = 0 + 1 \; ... |0 - 1| = 1 \qquad\qquad P \text{ state}$$
$$S = s_1 + s_2 \; ... |s_1 - s_2| = 1 \text{ and } 0, \qquad \text{i.e., triplets and singlets}$$

Thus from an sp configuration one gets a ^1P and ^3P.

The J values

For the ^1P state: $J = L + S \; ... |L - S| = 1 + 0 \; ... \; 1 - 0 = 1$ and thus the sate is $^1P_1$.

For the ^3P state: $J = L + S \; ... |L - S| = 1 + 1 ... \; 1 - 1 = 2, 1, 0$ and thus $^3P_{0,1,2}$.

Though we have derived the state considering an LS coupling, in a very strong magnetic field l and s precess around the magnetic field independently without making a resultant J. Thus for a sp configuration we have to consider $l_1 = 0$ (s orbit) $m_{l1} = 0$; $m_{s1} = \pm 1/2$ and $l_2 = 1$ (p orbit) $m_{l2} = 0, \pm 1$; $m_{s2} = \pm 1/2$. These values are arranged in a table. The table is already given in Chapter 9 (*See* Table 9.3). The m_{l1} and m_{l2} are written on the left up and on the right down in the Table. The sum of $m_{l1} + m_{l2} = M_L$ is created and written in the table in the middle.

11.3 EQUIVALENT ELECTRONS: PAULI'S EXCLUSION PRINCIPLE

One of the problems in atomic spectroscopy is in the determination of L and S for a given electron configuration. We have already seen that when two electrons are l values l_1 and l_2, we can obtain the total angular momentum L as having values ranging with difference of one from $l_1 + l_2 ... | l_1 - l_2 |$. The spins of the two electrons s_1 and s_2 combine to form the resultant S and gives S as $s_1 + s_2 ... | s_1 - s_2 | = 1$ or 0. The $S = 0$ gives a singlet and $S = 1$ a triplet. And the states can be represented as

$$^{2S+1}L_J$$

where J has the values $L + S ... | L - S |$.

However the difficulties arise when the two electrons are belonging to the same shell, i.e., having the same n, l values. These are equivalent electrons. In this case the Pauli's exclusion principle comes into play whereby certain combinations of L and S are not allowed. This reduces the number of possible combinations and hence the number of states.

The Pauli's principle is already discussed in Chapter 9 (Section 9.8) and as it is an important principle in atomic spectroscopy it is again discussed in this chapter with examples for brevity.

The ground state of carbon atom is as we know: $1s^2 2s^2 2p^2$. The inner shells $1s$ and $2s$ are completely filled as $1s^2 2s^2$ and only the outer unfilled $2p$ shell with 2 electrons are responsible for the spectral properties. If the two p electrons did not belong to the same n (i.e., if they were non-equivalent) we could have derived the states easily as described earlier. In that case we could have obtained $L = 2,1,0$ and $S = 0$ and 1, thereby obtaining the sates $^1D, ^1P, ^1S$ and $^3D, ^3P, ^3S$. If we consider the ^3D from the above, we could see that in this case $m_{l1} = m_{l2} = 1$ leading to M_L originating from $m_{l1} + m_{l2} = 2$ and $m_{s1} = m_{s2} = 1/2$ leading to the M_s value $m_{s1} + m_{s2} = 1$ and if the two n values are also the same, it then violates the Pauli's principle.

The Pauli's exclusion principle put forward by Pauli in 1925 play a major role in the determination of electronic configuration in an atom and the subsequent complex spectra. In its simplest form the Pauli's exclusion principle states that no two electrons in the same atom can have all their quantum numbers the same.

In order to obtain states originating from equivalent electrons following Pauli's Exclusion principle the following method can be adopted. As an example consider two equivalent p electrons.

We start with all possible combinations of m_l and m_s for a single p electron

m_l =	1	0	−1	1	0	−1
m_s =	1/2	1/2	1/2	−1/2	−1/2	−1/2
	a	b	c	d	e	f

We can choose all possible values of the combinations *ab, bc*, etc. which will follow Pauli's exclusion principle.

In Table 11.3 out of all combinations, the combinations in triangle are the only one possible according to Pauli's principle.

Table 11.3   The combinations of m_l and m_s for a single *p* electron in the triangle

I	aa	ab	ac	ad	ae	af
II	ba	bb	bc	bd	be	bf
III	ca	cb	cc	cd	ce	cf
IV	da	db	dc	dd	de	df
V	ea	eb	ec	ed	ee	ef
VI	fa	fb	fc	fd	fe	ff

Considering the 15 combinations in the triangle we make combinations of the strong field m_l's and m_s's to obtain strong field values of M_L and M_S.

Thus

M_L =	ab	ac	ad	ae	af	bc	bd	be	bf	cd	ce	cf	de	df	ef
	1	0	2	1	0	-1	1	0	-1	0	-1	-2	1	0	-1
M_S =	1	1	0	0	0	1	0	0	0	0	0	0	-1	-1	-1

We can group them in the following manner.

$$M_L = 2 \quad 1 \quad 0 \quad -1 \quad -2;\ 2 \quad 1 \quad 0 \quad -1 \quad -2;\ 2 \quad 1 \quad 0 \quad -1 \quad -2 \quad L=2$$
$$M_S = 1 \quad 1 \quad 1 \quad 1 \quad 1;\ 0 \quad 0 \quad 0 \quad 0 \quad 0;\ -1 \quad -1 \quad -1 \quad -1 \quad -1 \quad S+1$$
$$M_L = 2 \quad 1 \quad 0 \quad -1 \quad -2;\ L=2$$
$$M_S = 0 \quad 0 \quad 0 \quad 0 \quad 0 \quad S=0$$
$$M_L = 1 \quad 0 \quad -1;\ 1 \quad 0 \quad -1;\ 1 \quad 0 \quad -1 \quad L=1$$
$$M_S = 1 \quad 1 \quad 1;\ 0 \quad 0 \quad 0 \quad -1 \quad -1 \quad -1 \quad S=1$$
$$M_L = 1 \quad 0 \quad -1 \quad L=1$$
$$M_S = 0 \quad 0 \quad 0 \quad S=0$$
$$M_L = 0 \quad 0 \quad 0 \quad L=0$$
$$M_S = 1 \quad 0 \quad -1 \quad S=1$$
$$M_L = 0 \quad L=0$$
$$M_S = 0 \quad S=0$$

These lead to ^3D, ^1D, ^3P, ^1P, ^3S and ^1S and are possible if the Pauli's principle is violated. However when we consider the equivalent electrons and the Pauli's principle only the 15 combinations as in the triangle of Table 11.3 are possible.

Thus we will have

I	M_L	=	1	0	2	1	0
	M_S	=	1	1	0	0	0
II	M_L	=	−1	1	0	−1	
	M_S	=	1	0	0	0	
III	M_L	=	0	−1	−2		
	M_S	=	0	0	0		
IV	M_L	=	1	0			
	M_S	=	−1	−1			
V	M_L	=	−1				
	M_S	=	−1				

These are similar to the groups

$$M_L = 1 \quad 0 \quad -1; \quad 1 \quad 0 \quad -1; \quad 1 \quad 0 \quad -1 \qquad L = 1$$

$$M_S = 1 \quad 1 \quad 1; \quad 0 \quad 0 \quad 0 \quad -1 \quad -1 \quad -1 \qquad S = 1$$

$$M_L = 2 \quad 1 \quad 0 \quad -1 \quad -2 \qquad L = 2$$

$$M_S = 0 \quad 0 \quad 0 \quad 0 \quad 0 \qquad S = 0$$

$$M_L = 0 \qquad L = 0$$

$$M_S = 0 \qquad S = 0$$

These give rise to ^3P, ^1D and ^1S and excludes ^3D, ^3S and ^1S.

Breit's Scheme for equivalent electrons

m_{l1}	1	0	−1	
2		1	0	1
1		0	−1	0
0	−1	−2	−1
	S	P	D	m_{l2}

m_{s1}	1/2	−1/2	
M_s	1	0	1
	0	−1	0
	sing	trip	m_{s2}

If the two p electrons are equivalent, M_L values on the diagonal cannot be taken with M_S values also lying on the diagonal as this will mean that $m_{l1} = m_{l2}$ and $m_{s1} = m_{s2}$ and we permit m_s values alike (i.e., $M_S = 1$ or -1), the values $M_L = 2, 0, -2$ crossed out by the diagonal line is forbidden. Since the values in the lower left array are identical with the upper right half, one of these groups is eliminated. Leaving out the left array the remaining values are

$$M_S = 1 \qquad M_L = 1 \quad 0 \quad -1$$

The case of two equivalent d electrons are already described in Chapter 9.

11.3.1 Breit's Scheme for More than two Equivalent Electrons

The Breit's scheme is well suited for two equivalent electrons. When the number of equivalent electrons are more than two the Breits scheme can still be used. Thus if we have 3 equivalent p electrons we have to consider all combinations of a,b,c,d,e, and f in such a way that no symbol appears twice in any selected combination. In the case of p^3 electrons this is accomplished in $^6C_3 = 20$ ways. We can write down the combinations of M_L and M_s and obtain the states. Spectroscopic terms for some equivalent electrons are given in Table 11.4.

Table 11.4   Spectroscopic terms for some equivalent electrons

Electron configuration	Number of combinations	Spectroscopic terms
ns^2	1	^1S
np, np^5	6	^2P
np^2, np^4	15	$^1S, {}^1D, {}^3P$
np^3	20	$^2P, {}^2D, {}^4S$
nd, nd^9	10	^2D
nd^2, nd^8	45	$^1S, {}^1D, {}^1G, {}^3P, {}^3F$
nd^3, nd^7	120	$^2P, {}^2D, {}^2F, {}^2G, {}^2H, {}^4P, {}^4F$
nd^4, nd^6	210	$^1S, {}^1D, {}^1G, {}^3F, {}^1I, {}^3P, {}^3D, {}^3F, {}^3G, {}^3H, {}^5D$
nd^5	252	$^2S, {}^2P, {}^2D, {}^2F, {}^2G, {}^2H, {}^2I, {}^4P, {}^4D, {}^4F, {}^4G, {}^6S$
nf, nf^{13}	14	^2F

11.3.2 Isoelectronic Sequences

Isoelectronic sequences is a term used to describe atoms having the same number of electrons. If we consider an atom in the periodic table having n electrons, it is followed by an element with electrons $(n + 1)$. If we knock an electron out from this atom, i.e., if the atom is ionized, it becomes an element with n number of electrons, only difference being in the nuclear charge Ze. Consider the element sodium with $Z = 11$. The following elements Mg, Al, Si, P having $Z = 12, 13, 14, 15$ can all be made isoelectronic with sodium by removing one electron from Mg, two electrons from Al, three from Si and four from P. The neutral atom sodium is called Na I, singly ionized Mg is called Mg II, doubly ionized Al is called Al III, triply ionized Si is called Si IV and so on. Paschen and Fowler in 1923 noticed similarities and identified the sodium-like doublets in Mg II, Al III and in Si IV. If we consider potassium with $Z = 19$, the following element calcium having $Z = 20$ can be made isoelectronic by removing one electron from it and is called singly ionized calcium or Ca II, the next element scandium with $Z = 21$ can be made isoelectronic with potassium by removing two electrons thus by doubly ionizing the element called Sc III and the next element titanium can be made isoelectronic by removing three electrons and is named Ti IV. We have already seen that He^+ is isoelectronic with hydrogen atom so also Li^{++}. Because each atom in such a series has the same number of electrons vis-à-vis same electron configuration, the energy levels and the spectra in these sequences will be similar, only difference being in the magnitude of the energy values originating from Z. The empirical rules derived are closely associated with the well known X-ray doublet law.

ENDNOTES

1. Kossel, W. and Sommerfeld, A. (1919). *Auswahl Prinzip and Verchiebungssatz bei Serienspektren, Verh.d. Deutsch. Phys. Gesselscheft.* 21, 240.

2. Hund, F., (1927). "Linienspektren and periodisches system der Element, (1927). Springer Verlag, Berlin," p. 221.

REFERENCES

Condon, U and Shortley, G.H. (1935, 1999). *The Theory of Atomic Spectra.* Cambridge University Press, Cambridge, UK.

Foot, C.J. (2005). *Atomic Physics.* Oxford University Press, Oxford, UK.

Herzberg, G. (1945). *Atomic Spectra and Atomic Structure.* Dover, New York.

Rai, D.K. and Thakur, S.N. (2005). *Atomic Spectra, Structure and Modern Spectroscopy.* Vaivaswat Publication, Varanasi.

White H.E., (1934). *Introduction to Atomic Spectra.* McGraw Hill, New York.

12

CHARACTERISTICS OF BROADENING OF SPECTRAL LINES

12.1 BROADENING OF SPECTRAL LINES

In any kind of transitions involving atomic or molecular energy levels or indeed in any kind of spectroscopy, we should be aware that the spectral lines occurring from transitions from one energy state to the other are not precisely "sharp". There is always a finite width to the observed spectral lines.

Thus it is known from many experimental results that the spectral lines from atomic or molecular vapours show an observable wavelength spread around the line centre. It is many times independent of the experimental system used to observe the spectrum. This is known as width of spectral lines or broadening of spectral lines. As we have seen earlier in the case of alkali spectra and in many other cases there can be sharp and diffuse lines in the spectra obtained from the same atomic source. Sharp lines are those where frequency spread is small and diffuse lines are those where frequency spread is large and is spread such that the line looks like diffuse. Thus narrow and broad lines can be simultaneously observed even from the same source. There are many reasons for the breadth of spectral lines. These originate from instrumental or from the conditions prevailing in the source itself. The study of the shape and width of spectral lines is a new interesting field. It gives information not only on the origin of the spectral lines but gives also valuable information regarding the environment in which the atoms are placed. It can also give information on many characteristics of the atoms themselves. We can for the time being disregard the breadth of spectral lines due to instrumental technique as they can be considered mainly man made. We shall here consider the broadening of lines due to conditions prevailing in the source itself. The known causes can be classified mainly due to the following reasons.

1. Natural line width
2. Doppler effect
3. External effects

    i. Collision broadening

    ii. Asymmetry and pressure shift

    iii. Stark effect width

Consider spectral lines as shown in Figure 12.1.

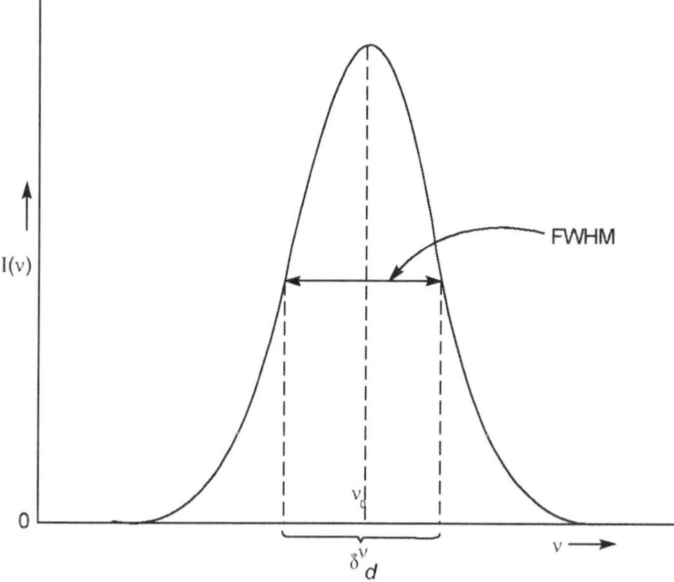

Figure 12.1   Intensity–frequency contour for the spectrum line

    It shows the variation of intensity of the spectral line as a function of frequency. Let us consider the intensity I_0 at the centre of the spectral lines corresponds to the frequency v_0 and the intensity drops out on either side of v_0. Let us draw a line from the point of the maximum intensity vertically down to the frequency axis and find the point where the intensity is half. There are two points on the line when the intensity is $I_0/2$ and we can draw a line parallel to the frequency axis which passes through these points. That means we can consider two points along the line on both sides of v_0 at which intensity can be considered as exactly one half of the maximum intensity. It need not be true always that these two points are always symmetrically situated. For the time being we can consider them as symmetrically situated on both sides of v_0. The separation between these two points is often referred to as the width of the line. It is the interval between the two points where the intensity drops to half its maximum value. It is also sometimes called the half width or the width of the line at half intensity. Or this is also called frequency width at half maximum (FWHM).

12.2 NATURAL LINE BROADENING

Natural line broadening is generally due to the finite life time of the atoms in the excited state and the consequent uncertainty in the energy values of the states involved in the transition. A typical lifetime for an atomic energy state is about 10^{-8} seconds, corresponding to a natural linewidth of about 6.6×10^{-8} eV.

Atoms radiate when energy is supplied to it from an external source. Otherwise even if the atom continues to radiate at the frequency $\omega_0 = 2\pi\nu_0$, the observed wave train of light is not monochromatic. Suppose for example a disturbance $f(t)$ oscillates at a fixed frequency ω_0, but the amplitude of $f(t)$ fall exponentially with time such that

$$f(t) = 0 \quad \text{for } t < 0 \qquad \text{and}$$
$$f(t) = e^{i\omega_0 t} e^{-\gamma t} \sqrt{2\gamma} \quad \text{for } t \geq 0$$

so that

$$\int_{\infty}^{\infty} |f(t)|^2 \, dt = 2\gamma \int_0^{\infty} e^{-2\gamma t} \, dt = 1 \tag{12.1}$$

The Fourier component of $f(t)$ is

$$F(\omega) = \left(\frac{\gamma}{2\pi}\right)^{\frac{1}{2}} \int_0^{\infty} e^{-i(\omega-\omega_0)t} e^{-\gamma t} \, dt = \sqrt{\frac{\gamma}{\pi}} \, \frac{(-i)}{(\omega_0 - \omega) + i\gamma} \tag{12.2}$$

$$I(\omega) = |F(\omega)|^2 = \frac{\gamma}{\pi} \, \frac{1}{(\omega_0 - \omega)^2 + \gamma^2} \tag{12.3}$$

The general shape of a distribution as given in the above equation is similar to the Figure 12.1.

The intensity will drop to its half value when the two terms in the denominator are equal.

$$(\omega_0 - \omega)^2 = \gamma^2$$
$$(2\pi\nu_0 - 2\pi\nu)^2 = \gamma^2$$
$$4\pi^2 (\nu_0 - \nu)^2 = \gamma^2$$
$$(\nu_0 - \nu)^2 = (1/4\pi^2)\gamma^2$$
$$\nu_0 - \nu = (1/2\pi)\gamma = \gamma / 2\pi \tag{12.4}$$

According to classical electromagnetic theory a vibrating electric charge is continuously damped by the radiation of energy following the expression

$$E = E_0 \, e^{-\gamma t} \tag{12.5}$$

And the amplitude $A = A_0 \, e^{-(\gamma/2)t}$ $\tag{12.6}$

where γ is given by $\gamma = \dfrac{2}{3}\dfrac{e^2}{mc^3}\omega_0^2 = \dfrac{8\pi^2 e^2 v_0^2}{3mc^3}$ $\tag{12.7}$

Therefore the natural half intensity is $\Delta v = \gamma / 2\pi = \dfrac{4\pi e^2 v_0^2}{3mc^3}$ $\tag{12.8}$

The half-intensity breadth in terms of wavelength is $\Delta\lambda = \dfrac{2\pi c}{\omega_0^2}\gamma = \dfrac{4\pi e^2}{3mc^2} = 1.16 \times 10^{-12}$ cm $\tag{12.9}$

$$\lambda = \frac{c}{v_0}, \Delta\lambda = \frac{c}{v_0^2}\Delta v = c.\frac{4\pi^2}{\omega_0^2}\Delta v = c.\frac{4\pi^2}{\omega_0^2}\frac{\gamma}{2\pi} = \frac{2\pi c}{\omega_0^2}\gamma$$

We can also derive the natural breadth of a spectral line by quantum mechanical means. According to quantum mechanics the energy level of an atom is to be regarded as spread following the uncertainty principle

$$\Delta E \cdot \Delta t = h / 2\pi \tag{12.10}$$

If we consider Δt as the mean life τ of the state, then $\Delta E \cdot \tau = h / 2\pi$

From the above expressions, we could see that if the life time is large ΔE is small and hence the spread or uncertainty in the energy level is small and thereby the line will be sharper (See Figure 12.2). It means for transitions from metastable states, the spectral lines will be sharper. The quantum mechanical derivation also leads to the same order of magnitude for the breadth of spectral lines though they are different for different levels. The frequency distribution $I(G)$ of an energy state of an atom on the basis of quantum mechanics may be written as

$$I(G) = \frac{\gamma_n}{2\pi}\left[\frac{1}{4\pi^2\left[(G_n - G)^2 + \left(\dfrac{\gamma_n}{2}\right)^2\right]}\right] \tag{12.11}$$

where G corresponds to the term values.

$$\gamma_n = 3\sum_{m<n} f_{nm}\left(\frac{g_m}{g_n}\right)\left(\frac{8\pi^2 e^2}{3mc^3}\right)\left[\frac{(G_n - G_m)^2}{mc^2}\right] \tag{12.12}$$

where f_m is the oscillator strength for the transition $n \rightarrow m$, g_n and g_m are the weight factors or quantum weights (degeneracies) which is $(2J+1)$ of the levels with quantum number J and the summation extends over all terms with $G_m < G_n$.

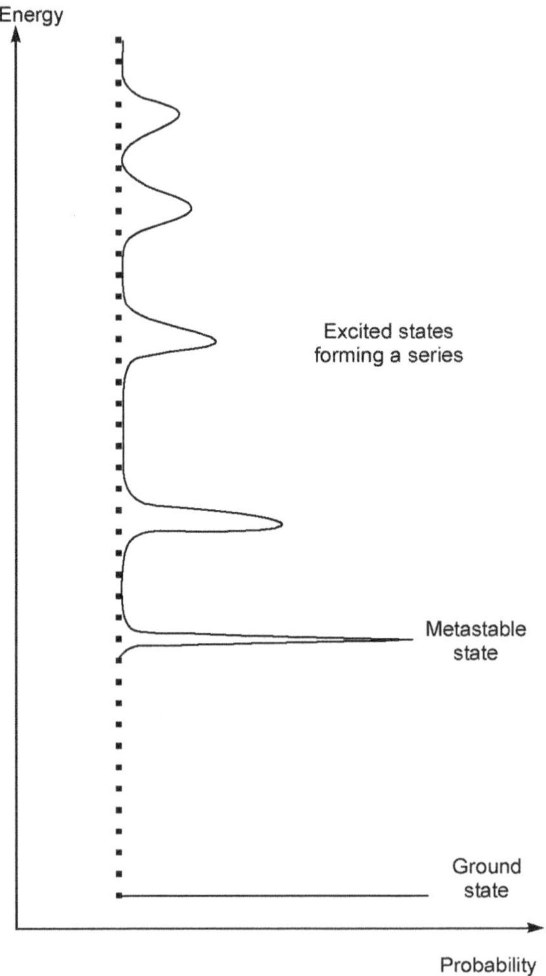

Figure12.2 Width of atomic energy levels

With the above value of γ_n the half-intensity breadth is given by

$$\Delta v = \gamma_n / 2\pi = 3\frac{1}{2\pi}\sum f_{nm}\left(\frac{g_m}{g_n}\right)\left(\frac{8\pi^2 e^2}{3mc^3}\right)\left[\frac{(G_n - G_m)^2}{mc^2}\right]$$

$$= 3\sum f_{nm}\left(\frac{g_m}{g_n}\right)\left(\frac{4\pi e^2}{3mc^3}\right)\left[\frac{(G_n - G_m)^2}{mc^2}\right] = \frac{1}{2\pi\tau} \tag{12.13}$$

F London[1] has shown that the sum of the oscillator strength associated with any state for all transitions into and out of a given state is one, i.e., $\Sigma f = 1$. If $\gamma_n / 2\pi$ and $\gamma_m / 2\pi$ are the respective half-intensity breadths of the two states, the intensity distribution of the observed spectral line can be shown to be

$$I_{nm}(\nu) = \frac{(\gamma_n + \gamma_m)}{2\pi}\left[\frac{1}{4\pi^2(\nu_{nm} - \nu)^2 + \left(\dfrac{\gamma_n + \gamma_m}{2}\right)^2}\right] \qquad (12.14)$$

where ν_{nm} refers to the frequency of the centre of the line.

The intensity drops to half its value when

$$4\pi^2(\nu_{nm} - \nu)^2 = \left(\frac{\gamma_n + \gamma_m}{2}\right)^2 \qquad (12.15)$$

Line width at half its maximum intensity is then

$$(\nu_{nm} - \nu) = \frac{\gamma_n + \gamma_m}{4\pi} \qquad (12.16)$$

As ν_{nm} refers to the frequency of the centre of the line the width of the line at half intensity is $2 \times \dfrac{\gamma_n + \gamma_m}{4\pi}$.

Thus line width is $\Delta\nu = \dfrac{\gamma_n + \gamma_m}{2\pi} = \delta_1 + \delta_2 \qquad (12.17)$

Thus it is the sum of the half-intensity breadths of the initial and final states (*See* Figure 12.3). Spectral lines with larger transition probability are relatively broader than that of smaller transition probabilities. If the transition is to the ground state the width of that level is negligible due to large life time and hence the width mainly depends on the width of the upper level.

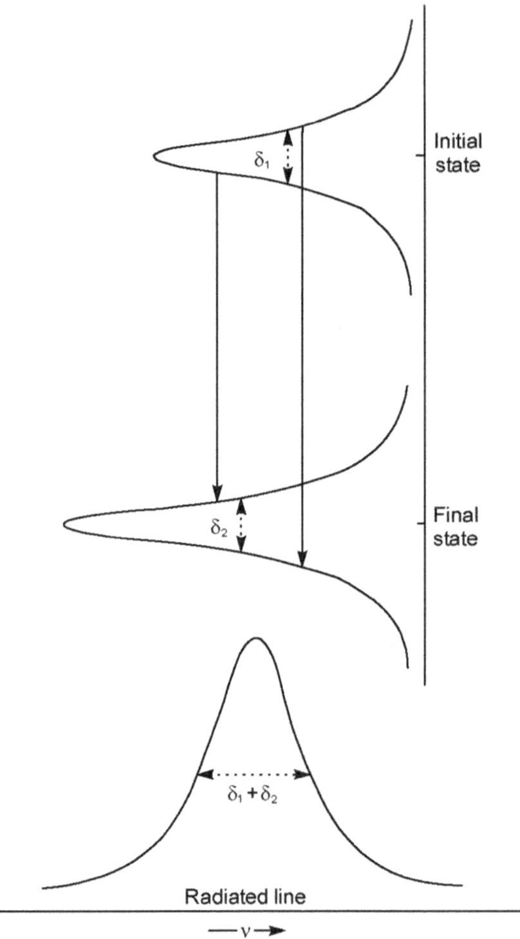

Figure 12.3   Relation between the natural breadth of a spectrum line and the natural breadths of the corresponding energy levels

12.3 Doppler Line Width

The Doppler line width originates from the distribution of frequencies occurring from the random motion of atoms as in the case in kinetic theory of gases. The frequency change originating from the relative motions of the source and observer is known for a long time, which is known as the Doppler shift. Thus there is a change in the observed wavelength of the spectral line when the velocities of the stars and their emitting gases with respect to the observer are high, sometimes large shifts, due to this effect. It is best seen in large spectral shifts of radiation from stars and other planetary objects whose velocities are large with respect to the motion of earth. However, when the shift is small it ends up with a broadening of the line. Thus the random motions of atoms or molecules produce a net broadening rather than any apparent shift. One could expect that the broadening (1) increases with temperature and (2) decreases with increasing atomic or molecular weight.

Since the thermal velocities are non-relativistic, the Doppler shift in the angular frequency is given by the simple form

$$\omega = \omega_0 \left(1 \pm \frac{u}{c}\right)$$

where ω_0 is frequency of an atom at rest

$$\frac{\omega - \omega_0}{\omega_0} = \frac{u}{c} \tag{12.18}$$

If the emitting atom is moving with a velocity v along a direction making an angle θ with the direction of observation, the change in frequency is given by the classical expression

$$\frac{\Delta v}{v_0} = \frac{v - v_0}{v_0} = \frac{v\cos\theta}{c} = \frac{u}{c} \tag{12.19}$$

where v_0 is the frequency of the line when $v = 0$ and v is the observed frequency. $u = v\cos\theta$ is the component of the velocity in the direction of observation and c is the velocity of light.

From the Boltzmann distribution, the number of atoms with velocity v in the direction of the observed light is given by

$$n(v)dv = N\sqrt{\frac{m_0}{2\pi kT}}e^{-m_0 v^2/2kT}\,dv$$

where, N = total number of atoms

$$m_0 = \text{atomic mass} \tag{12.20}$$

The distribution of radiation around the centre frequency is then given by

$$I(\omega) = I_0 \exp\left[\frac{-m_0 c^2(\omega_0 - \omega)^2}{2kT\omega_0^2}\right] \tag{12.21}$$

If we assume a Maxwellian distribution of velocities, the probability of velocity lying between u and $u + du$ is

$$d\omega = \sqrt{\frac{M}{2kT}}\exp\left[-\frac{M}{2kT}u^2\right]du \tag{12.22}$$

where M is the mass of the atom, k is the Boltzmann constant and T is the absolute temperature.

Substituting the value of u from equation (12.19), we get

$$I(\nu) = I_0 \exp\left[-\frac{M}{2kT}\frac{c^2}{\nu_0^2}(\nu - \nu_0)^2\right] \qquad (12.23)$$

where I_0 is the constant of proportionality.

The maximum intensity I_0 is at $\nu = \nu_0$. To find the two points at the half-intensity breadth, the exponential term in equation 12.23 is set to one half. Or we can find out $I_0/2$.

$$I(\nu) = I_0 \exp\left[-\frac{M}{2kT}\frac{c^2}{\nu_0^2}(\nu - \nu_0)^2\right] \qquad (12.24)$$

$$\frac{I_0}{2} = I_0 \exp\left[-\frac{M}{2kT}\frac{c^2}{\nu_0^2}(\nu - \nu_0)^2\right] \qquad (12.25)$$

$$\frac{1}{2} = \exp\left[-\frac{M}{2kT}\frac{c^2}{\nu_0^2}(\nu - \nu_0)^2\right] \qquad (12.26)$$

$$-\log_e 2 = \left[-\frac{M}{2kT}\frac{c^2}{\nu_0^2}(\nu - \nu_0)^2\right] \qquad (12.27)$$

$$\frac{\nu - \nu_0}{\nu_0}c = \pm\sqrt{\frac{\log_e 2}{M/2kT}} = \pm\sqrt{\frac{2kT}{M}\log_e 2} \qquad (12.28)$$

The full width at half intensity maximum is given by

$$\mathrm{FWHM} = \frac{2\nu_0}{c}\sqrt{\frac{2kT}{M}\log_e 2} = 1.67\frac{\nu_0}{c}\sqrt{\frac{2kT}{M}} = \frac{\nu_0}{c}\sqrt{\frac{8kT}{Mc^2}\log 2}$$

$$= 2.92 \times 10^{-20}\,\nu_0\left(\frac{T}{M}\right)^{\frac{1}{2}} \qquad (12.29)$$

$\Delta\nu/\nu = \Delta\lambda/\lambda$ and hence the half-intensity breadth in terms of wavelength is

$$\Delta\lambda = \frac{2\lambda_0}{c}\sqrt{\frac{2kT}{M}\log_e 2} = 1.67\frac{\lambda_0}{c}\sqrt{\frac{2kT}{M}} \qquad (12.30)$$

This can also be written as

$$\frac{\Delta\lambda}{\lambda_0} = 2\sqrt{2\ln 2\,\frac{2kT}{Mc^2}} \tag{12.31}$$

$$\Delta\omega_{\text{Doppler}} = \frac{2\omega_0}{c}\sqrt{2\ln 2\,\frac{kT}{M_0}} \tag{12.32}$$

Thus the Doppler broadening is (1) proportional to the square root of the temperature (2) proportional to the frequency of transition (3) inversely proportional to the square root of the atomic or molecular weight.

For atomic spectra in the visible and UV, the limit on resolution is often set by Doppler broadening.

It can be seen that the Doppler width is larger in optical regions than in microwave or radio frequency regions.

At ordinary temperatures the natural width is comparatively less than the Doppler broadening.

A consideration of a second order Doppler broadening is possible in terms of relativistic consideration, the magnitude of which is very less, generally 10^{-4} to 10^{-5} order less than that of first order Doppler broadening. The relativistic effect can be included by considering the relative motion between the atomic source and the observer. The well known "time dilation" theory of special theory of relativity is considered and it is a consequence of moving from a frame of reference in which the atom is at rest to one with respect to the observer at rest. This effect has also been measured in the laboratory.

12.4 COLLISIONAL WIDTH

While natural line width depends on the atom, the Doppler and collisional broadening depend on the environment of the atomic source. Thus the collisional broadening is due to external cause and is called by different names like collision damping, collisional broadening or pressure broadening. For a gas at a given pressure, the radiating atom interacts with its neighbours via collision and this affects the emission line width considerably. Sometimes it can even lead to line shifts. Thus it is the effect produced by the collision of two atoms, one in the process of emitting or absorbing radiation. The perturbation caused by collisions depends on the nature of the colliding particles, period of collision, the period between the collisions and also on the density of the gases. The density depends on the pressure and hence it is sometimes termed as pressure broadening. It was Michelson (1895)[2] and Lorentz (1905)[3] who suggested the phenomena based on the assumption that if, during the time an atom is emitting or absorbing radiation of frequency ν_0 it collides elastically with another atom, the phase and amplitude of the radiation have a chance of undergoing a considerable change.

We shall here present a simple classical picture of collision broadening. Let us consider the pressure of the gas is low so that the density of the atoms is low. Thus the time in which the atom is uninfluenced by it's neighbour is considerably longer compared to the time in which it is under the influence of the latter. That means the collisional time is smaller than that of the mean time between two collisions. We also assume that with every impact the oscillations are momentarily interrupted only to resume the same frequency again with possibly a new phase and amplitude (*See* Figure 12.4). We consider the radiation to be an undamped oscillation which can be represented by the form

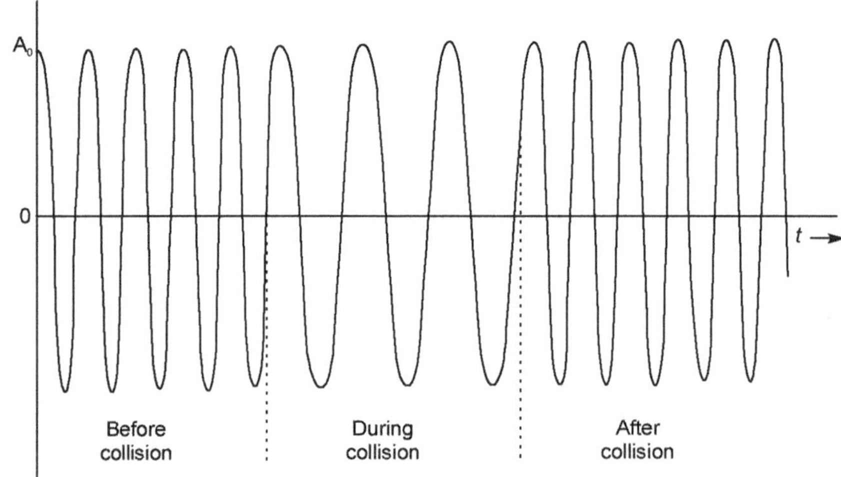

Figure 12.4 Schematic diagram of the change in frequency at the time of collision between two atoms

$$x = A_0 \cos(\omega_0 t + \phi) \quad \text{where} \quad \omega_0 = 2\pi v_0 \tag{12.33}$$

We assume that the oscillation is suddenly stopped after an interval of time τ. Thus τ is the time between the beginning of the radiation and the occurrence of collision. For each atom τ will be different and a probability distribution of which τ will lie between τ and $\tau + d\tau$ is

$$dw = \frac{1}{\tau_0} \exp\left(-\frac{\tau}{\tau_0}\right) d\tau \tag{12.34}$$

where τ_0 is the mean flight time between two collisions.

A Fourier analysis of the set of wave trains will give the intensity as a function of frequency as

$$I(v) = I_0 \left[\frac{1}{(\omega_0 - \omega)^2 + \left(\dfrac{1}{\tau_0}\right)^2} \right] = I_0 \left[\frac{1}{4\pi^2(v_0 - v)^2 + \left(\dfrac{1}{\tau_0}\right)^2} \right] \tag{12.35}$$

The intensity drops to its half value when

$$\omega_0 - \omega = \frac{1}{\tau_0} = 2\pi\,(\nu_0 - \nu) \tag{12.36}$$

Full width at half-intensity maximum is then $\text{FWHM} = \dfrac{1}{\pi\tau_0}$ (12.37)

From kinetic theory of gases the average velocity of an atom is given by

$$v_0 = \sqrt{\frac{8kT}{\pi M}} \tag{12.38}$$

The mean free path $l_0 = v_0 \tau_0$.

And from the Maxwellian distribution of velocities, the mean free path can be written as

$$l_0 = \frac{kT}{\sqrt{2\pi d^2 P}} \tag{12.39}$$

where d is the collision diameter which is the effective separation between the two atomic centres at the closest approach. P is the pressure. From the expressions of v_0, l_0 and τ_0 the half intensity breadth is obtained as

$$\text{FWHM} = \frac{4d^2 P}{\sqrt{\pi M kT}} \tag{12.40}$$

A comparison of Doppler half-intensity breadths with collision half-intensity breadths shows that at room temperature and normal atmospheric pressure the two are of the same order for visible light. In the microwave region the Doppler width is comparatively lower than the collisional width.

12.5 ASYMMETRY AND PRESSURE SHIFT

In the earlier example of collision broadening we have assumed that the collision time is small compared to the mean time between collisions. This, however, is not true at high gas pressures. In such conditions the interacting atoms interact each other for a longer period than compared to the time they are free. Also since collision damping takes into account only those portions of the radiation before the collision and after collision, we have to consider the radiations that take place during collision (*See* figure 12.4). The frequency of oscillation changes suddenly at the time of collision and regains it after collision. However at the centre of collision the situation is different. Asymmetry and pressure shift are possible at the centre. Jablonski had proposed a qualitative explanation of the line broadening in this case[4]. When a foreign atom approaches the emitting atom, the energy levels of the radiating atom are chiefly altered due to polarization effects. Due to this polarization the excited state of the emitting atom would be lowered in energy by a larger amount than the ground state at the distance of the closest approach. The net effect of the transition would be a shift towards lower frequency or a red shift. The energy levels can be well represented by the potential curves for a diatomic molecule as shown in figure 12.5. The change in energy is given by $\Delta E = -C / r^6$. C is given by London's formula

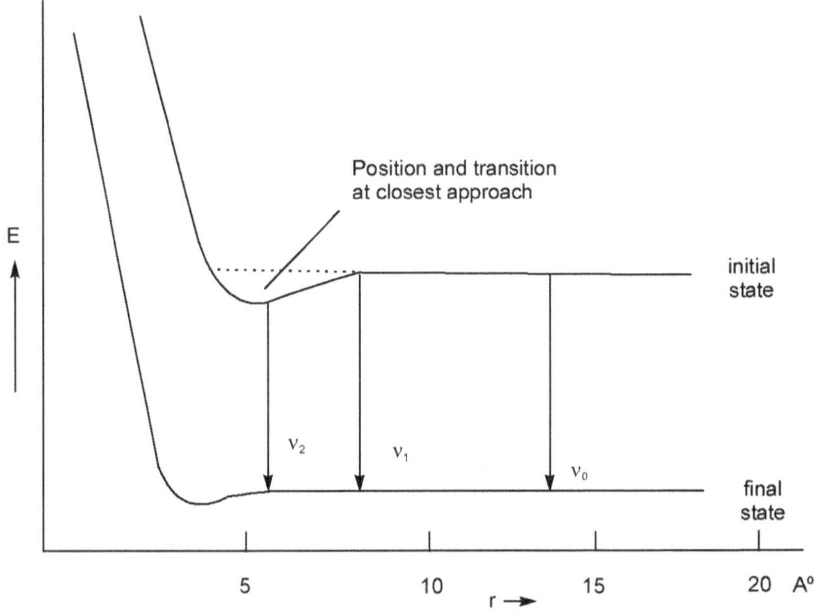

Figure 12.5   Change in energy of atomic levels due to close approach of two atoms at high pressure

$$C = \frac{3}{m^2}\left(\frac{eh}{2\pi}\right)\sum_{n'k'} \frac{f_{nn'}f_{kk'}}{(E_{n'} - E_n)(F_{k'} - F_k)(E_{n'} + F_{k'} - E_n - F_k)} \tag{12.41}$$

where E and F stand for the energy levels of emitting atom and the colliding atom, $f_{nn'}$ and $f_{kk'}$ are f values for the transitions $E_n \to E_{n'}$ and $F_k \to F_k'$ respectively. The collision diameter is of the form $d \approx (C / v)^{1/5}$, v being the mean velocity.

At normal temperatures and pressure, the collision times are less and the broadening and shift are less. However, with increasing pressure, the mean collision time increases and the time between two collisions decreases and as a result the line is shifted to the red and is asymmetrically broadened. As can be seen from the potential curve in the Figure 12.6, the transition frequency at the time the atoms are at the closest approach than at larger distances or when the atoms are separated. The result is that the line is spread towards the smaller frequency side (to the longer wavelength) than it is on the other side.

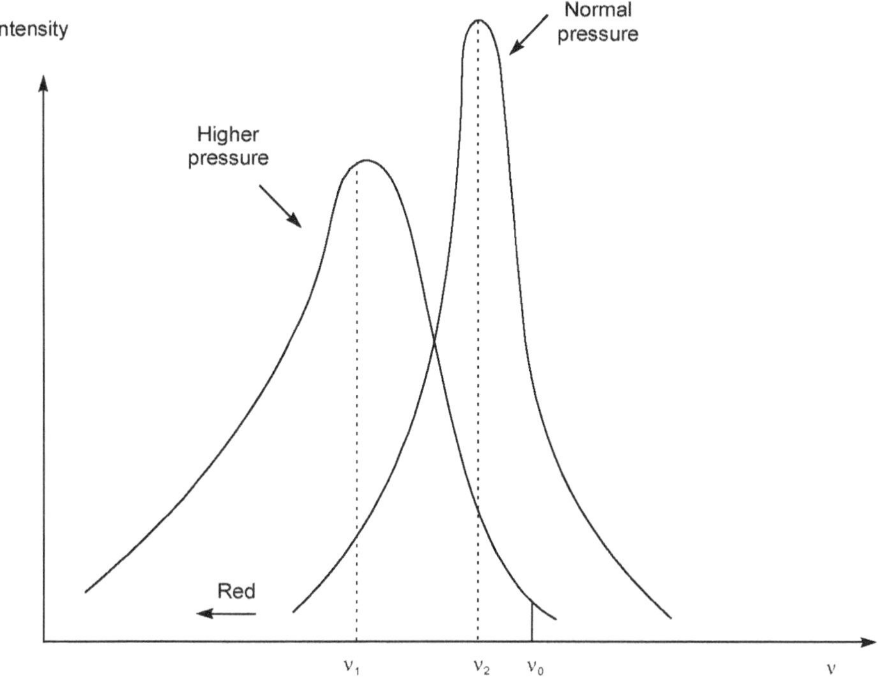

Figure 12.6   Effect of pressure on atomic line shape. v_0 is line centre without pressure broadening

12.6 STARK BROADENING

In an arc source many ions are produced which with collision with other atoms give rise to an electric field. In addition to it the dipole and quadrupole moments of the gas atoms or molecules can also produce strong interparticle electric fields. These kind of electric fields produce Stark effect and thereby a broadening of lines. The broadening is many times comparable to the collision broadening or Doppler broadening. The calculation of this type of Stark effect broadening is theoretically difficult, as one would be confronted with the problem of continually changing inhomogeneous electric fields. The electric field is large at the closest approach and decreases as the atoms get apart and hence the exact calculation of the field is difficult. However, the following expressions are derived due to the stark field originating from the three types of interactions mentioned above.

1. Average field due to the ions $F = a_1 \, e \, n^{2/3}$

2. Average field due to dipole moments $F = a_2 \, \mu \, n$

3. Average field due to quadrupole moments $\quad F = a_3 \, q \, n^{4/3}$ $\hfill$ (12.43)

where a's are constants, e is the ionic charge, μ the dipole moment and n is the number of corresponding particles per cubic centimetre. The total spread of the line is determined by the outermost component due to the Stark effect. The total spread of the lines is

$$\mathrm{FWHM} = A_{max} \qquad\qquad F(12.43)$$

where A_{max} depends on the magnetic quantum numbers (electric field quantum numbers) of the outer levels of the two combining states and F represents the average electric field. For the three types of electric fields mentioned above the line widths are given by

For electric field due to ions $\Delta v_i = 3.25 \, A_{max} \; e \; n^{2/3}$

For electric field due to dipoles $\Delta v_d = 4.54 \, A_{max} \; \mu \; n$

For electric field due to quadrupoles $\Delta v_q = 5.52 \, A_{max} \; q \; n^{4/3}$ $\qquad\qquad$ (12.44)

Using a high-voltage condensed spark discharge, Finkelnberg[5] studied the Stark broadening in the spectrum of atomic hydrogen at various pressures. H_α, H_β and H_γ were found to be symmetrically broadened, the enormous breadth were attributed to the Stark effect form the enormous Stark field originating from the high ion density in the spark discharge. Another example of Stark broadening in a diatomic molecule TlF is shown in Figure 12.7. where an external electric field is applied in order to determine the electric dipole moment in the molecule[6]. In this case the broadening is not from any internal magnetic or electric field, but from the non-homogeneity of an applied external electric field.

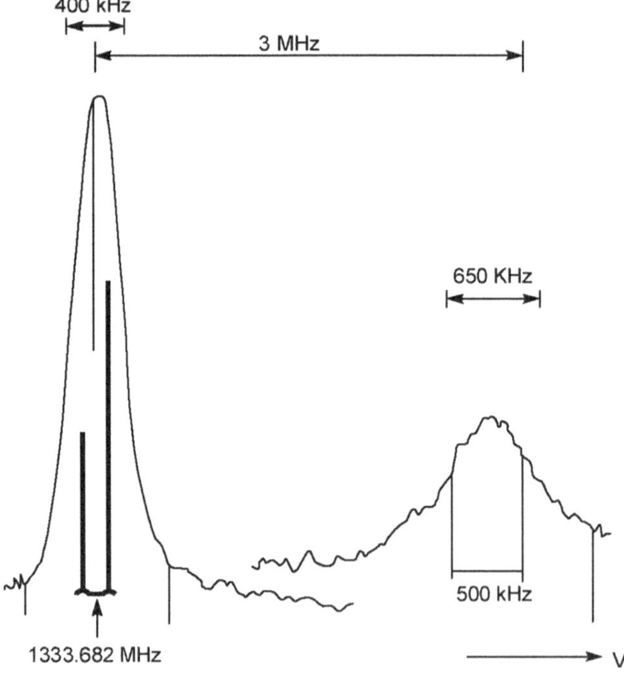

Figure 12.7   Stark effect in J = 1 – 0 rotational transition of ^{205}Tl ^{19}F molecule. On the left:-field-free transition. Thick straight lines show hyperfine components due to Thallium nucleus (I = 1/2) which are not resolved. On the right: Stark-shifted line. Field 180 V/cm, Temperature 450°C, Averaging time 4 minutes

12.7 RESONANCE WIDTH

Broadening of lines can also arise from an interaction arising between an atom in the excited state and an atom in the ground state. In this case there are two types of interactions.

1.   When the atom in the excited state emit a photon to transfer to the ground state, the atom in the ground state absorbs this photon and tries to be excited to the upper state. This is called self-absorption. This is easily visualized in gas discharges where the radiation has to pass through the plasma and the atoms in the lower state are absorbed. The absorption is near the centre of the line profile, which produces a modified line profile and an apparent line broadening.

2.   The atom in the excited state may be transferred to the ground state without a radiative transfer, simply by energy exchange. We can consider the two atoms interact via their dipoles and the average change suffered is

$$\left|\Delta \nu\right| = \frac{e^2}{16\pi^2\, m\nu_0}\, f\,\frac{1}{r^3}(e^2/16\pi^2\, m\nu_0) f\, 1/r^3 \tag{12.45}$$

where r is the separation between the two atoms, m is the mass of the electron, e its charge, ν_0 is the Bohr frequency of the transition and f is the oscillator strength. The full width at half maximum is calculated to be

$$\text{FWHM} - \frac{e^2}{4\pi m\nu_0}\, f\, N \quad \text{where } N \text{ is the density of the emitting atoms.}$$

ENDNOTES

1.   London, F. *Z.f.Phy*.39,526,1931.

2.   Michelson, A.A., *Astrophys.* J. 2,251 (1895)

3.   Lorentz, H.A. (1904).*Theory of electrons*.Akad Wetenschap. Amsterdam 13,346

4.   Jablonski, A. *Z.f.Phyik*, 70,723,1931)

5.   Finkelnburg W, (1931). *Z. f. Phys.* 70, 375

6.   Nair, K. P. R. and Hoeft, (1983). J. *Chem. phys* Lett. 96, 348

REFERENCES

Griem, H.R. (1994). *Spectral line Broadening by Plasmas*. Academic Press, New York.

Griem, H.R. (1997). *Principles of Plasma Spectroscopy*. Cambridge University Press, Cambridge.

Rai, D.K. and Thakur, S.N. (2005). *Atomic Spectra, Structure and Modern Spectroscopy*. Vaivaswat Publication, Varanasi.

Sobel'man, I.I., Vainshlein, L.A. Yukov, E.A. (1995). *Excitation of Atoms and Broadening of Spectral lines*, Springer-Verleg Heidelberg.

White, H.E. (1934). *Introduction to Atomic Spectra*. McGraw-Hill Book Co., New York.

13

EFFECT OF NUCLEAR SPIN ON ELECTRONIC MOTION: HYPERFINE INTERACTION

13.1 HYPERFINE STRUCTURE

The hyperfine structures arise due to the interaction of nuclear spin with the orbital motion and spin motion of the electron and results in the splitting of the spectral lines.

In terms of energy levels, either the initial state or the final state or both must be split or shifted by this interaction. Earlier it was thought that the hfs originated as a consequence of different isotopes, but as the hyperfine structure is observed even in atoms with only one isotopic species, this hypothesis was abandoned. The attempts to attribute the hfs to the presence of a small magnetic moment associated with the nucleus were first proposed by Pauli and independently by Russel[1]. Thus it was established that the hfs is due to an interaction of magnetic moment associated with the nuclear spin and the magnetic moment associated with the electron due to the orbital and spin motion.

The experimental observation of hfs in bismuth by Back and its explanation by Goudsmit[2] revealed that a new vector must be added to the vector atom model described earlier in Chapter 7, this time the nuclear spin vector. The model is as shown in Figure 13.1(c). For comparison, ordinary electron coupling, viz., spin-orbit coupling and the same in a magnetic field is again shown in Figure 13.1 (a) and (b) respectively.

In this model the total quantum vector J^* of the atom (for one electron system j^*) representing the total angular momentum due to all electrons is coupled with the nuclear quantum vector I^* having the mechanical moment of the nucleus $I*\dfrac{h}{2\pi}$ to form a resultant vector F^*. The $F*\dfrac{h}{2\pi}$ having the mechanical moment F^* represents in principle the total angular moment of the atom in place of J^*.

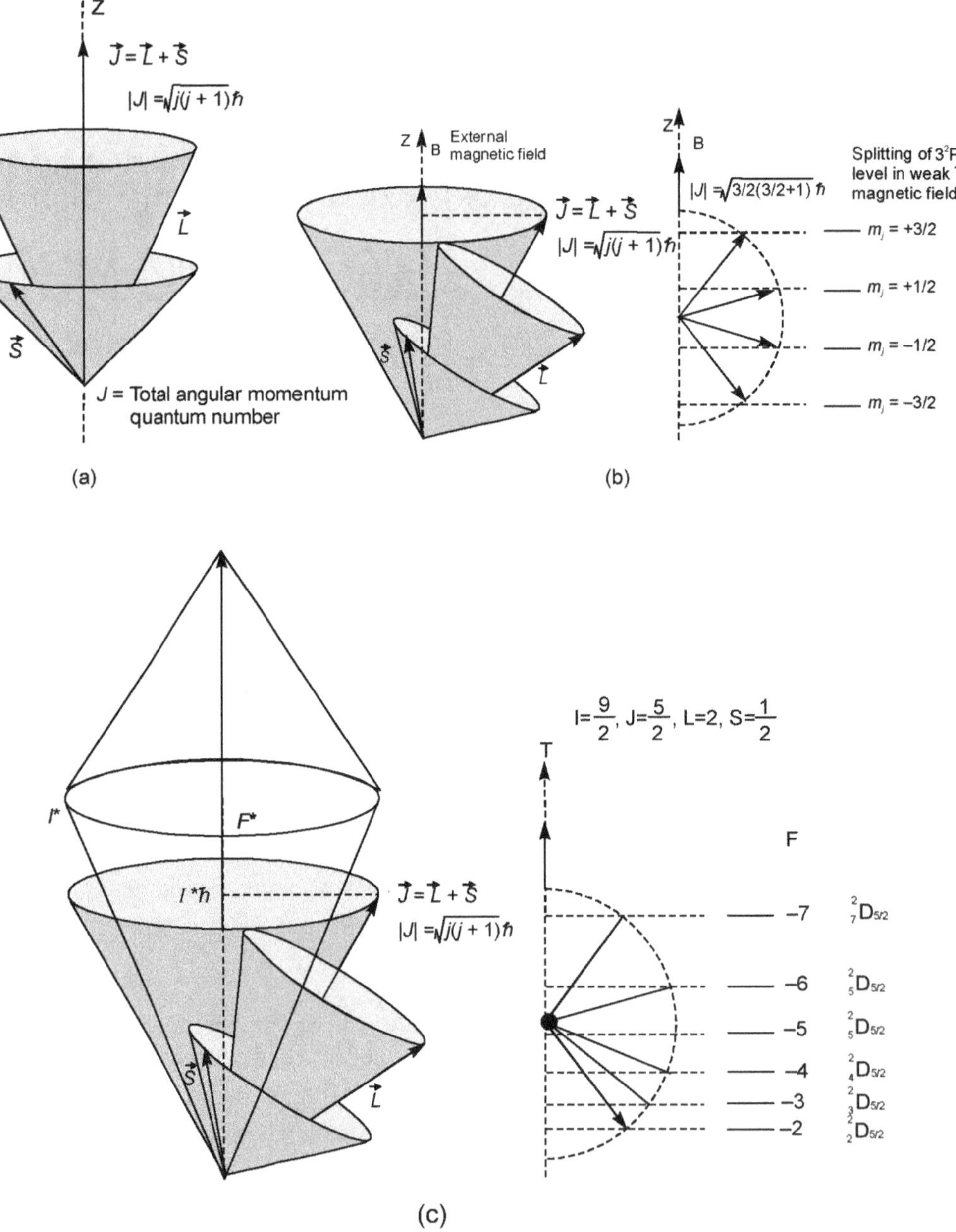

Figure 13.1 Vector model of atom under various interactions

The interaction energy between $I*$ and $J*$ is proportional to the cosine of the angle between them is just as in the case of the interaction between $L*$ and $S*$. The interaction energy is

$$\Gamma = A' I * J * \cos (I * J*) = A' \frac{F*^2 - I*^2 - J*^2}{2} \tag{13.1}$$

where A' is a constant which represents a measure of the strength of the coupling between $I*$ and $J*$. F takes values differing from unity from $I - J$ to $I + J$ if $I \geq J$ or $J - I$ to $J + I$ if $J \geq I$. Thus by knowing the number of splitting of levels one could determine the nuclear spin. Vector diagram and splitting for $I = \frac{9}{2}$, $J = \frac{5}{2}$, $L = 2$ and $S = \frac{1}{2}$ is shown in Figure 13.1(c). $L = 2$ means that it corresponds to a D state and since $S = \frac{1}{2}$ it refers to a doublet state and thus the state is $^2D_{5/2}$ where the subscript refers to J. Thus for $J = 5/2$, $I = 9/2$, F can have values from $I - J$ to $I + J$, thus $F = 2, 3, 4, 5, 6, 7$, all together $(2J + 1)$ levels. It can be shown that the splitting is $(2J + 1)$ if $J \leq I$ and it is $(2I + 1)$ if $I \leq J$. The Γ_F values are $-\frac{55}{4}A'$, $-\frac{43}{4}A'$, $-\frac{27}{4}A'$, $-\frac{7}{4}A'$, $\frac{17}{4}A'$ and $\frac{45}{4}A'$ the difference between the levels are $3A', 4A', 5A', 6A'$ and $7A'$. It can be seen that they are proportional to the larger of the F value.

The observed interval for the $^2D_{5/2}$ in bismuth were found to be $0.563, 0.491, 0.385, 0.312, 0.256$ cm^{-1}. The ratios are closely $7 : 6$, $5 : 4$ and $4 : 3$. The nomenclature for the levels are $^2_2D_{5/2}, ^2_3D_{5/2}, ^2_4D_{5/2}, ^2_5D_{5/2}, ^2_6D_{5/2}, ^2_7D_{5/2}$ in LS coupling. In jj coupling the $^2_7D_{5/2}$ level is written as $7\left(\frac{1}{2}, \frac{3}{2} \cdot \frac{3}{2}\right)_{5/2}$. As in the case of fine structure splitting, in hfs also some splittings are found to be normal, i.e., the smallest F value lie deepest and in others they are inverted.

Fermi, Hargreeves, Breit and Goudsmit[3] have made theoretical calculations on the interaction of the nuclear spin with an electron by semi-classical treatment. In order to obtain the interaction energy we have followed the semi-classical treatment as given by White.

We have already seen in Section 7.5.1 that the electric field at the nucleus by the electron at a distance r is

$$E = \frac{e}{r^3}\overline{r} \tag{13.2}$$

The magnetic field at the nucleus due to the orbital motion of the electron

$$H = \frac{E \times v}{c} = \frac{e}{cr^3}\overline{r} \times \overline{v} \tag{13.3}$$

From Bohr's relation $m\,r \times v = l * \dfrac{h}{2\pi}$

$$r \times v = \frac{1}{2\pi m} l * h \tag{13.4}$$

$$B = \frac{e}{mc} \cdot l * \frac{h}{2\pi}\left(\frac{1}{r^3}\right) \tag{13.5}$$

where m is the mass of the electron and e its charge and r the electron–nuclear distance. Since r is not constant in any orbit $\left(\dfrac{1}{r^3}\right)$ must be averaged. The nuclear spin with its mechanical moment $I * \dfrac{h}{2\pi}$ and a magnetic moment μ_1 carries out Larmor precession around the field and the angular velocity of the Larmor precession is given by

$$\frac{\mu_I}{P} = \frac{\mu_I}{I\hbar} = g_I \frac{e}{2mc} \tag{13.6}$$

where g_I is called the nuclear g factor. If the nuclear magnetic moment is due to the proton in an orbit, the value of g_I is $\dfrac{1}{1838}$ of an orbital motion and if it is due to a spinning proton, its value is anywhere between 4.5 and 7 times of the value of the orbital magnetic moment.

We have already seen in Chapter 7 that the angular velocity of the Larmor precession is equal to the field strength B times the ratio between the magnetic and mechanical moments:

$$\omega_L = B \cdot \frac{\mu}{p} = B \frac{e}{2mc} \tag{13.7}$$

or more exactly $\omega_L = B\, g_I \dfrac{e}{2mc}$

The precession frequency

$$\nu_L = \frac{B}{2\pi} \cdot \frac{\mu}{p} = B \cdot g_I \frac{e}{4\pi mc} \tag{13.8}$$

Thus in the present case

$$\omega_L = B\, g_I \frac{e}{2mc} = \frac{e}{mc} l * \frac{h}{2\pi}\left(\frac{\overline{1}}{r^3}\right) g_I \frac{e}{mc}$$

$$= g_I \frac{e^2}{2m^2c^2} \frac{l * h}{2\pi}\left(\frac{\overline{1}}{r^3}\right) \tag{13.9}$$

The interaction energy due to Larmor precession is given by the product of the angular frequency and projection of the mechanical moment (in this case nuclear mechanical moment) on $l*$, i.e.,

$$W_{I,l} = g_I \frac{e^2}{2m^2c^2} \frac{l*h}{2\pi} \left(\overline{\frac{1}{r^3}} \right) I* \frac{h}{2\pi} \overline{(\cos I*l*)} \tag{13.10}$$

where $\overline{(\cos I*l*)}$ is the average of the cosine and is given by

$$\overline{(\cos I*l*)} = \cos (I*j*) \cos (l*j*)$$

Thus $$W_{I,l} = g_I \frac{e^2}{2m^2c^2} \frac{l*h}{2\pi} \left(\overline{\frac{1}{r^3}} \right) I* \frac{h}{2\pi} \cos (I*j*) \cos (l*j*) \tag{13.11}$$

Since the angles are fixed the cosine terms can be easily calculated. The above equation gives the interaction energy arising from the orbital motion of the electron and the nuclear spin.

Now we will have to calculate the interaction energy arising from the electron spin and the nuclear spin. We can consider the two spins as two small magnetic dipoles and the interaction energy between two magnetic dipoles having magnetic moments μ_l and μ_s in classical electromagnetic theory is

$$W_{I,s} = \frac{\mu_I \mu_s}{r^3} [\cos (\mu_I \mu_s) - 3 \cos (\mu_I r) \cos (\mu_s r)] \tag{13.12}$$

We have already seen that

$$\mu_I = g_I \frac{e}{2mc} I* \frac{h}{2\pi}$$

$$\mu_s = -2 \frac{e}{2mc} s* \frac{h}{2\pi} \tag{13.13}$$

Thus $[\cos (\mu_I \mu_s) - 3 \cos (\mu_I r) \cos (\mu_s r)]$

$$= -\frac{1}{2} \cos (I*j*) [\cos (j*s*) - 3 \cos (j*l*) \cos (s*l*)] \tag{13.14}$$

And hence

$$W_{I,s} = + g_I \frac{e}{2mc} I* \frac{h}{2\pi} 2 \frac{e}{2mc} s* \frac{h}{2\pi} \left(\overline{\frac{1}{r^3}} \right) \frac{1}{2} \cos (I*j*) \times$$

$$[\cos (j*s*) - 3 \cos (j*l*) \cos (s*l*)] \tag{13.15}$$

The total interaction energy is thus

$$\Gamma_F = W_{I,l} + W_{l,s} = A' \, I * J * \ \cos \ (I * J*) = A' \frac{F*^2 - I*^2 - J*^2}{2}$$

$$A' = g_I \frac{e^2 h^2}{8\pi^2 m^2 c^2}\left(\overline{\frac{1}{r^3}}\right)\left[\frac{l*}{j*}\cos(l * j*) + \frac{s*}{2j*}\cos(s * j*) - \frac{3s*}{2j*}\cos(j * l*)\cos(s * l*)\right] \qquad (13.16)$$

The quantum mechanical value of $\left(\overline{\dfrac{1}{r^3}}\right) = \dfrac{R\alpha^2 ch Z^4}{a_1^3 n^3 l\left(l + \dfrac{1}{2}\right)(l+1)}$ $\qquad (13.17)$

where a_1 is the radius of the first Bohr orbit which is

$$a_1 = \frac{h^2 \varepsilon_0}{\pi m e^2} \ \text{and} \ \alpha = \frac{e^2}{2\varepsilon_0 hc}$$

and hence

$$A' = g_I \frac{Rhc\alpha^2 Z^4}{n^3 l(l + \frac{1}{2})(l+1)}\left[\frac{l*}{j*}\cos(l * j*) + \frac{s*}{2j*}\cos(s * j*) - \frac{3s*}{2j*}\cos(j * l*)\cos(s * l*)\right] \qquad (13.18)$$

It has been later shown by Fermi and others and Goudsmit from a classical treatment that the terms in brackets must be replaced by $l*^2/j*^2$.

Thus $\qquad A' = g_I \dfrac{Rhc\alpha^2 Z^4}{n^3 l\left(l + \dfrac{1}{2}\right)(l+1)} \dfrac{l(l+1)}{j(j+1)} = g_I \dfrac{Rhc\alpha^2 Z^4}{n^3 \left(l + \dfrac{1}{2}\right)j(j+1)}$ $\qquad (13.19)$

Dividing by hc A' can be written in cm^{-1}.

Hence $\qquad A' = g_I \dfrac{R\alpha^2 Z^4}{n^3 \left(l + \dfrac{1}{2}\right)j(j+1)} \ \text{cm}^{-1}$ $\qquad (13.20)$

From expression (13.6) we have seen that $\dfrac{\mu_I}{I} = g_I \dfrac{eh}{4\pi mc}$

The value of g_I is expected to be $\dfrac{1}{1838}$ of that of an orbital electron. Thus if g_I is to be expressed

in nuclear magneton $\dfrac{eh}{4\pi Mc}$, g_I should be divided by the ratio of the mass of proton to the mass of

electron which is $\dfrac{1}{1838}$.

Hence $$A' = \frac{g_I}{1838} \frac{R\alpha^2 Z^4}{n^3 (l+\frac{1}{2}) j(j+1)} \, \mathrm{cm}^{-1} \tag{13.21}$$

13.2 NORMAL AND INVERTED HYPERFINE STRUCTURES

The hyperfine splitting is said to be normal if the level with the lowest F lies lowest in energy. The hyperfine splitting with the reverse order is said to be inverted. That means in an inverted state the level having the largest F value has the least energy. If we have two fine structure components $j_1 = l + s$ and $j_2 = l - s$, by interaction with the nuclear spin we obtain the hyperfine components as

$$F_1 = j_1 + I = l + s + I$$
$$F_1' = j_1 - I = l + s - I$$
$$F_2 = j_2 + I = l - s + I$$
$$F_2' = j_2 - I = l - s - I$$

We have already seen that the orbital motion of electron produces a magnetic field which is opposite in direction to its mechanical moment. The most stable state of the nucleus in this field where the nuclear magnetic moment is parallel to the electron orbital magnetic field. Since the nucleus has positive charge, its mechanical and magnetic moments are in the same direction and is in the direction of the orbital magnetic moment of the electron. The most stable state is the one in which I and l are oppositely directed. Thus the most stable states are $F_1' = j_1 - I = l + s - I$ and $F_2' = j_2 - I = l - s - I$ on their respective groups.

If we consider the electron spin–nuclear spin interaction, we can consider them as two tiny magnets acting on each other at some distance from one another. We can consider that these tiny magnets are moving in a magnetic field arising from the orbital motion and the stable position of these tiny magnets are when their magnetic moments are oppositely directed. We have already seen that since the electrons are negatively charged their magnetic moments and the mechanical moments are oppositely directed. Since the nucleus is positively charged its magnetic and mechanical moments are in the same direction. And hence the stable configuration is the one in which the two mechanical moments $\vec{I}$ and $\vec{S}$ are in the same direction. The effect is that the energy of the level F_1 decreases and that of F_1' increases. There is a tendency to invert. On the other hand the energy of F_2 tends to increase and that of F_2' tends to decrease, thereby the separation of the hyperfine components tends to increase.

13.3   ZEEMAN EFFECT WITH HYPERFINE STRUCTURE

The Zeeman effect in hyperfine structure was first studied by Back and Goudsmit[4]. They had observed the effect similar to the Paschen–Back effect. However, we can construct a vector model in a weak magnetic field as shown in Figure 13.2.

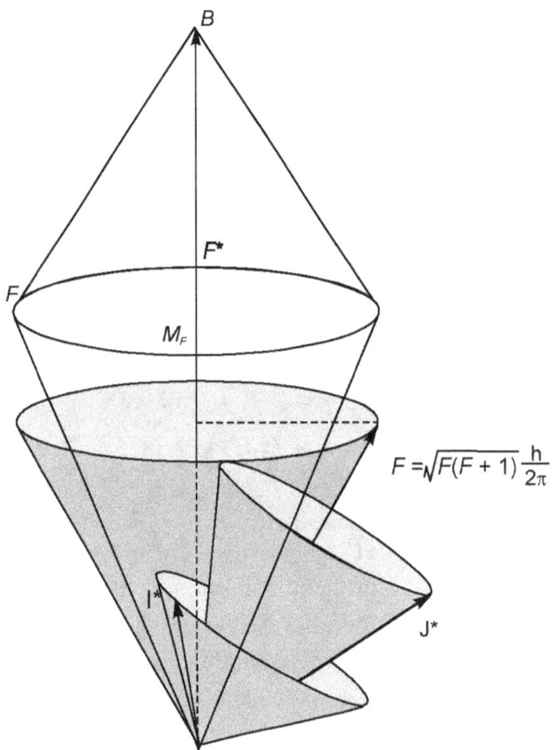

Figure 13.2   Vector model in a weak magnetic field

In an atom without the influence of the nuclear spin we have seen that the total angular momentum J^* precess around the magnetic field. But with the presence of a nuclear spin I^* we have to consider the total angular momentum of the atom as $F = I \pm J$ and F in turn process around the applied magnetic field.

Hence instead of

$S,\ L,\ J,\ g_s,\ g_L,\ g_J,\ M_S,\ M_L,\ M_J$ we have

$I,\ J,\ F,\ g_I,\ g_J,\ g_F,\ M_I,\ M_J,\ M_F$

When F precess around the magnetic field the projection of the mechanical moment $F * \dfrac{h}{2\pi}$ on the magnetic field B is $M_F \dfrac{h}{2\pi}$. The magnetic quantum number M_F takes values from $-F$ to $+F$. In order to calculate the energy levels classically we have to obtain the magnetic moment of the atom as a whole. It is the sum of the electron contribution μ_J and the magnetic moment of the nucleus μ_I. We have already seen that

$$\mu_J = g_J \, J * \frac{eh}{4\pi mc} \quad \text{and} \quad \mu_I = g_I \, I * \frac{eh}{4\pi mc} \tag{13.22}$$

The total magnetic moment along the magnetic field is then

$$\mu_F = \mu_J \cos \, (J * F*) - \mu_I \cos \, (I * F*)$$

As it is the mechanical moments which are quantized and as μ_I and $I*$ are parallel to each other in the same direction and however, μ_J and $J*$ are oppositely directed, the minus sign in the second term occurs in the resultant magnetic moment. From expression

$$\mu_F = [g_J \, J * \frac{eh}{4\pi mc} \cos \, (J * F*) - g_I \, I * \frac{eh}{4\pi mc} \cos \, (I * F*)]$$

$$\mu_F = [g_J \, J * \cos \, (J * F*) - g_I \, I * \cos \, (I * F*)] \, \frac{eh}{4\pi mc} \tag{13.23}$$

We can replace the bracketed quantity as

$$g_F \, F* = [g_J \, J * \cos \, (J * F*) - g_I \, I * \cos \, (I * F*)] \tag{13.24}$$

Hence from equation

$$\mu_F = [g_J \, J * \cos (J * F*) - g_I \, I * \cos \, (I * F*)] \frac{eh}{4\pi mc} = gF \, F * \frac{eh}{4\pi mc} \tag{13.25}$$

We can evaluate the cosine terms from the vector model which give

$$\cos \, (J * F*) = \frac{F *^2 + J *^2 - I *^2}{2F *^2} \quad \text{and} \quad \cos(I * F*) = \frac{F *^2 + I *^2 - J *^2}{2F *^2} \tag{13.26}$$

$$\text{Thus } \, g_F = g_J \, \frac{F *^2 + J *^2 - I *^2}{2F *^2} - g_I \, \frac{F *^2 + I *^2 - J *^2}{2F *^2} \tag{13.27}$$

The nuclear g factor g_I is a number which is the ratio of the magnetic moment of the nucleus in nuclear magneton and the mechanical moment of the nucleus in quantum units $h/2\pi$. The nuclear g factor is different for different nuclei and is a small quantity. Thus we can write the expression 13.27 as

$$g_F = g_J \frac{F*^2 + J*^2 - I*^2}{2F*^2} \tag{13.28}$$

As in the case of electrons we can write

$$\frac{\mu_F}{\dfrac{F*h}{2\pi}} = g_F \frac{e}{2mc} = g_J \frac{e}{2mc} \frac{F*^2 + J*^2 - I*^2}{2F*^2} \tag{13.29}$$

According to Larmor's theorem the angular velocity arising out of the precession around B is

$$\omega_F = B\, g_F \frac{e}{2mc} \tag{13.30}$$

The energy of precession $\quad \Delta W = \omega_F F* \dfrac{h}{2\pi} \cos(F*B) = B\, g_F \dfrac{e}{2mc} \cos(F*B)$

$F* \dfrac{h}{2\pi} \cos(F*B) = M_F \dfrac{h}{2\pi}$ and hence

$$\Delta W = B g_F \frac{e}{2mc} M_F \frac{h}{2\pi} = g_F M_F\, B \frac{eh}{4\pi mc} \tag{13.31}$$

In wave numbers it is $\quad \dfrac{\Delta W}{hc} = g_F M_F \dfrac{eB}{4\pi mc} = g_F\, M_F\, L\ \text{cm}^{-1} \tag{13.32}$

The selection rules are $\Delta M_F = 0$ $\ p$ components

$$\Delta M_F = \pm 1 \ \ s \text{ components}$$

The energy level diagram of $^2P_{3/2} - {}^2S_{1/2}$ is shown in Figure 13.3. Here the nuclear spin is 1/2. In $^2P_{3/2}$, the F components are $F = J + I \ldots |J - I| = 2,1,0,-1,-2$. For $^2S_{1/2}$, the F values are 1, 0, −1. In computing $g_J M_F$, the g_J factor for $^2P_{3/2}$ is 4/3 which gives $g_F = 1$ for $F = 2$ and $g_F = 5/3$ for $F = 1$. Similarly the g_J factor for $^2S_{1/2}$ is 2 which gives $g_F = 1$ for $F = 1$ and $g_F = 0$ for $F = 0$.

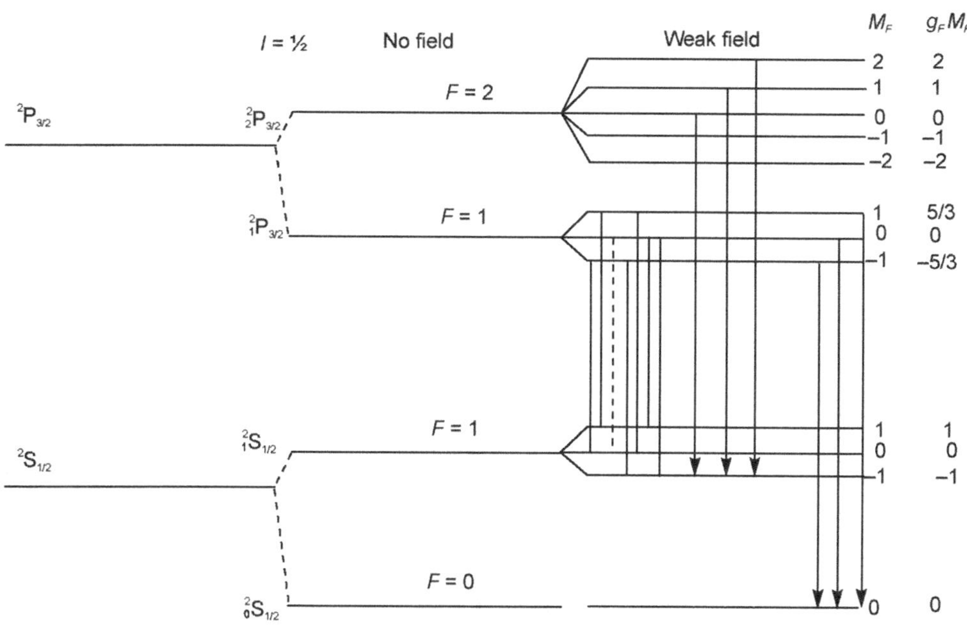

Figure 13.3   Zeeman effect in hyperfine structure $^2P_{3/2} - {}^2S_{1/2}$

13.4 ZEEMAN EFFECT IN STRONG MAGNETIC FIELD (BACK–GOUDSMIT EFFECT)

This is similar to the Paschen–Back effect. The total angular momentum of the electron J and the nuclear spin angular momentum I precess around the applied electric field independently instead of their resultant F as in the previous case of low magnetic field. Thus the coupling between the nuclear moment I^* and the electron angular momentum J^* is broken down and they precess around B independently as shown in Figure 13.4.

The magnetic energy is the sum of (1) the energy due to the interaction between I^* and the magnetic field B, (2) the energy due to the interaction between J^* and the magnetic field B and (3) the energy due to the interaction between I^* and J^*. According to Larmor's theorem we can write the angular velocity as

$$\omega_L = B\frac{\mu}{p} = B\,g_I\,\frac{e}{2mc} \quad \text{and} \quad \omega_J = -B\,g_J\,\frac{e}{2mc} \tag{13.33}$$

The energy contribution

$$\Delta W_{I,B} = -B\,g_I\,\frac{e}{2mc}\,I^*\frac{h}{2\pi}\cos\,(I^*B) = -B\,g_I\,\frac{e}{2mc}\,M_I\,\frac{h}{2\pi} \tag{13.34}$$

$$\Delta W_{J,B} = B\, g_J \frac{e}{2mc}\, J * \frac{h}{2\pi} \cos(J * B) = B\, g_J \frac{e}{2mc}\, M_J \frac{h}{2\pi} \tag{13.35}$$

The sum of these energies gives the main energy shift from the zero field lines. It can be written as

$$\Delta W = (-\,g_I\, M_I + g_J\, M_J)\, B \frac{eh}{4\pi mc} \tag{13.36}$$

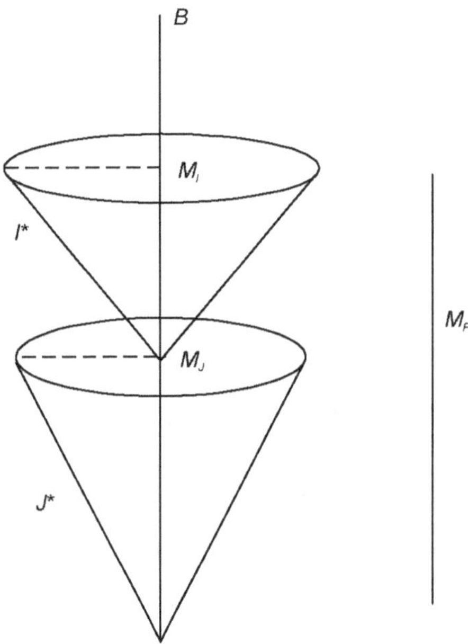

Figure 13.4   Back–Goudsmit effect

And in cm⁻¹, it is

$$\frac{\Delta W}{hc} = \Delta T = (-\,g_I\, M_I + g_J\, M_J)\, B\, \frac{e}{4\pi mc^2}\,\mathrm{cm}^{-1} = (-\,g_I\, M_I + g_J\, M_J)\, L$$

$$\text{where}\quad L = \frac{eB}{4\pi mc^2} \tag{13.37}$$

The third interaction is similar to the spin–orbit interaction which we have discussed earlier and the energy can be written as

$$\Gamma = A'\, I * J * \overline{\cos(I * J*)}\ \mathrm{cm}^{-1} \tag{13.38}$$

$\overline{\cos(I * J^*)}$ is the average of cosine between I^* and J^* and is given by

$\overline{\cos(I * J^*)} = \cos (I * B) \cos (J * B)$ and hence

$$\Gamma = A' I * J * \cos(I * B)\cos(J * B) = A' I * \cos (I * B) J * \cos (J * B) \tag{13.39}$$

$$= A' M_I M_J$$

The total energy change now is

$$\Delta T = (- g_I M_I + g_J M_J) L + A' M_I M_J \tag{13.40}$$

The g_I values for the nuclei are generally very small compared to g_J and hence the first term is negligibly small. Hence

$$\Delta T = g_J M_J L + A' M_I M_J \ \text{cm}^{-1}$$

The selection rules are $\Delta M_I = 0$

$$\Delta M_J = 0 \ \ \text{for } p \text{ components}$$

$$\Delta M_J = \pm 1 \ \ \text{for } s \text{ components} \tag{13.41}$$

Generally $M_F = M_I + M_J$ in all field cases.

13.5 RELATIVISTIC APPROACH: FINE AND HYPERFINE STRUCTURE

The Hamiltonian under a coulomb potential can be written as

$$H_o = \frac{p^2}{2m} - \frac{Ze^2}{r} \tag{13.42}$$

The eigen states can be grouped into shells which are labelled by the principal quantum number n which takes non-negative integer values. The energy of the eigen states was derived as

$$E = -\frac{Z^2 m e^4}{2\hbar^2 n^2} \tag{13.43}$$

where n is $(2l + 1)$ degenerate and hence each shell n has subshells of orbital angular momentum $l = 0, 1, 2 \ldots (n-1)$.

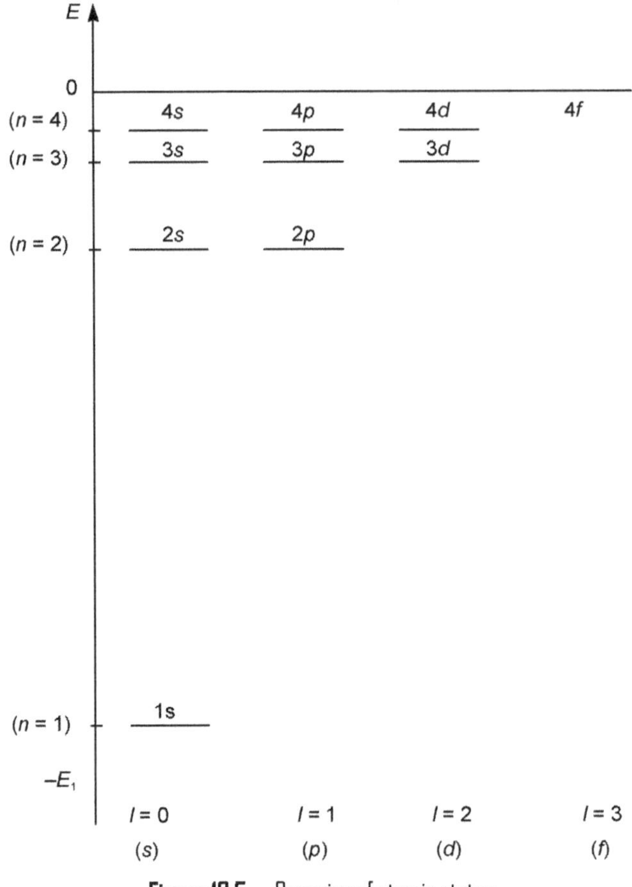

Figure 13.5　Grouping of atomic states

This is a simple picture of an atom in the first approximation. The basic structure of the hydrogen atom can be explained by this treatment of the theory. In order to get a better picture of the atom we have to consider the following:

1. relativistic effects
2. electron spin
3. nuclear spin

The first two items combined contributes to what are known as the fine-structure terms in the atomic Hamiltonian, while the third gives rise to hyperfine terms.

13.5.1　The Fine-Structure Hamiltonian

In relativistic quantum theory, since the fine structure constant, the coupling constant between the charged matter and the electromagnetic field, $\alpha = \dfrac{e^2}{\hbar c} \approx \dfrac{1}{137}$ is a small quantity, perturbation theory could be applied. For the hydrogen atom α appears as the ratio between the electron velocity and the speed of light, i.e., $\dfrac{v}{c}$. This indicates that the effect brings relatively small contribution.

A more realistic relativistic equation following Dirac's theory can be used to correct the basic Coulomb Hamiltonian.

$$H = m_e \, c^2 + \left[\frac{p^2}{2m_e} + V(r) \right] - \left[\frac{p^4}{8m_e^3 c^2} \right] + \left[\frac{1}{2m_e^2 c^2} \frac{1}{r} \frac{dV(r)}{dr} \vec{L}.\vec{S} \right] + \left[\frac{\hbar^2}{8m_e^2 c^2} \nabla^2 V(r) \right] + \dots$$

$$= m_e \, c^2 + H_o + W_{mv} + W_{SO} + W_D + \dots \tag{13.44}$$

where m_e is the electron mass. It will be appropriate to take reduced mass, however, we consider here only the electron mass. The rest-mass energy term $m_e \, c^2$ shifts all energy levels equally, and therefore has no effect on the hydrogen spectrum per sec. The next term in square brackets reproduces the Coulomb Hamiltonian. The third term $W_{mv} = -\left[\dfrac{p^4}{8m_e^3 c^2} \right]$ represents relativistic variation of the electron mass with velocity and is the first relativistic correction.

We already know the relativistic expression for energy $E = -c\sqrt{p^2 + m^2 c^2}$

We can expand this in terms of $\dfrac{p}{mc}$ and obtain

$$E \approx mc^2 \left(1 + \frac{p^2}{2m^2 c^2} - \frac{p^4}{8m^4 c^4} \right) \approx mc^2 \left(\frac{p^2}{2m} - \frac{p^4}{8m^3 c^2} \right) \tag{13.45}$$

Using $\alpha = \dfrac{v}{c} = \dfrac{p}{mc}$, we can estimate the first relativistic correction

$$W_{mv} = -\frac{\left[\dfrac{p^4}{8m_e^3 c^2} \right]}{\dfrac{p^2}{2m}} - \frac{4p^2}{4m^2 c^2} \approx \alpha^2 \approx \left(\frac{1}{137} \right)^2 \tag{13.46}$$

Evidently this is a small correction.

The spin–orbit interaction term is $\left[\dfrac{1}{2m_e^2 c^2} \dfrac{1}{r} \dfrac{dV(r)}{dr} \vec{L}.\vec{S} \right]$ $\tag{13.47}$

When the electron moves in an orbit, it moves rapidly through the electrostatic field created by the proton. The magnetic field in terms of v/c, $B = -\dfrac{1}{c^2}\vec{v} \times \vec{E}$.

Magnetic moment due to the electron spin $\vec{\mu} = \gamma \vec{S} = \dfrac{q}{m}\vec{S}$.

This interacts with the magnetic field B which yields an energy correction $W = -\vec{\mu}.\vec{B}$.

The electrostatic field seen by the electron is $-\dfrac{1}{q}\dfrac{1}{r}\dfrac{dV(r)}{dr}\dfrac{\vec{r}}{r}$.

$V(r) = -\dfrac{e^2}{r}$ is the coulomb potential. Hence $B = \vec{B}\dfrac{1}{qc^2}\dfrac{1}{r}\dfrac{dV(r)}{dr}\dfrac{\vec{p}}{m}\times\vec{r}$

$\vec{p}\times\vec{r} = -\vec{L}$ and hence the quantum mechanical Hamiltonian

$$W = -\frac{q}{m}\vec{S}.\frac{1}{qc^2}\frac{1}{r}\frac{dV(r)}{dr}\frac{\vec{p}}{m}\times\vec{r} = \frac{1}{m^2c^2}\frac{1}{r}\frac{dV(r)}{dr}\vec{S}.\vec{L} = \frac{e^2}{m^2c^2}\frac{1}{r^3}\vec{S}.\vec{L} \qquad (13.48)$$

The actual spin–orbit terms differ by a factor of 1/2 which is believed to be due to the curved orbit of the electron. L and S are of the order of h and r is of the order of $a_o = \dfrac{\hbar^2}{me^2}$ and we estimate

$$\frac{W_{SO}}{H_o} = \frac{\left(\dfrac{e^2}{m^2c^2}\dfrac{1}{r^3}\vec{S}.\vec{L}\right)}{\dfrac{e^2}{r}} = \frac{\hbar^2}{2m^2c^2r^2} \approx \frac{e^4}{\hbar^2c^2} = \alpha^2 \qquad (13.49)$$

Finally the $W_D = \left[\dfrac{\hbar^2}{8m_e^2c^2}\nabla^2 V(r)\right]$ and is called Darwin term and is difficult to estimate. We can derive the term in terms of the atomic wave function and the operator is $<W_D> = \dfrac{\pi e^2\hbar^2}{2m^2c^2}|\Psi_o|^2$.

$|\Psi_o|^2$ is of the order of a_1^{-3} and hence we estimate

$$\frac{W_D}{H_o} \approx \frac{\pi e^2\hbar^2}{2m^2c^2a_1^{-3}}\bigg/ -\frac{e^2}{a_1} \approx \frac{\hbar^2}{m^2c^2a_1^2} = \frac{\hbar^2}{m^2c^2}\frac{m^2e^4}{\hbar^4} = \frac{e^4}{\hbar^2c^2} = \alpha^2 \qquad (13.50)$$

The overall Hamiltonian is thus broken down into

$$H = mc^2 + H_o + W_f$$

where $\quad H_0 = \dfrac{p^2}{2m} + V(r)$

$$W_f = W_{mv} + W_{SO} + W_D$$

$$\frac{H_0}{mc^2} \approx \frac{e^2/a_0}{mc^2} = \frac{e^2}{mc^2}\frac{me^2}{\hbar^2} = \frac{e^4}{\hbar^2 c^2} = \alpha^2 \tag{13.51}$$

$$\frac{W_f}{H_o} \approx \alpha^2$$

The hydrogen nuclear magnetic moment operator can be written as

$$\vec{\mu}_I = \frac{g_p \mu_N}{\hbar} \vec{I}$$

where I is the nuclear spin vector operator g_p is proton g factor and is equal to 5.585. μ_N is the

nuclear magneton and $\mu_N = \dfrac{q_p \hbar}{2m_p}$. It is to be remembered that μ_N is smaller than Bohr magneton by

a factor of $m_e/m_p = 1/1873$ and $q_p = -q_e$. The electron magnetic moment operator is $\vec{\mu}_s = \dfrac{g_e \mu_B}{\hbar} \vec{S}$

with $g_e = 2$.

The hyperfine Hamiltonian is

$$W_{hf} = -\frac{\mu_0}{4\pi}\left\{ \frac{q}{m_e r^3}\vec{L}.\vec{\mu}_I + \frac{1}{r^3}[3(\vec{\mu}_s.\hat{n})(\vec{\mu}_I.\hat{n}) - \vec{\mu}_s\vec{\mu}_I] + \frac{8\pi}{3}\vec{\mu}_s.\vec{\mu}_I\partial(\vec{r}) \right\}$$

where $\hat{n}$ is the unit vector.

The first term of W_{hf} reflects the interaction of the nuclear magnetic moment with the magnetic field $(\mu_0/2\pi)(q_L/m_e\ r^3)$ created by the orbiting electron. The second term represents the magnetic dipole–dipole interaction between the nuclear and electronic spins. The third term, known as Fermi's 'contact term,' has to do with the internal magnetic structure of the proton. The important thing to note about the hyperfine Hamiltonian is that it leads us to consider coupling of angular momenta between the nuclear spin I and the total electron angular momentum J, yielding $\vec{F} = \vec{I} + \vec{J}$.

The reason for considering such a coupling scheme (as opposed, e.g. to coupling S and I first and then adding L) is that the spin–orbit coupling is so much larger than the hyperfine interaction.

ENDNOTES

1. Pauli, W. *Z.Naturforsch*. 12, 741 (1924); Russel, H.N., Meggers, W.F. and Burns, K. *J.Opt.Soc.Amer*. 14, 449, 1927.

2. Goudsmit, S. and Back, E. *Z.f Phys*. 43, 321, 1927; 47,174, 1928.

3. Fermi, E. *Z. f. Phys*. 60, 320, 1930; *J. Hargreaves, Proc. Roy. Soc.* A, 127, 141, 1930; G. Breit, *Phy.Rev*. 37, 51, 1931; Goudsmit, S.A. *Phys.Rev*. 37, 663 1931.

4. Back, E. and Goudsmit, S.A. *Zeits.f.Phys*. 47 (1928) 174.

REFERENCES

Cowan, R.D. (1981). *The Theory of Atomic Structure and Spectra*. University of California Press, Berkeley.

Foot, C.J. (2005). *Atomic Physics*. Oxford University Press, Oxford, U.K.

Rai, D.K. and Thakur, S.N. (2005). *Atomic Spectra, Structure and Modern Spectroscopy*: Vaivaswat Publication, Varanasi.

Shore, B.W. and Menzel, D.H. (1968). *Principles of Atomic Spectra*. John Wiley, New York.

White, H.E. (1934). *Introduction to Atomic Spectra*. McGraw Hill., New York.

Woodgate, G.K. (1992). *Elementary Atomic Structure*. Clarendon Press, Oxford, U.K.

14

EXPERIMENTAL TECHNIQUES IN SPECTROSCOPY

14.1 LIGHT SOURCES IN ABSORPTION AND EMISSION SPECTROSCOPY

Though it is difficult to describe all experimental techniques available for spectroscopic work, some of the important ones are given in this chapter.

The phenomena used for optical measurements are:

1. Absorption
2. Emission
3. Luminescence (fluorescence, phosphorescence, chemiluminescence)
4. Scattering

And almost all optical instruments share similar design. The basic parts are:

1. stable radiation source
2. transparent sample holder
3. wavelength selector
4. radiation detector
5. signal processor

The basic parts of an optical instrument is shown in Figure 14.1.

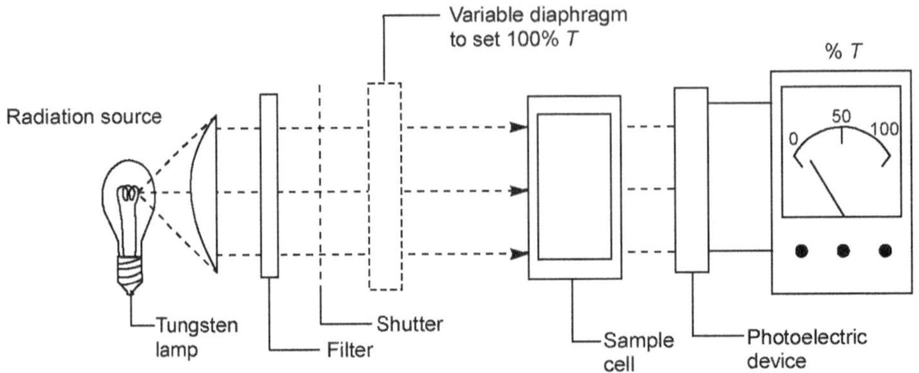

Figure 14.1   The basic parts of an optical instrument

If I_0 is the intensity of radiation on a sample in a container having a thickness of b, and I is the intensity after the transmission as shown in Figure 14.2, the transmittance $T = I / I_0$ and the percentage transmittance is then $T = (I / I_0) \times 100$.

The absorbance $A = -\log T = -\log (I / I_0) = \log (I_0 / I)$ (14.1)

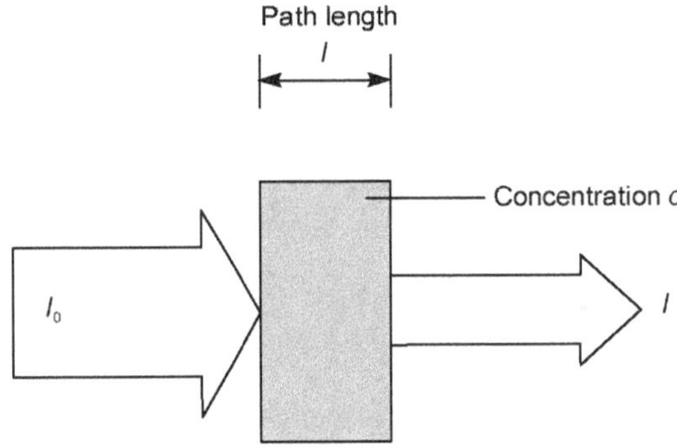

Figure 14.2   The absorbance spectrophotometry

As an example if a sample has 100% transmittance

Examples

If $T = 1.00$ (100 % T), $A = 0.00$

If $T = 0.10$ (10 % T), $A = 1.00$ and

If $T = 0.001$ (0.1 % T), $A = 3.00$

The basis for absorbance spectrophotometry is Beer's Law.

It is based on $A \propto c$ and $A \propto l$ where A is absorbance and c is concentration of the sample and l is the length of the sample cell.

Hence $A \propto l \cdot c$

$$A = a \cdot l \cdot c$$

The proportionality constant a is called absorptivity and the unit is $\text{L/g} \cdot \text{cm}$. If the unit of concentration is M (mol/L) then the absorptivity is molar absorptivity ε.

$$A = \varepsilon \cdot l \cdot c \ \ \text{L/mol} \cdot \text{cm}$$

Radiation sources generally used at various wavelengths are given in Table 14.1 and Table 14.2 gives the various kind of detectors.

14.1.1 Beer–Lambert Law

The law states that there is a logarithmic dependence between the transmissivity (or transmittance) T of light passing through a substance and the product of the absorption coefficient of the substance a, and the distance the light travels through the material (i.e., the path length), l. The absorption coefficient can be written as the product of molar absorptivity of the absorber ε and the concentration c of the absorbing species or it can also be the product of the absorption cross section σ, and the number density N of the absorbers.

For liquids, these relations are written as

$$T = \frac{I}{I_0} = 10^{\alpha l} = 10^{-\varepsilon lc} \tag{14.2}$$

whereas for gases, they are written in the form

$$T = \frac{I}{I_0} = 10^{-\alpha' l} = e^{-\sigma lN} \tag{14.3}$$

where I_0 and I are the intensity (or power) of the incident light and the transmitted light respectively. The transmission (or transmissivity) is expressed in terms of an absorbance which for liquids is defined as

$$A = -\log_{10}\left(\frac{I}{I_0}\right) \tag{14.4}$$

and for gases,

$$A' = -\ln\left(\frac{I}{I_0}\right) \tag{14.5}$$

It shows that the absorbance becomes linear with the concentration (or number density of absorbers) according to

$$A = \varepsilon lc = \alpha l$$

Table 14.1   Radiation sources

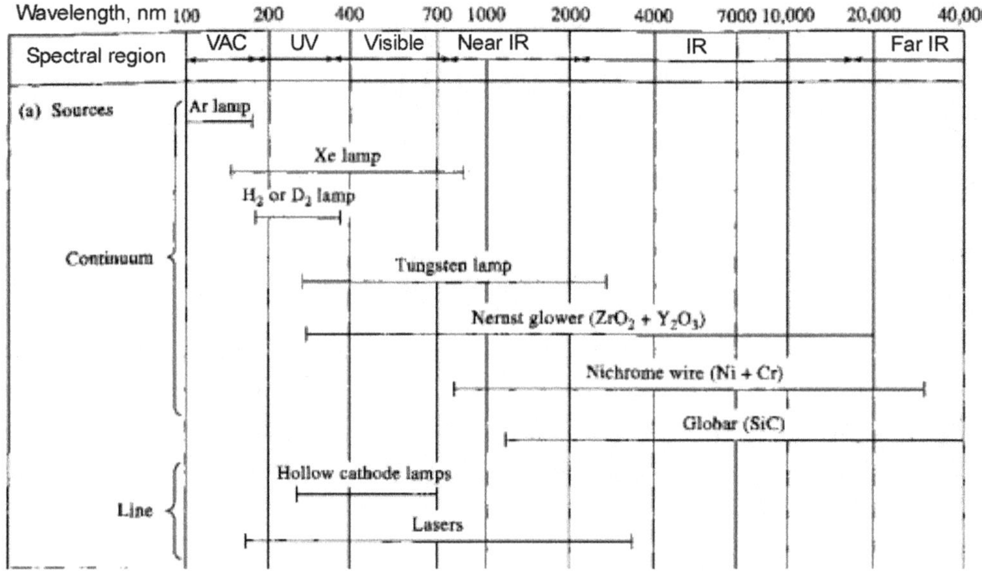

Table 14.2   Detectors

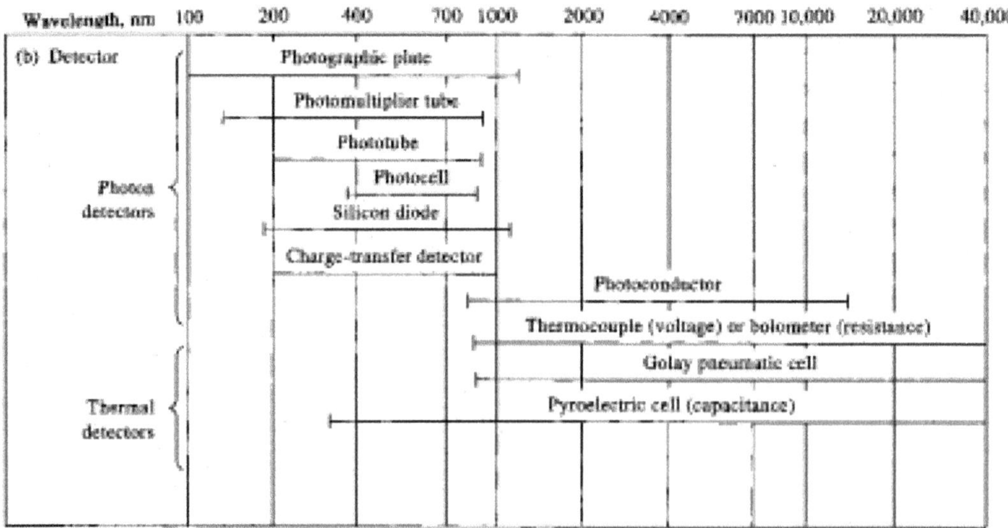

Heated solid globar and nichrome wires are suitable for the region (1–40 μm) as continuous source, tungsten lamp for (300–3000 nm), Quartz tungsten halogen (QTH) lamp for the region (200–3000 nm) and D_2 lamp or Hg/Xe arc lamp for (160–400 nm).

The degree of excitation of atoms depends on the temperature of the source in which atoms are excited. This is how we could study the temperature and abundance of atoms and molecules in astrophysical objects and interstellar media. The method of excitation largely affects the width of spectral lines. At high temperatures one could expect large Doppler Effect and at high densities the pressure broadening becomes important. If we consider sources which are not in thermal equilibrium such as low pressure glow discharges, the gas kinetic temperature may be close to the room temperature though electron temperature may be very high, thereby reducing the Doppler broadening. We shall here describe some of the emission sources, which are generally used in atomic spectroscopy.

14.1.2 Flames

This is a source of thermal excitation and corresponds to a temperature of around 2000° K. The source is mainly used in visible spectroscopy.

14.1.3 Arcs

The mechanism of excitation is thermal. Iron and copper arcs are common in experimental spectroscopy. Iron and copper arc spectra are used as standard for wavelength vis-à-vis frequency measurements and are used for calibration in visible and ultraviolet spectroscopy. A typical arc runs with 2 to 5 amperes of current and at 60–100 volts and the operating temperatures are around 5000° K. There are a number of varieties of arcs depending on the experiments and the sample. For example, for powder samples cavity electrodes are used. Water-cooled electrodes are also used to reduce the temperatures of the source.

14.1.4 Sparks

The mechanism of excitation is electrical rather than thermal in this case. High voltage sparks with gas at low pressures between the two electrodes are used. The voltage is generally 10 to 40 KV per centimetre of the spark gap. Ionization of gases occurs at various degrees in the atoms of the metal forming the electrodes or the gas through which discharge passes. The degree of excitation can be controlled by adjusting the capacitance and inductance in the electric circuit and also by varying pressure of the gas.

14.1.5 Glow Discharges

A glow discharge is a kind of plasma. In fact the glow discharge owes its name to the very reason that plasma is luminous. The plasma consists of ionized gas with equal concentrations of positive and negative charges and a large number of neutral species. The excitation in a glow discharge is accomplished by electron bombardment. When a potential difference is applied between the two electrodes in a gas the glow can be produced and maintained. The potential drops rapidly close to the cathode, vary slowly in the plasma, and changes again close to the anode. Atoms are excited in the positive column of the glow discharge at low pressures generally few mm of pressure using few kilovolt of potential difference with small discharge currents, generally 1 amp or less. Electrons that are emitted from the cathode are accelerated away from the cathode, and give rise to collisions with the gas atoms or molecules and thereby giving excitation, ionization, energy transfer, dissociation, etc. to the atoms or molecules present. The gas discharges and the excitation collisions can produce atoms or molecules in the excited states, which can decay to lower levels by the emission of light. The emission of light is responsible for the characteristic name of the "glow" discharge. The ionization collisions create ion–electron pairs. As the ions thus produced have a net positive charge they are accelerated toward the cathode, where they release secondary electrons. These electrons are accelerated away from the cathode and then can give rise to more ionization collisions. In its simplest way, the combination of secondary electron emission at the cathode and ionization in the gas, gives rise to a self-sustained plasma. In glow discharge experiments some times hollow cathodes are used. A hollow cathode glow discharge tube is shown in Figure 14.3.

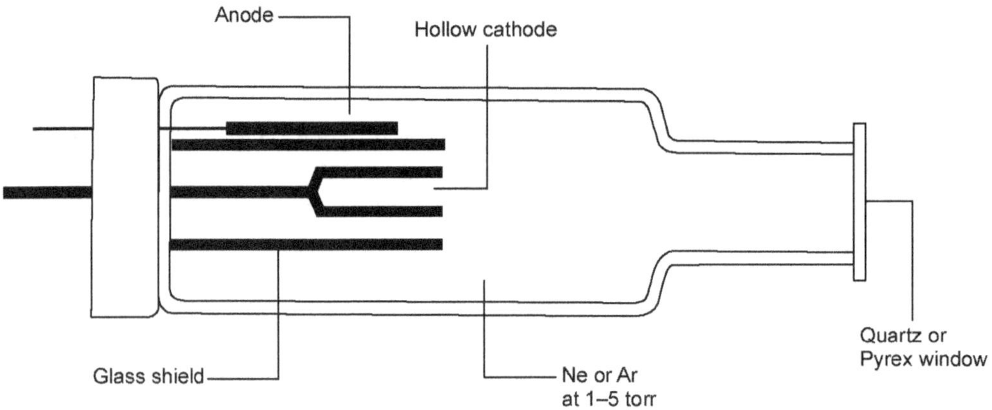

Figure 14.3   A hollow cathode glow discharge

Hollow cathode techniques are usually used with materials deposited on the inside walls of the hollow cathode. This method is used in high resolution studies with cooled cathodes which could maintain the glow discharge for sufficiently long time. The spectra of ions can also be excited in hollow cathodes. Glow discharges by radiofrequency excitation or by microwaves are also increasingly used. They do not require any electrodes. They work on the change in the orientation of magnetic dipoles. A microwave plasma source is shown in Figure 14.4(a). As they do not have any electrodes they are also called electrodeless excitation[1]. The electrodeless discharges are increasingly being used in the production of short-lived molecules[2]. An electrodeless microwave source used by Nair and Hoeft for the production of short-lived molecules is shown in Figure 14.4(b).

Microwave plasma source

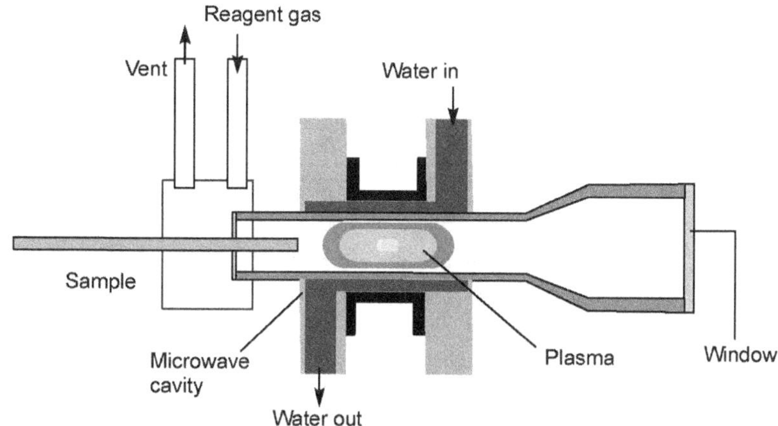

Figure 14.4 (a)   Excitation sources

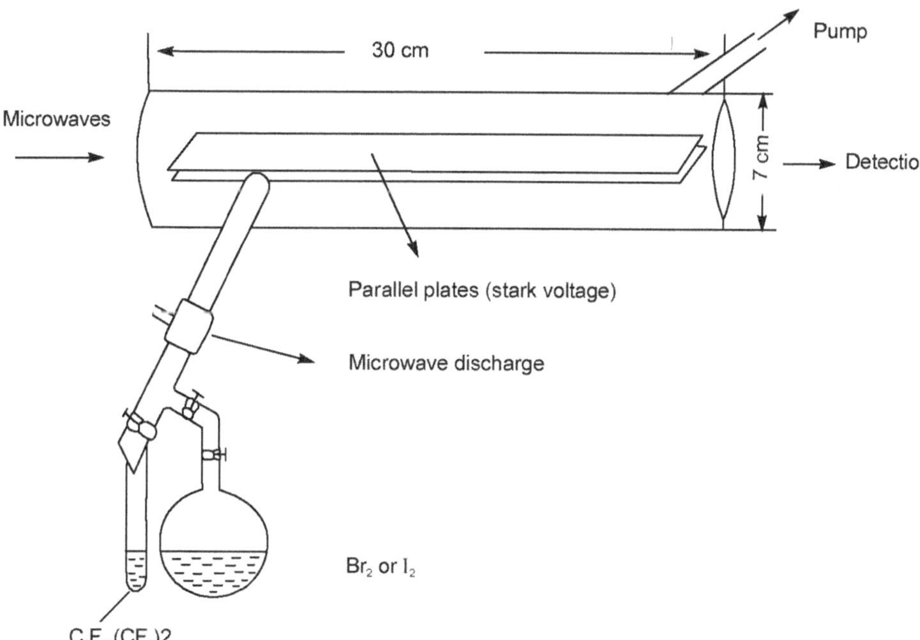

Figure 14.4 (b)   Experimental set-up for the production and study of BrF and IF molecules used by Nair and Hoeft for microwave rotational spectroscopy

Lasers are increasingly used as light source. Lasing medium can be solid (Nd:YAG, semiconductor diode laser AlGaAs), gas (noble gas Ar$^+$, He/Ne, CO$_2$, N$_2$) or liquid (dye). Such a system is shown in Figure 14.5 and the energy level for three- and four-level systems are given in Figure 14.6.

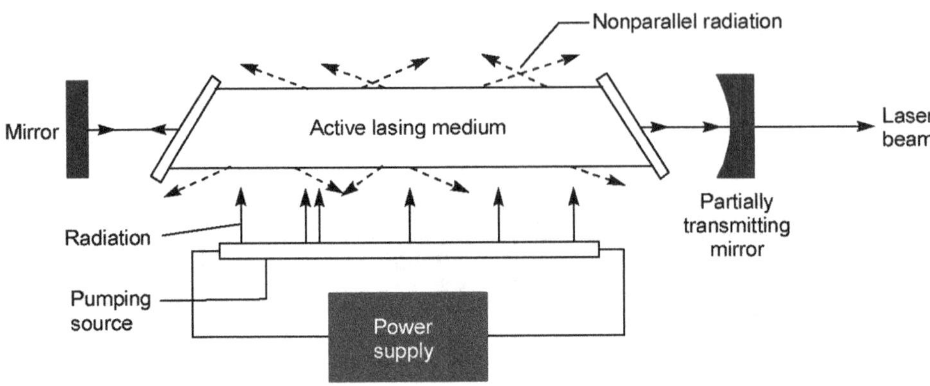

Figure 14.5   Laser

It needs population inversion for lasing. It is easier to produce population inversion in 3- or 4-levels rather than in a two-level system.

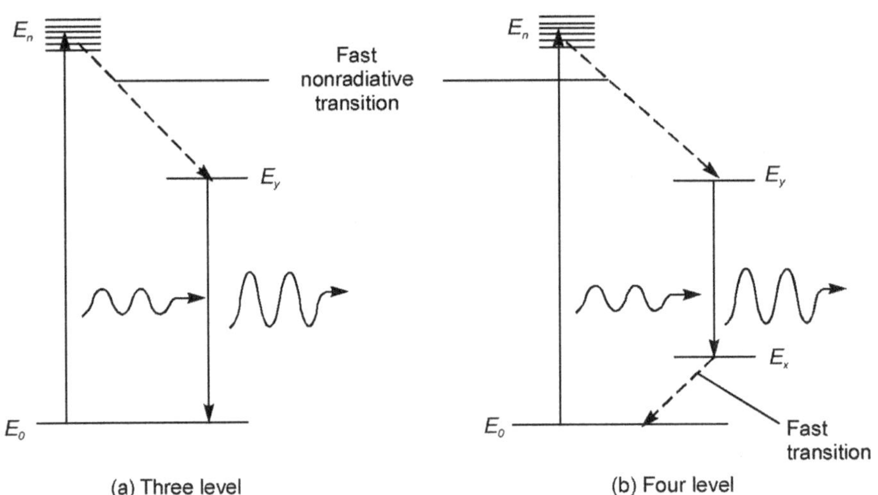

Figure 14.6   Population inversion in three- or four-level system

The advantages of laser sources are

1.   they are intense
2.   monochromatic (line sources)
3.   they can be pulsed (10-15-10-6 s) or continuous wave (cw)
4.   they are coherent
5.   they have small beam divergence

14.1.6 Shock Tubes

In a shock tube there are two regions of a metal tube where the gas is kept under different pressures. A diaphragm separates the two regions of low pressure and high pressure. The gases in the two regions need not be of the same chemical composition. At a certain time we can break the diaphragm which will then produce a shock wave that travel down the low pressure section of the tube. The shock wave then increases the pressure and temperature of the gas which induces a flow in the direction of the shock wave. When the incident shock wave reaches the end of the shock tube, it reflects back into already heated gas. This results in further rise in temperature, pressure and density. This arrangement creates a high temperature-high pressure reaction zone to which the driven gas is subjected. It is possible to quench the reaction by using a 'dump tank' which takes away the shock wave. The gas samples are then collected from the tube and can be used for further studies. Other than the study of high temperature high pressure samples, shock tubes are also used for other studies like combustion and aerodynamics.

14.1.7 Pinch Discharge

In this case the magnetic field connected with a pulse of high current squeezes a column of ionized gas towards the axis in very short durations. When the collapse is completed, the ordered radial velocities become random velocities by the process of collisions resulting in a narrow beam of gas with electron temperatures of the order of 10^5 K and with electron densities of 10^{17}/cc which lasts for a few microseconds. The gas can then be studied spectroscopically or by other means.

14.1.8 Laser Excitation

A high power laser directed to a small region of target material can be used to produce electron temperatures of the order of 50–500 eV and electron densities of the order of 10^{20}/cc lasting for few nanoseconds. The laser excitation is very useful in studying ionized metal atoms.

14.1.9 Beam Foil Source

A beam of ions accelerated in a van de Graaf accelerator is impinged on a thin target foil and the beam emerges as a mixture of ions of various degrees of ionization and in various states of excitation. When the excited ions return to the lower states or to the ground state, spectral lines are emitted characteristics of the ions.

14.2 ABSORPTION SPECTROSCOPY

By sending a continuous beam of light through a cell, which contains the substance under investigation, absorption features of the sample can be spectroscopically studied. The absorption cell should be equipped with suitable windows, which transmit the radiation, and when the substance under study absorbs the continuous radiation which is passed through it, the transmitted radiation is deficient in intensities at precisely the wavelengths absorbed by the atoms in the cell. The absorption cell, which contains the sample, is heated or is generally placed in a furnace so that the substance under investigation is in the vapour state. The furnace is heated to such a temperature so that the absorption cell is filled with sufficient gas of the sample. The window materials used are quartz for the ultraviolet and visible regions and glass is used for the visible region. For the infrared regions alkali halide crystals are used. The sources of background radiations are generally tungsten filament or high pressure Xenon arc for the absorption region 200–3000 nm and hydrogen discharge lamp for 160–400 nm region. Flash tubes are also used as sources of continuous radiation in the far ultraviolet region.

14.2.1 Dispersing Element

For obtaining the spectrum a dispersing element is necessary. It is achieved by refraction, diffraction or by interference. Prisms, gratings or interferometers are used for this purpose. The important characteristics are their resolving power, spectral range and light gathering power.

The resolving power of the prisms is usually smaller than those of the gratings. With a base of approximately 10 cm one could arrive at a value of 10^4 as the order of the resolving power of a prism.

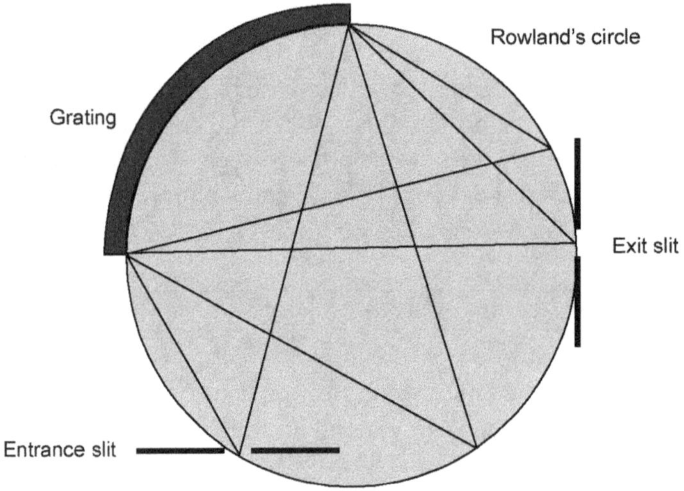

Figure 14.7   Monochromator system

The monochromator and spectrograph system consists of a dispersing element, an entrance slit and an exit slit where the detection is made. There are many modifications and numerous configurations by which this may be achieved. In order to get good focusing at the entrance slit and the exit slit, the Rowland's circle principle should be obeyed (*See* Figure 14.7).

The essential features of a monochromator are

- ✧ Entrance slit
- ✧ Collimating lens or mirror
- ✧ Dispersing element (prism or grating)
- ✧ Focusing lens or mirror
- ✧ Exit slit

A typical prism monochromator is shown in Figure 14.8.

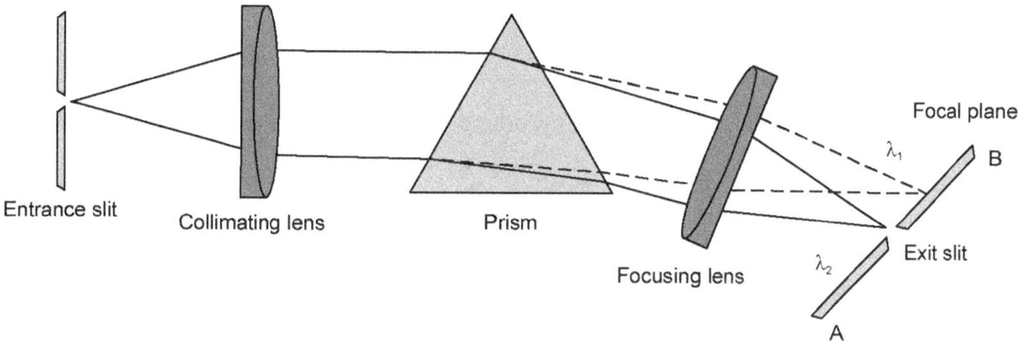

Figure 14.8   A typical prism monochromator

For a prism

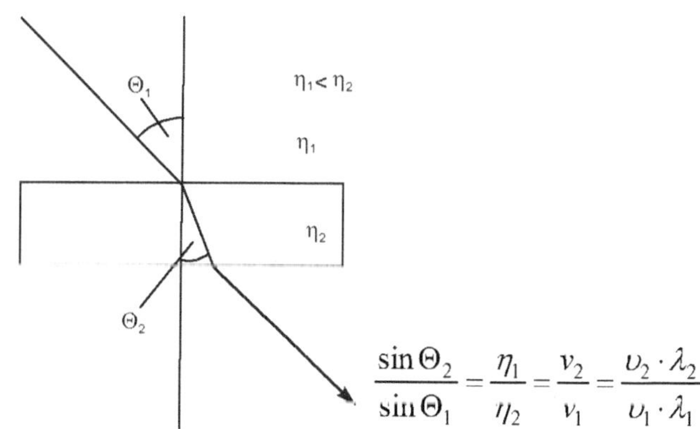

$$\frac{\sin\Theta_2}{\sin\Theta_1} = \frac{\eta_1}{\eta_2} = \frac{v_2}{v_1} = \frac{\upsilon_2 \cdot \lambda_2}{\upsilon_1 \cdot \lambda_1}$$

Figure 14.9   Snell's law

Short wavelengths are refracted more. The dispersion of light in a prism spectrometer is shown in Figure 14.10.

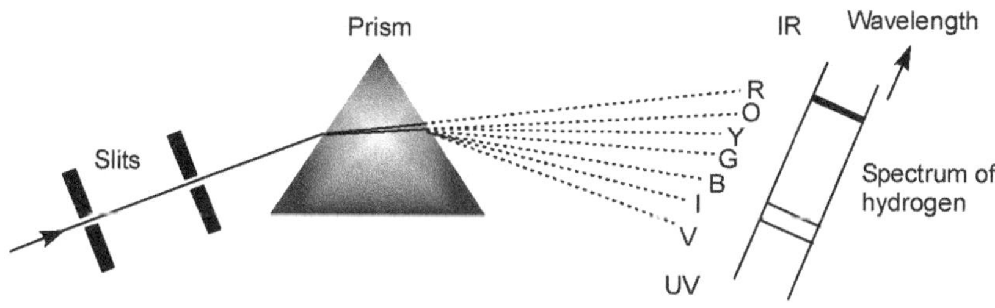

Figure 14.10   Dispersion of light in a simple spectrometer

14.2.2 Quality of Monochromators

1. **Spectral purity**   In order to reduce scattered or stray light in exit beam, the entrance and exit windows, they are kept in a dust and light-tight housing and the interior is coated with light absorbing paint.

2. Dispersion is the ability to separate small wavelength differences. Linear dispersion or reciprocal linear dispersion refers to variation in λ across the focal plane. Thus

$$D = \frac{dy}{d\lambda} \tag{14.6}$$

$$D^{-1} \frac{dy}{d\lambda} = \frac{d}{nF} \tag{14.7}$$

Here F is focal length and D^{-1} has units nm/mm or A/mm.

3. Light gathering is light collection efficiency. It is called

f number

$$f = \frac{F_{\text{collimating-mirror}}}{dia_{\text{collimating-mirror}}} \tag{14.8}$$

4. Spectral bandwidth is the range of wavelengths exiting in the monochromator. It is related to dispersion and entrance/exit slit widths

$$\text{Effective bandwidth} = \frac{\text{bandwidth}}{2} = \frac{\Delta\lambda}{\Delta y} = D^{-1}. \tag{14.9}$$

14.2.3   Littrow Type of Prism Spectrograph

This is one of the common forms of prism spectrometers. It generally consists of a littrow prism (a 30° prism silvered on the back side and a long focus lens. The ray of light coming through a slit passes through the achromatic lens which provides a parallel beam of light towards the prism face. After reflection at the back surface of the prism the emergent light of different wavelengths passes through the prism again and are focused at different points on the photographic plate. Since the light passes through the prism two times the dispersion and resolution obtained is equivalent to that of light dispersed by an unsilvered 60° prism. If the spectrograph is to be used for ultraviolet studies the optical components are made up of quartz since the glass optics will only transmit visible light. For infrared region alkali halide crystals are used for prism with other suitable materials for the infrared region. The photographic plates are nowadays replaced by CCD cameras. A Littrow-type prism spectrograph is shown in Figure 14.11.

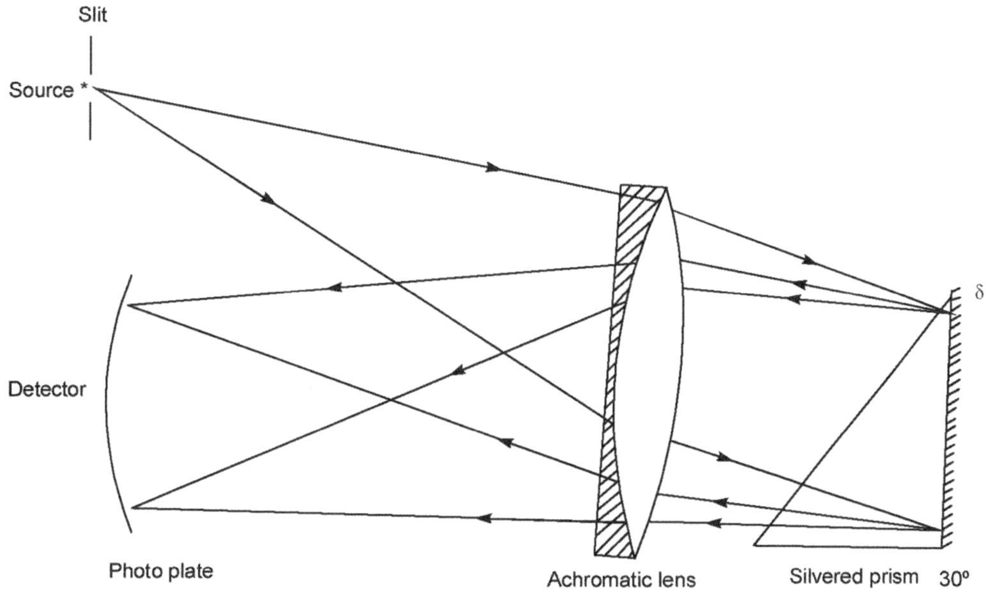

Figure 14.11   Littrow mounting of a prism spectrograph

Plane gratings with suitable blaze angles can be used instead of prism. In Littrow mounting the prism is replaced by the grating. To change the spectral region the grating can be rotated.

14.2.1 Grating Spectrometers

Diffraction gratings are used nowadays in most modern instruments. The principle behind a grating spectrometer is depicted in Figure 14.12.

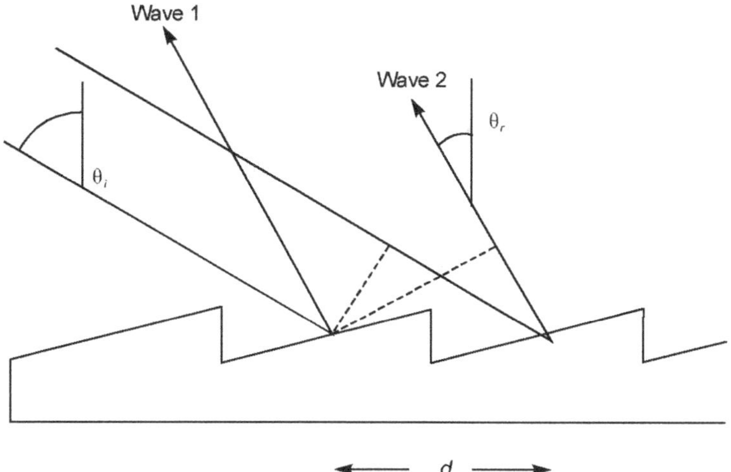

Figure 14.12   Echellette grating

It can be seen from Figure 14.12 that the extra pathlength travelled by wave 2 must be $n\lambda$ for constructive interference

$$n\lambda = d(\sin\theta_i + \sin\theta_r) \tag{14.10}$$

For UV closely-spaced parallel lines 1000–2000/mm,are grooved whereas for IR they are 10–200/mm. They are called blazing and there is a particular angle in which they are blazed and is called blazing angle.

Example

For $\theta_i = 30°$, $\theta_r = 45°$ and grating ruled at 2000 lines/mm (blazes)

$$n\lambda = d(\sin\theta_i + \sin\theta_r) = \frac{1\text{mm}}{2000}(\sin 30° + \sin 45°)$$

$$= 6.03 \times 10^{-7} \text{ m or 603 nm} \tag{14.11}$$

At 603 nm region one obtains first order spectrum, at $\dfrac{603 \text{ nm}}{2} = 301.5$ nm we obtain and at

2nd order spectrum and at $\dfrac{603 \text{ nm}}{3} = 201$ nm the 3rd order spectrum is obtained.

Thus the only drawback is that higher order diffraction will give different λ's at same angle. To overcome this, filters can be used to reduce multiple order intensity.

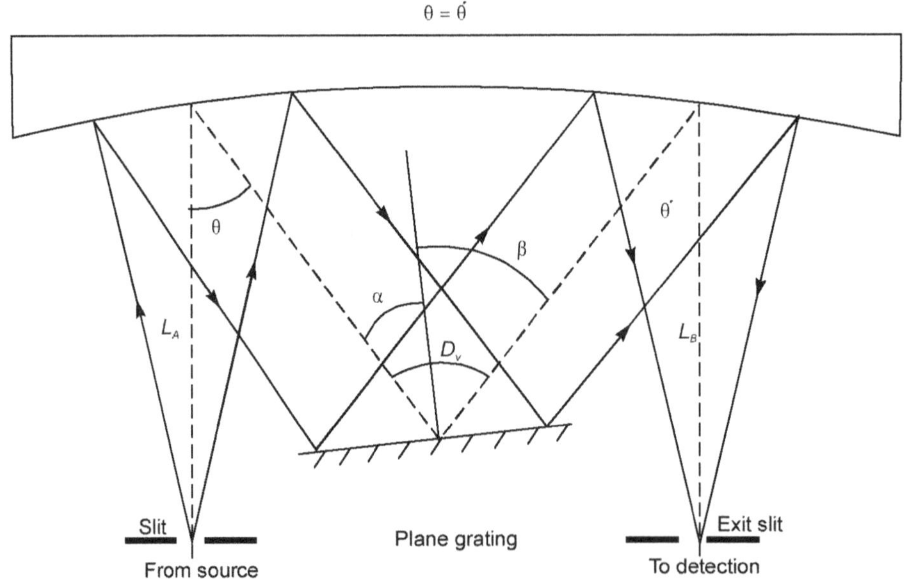

Figure 14.13   Plane grating configuration

In the grating spectrometers, the Littrow mounting is almost the same, with the prism being replaced by gratings. Plane gratings are now commercially available with high quality perfection. Plane grating systems (PGS) and Aberration corrected holographic grating(ACHG) systems are commercially available.

The other commonly employed mounting is Fastie–Ebert mounting which generally consists of one large spherical mirror and one plane diffraction grating (*See* Figure 14.13). A portion of the mirror first collimates the light, which will fall upon the plane grating. A separate portion of the mirror then focuses the dispersed light from the grating into images of the entrance slit in the exit plane.

It is a commonly used design, but has limited ability to maintain image quality off axis due to system aberrations such as spherical aberration, coma, astigmatism, and a curved focal field.

14.2.5 Czerny–Turner Configuration

The Czerny–Turner (CZ) monochromator consists of two concave mirrors and one plane diffraction grating (*See* Figure 14.14). The design is almost the same as in Ebert mounting, but uses two mirrors in the place of the single spherical mirror. The first mirror M_1 collimates the light source and second one M_2, focus the dispersed light from the grating, the geometry of the mirrors are flexible.

By using an asymmetrical geometry, a Czerny–Turner configuration may be designed to produce a flattened spectral field and good coma correction at one wavelength. In this type of mounting also spherical aberration and astigmatism may be prevalent at all wavelengths.

It is also possible to design a system that may accommodate very large optics.

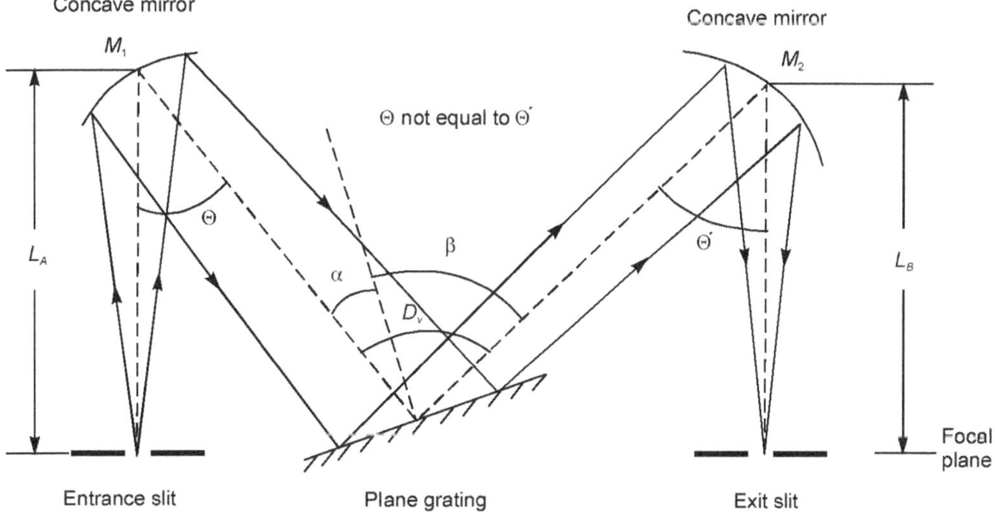

Figure 14.14   Czerny–Turner configuration

14.3 FABRY–PEROT INTERFEROMETER

This interferometer makes use of multiple reflections between two closely spaced partially silvered surfaces. Part of the light is transmitted each time the light reaches the second surface, resulting in multiple offset beams which can interfere with each other. The large number of interfering rays produces an interferometer with extremely high resolution, somewhat like the multiple slits of a diffraction grating increase its resolution (*See* Figure 14.15).

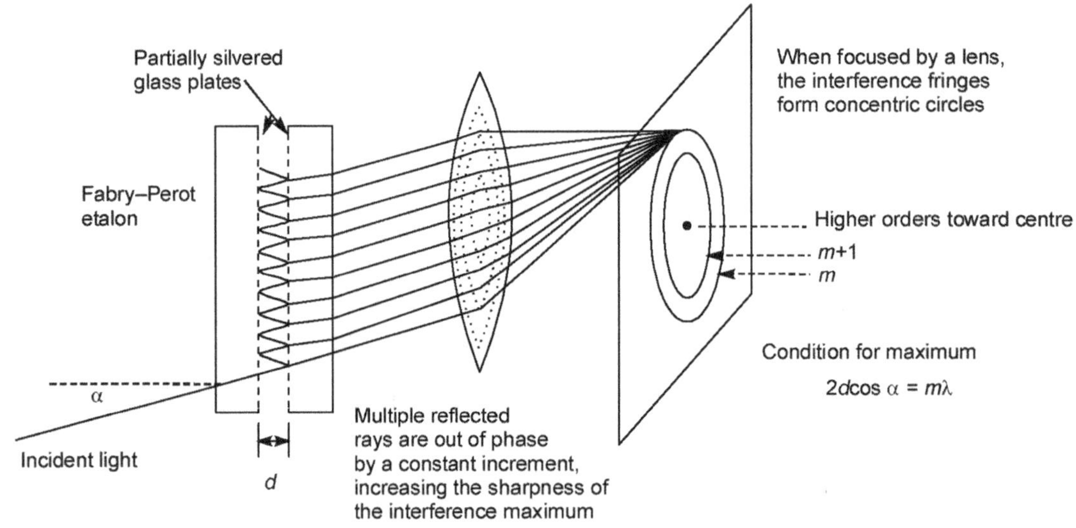

Figure 14.15   Fabry–Perot interferometer

14.3.1 Fabry–Perot Geometry

Two optically flat glasses or quartz plates, each partially silvered on one face on rigid frames separated by distance d as shown in Figure 14.16. The two surfaces are made parallel with high degree of precision. Light from the source S passing through the interferometer undergoes reflections back and forth and the emerging parallel rays are brought together to interfere in the local plane of a lens. Thus the Fabry–Perot interferometer makes use of multiple reflections which follow the interference condition for thin films. The net phase change is zero for two adjacent rays.

The path difference between successive rays is $2\mu d \cos \alpha$ and for a given angle α one obtains a bright circular fringe on the screen if

$$2\mu d \ \cos \ \alpha = n\lambda \qquad (14.12)$$

where n is a whole number.

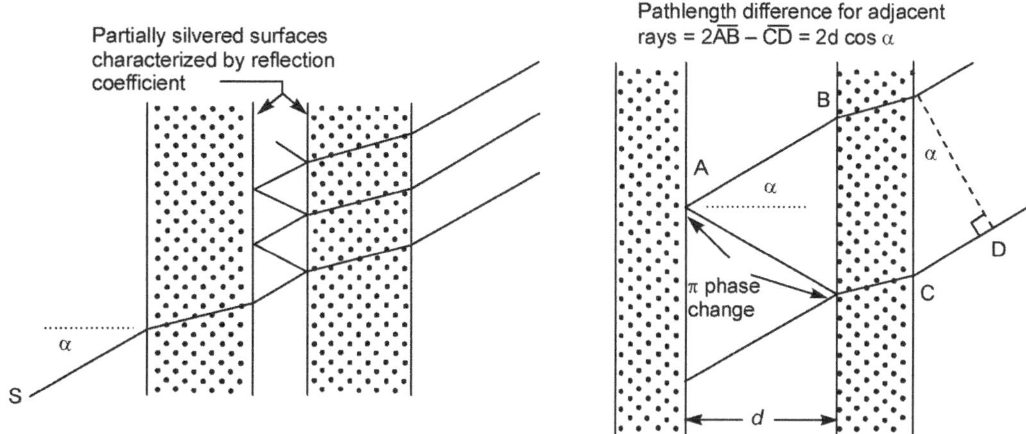

Figure 14.16   Fabry-Perot geometry

14.3.2 Interference Condition for Thin Films

Interference condition for thin films are pictorially represented in Figure 14.17.

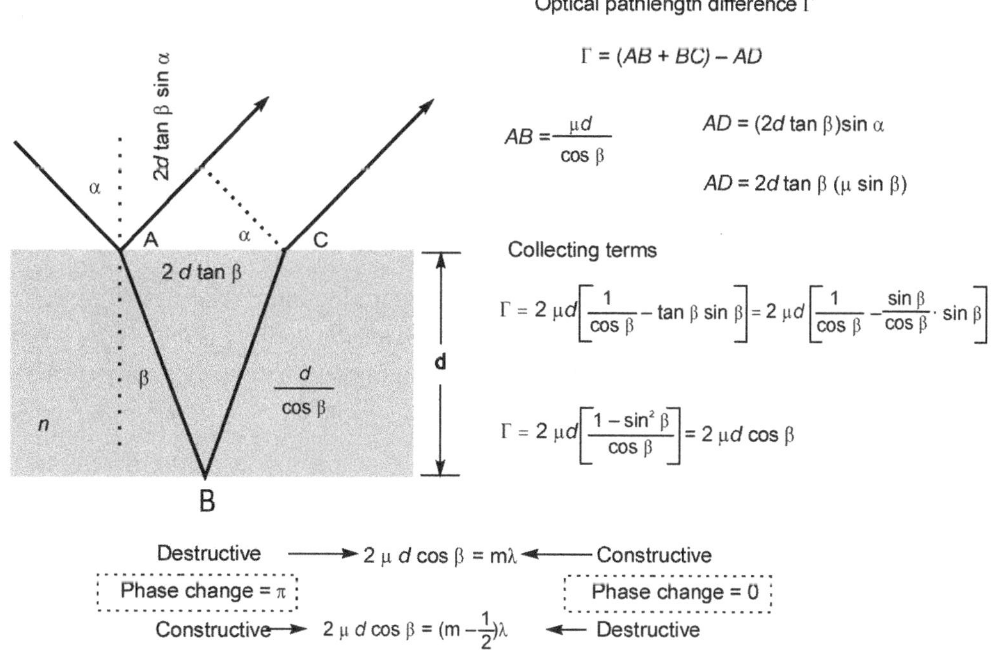

$$\Gamma = (AB + BC) - AD$$

$$AB = \frac{\mu d}{\cos \beta}$$

$$AD = (2d \tan \beta)\sin \alpha$$

$$AD = 2d \tan \beta \,(\mu \sin \beta)$$

Collecting terms

$$\Gamma = 2\,\mu d\left[\frac{1}{\cos \beta} - \tan \beta \sin \beta\right] = 2\,\mu d\left[\frac{1}{\cos \beta} - \frac{\sin \beta}{\cos \beta}\cdot \sin \beta\right]$$

$$\Gamma = 2\,\mu d\left[\frac{1 - \sin^2 \beta}{\cos \beta}\right] = 2\,\mu d \cos \beta$$

Figure 14.17   Interference condition for thin films

In the above expressions m is a whole number and λ is the wavelength of light falling on the interferometer and μ is the refractive index of the medium between the two parallel plates. For those rays for which α is such that the path difference between the successively reflected rays is $\left(m+\dfrac{1}{2}\right)\lambda$, dark circular fringes are obtained. The number m is called the order of interference.

$$m = 2\mu d \cos \alpha / \lambda \quad \text{and expressing in wavenumber units} \quad m = 2d \, v \cos \alpha \tag{14.13}$$

For very small angles α, $\quad m = 2d \, v$ \hfill (14.14)

$$\Delta m = 2d \, \Delta v$$

When $\Delta m = 1$, the free spectral range of the Fabry–Perot interferometer is

$$\Delta v = \frac{1}{2d} \tag{14.15}$$

Figure 14.18 shows a Fabry–Perot interferometer with curved cavity mirrors with reflection coefficient $R > 0.995$ or 99.5%. The curved surface mirrors are coated for high reflectivity while the outer flat surfaces are anti-reflection coated.

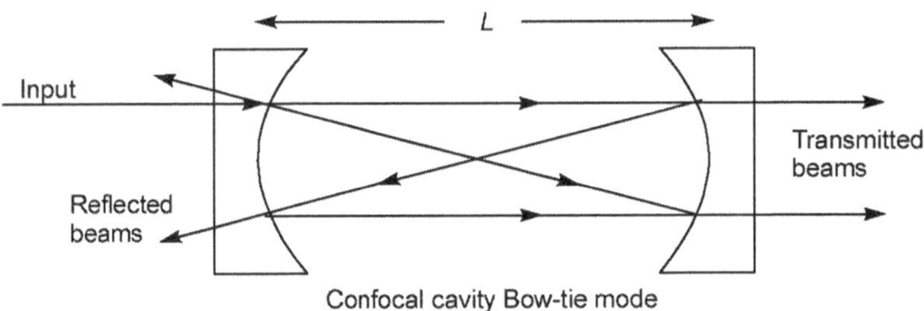

Figure 14.18   The Fabry–Perot cavity

The interferometer consists of a 0.5 inch inner diameter aluminum tube with high reflectivity curved cavity mirrors at each end. The mirrors are mounted in adjustable lens tubes by which we can control the cavity length. The optimum cavity length is roughly 20 cm giving a free spectral range of about 380 MHz.

The curved surface mirrors "trap" light in a stable, sometimes referred to as bow-tie, mode. Light intensity builds up inside the cavity whenever the round trip path length of the laser beam is equal to an integral number of wavelengths. These are called longitudinal modes. The frequency separation between the longitudinal modes is called the "free spectral range" of the cavity and is given by the equation: $\Delta f = c / 4L$.

There are also transverse modes in an optical cavity. They are characterized by differences in the intensity of the light within a cavity in directions transverse to the direction of propagation.

Usually, each transverse mode has a different wavelength. However, in a configuration called a confocal cavity, where each mirror has the same radius of curvature and the cavity length is equal to the radius of curvature, all transverse modes become degenerate and resonant at the same frequency.

The Fabry–Perot cavity is used primarily to calibrate the laser sweep and to observe sidebands on the current modulated laser output.

14.4 DETECTORS

Traditionally photographic plates were used for photographing the spectra. It had the advantage of recording a large region of the spectrum over and above the advantage of exposing the plate for a longer period and is a method for conveniently making time integration for very weak lines. Photoelectric detectors increasingly replace photographic plates nowadays. These are more useful when time resolution is needed as in the case of lifetime measurements of the excited states. For line shape measurements these kinds of detectors are more useful. To resolve two wavelengths on a photomultiplier a slit is used in front of it and in order to control the exposure time in a photographic plate a shutter is used.

14.5 INFRARED ABSORPTION SPECTROSCOPY

Infrared spectroscopy is most probably the widely used instrumental technique in analytical chemistry today. Many properties of infrared technique contribute in the universal use of infrared spectroscopy, viz., the easiness in handling the spectrometer, the wide range of facility in classifying the compounds according to the chemical class or chemical groups, the ability to study various states of the compound and the ease in identifying the compound with already available library data of frequencies. For a qualitative and quantitative analyses, infrared spectroscopy is widely used. Though IR is used mainly in molecular spectroscopic studies it will not be out of place if we describe a spectrometer here.

The source for the infrared can easily be obtained by electrically heating an inert solid to a temperature in the range 1500–2000 K. The heated material will then emit infrared radiation.

The Nernst glower is a commonly used infrared source. It is a cylinder (1–2 mm diameter and approximately 20 mm long) of rare earth oxides. A current is passed through the cylinder after two platinum wires are sealed to the ends. The Nernst glower can reach temperatures of 2200 K.

The globar is another source of IR radiation. It is a silicon carbide rod (5 mm diameter, 50 mm long) which is electrically heated to about 1500 K. Water cooling of the electrical contacts is needed to prevent arcing. The spectral output is comparable with the Nernst glower.

The incandescent wire is another infrared source. It is a tightly wound coil of nichrome wire, electrically heated to 1100 K. It produces a lower intensity of radiation than the Nernst or globar sources, but has a longer working life.

There are mainly four categories of IR detectors:

Thermal They are sensitive in IR region ($\lambda > 750$ nm).

Thermocouples These are junction thermometers which consists of a pair of junctions of metals of different temperature coefficients. A piece of bismuth fused to either end of a piece of antimony will serve the purpose. The potential difference (voltage) between the junctions changes according to the difference in temperature between the junctions.

Pyroelectric detectors These are devices which exhibits piezoelectric effect. In this case pyroelectric materials are used as detectors. This kind of detector is made from a single crystalline wafer of a pyroelectric material, such as triglycerine sulphate. In any dielectric material when an electric field is applied across it, electric polarization occurs. However, in pyroelectric material the polarization persists even when the field is removed. The degree of polarization is also temperature dependant. The pyroelectric material is sandwiched between two electrodes and a temperature dependent capacitor made. When the IR radiation falls on the detector, there will be a change in capacitance of the material. Pyroelectric detectors have a fast response time. These kinds of detectors are increasingly used in FT-IR spectrometers. Yet another detector—the bolometers are resistance thermometers.

Photoconducting or photoelectric detectors It consists of a film of semiconducting material such as the mercury cadmium telluride deposited on a glass surface and is sealed in an evacuated envelope. Absorbing IR radiations the nonconducting valence electrons receive energy and goes to a higher, conducting state. The electrical resistance of the semiconductor then decreases. These detectors have better response characteristics than pyroelectric detectors and are used in FT-IR instruments.

14.6 PHOTON TRANSDUCERS

They are photovoltaic cells and are metal-semiconductor-metal sandwiches that produce voltage when irradiated with light (350–750 nm). Such a detector is shown in Figure 14.19.

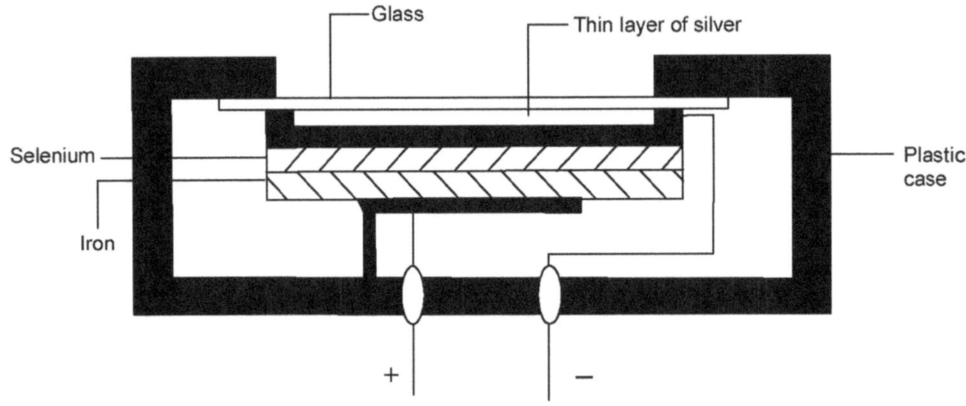

Figure 14.19 Photon transducers

Phototubes   They are tubes in which electrons produced by irradiation of cathode travel to anode. λ response depends on cathode material (200–1000 nm). It is shown pictorially in Figure 14.20.

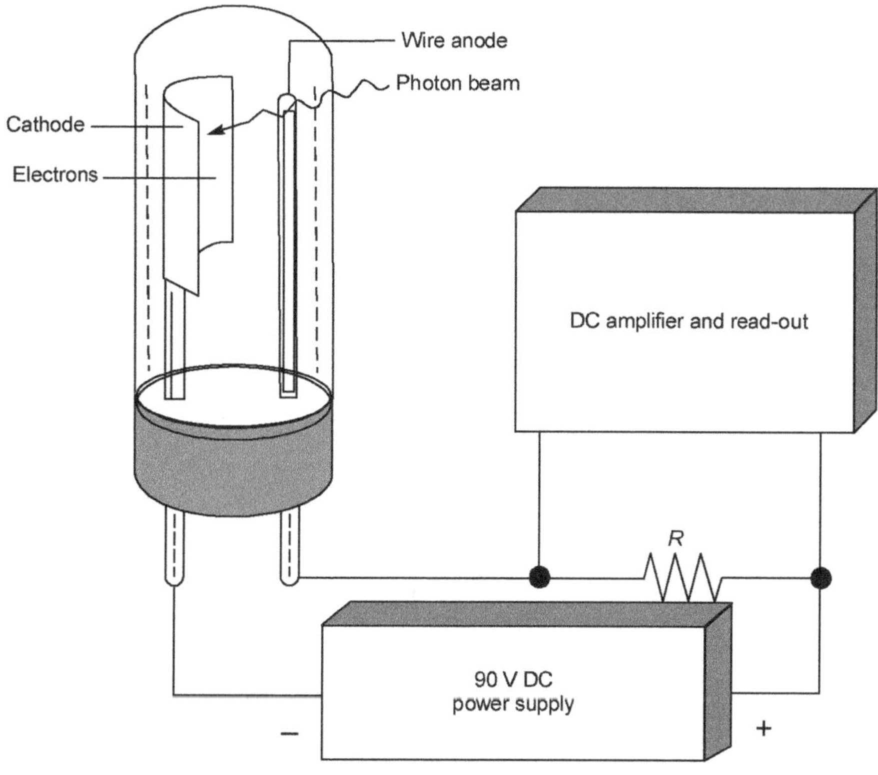

Figure 14.20   Phototube

Photomultiplier tube (PMT)   In PMT irradiation of cathode produces electrons, and a series of anodes (dynodes) increases gain to 10^5–10^7 electrons per photon. This is shown in Figure 14.21.

Photodiode arrays—(multichannel transducer)   Photon striking n-type Si creates free electrons which travel to p-type Si. In many cases, dark current reduced by cooling transducer (250 K to 1.5 K) reduces thermal excitation of electrons.

Radiation transducers have high sensitivity, low noise, wide wavelength response, linear output and low dark current.

Dispersive infrared spectophotometers   These spectrometers are usually double-beam recording instruments, employing diffraction gratings for dispersion of radiation.

Radiation from the source is passed between the reference and sample paths. Often, an optical null system is used. That means the difference spectrum is made and is used as base line. The detector only responds if the intensity of the two beams is unequal. If the intensities are unequal, a light attenuator restores equality by moving in or out of the reference beam. The recording pen is attached to this attenuator. Double-beam IR spectrometers are shown in Figures 14.22 and 14.23.

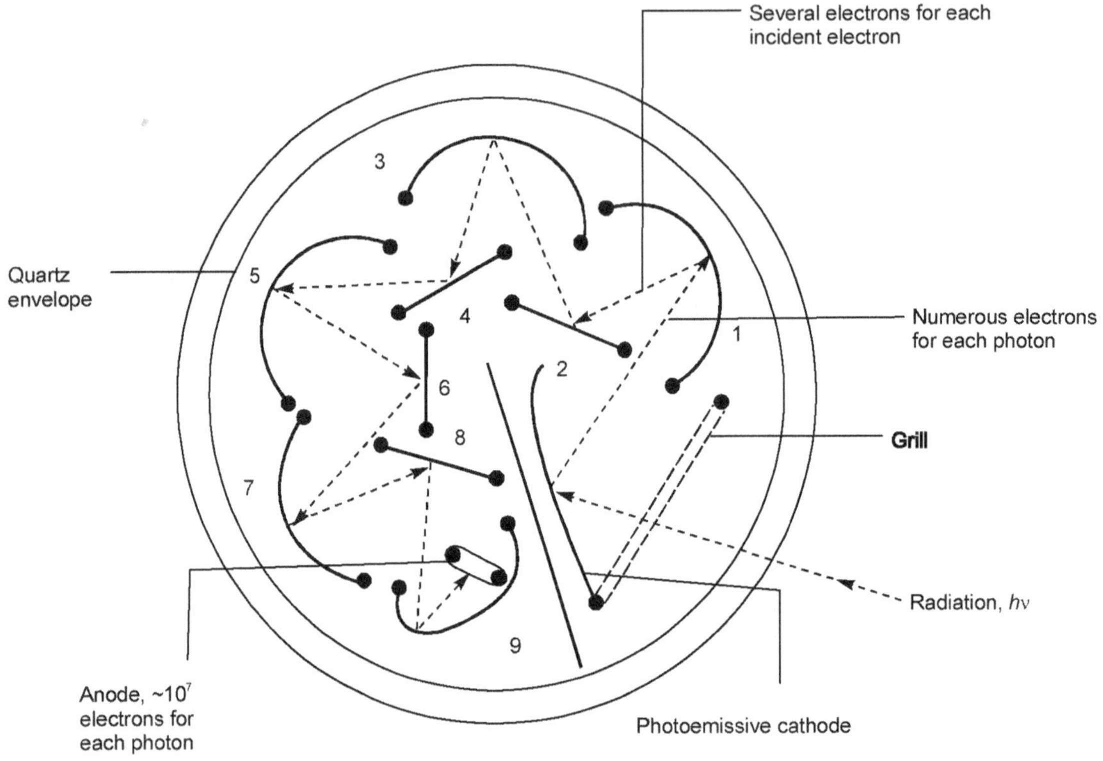

Figure 14.21 Photomultiplier tube (PMT)

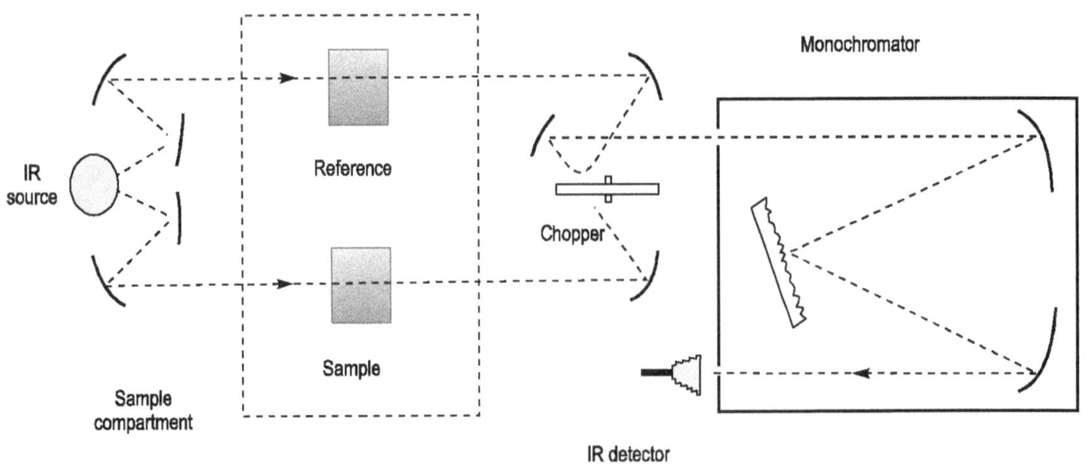

Figure 14.22 IR double-beam grating spectrometer

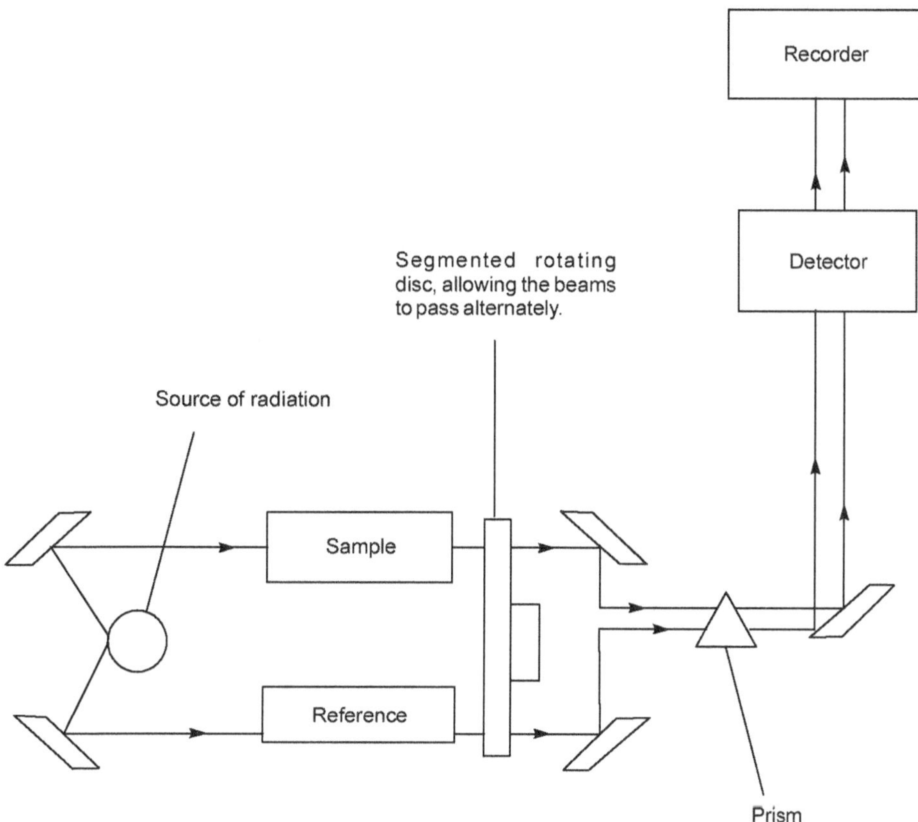

Figure 14.23   IR double-beam prism spectrometer

14.7 FOURIER-TRANSFORM SPECTROMETERS

Dispersive IR instruments operate in the frequency domain. There are, however, advantages to be gained from measurement in the time domain followed by computer transformation into the frequency domain. This is done by Fourier-transform of the signal.

Fourier-transform spectroscopy is the amalgamation of the old optical instrument, still old mathematical principles with modern computer approach. The basic old instrument is the Michelson's interferometer which was conceived in 1857 and the mathematical principle is that of the French mathematician Baron Jean Baptiste Fourier by which a complex summation of cosine waves, an interferogram, is transformed into an optical spectrum. The mathematical manipulations were very tedious and wieldy and time consuming. However, the FT spectrometers gained momentum with the availability of digital computers. It was P.D. Fellgett who first carried out a numerical analysis of Fourier-transform interferogram.

The principle behind FT method is the fact that any waveform can be shown in one of two ways; either in frequency domain or time domain (*See* Figure 14.24).

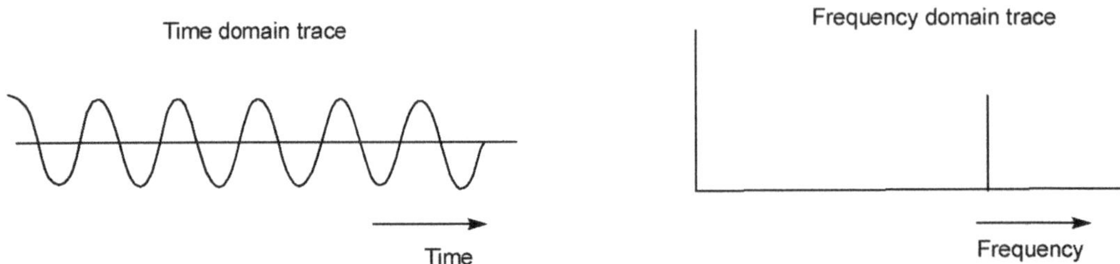

Figure 14.24   Waveform in Fourier-transform method

To record a spectrum in the time domain, radiation is allowed to fall on a detector and its response over time is recorded. In practice, no detector can respond quickly enough (the radiation has a frequency greater than 10^{14} Hz). This problem can be solved by using interference to modulate the IR signal at a detectable frequency. The Michelson interferometer is used to produce a new signal of a much lower frequency which contains the same information as the original IR signal. The output from the interferometer is an interferogram.

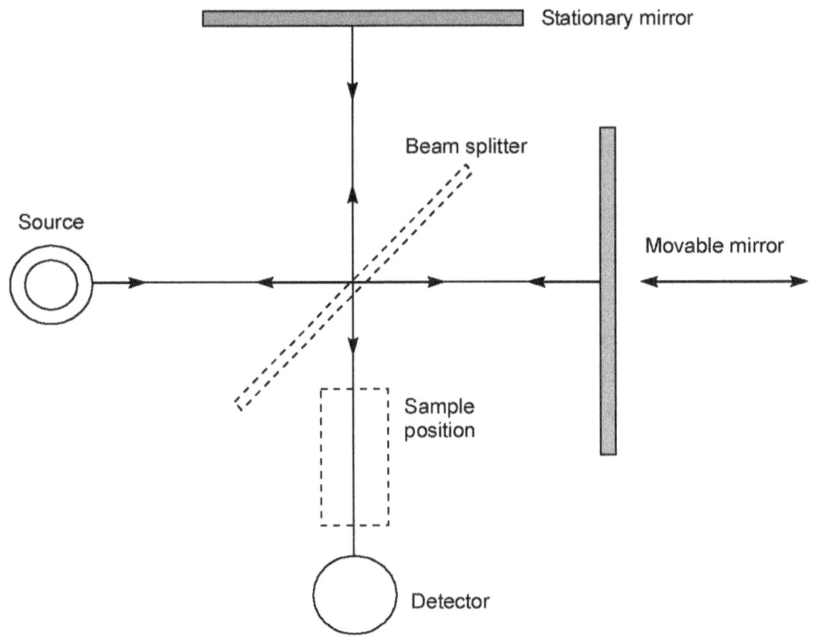

Figure 14.25   The Michelson interferometer

Thus essentially the Fourier-transform spectrometer is a version of the Michelson interferometer. The essential parts of a Michelson interferometer are two mirrors and a beam splitter. The beam splitter transmits half of all incident radiation from a source to a moving mirror and reflects half to a stationary mirror. Each component reflected by the two mirrors returns to the beam splitter where the amplitudes of the waves are combined. A schematic diagram is shown in Figure 14.25.

Radiation which falls on the stationary mirror and then back to the splitter travels a fixed distance. The path of the other half of the radiation from the source passes through the splitter and is reflected back by the movable mirror. Therefore the pathlength of this beam is variable. The two reflected beams recombine at the splitter, and they interfere.

If the two mirrors are equidistant from the beam splitter, the amplitudes combine constructively. But if the movable mirror is moved a distance equal to $\lambda/4$, the beam to and from that mirror travel an extra distance of $\lambda/2$. The emerging beam will be the result of two 180° out-of-phase beams that recombine destructively. We have here assumed a monochromatic incident radiation. The detector sees a cosine signal whose amplitude is a function of mirror distance.

For any one wavelength, interference will be constructive if the difference in path lengths is an exact multiple of the wavelength. If the difference in path lengths is half the wavelength then destructive interference will occur. If the movable mirror moves away from the beam splitter at a constant speed, radiation reaching the detector goes through a steady sequence of maxima and minima as the interference alternates between constructive and destructive phases.

We can extend the simple treatment to the case where the source is polychromatic. The signal received at the detector is a summation of all the interferences as each wavelength interfere constructively or destructively with every other component. The resulting signal is an interferogram which is a complex pattern of light amplitude vis-à-vis energy as a function of distance travelled by the mirror. The relationship between the intensity of the inteferogram as a function of the mirror travelled $I(x)$ and the intensity of the source as a function of frequency of radiation $I(v)$ is given by a cosine Fourier-transform

$$I(x) = \int_{-\infty}^{+\infty} I(v)\cos(2\pi x v)dv$$

And the inverse transform

$$I(v) = \int_{-\infty}^{+\infty} I(x)\cos(2\pi v x)dx$$

where $I(x)$ is the amplitude of the interferogram, x is the mirror displacement and $I(v)$ is the intensity of the source as a function of frequency v of the spectrum. By means of this pair of equations we obtain a relationship between the interferogram $I(x)$ and the spectrum $I(v)$. The spectrum can be accomplished by computers.

Because all wavelengths emitted by the source are present, the interferogram is usually complicated. Normally a frictionless bearing is used with electromagnetic drive to move the mirror.

The position of the mirror is measured by a laser shining on a corner of the mirror. A simple sine wave interference pattern is produced. Each peak indicates mirror travel of one half the wavelength of the laser. The accuracy of the frequency measurement depends on the accuracy of the mirror positions. In the FT-IR instrument, the sample is placed between the output of the interferometer and the detector. The sample absorbs radiation of particular wavelengths. Therefore, the interferogram contains the spectrum of the source minus the spectrum of the sample. An interferogram of a reference (sample cell and solvent) is needed to obtain the calibration of the spectrum of the sample.

After an interferogram has been collected, a computer performs a Fast Fourier-transform, which results in a frequency domain trace (i.e., intensity vs. wavenumber).

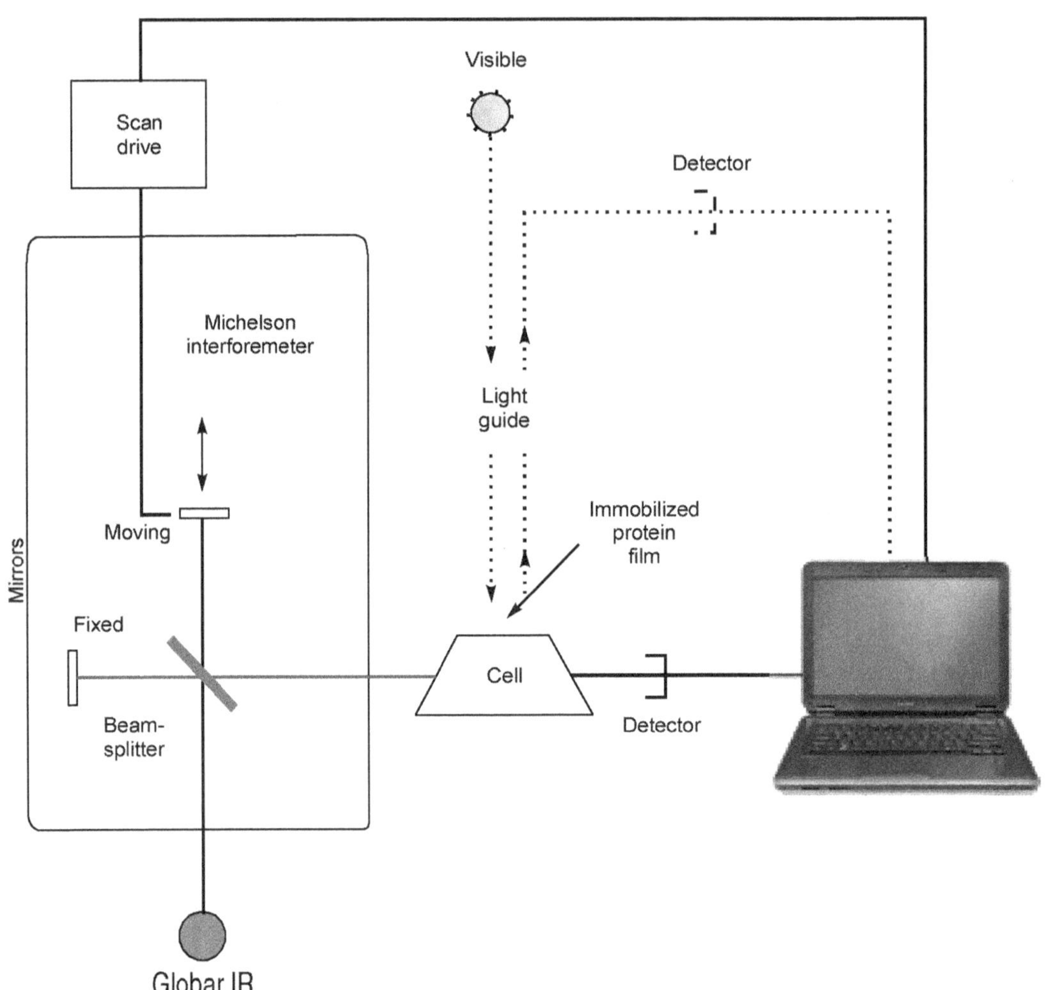

Figure 14.26(a)   Simple FT spectrometer

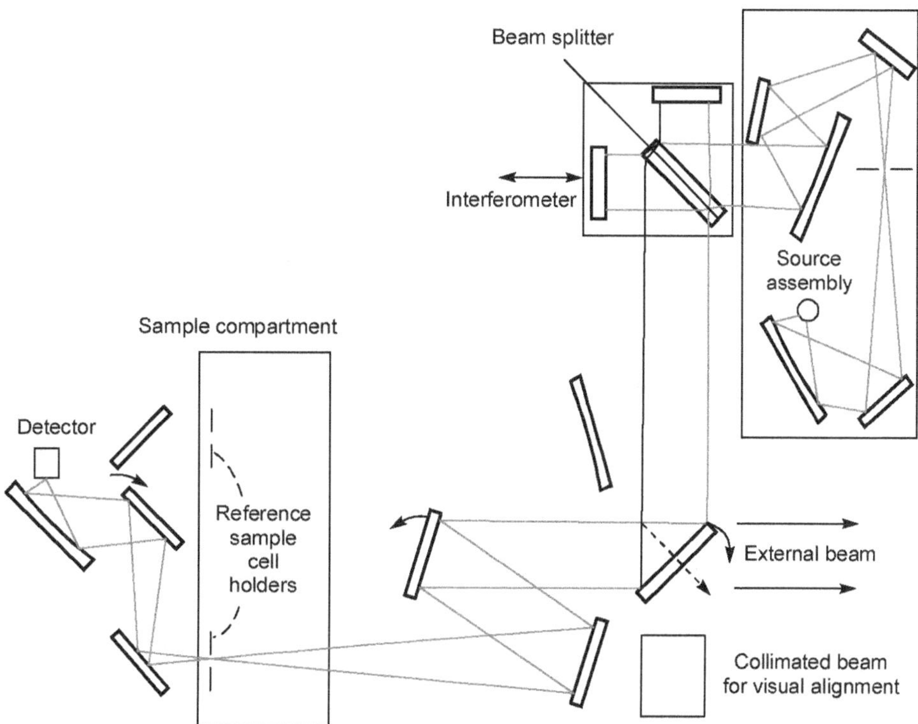

Figure 14.26 (b)   FT spectrometer with reference sample cell

The detector used in an FT-IR instrument must respond quickly because intensity changes are rapid (the moving mirror moves quickly). Pyroelectric detectors or liquid nitrogen cooled photon detectors must be used. Thermal detectors are too slow in this case.

To achieve a good signal to noise ratio, many interferograms are obtained and then averaged. This can be done in less time than it would take a dipersive instrument to record one scan.

Advantages of Fourier transform-IR over dispersive-IR are

◇   Improved frequency resolution

◇   Better frequency reproducibility (older dispersive instruments must be recalibrated for each session of use)

◇   Higher energy throughput

◇   Faster operation

◇   Computer-based recording (allowing storage of spectra and facilities for processing spectra)

◇   Easily adapted for remote use (such as diverting the beam to pass through an external cell and detector)

Two simple FT spectrometers are shown in Figures 14.26a and 14.26b.

14.8 MICROWAVE AND RADIO FREQUENCY MEASUREMENTS

Rotational spectroscopy or microwave spectroscopy studies the absorption or emission of electromagnetic radiation typically in the microwave region of the electromagnetic spectrum. These are caused by rotational motion of molecules and are associated with a corresponding change in the rotational quantum number of the molecule. The development of microwave technology for RADAR during World War II has essentially made it possible to use this technique in spectroscopy. Rotational spectroscopy is only really practical in the gas phase where the rotational motion is quantized. In solids and liquids the rotational motion is usually quenched due to collisions.

Rotational spectrum from a molecule (to first order) requires that the molecule have a permanent dipole moment. It is this dipole moment that enables the electric field of the light ,in this case microwaves, to exert a torque on the molecule causing it to rotate more quickly (inexcitation) or slowly (in de-excitation). Diatomic molecules such as oxygen (O_2), hydrogen (H_2), etc. do not have a dipole moment and hence have no pure rotational spectrum. However, electronic excitations can lead to asymmetric charge distributions and thus provide a net dipole moment to the molecule. Under such circumstances, these molecules will exhibit a rotational spectrum, i.e., in the electronic transitions.

Microwave and radiofrequency spectroscopic techniques are used for the measurement of energy changes between 10 cm^{-1} and a few thousandths of a cm^{-1}. The Zeeman and Stark splittings and hyperfine structures can be observed at wavelength regions between 1 mm to 10 cm. All these effects can also be observed in the optical region as well, however splittings are very small compared to large frequencies in the optical region. A typical hyperfine splitting in a light atom of few tenths of a cm^{-1} represents a difference of one part in 5×10^5 in the visible region whereas it is in the microwave region. The limit in the optical region is very close to the Doppler width.

The main advantage is that we can directly measure the rotational transitions in molecules or hyperfine or other small transitions in molecules or in atoms rather than indirectly measuring these differences as by optical spectroscopy. The radiative transitions between the fine or hyperfine levels depend on the transition moment, electric dipole, magnetic dipole or electric quadrupole. Magnetic interactions are more important in atoms whereas electric interactions are important in molecules. Any atom with resultant electronic angular momentum has a magnetic moment. However most stable molecules are closed shell molecules in the ground state having no magnetic moment due to the pairing of electrons to give zero nuclear spin and orbital angular momentum. As stated earlier only molecules having a permanent electric dipole moment exhibit microwave spectrum.

The most common study by microwave spectroscopy is based on the electric dipole transitions. Magnetic dipole transitions between sublevels in a magnetic field are also studied by this spectroscopy. The commonly used microwave spectrometers use stark modulation technique. The details of the instrumentation are given in the book by the author[3]. A typical Stark modulation spectrometer[5] is given in Figure 14.27. Various modifications in instrumentation have been achieved nowadays. A molecular beam Fourier-transform microwave spectrometer fabricated at the University of

Hannover by Professor Grabow[4] is an example which gives Doppler free resolutions and is very suitable for measuring small splittings.

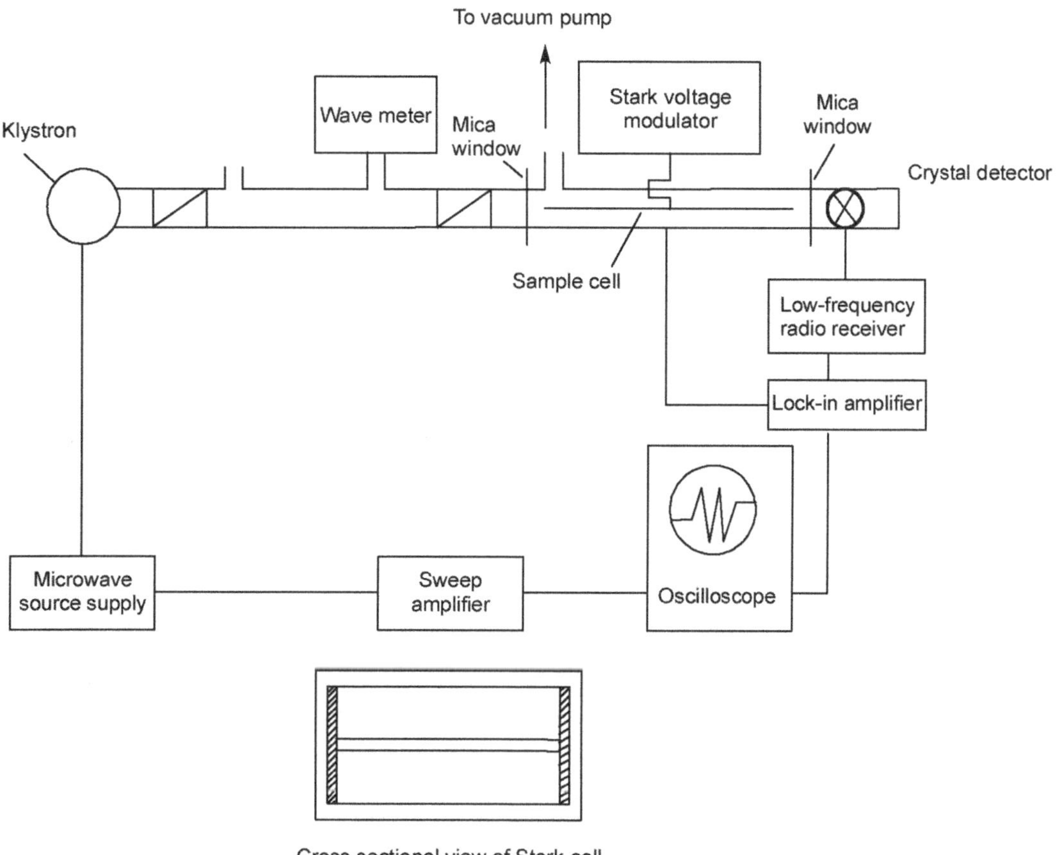

Figure 14.27   A typical Stark modulation spectrometer

14.9 MODERN TRENDS IN EXPERIMENTAL SPECTROSCOPY

In order to measure small splittings in the spectra, it is necessary to fabricate and build new experimental set-ups. These are very high resolution instruments. We have seen that a transition in the optical region involves frequency intervals of the order of 3×10^{15} Hz, the fine structure interval are of the order of 10^{11} Hz, hyperfine structures are of the order of 10^8 Hz, Doppler broadening in the optical region is of the order of 10^9 Hz, natural line width and Lamb widths are of the order $10^8 - 10^9$ Hz.

We can consider the spectroscopy beyond the Doppler limit is very high resolution though the definition is vague. The resolution in classical spectroscopy involving grating instruments is of the order of 10^5 to 10^6 and is Doppler limited. Many improvement can be made by using atomic or molecular beams as the source and instrumental resolution can be improved by multiple beam interferometer and in the following sections we describe some of the modern trends in the direction of high resolution spectroscopy.

14.9.1 Double Resonance Technique

By measurements in the microwave region one could obtain very high resolution spectra which can resolve many fine and hyperfine splittings. The large Doppler shift usually observed in the optical region is considerably reduced in the microwave region.

The double resonance experiments had been first performed in the microwave region. An atomic or molecular system can be produced to have a large population in one of the substates. This can be accomplished by optical pumping, microwave pumping by high power microwave source, or by state selection using a magnetic field to separate atoms or molecules in a particular state by using its symmetry character as in the case of Stern and Gerlach experiment or in the case of ammonia maser. Resonance from this state to other states are then accomplished by another microwave source. The pumping source must be strong and the probe source can be comparatively less in power. A unique laser–microwave double resonance had been described by Kindt, Ernst, Nair and Torring[6].

14.9.2 Atomic Beam Technique

In this experiment the source is an atomic beam and it is passed through a magnetic field of various strengths. The transitions among hyperfine levels of free atoms are measured in these kinds of experiments. The experimental set-up is as shown in Figure 14.28. The molecular beam is passed through three different sections of the magnetic field P, Q and R where the magnetic fields in the P and the R are of equal length and have equal and opposite field gradients. The field gradient in the Q region is uniform. An atom leaving the source at an angle is deflected up by the field in the region P and then it is deflected down by an equal amount by the field in the region R and thus it meets the axis again. If μ_J is the magnetic moment of the atom, the force in the z-direction acting on the atom in the inhomogeneous field B experiences a force

$$F_z = \mu_z \delta B / \delta z$$

where μ_z is the component of μ_j in the field direction. If μ_z is changed while the atom is in the region Q then the field in the region R will not bring the atoms to the axis and ultimately to the detector. However it can be accomplished if a radio frequency field is also applied in the region Q and the space quantization which determines μ_z is changed when the atom passes through this region. We can explain the phenomena considering the example of an atom in the state $J = 1/2$. μ_z has two possible values $\pm\mu_J$, in this case $+1/2$ or $-1/2$. Thus in a magnetic field the two μ_j's are as shown in Figure 14.29.

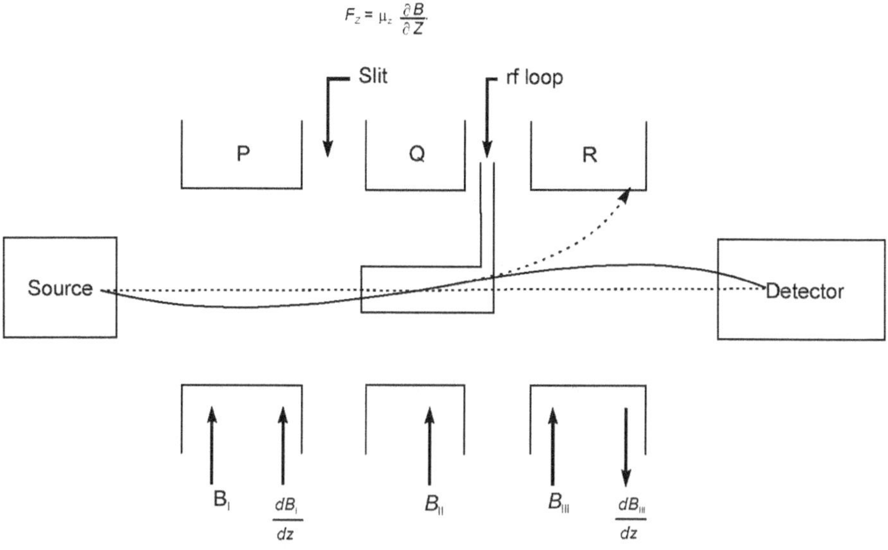

Figure 14.28   Schematic experimental set-up for magnetic resonance in an atomic beam

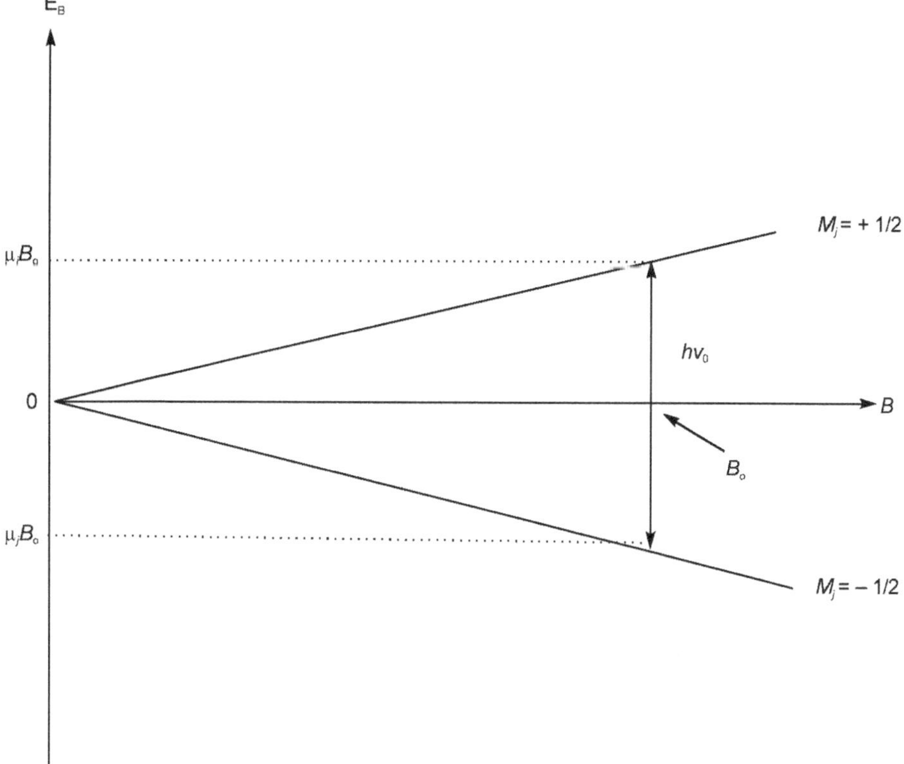

Figure 14.29   Splitting of magnetic sublevels of an atomic state $J = 1/2$ in a variable magnetic field

If a radio frequency v_0 is applied, then $hv_0 = 2\mu_J\,B_{II} = 2(1/2)(g_J\,\mu_B\,B_{II}) = g_J\mu_B\,B_{II}$

where g_J is the Lande g-factor, μ_B is the Bohr magneton and B_{II} is the magnetic field in the region II.

A number of atoms with ground state in the $^2S_{1/2}$ and $^2P_{1/2}$ have been investigated by this technique and transitions amongst the hyperfine levels were studied.

14.9.3 Laser-induced Fluorescence Spectroscopy

Laser-induced fluorescence (LIF) is widely used to prepare a large number of atoms in specific energy states by selectively pumping. The extreme high intensities together with wide tenability and pulsed operation of the dye lasers make it easy to tune to any desired frequency. Fluorescence can be obtained from these selectively excited states thereby reducing the otherwise thick spectra. The atom in the atomic beam can re-emit spontaneously after absorption of the laser photon.

14.9.4 Detection of Recoil Effect

By absorbing a photon from the laser beam, an atom in the atomic beam receives a momentum hv/c and undergoes a recoil. If Δv is the increase in velocity of the atom in a direction perpendicular to the atomic beam then from the conservation of momentum

$$M\Delta v = \frac{hv}{c}$$

$$\Delta v = \frac{hv}{Mc} \tag{14.17}$$

where M is the atomic mass. The mean thermal velocity of atoms along the beam is $v_{thermal} = \sqrt{\dfrac{kT}{M}}$ where k is the Boltzmann constant and T is the beam temperature.

The angle of deflection due to the recoil effect is then $\theta = \Delta v / v_{thermal} = \sqrt{\dfrac{(hv)^2}{kTMc^2}}$

For an optical beam $hv = 1\,\text{eV}$, $Mc^2 = 10^{10}\,\text{eV}$, $KT = 10^{-2}\,\text{eV}$ and hence $\theta = 10^{-4}$ radian

This may correspond to a lateral displacement of 0.1 mm of the collimated beam at a distance of 1 metre from the crossing point of the atomic beam and the laser beam. Detector at the appropriate position detects this. The laser frequency can be tuned and this effect can be noticed by a dip in the detector current when the laser frequency is at the resonance while sweeping the frequency.

14.9.5 Doppler-free Saturation Spectroscopy

By this method one could obtain very high resolution. With the advent of narrow line width tunable dye lasers one could obtain very high resolution in optical spectra and could obtain very small splittings resolved which give subtle details of the atomic and molecular interactions. With the help of tunable dye lasers it was possible to carry out measurements on the Balmer lines of the hydrogen atom which were then used to obtain highly accurate values of Rydberg constant R which is determined to be $R = 10973731.573(3)$ m^{-1}. High resolution laser measurements made on hydrogen and other atoms were able to eliminate the Doppler broadening of spectral lines caused by the random, thermal motion of the absorbing or emitting atoms at ordinary temperatures.

Saturated absorption laser spectroscopy employs two laser beams derived from a single laser to insure that they both have the same frequency. The beams are arranged to propagate in opposite directions and overlap one another as they pass through the target. One beam, the "pump" beam, is more intense. To perform saturation spectroscopy, one monitors the amount of absorption that the probe beam undergoes as it passes through the sample.

Let us consider two energy states, a lower level $|g\rangle$ and an upper level $|e\rangle$. Let us now use a monochromatic, tunable, highly directional, laser beam to excite atoms from lower levels, $|g\rangle$, to excited levels $|e\rangle$. When the laser frequency ν_L^0 exactly coincides with an atomic transition ν_o^0 absorption from the laser beam is maximum. The resonance is sharp in this case and the line is in a small range of frequency.

However we know that the atoms are not at rest and the random thermal motion of atoms at room temperature complicates the simple picture we mentioned. As the atoms move with different speeds along the pump-probe axis they absorb laser light of different frequencies because of the Doppler effect. Specifically, an atom that absorbs light at frequency ν_0, when at rest, will absorb laser light of frequency ν_L given by

$$\nu_L = \nu_o(1 \pm \text{v}/c) \tag{14.18}$$

The plus sign is valid when the atom moves with speed v towards the laser beam and the minus sign is valid when the atom moves away from the laser light source.

Thus if the emitting atom is moving with a velocity v along the direction of probe laser, the Doppler shifted frequency is given by the classical expression

$$\tag{14.19}$$

where v_p is the velocity component along the probe laser. This should be the resonance frequency ν_0 for the atoms in question.

Thus $\nu_p = \nu\,[1 - (\text{v}_p/c)] = \nu_0$

$$\nu_p = (\nu - \nu_0)\,\frac{c}{\nu} \tag{14.20}$$

Therefore, an ensemble of atoms moving at different thermal speeds will absorb light over a range of laser frequencies determined by the atomic velocity distribution. Thus it gives a broadening of absorbed frequencies or in other words we obtain a broadened line. This is called Doppler broadening.

The intensity absorbed from the probe laser will be proportional to the population of the atoms in the two states and is given by

$$\Delta I_p \ \alpha \ N_g \ (v_p) - N_e (v_p) \tag{14.21}$$

The atomic vapour in thermal equilibrium can be considered to have a Maxwellian velocity distribution and the population difference $N_g (v_p) - N_e (v_p)$ is proportional to the probability distribution $W(v)$ which is a Gaussian curve having a half maximum width $\Delta v = \sqrt{\dfrac{kT}{M}}$ where M is the atomic mass and T is the absolute temperature.

The absorption line will have thus an intensity width of

$$\Delta v_D = (v / c) \sqrt{\dfrac{kT}{M}} \tag{14.22}$$

This is the usual Doppler width. If the atomic beam is subjected to another laser beam of the same frequency in the opposite direction of the probe beam (usually this is achieved by the same laser by using beam splitters and reflectors) the atomic beam is under the influence of two equal beams passing in opposite directions as shown in Figure 14.30 (a). The second beam is called saturating beam or the pump beam. The atom with velocity v along the probe beam will "see" the saturating laser beam with a Doppler shifted frequency

$$v_s = v[1 + v_s / c] \tag{14.23}$$

The pump laser beam is thus in resonance with the group of atoms with velocity v_s.

$v_s = (v_s - v)c / v = v_0$. This should be the resonance frequency v_o for the atoms in question.

Thus

$$v_s = (v_s - v)c / v = v_0 \tag{14.24}$$

The pump beam generally excites the atoms with velocity v_s with a large transition rate thus the excited state gets saturated. Thus the number of atoms $N_g (v_s)$ decreases and $N_e (v_s)$ increases. g and e stands for ground and excited states. As a consequence the population difference $N_g (v_s) - N_e (v_s)$ exhibits a dip or a "hole" in the velocity distribution at $v = v_s$. Thus the pump beam or generally called as saturating beam "burns" a "hole" in the velocity distribution and it occurs at $v = v_s$.

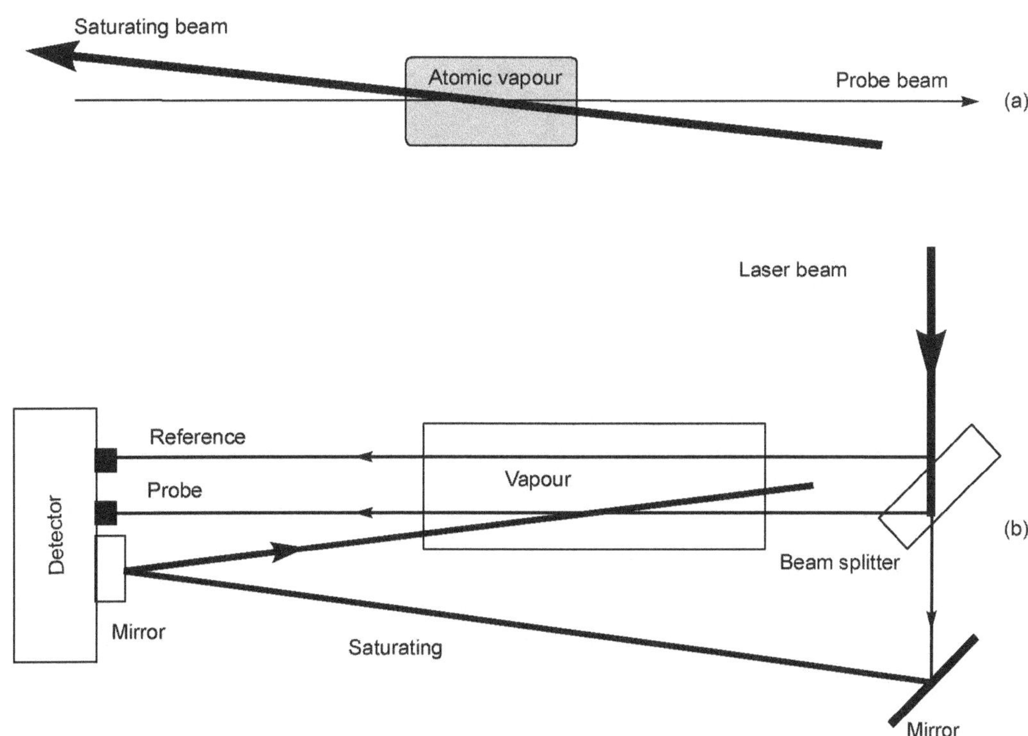

Figure 14.30   Experimental set-up for Doppler-free spectrum

The effects of Doppler broadening can be overcome in a pump-probe saturated absorption measurement scheme such as the one shown in Figure 14.30 (b). The probe and pump laser interact with the different groups of atoms with different velocity distribution. The pump (saturating) laser interacts with atoms with $v = v_s$ and the probe beam interacts with atoms with velocity $v = v_p = -v_s$. The saturating beam burns a hole at $v = v_s$ and the absorption due to the probe beam is not affected by the presence of the pump beam. Consider the familiar Maxwell distribution of atomic velocities shown in Figure 14.33(a), where the number of ground state atoms is plotted against the component of atom velocity along the pump-probe axis. It is useful to distinguish the three cases of Figure 14.33. separately.

(a) $v_L < v_o$. In Figure 14.33(a), atoms moving toward either the pump or probe beams see the pump or probe beam as blue-shifted in the atom's rest frame (*See* Figure 14.31, and the number of ground state atoms at a speed v given by

$$v = \pm c \, (v_L / v_o - 1) \tag{14.25}$$

is reduced because the laser frequency is not in the rest frame of these atoms.

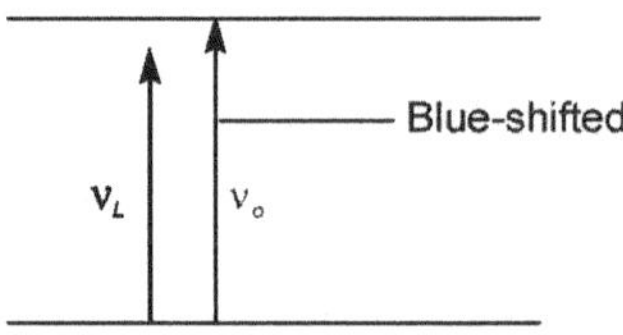

Figure 14.31 v_L and v_o in the case $v_L < v_o$

For example if we consider the probe beam and if the atoms are moving in the direction of the probe beam the Doppler shift is

$$v_L = v_o(1 + v/c) \tag{14.26}$$

and hence $v = c\,(v_L / v_o - 1)$

It is true for other atoms moving in the direction of the pump beam. It should be noted that the probe beam and the pump beam encounter different atoms depending upon the atoms' speed.

It is important to recognize in this case that because a Doppler shift of a particular sign is required for absorption, an atom that absorbs light from one beam, cannot absorb light from the other beam. The pump and probe beams therefore interact with entirely different groups of atoms within the cell. And if the pump beams pump the atoms it will be doing so only for certain atoms which "see" the frequency matching with the Doppler shifted frequency. The transparency of the cell as seen by the probe beam is therefore unaffected by the presence of the pump beam.

(b) $v_L > v_o$. Atoms moving away from either the pump or probe beams with a speed given by Eqution 14.27 will see the pump or probe beams as red-shifted to frequency (*See* Figure 14.32) v_o in the atom's rest frame and the number of ground state atoms at these speeds will be reduced, as shown in Figure 5(c). Again, because a Doppler shift of a specific sign is required for absorption, the pump and probe beams interact with entirely different atoms in the cell. Probe beam absorption in the cell is entirely unaffected by the pump beam.

$$v_L = v_o(1 - v/c) \tag{14.27}$$

and hence $v = -c\,(v_L / v_o - 1)$

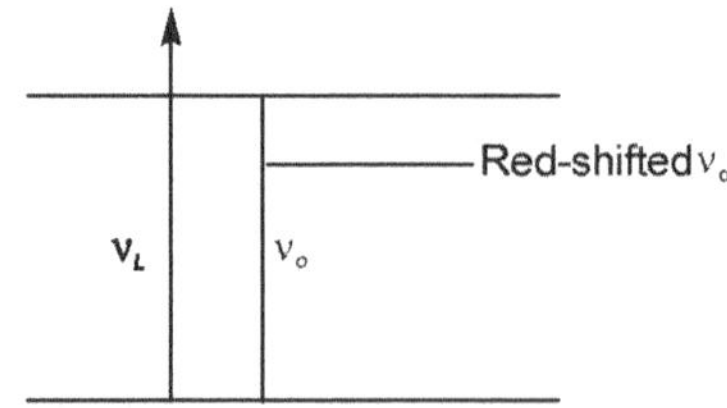

Figure 14.32 v_L and v_o in the case $v_L > v_o$

Consider next the case of a laser beam with frequency $v_L = v_o$. Figure 14.33(b) indicates what happens when such a beam passes through a gaseous sample. Only the $v = 0$ group in the ground state interacts with this beam, and hence this group is the only one that suffers depopulation due to promotion of atoms into the excited state. One says that the laser has burned a "hole" in the middle of the ground-state distribution since fewer $v = 0$ atoms remain to absorb laser light. Such hole-burning or depopulation is central to saturated absorption laser spectroscopy. In the two-level case considered here, this phenomenon requires considerable laser power and/or a relatively long excited state lifetime τ. The long lifetime helps because it delays the atom's return to the original state.

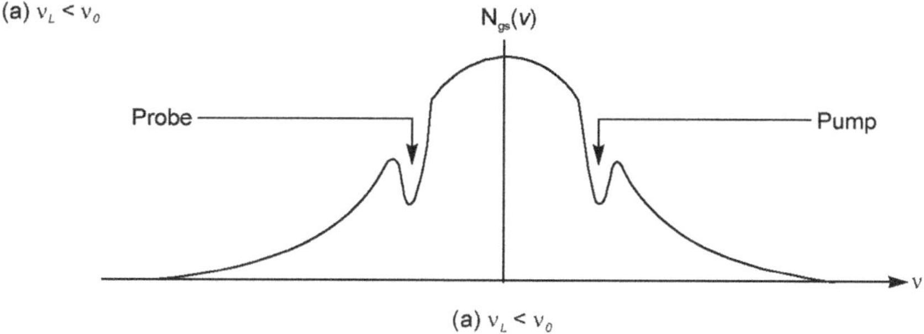

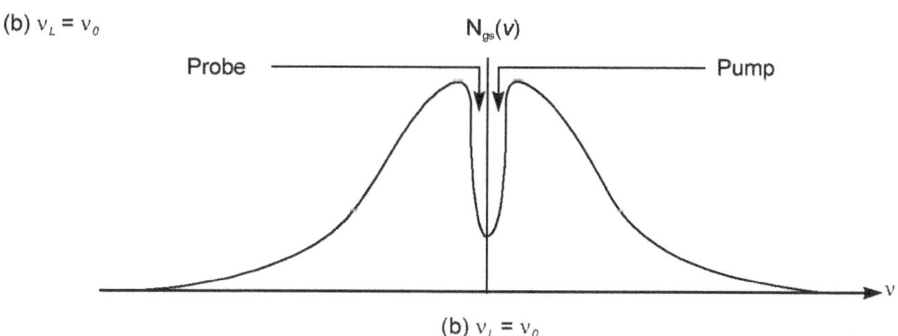

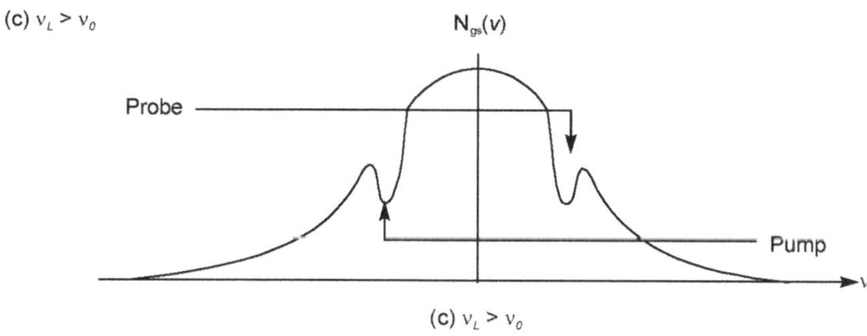

Figure 14.33 Absorption of pump and probe beams by ground state atoms

The modulation of the transmitted probe beam by the chopped pump beam occurs only when the pump and probe beams interact with the same atoms. Because of the Doppler effect this condition only occurs for atoms near $v = 0$. The lineshape of the probe signal detected by the lock-in detector is therefore determined by the natural linewidth of the atomic transition and not by the effects of Doppler broadening.

The lineshape exhibits the sharp dip called a Lamb dip characteristic of "saturated absorption." A Lamb dip can be several orders of magnitude narrower than the Doppler-broadened spectrum on which it sits. Therein lies the reason why saturated absorption signals can resolve previously shrouded features and offer the prospect of increased spectroscopic accuracy and precision.

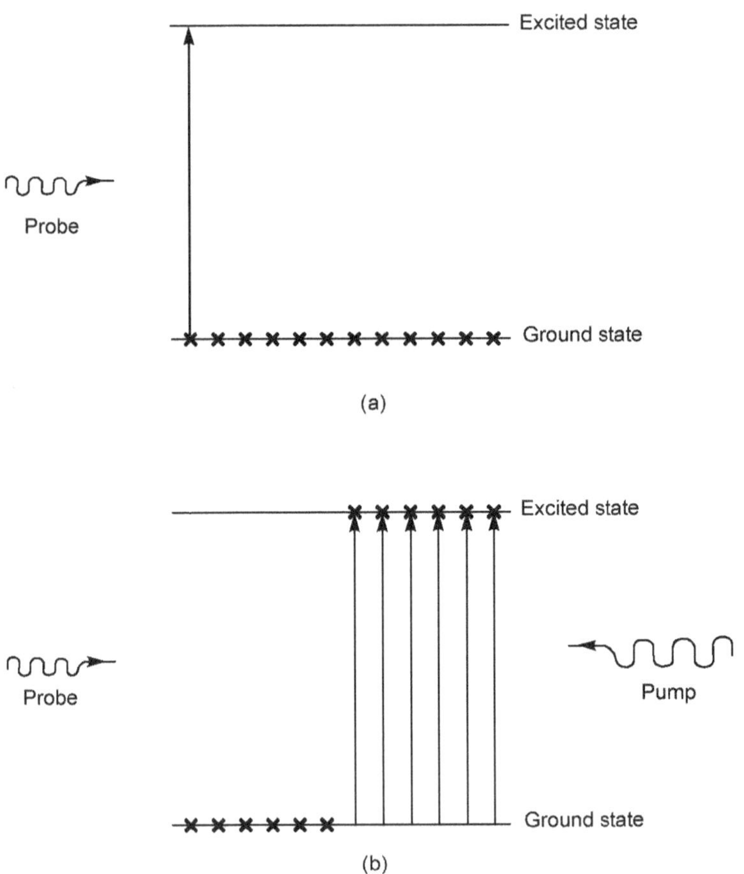

Figure 14.34   (a) Strong absorption of probe beam   (b) Reduced absorption of probe beam

Figure 14.34(a) shows strong absorption of probe beam by atoms in their ground states and Figure 14.34(b) shows reduced absorption of probe beam because pump beam has reduced the population of ground state atoms.

The experimental arrangement for Doppler-free saturated absorption spectroscopy of I_2 and Na is shown in Figure 14.35. A beam splitter BS is used to split the output beam from the dye laser is split into two beams. There are two beams from the same source now and the more intense beam is called pump beam and the less intense beam is called the probe beam, by which we will be probing the sample. The pump beam is chopped by a mechanical chopper, and, after passing through the window of the 300 MHz Fabry–Perot interferometer, traverses the I_2 or Na cell as in Figure 14.31. Note that the pump beam travels from right to left. The less intense beam, i.e., the probe beam, passes through the I_2 or Na cell from left to right, and through the slightly transmitting mirror M_3. It reaches then to a photodiode detector PD1.

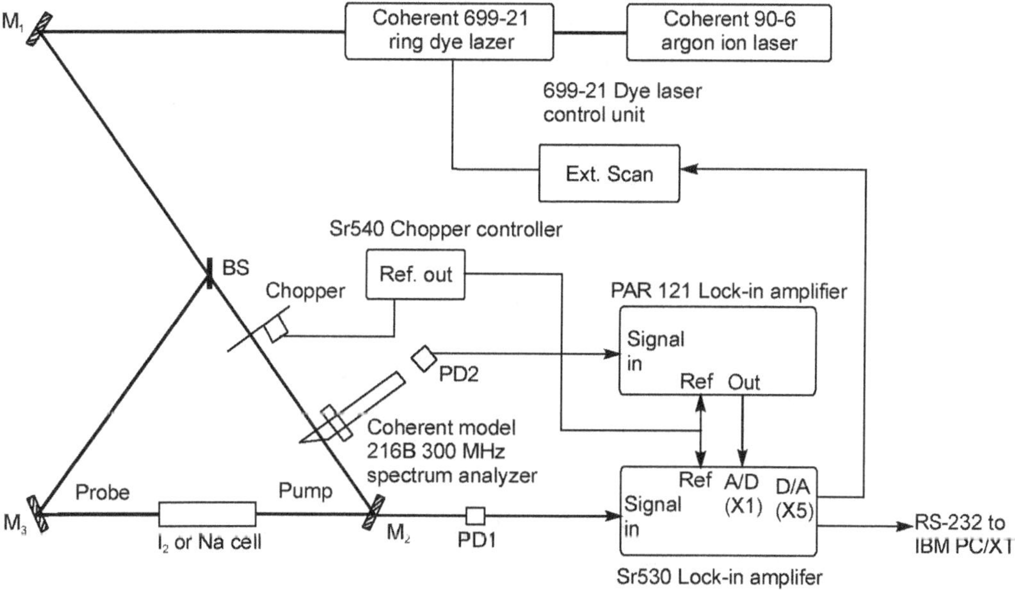

Figure 14.35   Apparatus for Doppler-free saturated absorption spectroscopy of I_2 and Na

We now consider that all the atoms in the Na cell are at rest. For clarity we will consider here that only two levels are present in the molecule. When the laser frequency ν_L is equal to the energy difference of the two levels the atoms absorb the frequency and make a transition to the upper state giving rise to an absorption frequency which is equal to the laser frequency. Thus when the laser frequency is tuned to an atomic absorption line of the Na atoms, the atoms get absorbed. As long as the pump beam is blocked by the chopper, the probe beam is absorbed by ground state atoms, as shown in Figure 14.34(a). However when we allow the pump beam to pass through the chopper, the pump intensity can be so great that a large fraction of the atoms are pumped to the excited state of the transition. As the intensity of the pump beam is large it produces what is called saturation of the transition, as depicted schematically in Figure 14.34(b). In this case when the probe beam passes through the atoms, it will see that all the atoms are in the excited state and hence there will not be any more absorption due to the probe beam. Thus the probe beam passing through the saturated medium will experience reduced absorption, due to the reduction in the number of ground state atoms available to absorb the probe beam or equivalently we can say that the probe beam experiences increased transparency. As we are using a chopper with a particular frequency, the pump beam is blocked and unblocked by the chopper. Thus the photodiode PD1 measures a modulated intensity at the chopping frequency that can be detected by a lock-in amplifier. The modulated intensity of the probe beam reflects the situation that the pump and probe beam can interact with the same atoms, here assumed to be at rest.

14.9.6 Optical Pumping

Optical pumping is a process in which electromagnetic radiation is used to raise (or "pump") electrons from a lower energy level in an atom or molecule to a higher one. It is a process used in high frequency spectroscopy which was developed by A. Kastler. It is very useful to study atomic energy states in an energy region which is not accessible by means of direct, optical observation. The technique was developed by Alfred Kastler in the early 1950's. Kastler was awarded the Nobel Prize for Physics for this in 1966.

Optical pumping technique   One can selectively populate or depopulate a state by this technique. The method was first proposed by Brossel and Kastler to change the population in the hyperfine levels of atoms. We shall here consider an atomic system with three energy levels E_1, E_2 and E_3 in increasing order of energy as shown in Figure 14.36. In order to populate the level E_2, we can proceed as follows: The first step is to pump the atoms from the level E_1 to E_3. This can be achieved by strong laser light. The next step is to bring the atoms from level E_3 to E_2 by spontaneous emission or radiationless transition. Thus the population ratio between the levels E_1 and E_2 is changed if there is no relaxation from level E_2 to E_1. Mercury, on which Brossel and Kastler did the first experiment, has $^1S_0$ as its ground state originating from the electron configuration $6s^2$. The transition from this state to the excited state $^3P_1$ originating from the $6s6p$ configuration gives rise to the strong intercombination line. Normally singlet–triplet transition is a forbidden transition. In a magnetic field the ground $^1S_0$ has no splitting though a shift is possible. The upper $^3P_1$ state splits into three components in the magnetic field, viz., $M = +1, 0, -1$. The Zeeman pattern of the 2537 Å $^1S_0 - {}^3P_1$ consists of three lines. We already know that the transition $\Delta M = 0$ is called π component and that with $\Delta M = 1$ is called σ component. This is shown in Figure 14.37.

In the experimental set-up, the mercury source is kept in a magnetic field and is illuminated by radiation from another mercury lamp after passing through a polarizer as shown in Figure 14.38. So that the polarizer allows only π radiation from the mercury lamp to pass. This excites the mercury atoms to the $M = 0$ level of the upper state and from this state only π radiation can be emitted from the source. If a detector is placed along the magnetic field and it receives only $\Delta M = 0$ components and no signal due to $\Delta M = \pm 1$ components (π components) will be detected. If a radio frequency field is applied perpendicular to the direction of detection and is tuned to the transition frequency $M = 0$

and levels of the excited state then the magnetic dipole transitions to $\Delta M \pm 1$ to the ground state can be observed. Thus the σ components can be detected. The experiment is also called sometime as radiofrequency double resonance experiment and is usually used to study the hyperfine structures in atoms.

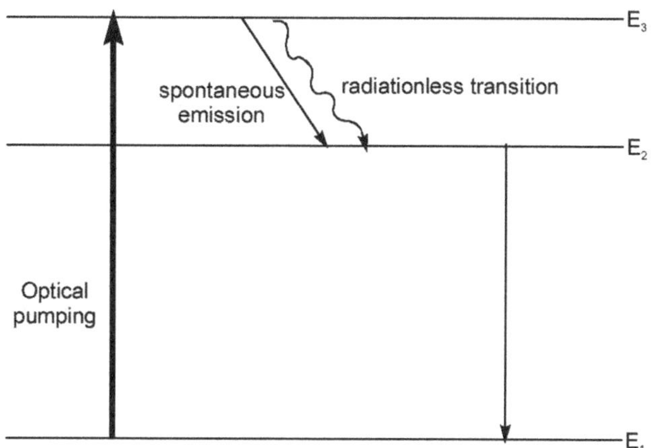

Figure 14.36   Schematic diagram of optical pumping

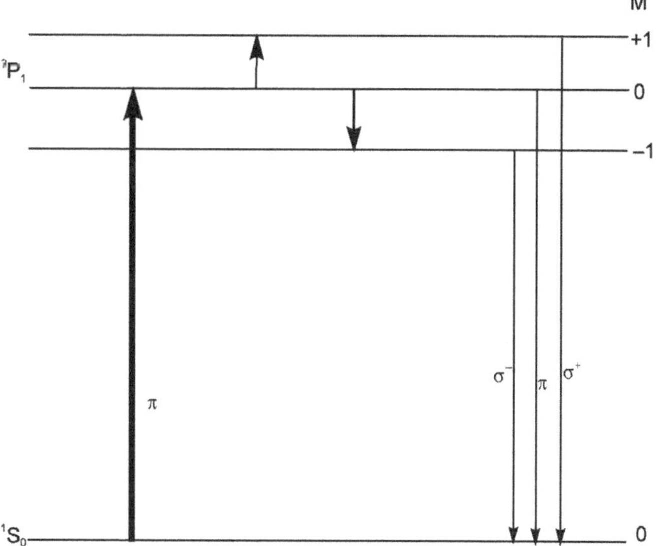

Figure 14.37   Mercury atom in a magnetic field

Transitions are induced in a low density atomic vapour by means of high-frequency irradiation. These transitions can be detected through a change in the optical absorption, which occurs during this process. With the help of high-frequency spectroscopy, it is possible to observe transitions between the Zeeman levels of hyperfine states in weak magnetic fields, where the spacing between neighbouring Zeeman states is less than 10^{-8} eV.

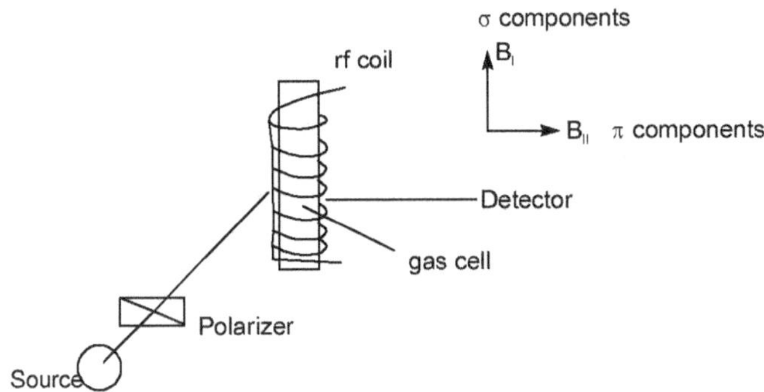

Figure 14.38   Experimental set-up for a double-resonance experiment

Optical pumping is used to achieve population inversion in laser medium.

Optical pumping is also used to cyclically pump electrons bound within an atom or molecule to a well-defined quantum state. An optical pumping experiment is commonly found in physics undergraduate laboratories, using rubidium gas isotopes and displaying the ability of radiofrequency to effectively pump and unpump these isotopes.

Optical pumping is a technique for producing spin alignment in a gas of suitable atoms. Circularly polarized light is applied at the frequency of an electronic transition from the ground state, causing transitions to an excited state. Each photon absorbed adds one quantum of angular momentum in the direction of the axis of the light beam to the system of atoms. However, atoms in the single Zeeman sublevel of the ground state with highest angular momentum projection cannot be excited because there is no excited level of higher angular momentum. Thus a surplus of atoms (if there were no relaxation process, all atoms) accumulates in this sublevel, producing a net macroscopic magnetic moment. This condition can be detected by the resulting increased transmission of the pumping light.

Rubidium is a fairly simple atom to study optical pumping There are many laboratories which use rubidium to demonstrate optical pumping. The electronic structure of rubidium closely approximates that of a hydrogen atom, so we can get a pretty good theoretical understanding. However, even in hydrogen-like atoms there are interesting effects that can be investigated spectroscopically using optical pumping.

Optical pumping can be used to polarize a gas which has magnetic dipole moments. The rubidium atom is a fairly simple atom. We know the electron configuration of rubidium is

Rb $1s^2\ 2s^2\ 2p^6\ 3s^2\ 3p^6\ 3d^{10}\ 4s^2\ 4p^6\ 5s$

Its electronic structure closely resembles that of a hydrogen atom, and hence we can get a pretty good theoretical understanding. Rubidium in its atomic state has just one valence electron and thus it can be well approximated by a one-electron-atom model. Its nuclear properties are different from hydrogen, however, and this will give it a different energy-level structure. We know that the structure of these two isotopes is same except for their nuclear spins. There are two commonly occurring isotopes of rubidium in nature, ^{85}Rb, with a natural abundance of 72%, and ^{87}Rb, with an abundance of 28%. ^{85}Rb has a nuclear spin I which is equal to 5/2 and for ^{87}Rb nuclear spin $I = 3/2$. It means that over and above the energy levels split by electron spin–orbit interaction (fs), they also split by nuclear spin interaction (hfs). If we consider a crude model of the atom which neglects the fine structure and the hyperfine structure, the energy level diagram is as shown in Figure 14.39. We will consider Zeeman effect in such a system.

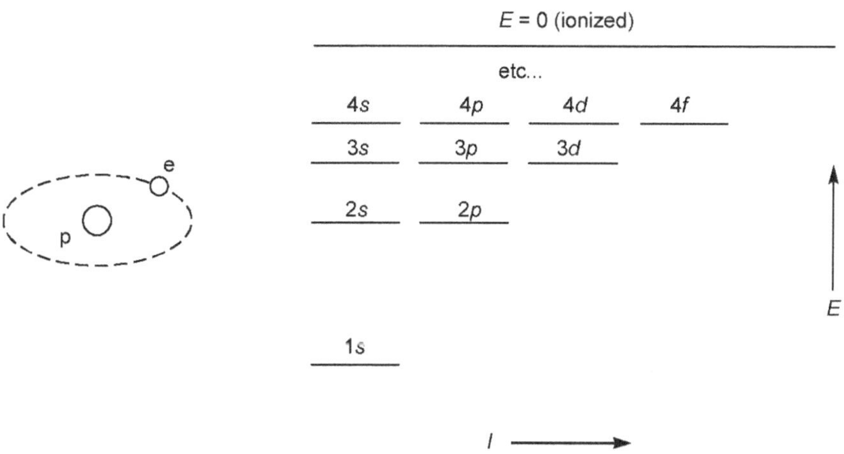

Figure 14.39   The energy-level diagram of a crude model of the atom

By taking into consideration the electron spin interaction the first three levels will be split as shown in Figure 14.40.

We have already seen that the fine structure arises from electron spin–orbit interaction and the level will be split into $2S+1$ components. The $2p$ level split into $^2P_{1/2}$ and $^2P_{3/2}$ components as shown in Figure 14.40. The notation is $^{2S+1}L_J$. S is the total electron spin and in this case it is ½. J is $(L + S)$ or $(L - S)$. If we now consider the interaction of nuclear spin I, each fine structrure level will be split into $F = (I + J), (I + J-1)...| I-J |$. The energy structure is given as in Figure 14.41(a) for $^2P_{1/2}$ and $^2P_{3/2}$. For $^2P_{3/2}$, F values for ^{87}Rb having nuclear spin $I = 3/2$ are $F = 3/2 + 3/2 = 3$ and up to $F = 3/2 - 3/2 = 0$, giving rise to $2J + 1$ components $F = 3,2,1$ and 0. Similarly for $^2P_{1/2}$, $F = 2$ and 1. A spectrum is shown in Figure 14.41(b). An experimental set for saturation spectroscopy is shown in Figure 14.42.[6]

Teachspin, Buffalo, NY produces instruments designed for teaching and the optical pumping system for rubidium is one of them.

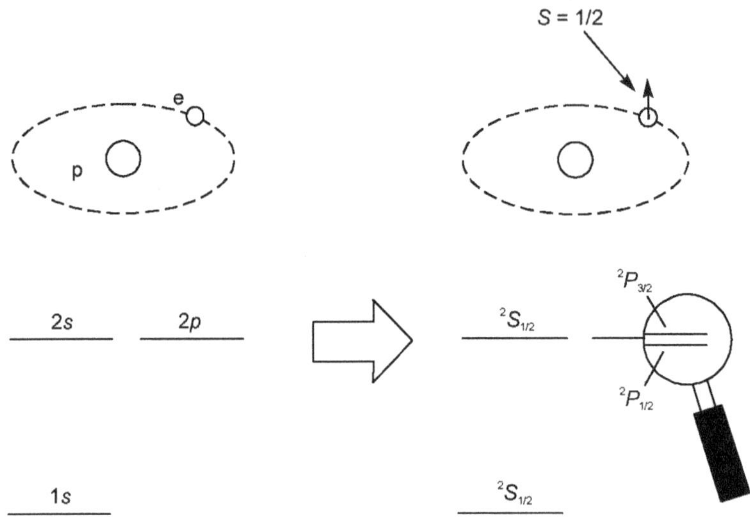

Figure 14.40   The splitting of the first three levels

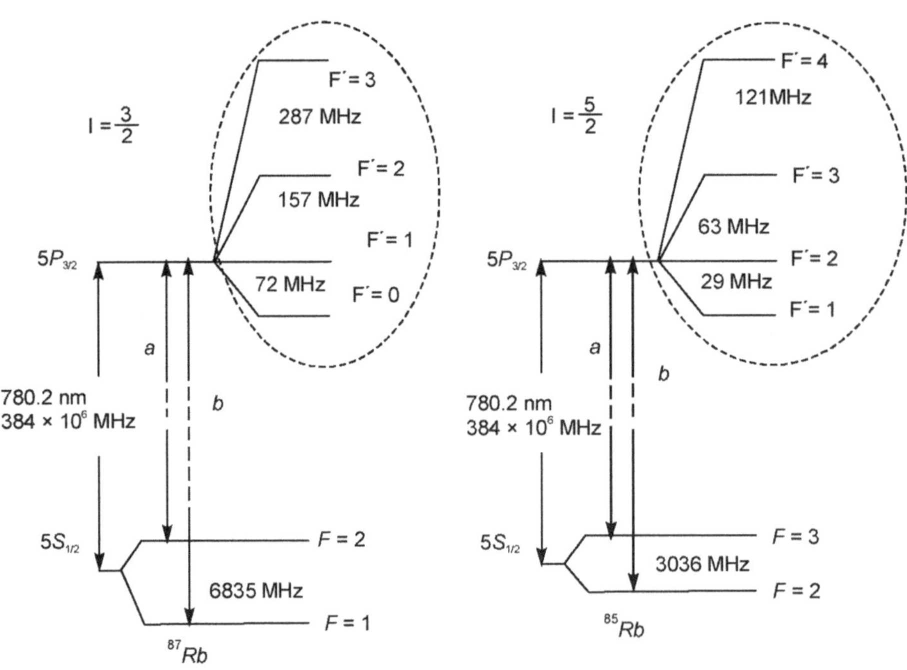

Figure 14.41(a)   The energy structure of ⁸⁷Rb and ⁸⁵Rb

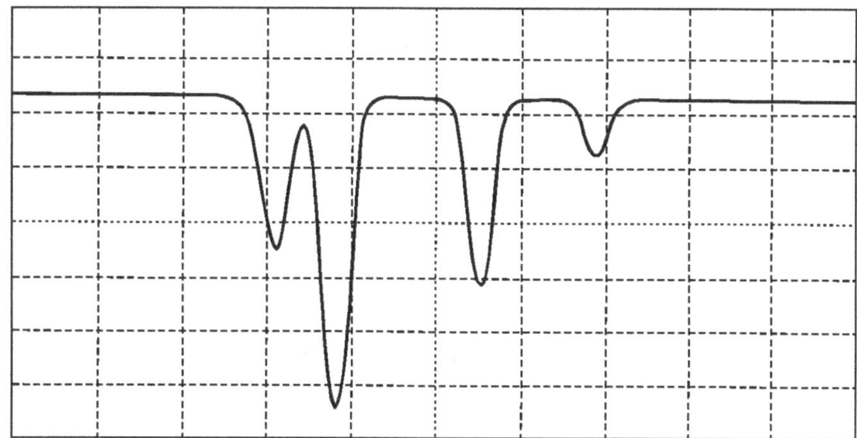

Figure 14.41(b)   The spectrum having lamb dips

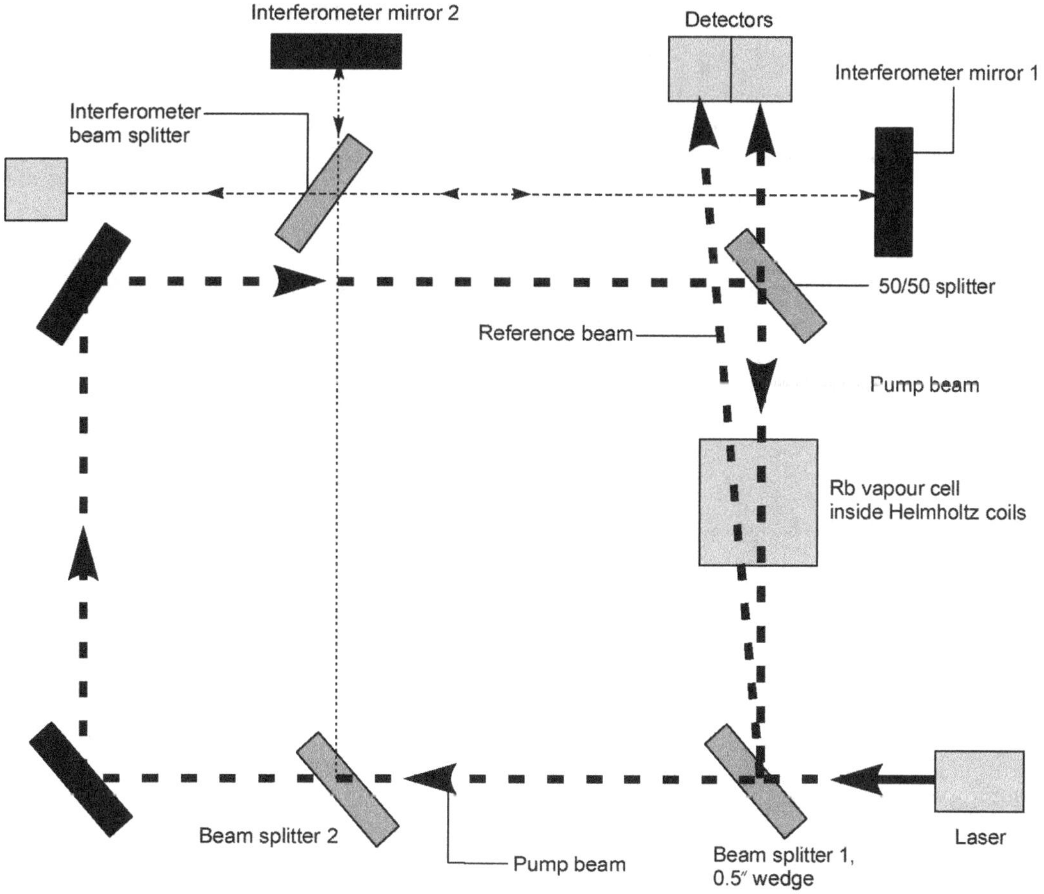

Figure 14.42   The experimental set-up of saturation spectroscopy

In the above set-up both pump and probe beam interact with the atoms having zero longitudinal velocity. The much stronger pump beam pumps the atoms and saturates the upper state and the probe beam detects the variation. It is possible to remove the Doppler background by using a reference beam and getting the difference between the signals of the probe beam and the reference beam. The beam splitter can divert a beam to a Fabry–Perot interferometer in order to measure the difference in frequencies more accurately and the Figure 14.43 shows a typical spectrum with interferometer fringes.

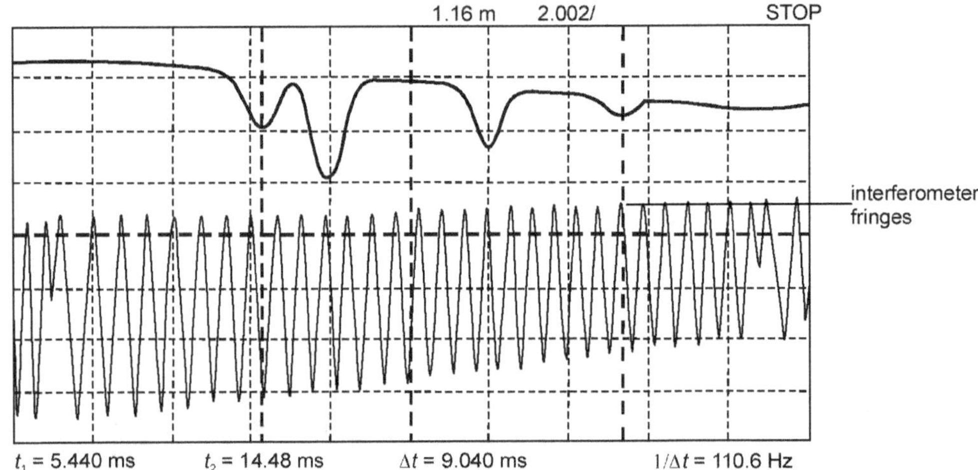

Figure 14.43   The spectrum with interferometer fringers

Placing the atom in a magnetic field we obtain again Zeeman splittings into different M components. There are two transitions each for the two isotopes designated as a and b. It should be noted from the spectra that the hyperfine structure in the excited state is not resolved in such kind of Doppler broadened spectrum. Note the ground state F values of both isotopes. They come from $F = I + J$ and $I - J$. In ^{87}Rb, $I = 3/2$ and in ^{85}Rb, $I = 5/2$ and hence $F = 1$ and 2 in ^{87}Rb, and $F = 2$ and 3 in ^{85}Rb for the ground state.

In a magnetic field the energy levels will be split into different M components.

Optical pumping in hydrogen   If we consider the $1s$ and $2p$ states of hydrogen atom and if we introduce a photon of light energy the electrons from the $1s\ ^2S_{1/2}$ level will be excited to the $2p\ ^2P_{1/2}$ level. The hyperfine levels of the $1s,\ ^2S_{1/2}$ state are close enough together that they are more or less equally populated. In the figure 14.44 they have been shown widely separated in order to have clarity of explanation. It means that their spacing is much less than k_BT, and the valence electron in any given atom has about an equal probability of being in any F state with any M. An electron in any one of these states will get excited by the photon into one of the (F, M) states in $^2P_{1/2}$, but there are only certain states it can go to because of the selection rules. For example, the electron's final M value, after it lands in the excited state, can't differ from its initial M value by more than one (The selection rule is $\Delta M = +1,\ 0, -1$). Or in other words a photon is not capable of changing M by more than one.

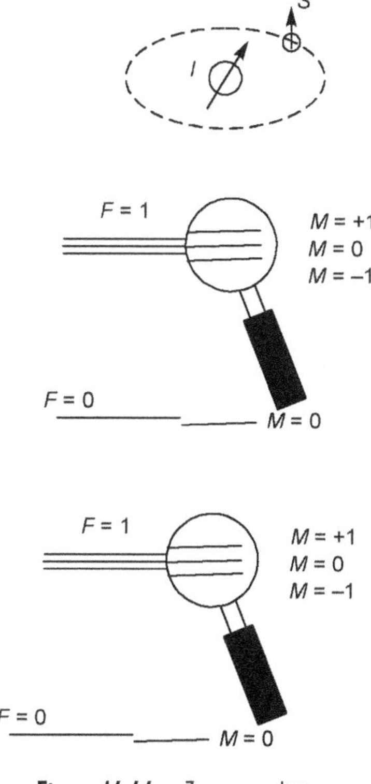

Figure 14.44   Zeeman splitting

Whether M changes by $+1$, -1, or 0 depends on the nature of the photon. If the applied magnetic field is parallel to the direction of propagation of the photon, then a right-circularly-polarized photon will always induce transitions that have $\Delta M = +1$. Left circularly polarized light will induce transitions with $\Delta M = -1$. The same thing is true for emission. An electron can fall from the $2p$, $^2P_{1/2}$ level into the $1s$, $^2S_{1/2}$ level emiting a photon with right or left circular polarization, depending on whether ΔM is $+1$ or -1. For emission, both cases are equally likely. For absorption, we control which kind of photon is incident, so we can control ΔM.

Let us now consider how the polarization of atoms work. If we shine with right-circularly polarized light and apply a magnetic field, along the same direction, to a gas of atoms, each absorption will force $\Delta M = +1$, whereas each emission event will have, on average, $\Delta M = 0$.

We shall here take the example of hydrogen atom with $I = \frac{1}{2}$ and $F = 1$ ($^2S_{1/2}$) and $F = 1$ ($^2P_{1/2}$).

By repeated absorption and emission the electrons can be brought to the highest M level of the ground state. If we apply a right circularly polarized light, $\Delta M = +1$ selection rule is followed and the electron gets into the $M = +1$ level of the upper state from the $M = 0$ sublevel of the lower state. This is the case with every absorption. The atoms re-emit a photon and on the average $M = 0$ transition occurs. After a few absorption emission processes all the electrons reach the highest

M level of the lower state as illustrated in Figure 14.40 in the case of a hydrogen atom. The same principle works for rubidium also only difference is that there can be many levels and transitions. In the case of hydrogen nuclear spin $I = \frac{1}{2}$ whereas in rubidium $I = 3/2$ (^{87}Rb) or $5/2$ (^{85}Rb). Thus with optical pumping the electrons can "walk" into the $1s$ state with the highest value of *M*, as shown in Figure 14.45.

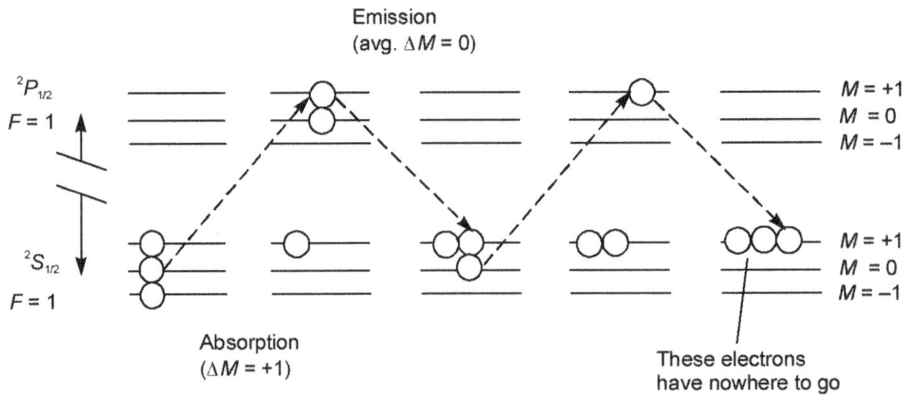

Figure 14.45   Basic principle of pumping the atoms to the highest *M* level of the ground state.

Optical pumping in rubidium   Once polarized, the gas will have a total magnetization which we could, in principle, measure. However, it is much more useful to look at how well the gas absorbs the photons. As long as absorption can occur, the gas will be partially opaque to our circularly-polarized light. Once all of the electrons are pumped into the highest *M* state of the lower state and the gas is thus polarized, no more absorption can occur. The pumped electrons have nowhere to go that would satisfy the $\Delta M = +1$ requirement. When it is in this state, the gas is transparent to incident photons.

The gas will remain transparent as long as the polarization is maintained. If we switch off the magnetic field, the polarization will be lost, and the gas will become opaque. Similarly, pumping, and hence absorption, can occur if we scramble the electrons in the ground state by applying an RF signal that is resonant with the Zeeman splitting. An energy level scheme for ^{87}Rb with nuclear spin $I = \dfrac{3}{2}$ is again shown in Figure 14.46 with Zeeman splittings of the levels.

An experimental set-up for optical pumping is shown in Figure 14.42. A beam from a diode laser is split by beam splitting into three beams: a saturation beam, a probe, and a reference beam. The probe and reference beams are of the same intensity and are passed through a low-density rubidium vapour cell. The transmitted intensities of the probe and reference beams are recorded by a differential photo-detector. This detector has the property of recording the amplified output difference between the two signals. The much more intense saturating beam enters the rubidium vapour cell from the opposite side and overlaps the probe beam.

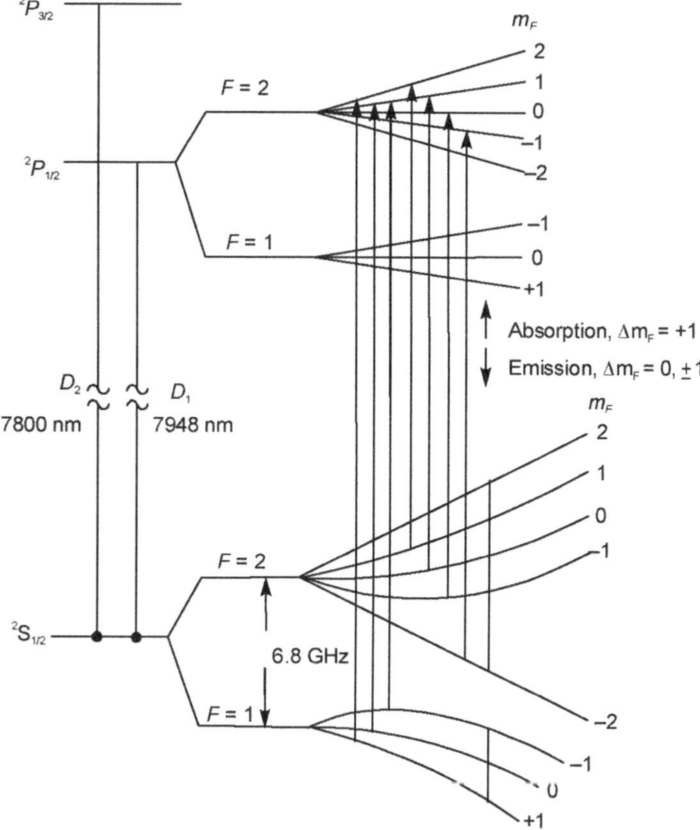

Figure 14.46 Energy level scheme for ^{87}Rb

If only the probe beam is propagating through the cell, we measure the full Doppler profile of the $5\,^2S_{1/2}$ to $5\,^2P_{3/2}$ transitions in ^{85}Rb and ^{87}Rb, as shown in the Figure 14.48.

If the saturating beam is on when the laser is resonant with a particular transition, then the probe and saturating beams will both excite atoms with approximately zero velocity along the axis of the incident beams. Since the saturating beam has a higher intensity than the probe beam, it will excite most of the atoms, allowing the probe beam to pass through virtually unabsorbed. As we are looking for signals from the probe beam, we observe a dip or a hole in the line profile which can be measured accurately. It is the Doppler-free transition frequency.

The reference beam, however, does not overlap the saturating beam and it is almost completely absorbed at resonance. If the transmitted intensity of the reference beam is subtracted from the transmitted intensity of the probe beam, then the difference in the intensities when plotted as a function of laser frequency shows the hyperfine structure transitions as a series of spikes. The hyperfine structure transitions in the case of rubidium is shown in Figure 14.48.

Another experimental set-up to demonstrate optical pumping is shown in Figure 14.49.

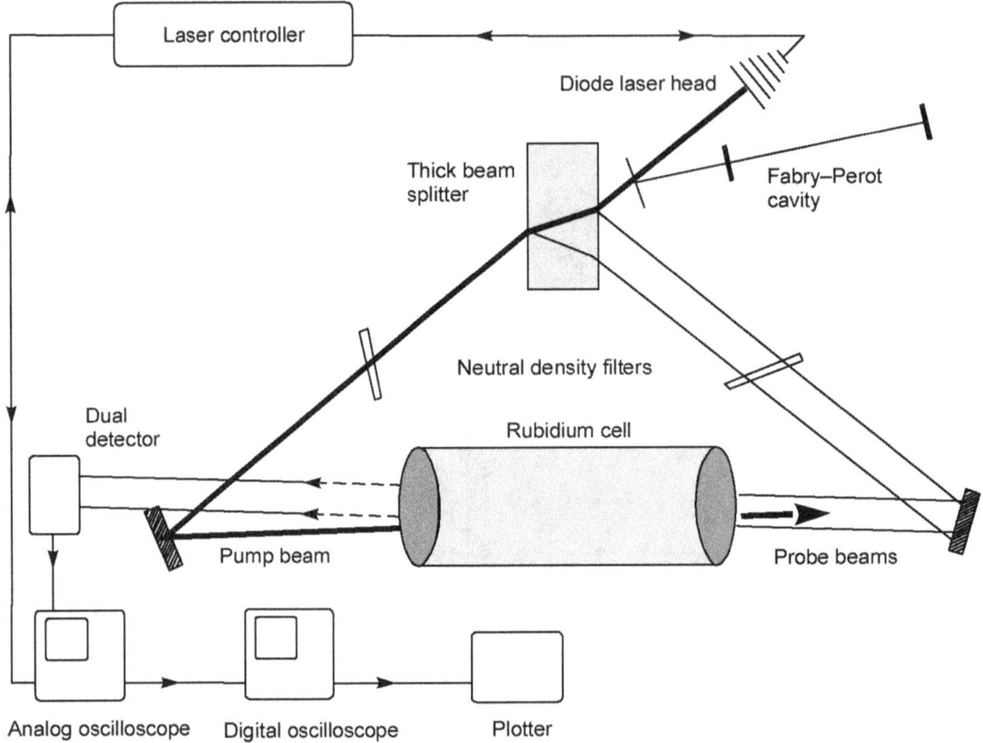

Figure 14.47   Optical pumping in rubidium

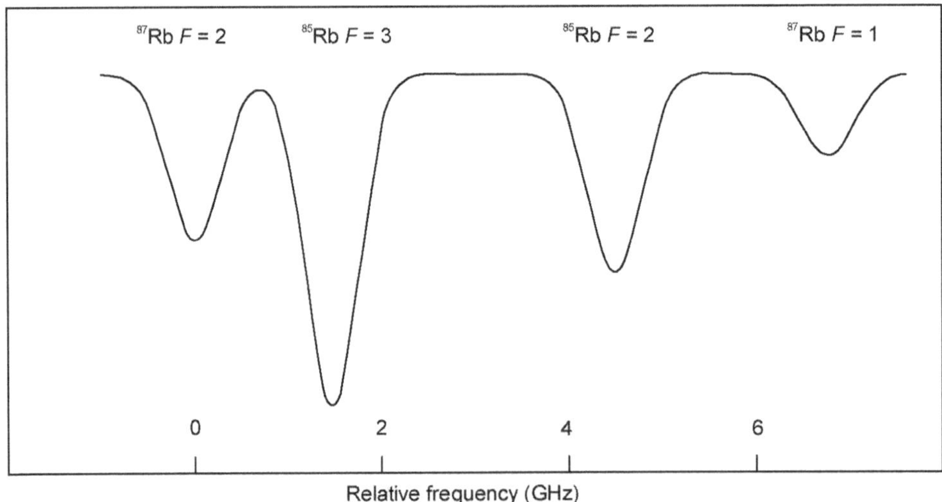

Figure 14.48   Hyperfine structure transitions

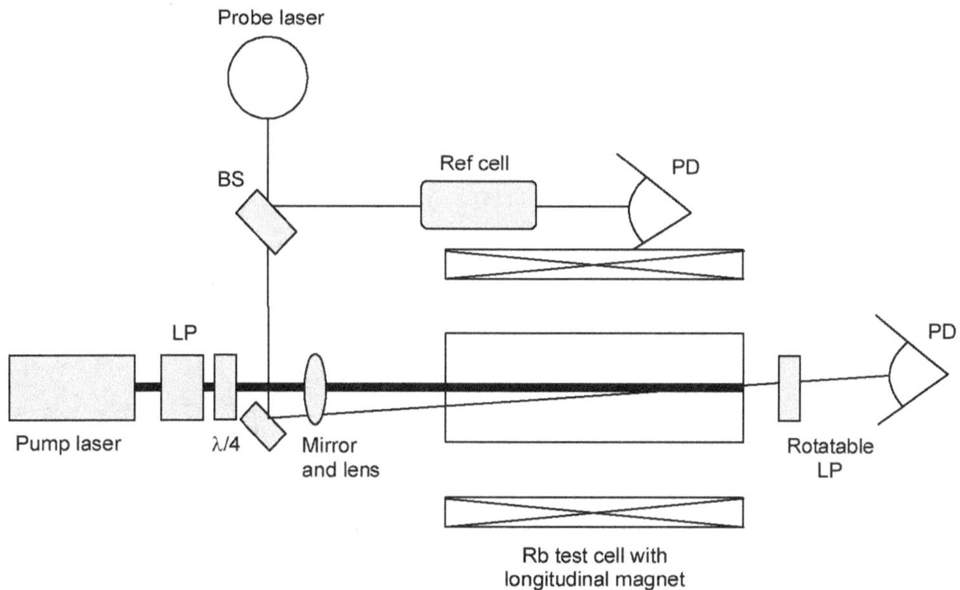

Figure 14.49   Studies of rubidium optical pumping

Apparatus   The apparatus is a 15 cm long stainless steel pipe with two 2 ¾″ conflat windows at the ends. A solenoidal electromagnet encompasses the tube. It is possible to control the pressure of various gases in the cell. The test cell has a base pressure of 1×10^{-4} Torr, produced with a diffusion pump. A diode laser is used as pump laser and passes through a linear polarizer and a quarter-wave plate before entering the cell. A single diode probe laser is split with a plain glass beam splitter. After the beam splitter, a reference beam goes through a small Rb reference cell into a photodetector. The probe beam, aligned to pass through the cross section of the pump beam, exits the test cell and passes through a rotatable linear polarizer before entering another photodetector.

We now consider the two quantum states $^1P_1$ and $^1S_0$ of an atom with a nuclear spin $\frac{1}{2}$. we already know that the total quantum number with nuclear spin is obtained as $F = (J + I)$, $(J + I - 1)$, ... $|J - I|$ where $J = (L + S)$, $(L + S - 1)$,...$|L - S|$.

In the case of $^1P_1$ state we obtain $F = \frac{3}{2}$ and $\frac{1}{2}$ and for $^1S_0$ we obtain $F = \frac{1}{2}$. Let us consider a transition between $F' = \frac{3}{2} \leftarrow F'' = \frac{1}{2}$ where F′ corresponds to upper state and F″ lower state. In a magnetic field the state having $F' = \frac{3}{2}$ will split into $(2F + 1)$ different M_F components having the values $+\frac{3}{2}, +\frac{1}{2}, -\frac{1}{2}, -\frac{3}{2}$ and the lower $F = \frac{1}{2}$ states to M_F components $+\frac{1}{2}$, and $-\frac{1}{2}$. The allowed dipole transition selection rules are $\Delta M_F = 0$ or ± 1

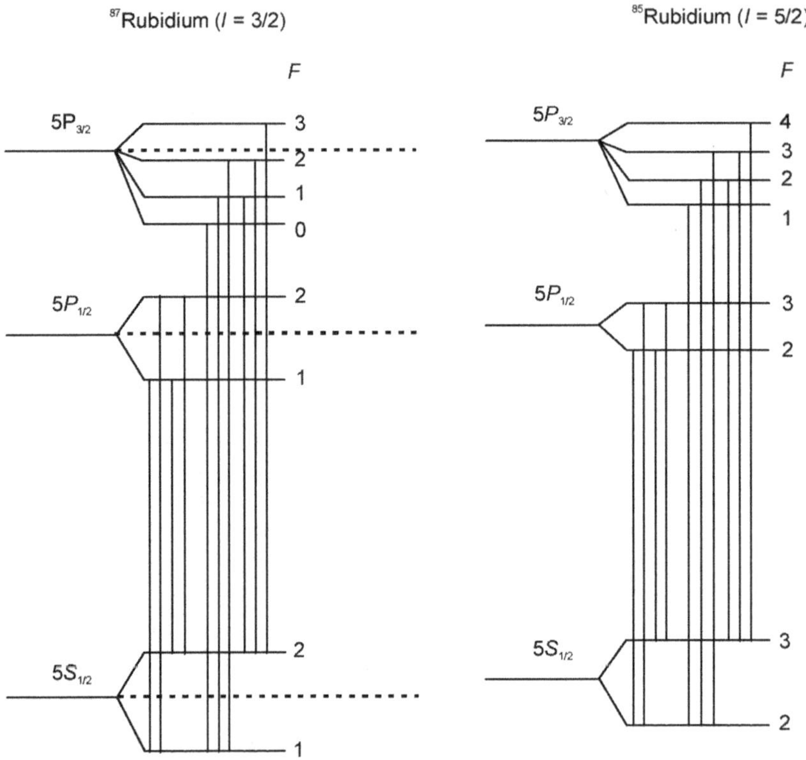

Figure 14.50   Energy level scheme for ^{87}Rb and ^{85}Rb

If there is no magnetic field, these levels are degenerate. The energy levels are shown in Figure 14.51

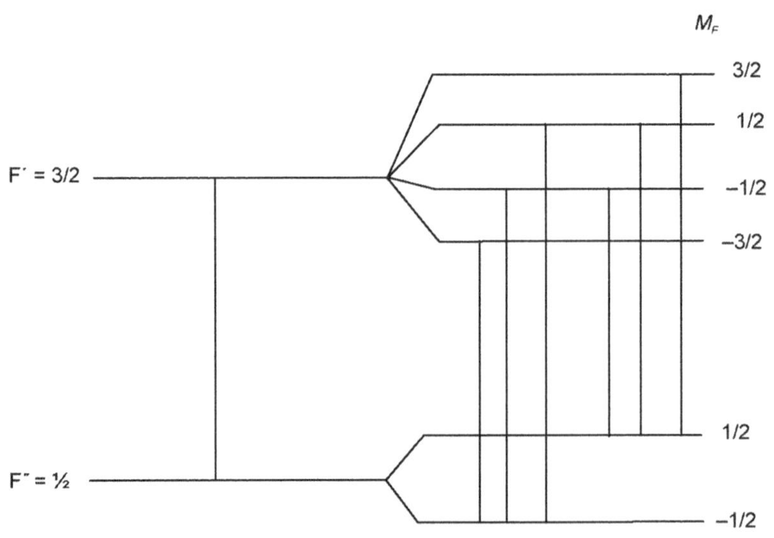

Figure 14.51   Hyperfine transitions in a magnetic field

In a field free case it can be shown in the following way.

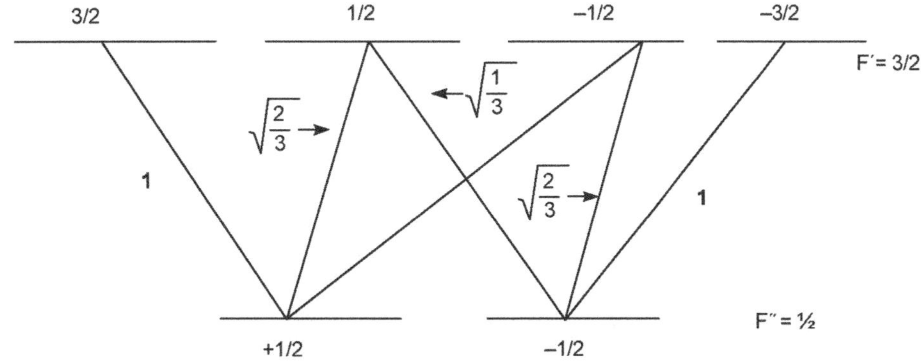

Figure 14.52   Hyperfine transitions in a field free case

Transitions between magnetic sublevels of the ground and excited states are dependent on the state of polarization of the incident light when the frequency of transition satisfies the Bohr's condition for transition

The selection rules are for LCP (σ^{+}) light $\Delta M = M' - M'' = +1$

For RCP (σ^{-}) light $\Delta M = M' - M'' = -1$

And for linearly polarized light $\Delta M = M' - M'' = 0$

The transition probabilities are proportional to the Clebsch–Gordon coefficients. The Clebsch–Gordon coefficients are indicated in the figure against the transitions. These are indicated by numbers on the inclined lines representing the transitions.

If a left circular polarized (LCP) σ^{+} light with its frequency equals to the transition $F' - F''$ falls on the atoms, the atoms in levels $M'' = \dfrac{1}{2}$ can have transitions to the state $M'' = \dfrac{3}{2}$ and the atoms in the state $M'' = -\dfrac{1}{2}$ can have transition to $M'' = \dfrac{1}{2}$. Here the selections rules $\Delta M = M' - M'' = +1$ is followed. Thus the transitions in this case are

$$M' = \frac{3}{2} \leftarrow M'' = \frac{1}{2}$$

$$M' = \frac{1}{2} \leftarrow M'' = -\frac{1}{2}$$

These excited atoms return to the lower state by spontaneous emission and the following rules are then followed.

1. the atoms excited to the $M' = \dfrac{3}{2}$ can make a transition to the $M'' = \dfrac{1}{2}$ (Selection rule

    $\Delta M = M' - M'' = +1$)

2. the atoms excited to the $M' = \dfrac{1}{2}$ can make a transition to the $M'' = \dfrac{1}{2}$ and also to

    $M'' = -\dfrac{1}{2}$ (Selection rule $\Delta M = M' - M'' = 0, +1$). The transition probabilities are in the

    ration 2 to 1 (square of the Clebsch–Gordon coefficient).

Thus it is clear that by this process of absorption and spontaneous emission, the number of atoms

going to the $M'' = -\dfrac{1}{2}$ level is comparatively smaller than the atoms going to the level $M'' = \dfrac{1}{2}$.
Thus after many absorption–spontaneous emission processes it is possible to deplete the level

$M'' = -\dfrac{1}{2}$. Thus after many cycles of absorption–spontaneous emission almost all atoms in the

$M'' = -\dfrac{1}{2}$ level will be transferred to $M'' = \dfrac{1}{2}$.

If suppose the atoms are interacted with a right circular polarized light, then the transitions with

the selection rules $\Delta M = M' - M'' = -1$ are absorbed. Thus the atoms in the $M'' = \dfrac{1}{2}$ make

transitions to $M' = -\dfrac{1}{2}$ and the atoms in $M'' = -\dfrac{1}{2}$ make transitions to $M' = -\dfrac{3}{2}$. These excited
atoms will make spontaneous emission in the following way.

1. The atoms excited to the $M' = -\dfrac{3}{2}$ can make a transition to the $M'' = -\dfrac{1}{2}$ (Selection rule

    $\Delta M = M' - M'' = -1$)

2. The atoms excited to the $M' = -\dfrac{1}{2}$ can make a transition to the $M'' = -\dfrac{1}{2}$ and also $M' = -\dfrac{3}{2}$

    to $M'' = -\dfrac{1}{2}$ (Selection rule $\Delta M = M' - M'' = 0, +1$). The transition probabilities are in the
    ration 2 to 1 (square of the Clebsch–Gordon coefficient).

After many cycles of absorption–spontaneous emission processes almost all atoms in the $M'' = \dfrac{1}{2}$

level will be transferred to $M'' = -\dfrac{1}{2}$ and thus the level $M'' = \dfrac{1}{2}$ will be depleted. The French
physicist A. Kastler obtained Nobel Prize in the year 1966 for demonstrating optical pumping
phenomena for the first time. Another type of optical pumping and microwave optical double resonance
was demonstrated by Torring and his group at Berlin Free University[7]. Such a scheme is shown in

Figure 14.53. They used the system to determine hyperfine structure and electric dipole moment in CaCl molecule.

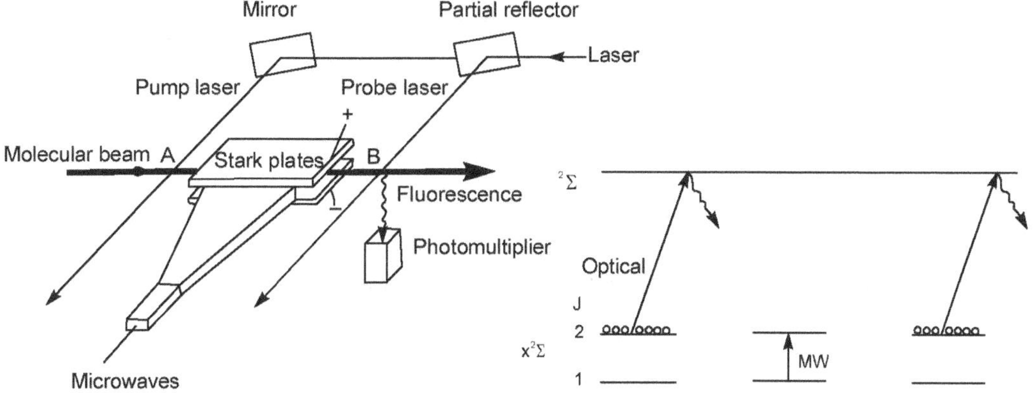

Figure 14.53   Laser-microwave double resonance experimental scheme

14.8.7 Two-photon Spectroscopy

We first consider an atom at rest and if a resonant absorption takes place, the transition frequency ν lies within the natural line width $\Delta\nu_n$. If ν_0 is the right frequency when there was no line width, then the observed frequency

$$\nu = \nu_0 \pm \Delta\nu_n \tag{14.28}$$

The intensity of the transition is

$$I = E^2 \left| <e|M|g> \right|^2 \tag{14.29}$$

where E is the electric field and $\left| <e|M|g> \right|^2$ is the dipole matrix element between the two states. This is the case of single photon absorption

In the case of two-photon absorption the resonance is the sum of the two photon frequencies and it lies within the natural line width of the transition, i.e.,

$$\nu_1 + \nu_2 = \nu_0 \pm \Delta\nu_n \tag{14.30}$$

and the intensity of the transition is

$$I = E^4 \left| \sum_i \left(\frac{\langle e|M|i\rangle\langle i|M|g\rangle}{h(\nu_1 - \nu_0)} + \frac{\langle e|M|i\rangle\langle i|M|j\rangle}{h(\nu_2 - \nu_0)} \right) \right|^2 \tag{14.31}$$

where the summation is over all possible intermediate states.

ν_0 is the transition $g \leftrightarrow i$. The selection rule for the two photon absorption is $\Delta l = 0$ or ± 2 whereas we already know that for single photon absorption it is $\Delta l = \pm 1$.

It can be seen that for two photon absorption much laser power is needed as the intensity is proportional to the fourth power of electric field. It can be shown that for two photon absorption E should be larger than that in one photon absorption by a factor of

$$\frac{\nu_L - \nu_{0i}}{\Delta \nu_n} \tag{14.32}$$

Since the atoms are not at rest and if the atom is subjected to two laser beams of the same frequency propagating in opposite direction then the +ve wave will be seen at a frequency

$$\nu_+ = \nu - \frac{v}{c}\nu \tag{14.33}$$

and the −ve wave at

$$\nu_- = \nu + \frac{v}{c}\nu \tag{14.34}$$

If we assume that the atom absorbs one photon from each wave, then the two photon absorption occurs with

$$\nu_+ + \nu_- = 2\nu = \nu_0 \tag{14.35}$$

However one should consider the fact that the possibility of two photon absorption from the same laser beam which leads to the condition

$$2\nu_\pm = 2\nu \pm \frac{2v}{c}\nu = \nu_0 \tag{14.36}$$

$$\nu = \frac{\nu_0}{2}\nu \pm \frac{v}{c}\nu_0 \tag{14.37}$$

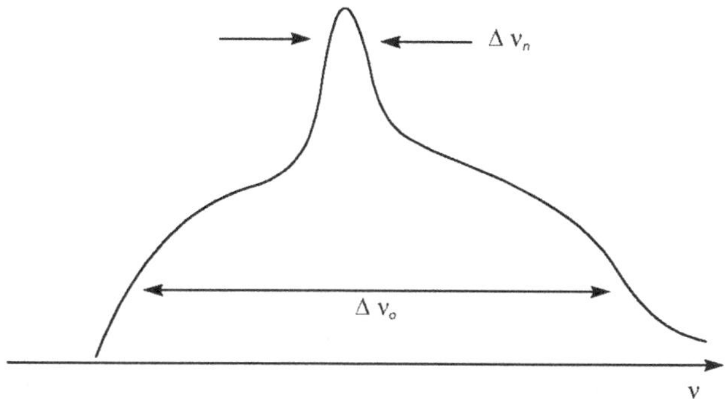

Figure 14.54 Doppler-free peak

It means that the first order Doppler effect does not completely vanish as seen by expression 14.35. As both of the above cases are possible the two photon absorption is generally seen as a narrow peak (Doppler-free peak) on the top of the Doppler broadened back ground as shown in Figure 14.54.

ENDNOTES

1. Bogaerts. A., Neyts, E., Gijbels, R. and Van der Mullen, J.J.A.M. (2002). Gas discharge plasmas and their applications, *Spectrochemica Acta*. 57, 609–658.

2. Nair, K.P.R. and Upadhya, K.N. (1966). *Nature*. 211, 1170.

   Nair, K.P.R. and Hoeff, J. (1979). *Chem Phys. Lett*. 60, 253

3. Nair, K.P.R. (2006, 2008). Atoms, Molecules and Lasers, Narosa (New Delhi) and Alpha Science International, Oxford.

4. Jens-Uwe Grabow and Walther, C. (2009). "Microwave Spectroscopy, Experimental Techniques. Laane, I (Ed.) *Frontiers in Molecular Spectroscopy*, Elsevier, Amsterdam.

5. Nair, K.P.R.. (1974). Science Reporter, CSIR, New Delhi.

6. Teachspin (2009). Instruments Designed for teaching, Inc. Buffalo, New York State, Reproduced with permission

7. Ernst, W.E., Kindt, S., Nair, K.P.R. and Torring, T. (1984). *Phys. Rev*. A29, 1158.

REFERENCES

Benumof, (1965). "Optical pumping theory and experiments." *Am. J. Phys*. 33, p. 151.

Bloom, A.L. (1960). "*Optical Pumping*." Scientific American. p. 72.

Carver, (1963). "Optical pumping." *Science*. 144, p. 599.

Claude Cohen-Tannoudji, Bernard Diu, and Frank Lalo"e, (1977). *Quantum Mechanics*. John Wiley and Sons, New York.

Davis, K.B. (1990). *The Hyperfine Structure of Molecular Iodine*, B.A. Thesis, Middlebury College.

Demtröder, W. (1982, 2003). *Laser Spectroscopy*. Springer-Verlag, New York.

Foot, C.J. (2005). *Atomic Physics*. Oxford University Press.

Grant, R. Fowles. (1989). *Introduction to Modern Optics*. Dover Publications, Inc., New York.

Hänsch, T.W., Shahin, I.S. and Schawlow, A.L. (1971). *Phys. Rev. Lett.* 27, 707.

Hänsch, T.W. (1987). *In Lasers, Spectroscopy and New Ideas*, (ed.) Yen, W.M. and Levenson, M.D. Springer-Verlag, New York. p. 3.

Hollberg, L. (1990). *In Dye Laser Principles with Applications*, (ed.). Duarte, F.J. and Hillman, L.W. Academic Press, New York. p. 185.

Jänsch, H. (1982). *Messung der Relativen Besetzungszahlen am 23Na Atomstrahl durch Selektives Optisches Anregen im Magnetfeld*, Diplomarbeit, Marburg University.

Johnston, T.F. (1987). *In Encyclopedia of Physical Science and Technology*, Vol. 14, Robert A. Meyers.(ed.). Academic Press, New York. p. 96.

Levenson, M.D. (1971). *Hyperfine Interactions in Molecular Iodine*, Ph.D. Thesis, Stanford University.

Optical pumping of Rubidium: Guide to the Experiment Instructor's manual, TeachSpin, Inc. (2002).

Rammamurti Shankar, (1980). *Principles of Quantum Mechanics*. Plenum Press, New York.

Sorem, M.S. (1972). *Spectroscopy by Saturated Fluorescence and Absorption in Molecular Iodine*, Ph.D. Thesis, Stanford University.

Ziock, K. (1969). *Basic Quantum Mechanics*. Wiley, New York.

Zhao, P., Lichten, W., Layer, H.P. and Bergquist, J.C. (1987). *Laser Spectroscopy VII*. Persson, W. and Svanberg, S. (eds.). Springer-Verlag, New York. p. 12.

15

MAGNETIC RESONANCE SPECTROSCOPY

There are two kinds of commonly used magnetic resonance techniques, one is electron spin resonance and the other is nuclear magnetic resonance. In the former it is mainly the electron spin, which takes part, and in the latter it is the nuclear spin.

An atom having a nonzero resultant angular momentum (the so called paramagnetic atoms or molecules) has a magnetic moment $\bar{\mu}_j$ and the angular momentum is $\bar{j}\hbar$. We have already seen that an interaction of the angular momentum (magnetic moment) with an external magnetic field gives Zeeman splitting of levels. And the splitting is

$$\Delta E = g_j \, B \, L$$

where L is the Bohr magneton unit with a value of 9.27×10^{-24} J per weber m^{-2}. g_j is the Lande g factor. The energy separation between the Zeeman levels is generally very small, of the order of 1 cm^{-1}. These transitions can be generally detected by microwave or radio frequency regions, the measurement of the absorbed energy becomes less sensitive. In thermal equilibrium the relative population of the two Zeeman levels is nearly equal. This means that the number of induced absorptions is almost equal to the induced emission. Moreover the magnetic dipole transitions are generally weak, atleast 10^{-5} times weaker than electric dipole transition. However these factors can be increased when a large concentration of absorbing samples are present and these are generally accomplished in liquid and solid phases and the absorptions are detected by ESR or NMR technique.

15.1 ELECTRON SPIN RESONANCE (ESR)

In the gas the atoms are almost free and in such a case the electronic angular momentum quantum number J is a good quantum number. However, in liquid or solid state, the situation is different. The electric field in the crystal can interact with the orbital angular momentum L strongly so that the coupling between L and S breaks. The L tries to precess along the internal magnetic field giving rise to $2L+1$ levels from $-L$ to $+L$. These levels are in principle $2S+1$ fold degenerate as the electron spin has no direct interaction with the crystalline electric field. However an external magnetic field removes this degeneracy splitting each of the $2L+1$ components into $2S+1$ components again. The magnetic dipole selection rule is $\Delta M_j = \pm 1$ and hence the transitions are allowed between the adjacent sublevels and the absorption frequency is given by the relation

$$h\nu_0 = g_s \, B \, L = 2 \, L \text{ for a magnetic field of 1 kilogauss}$$

In a second order approximation a coupling between the electron spin and the crystal field is also possible which may remove the $2S+1$ degeneracy even in the absence of an external magnetic field. The splitting is usually very small. The Figure 15.1 shows the splitting in a $^6S_{5/2}$ (example Mn^{++} atom). The magnetic sublevels are equally spaced and hence the transitions among them are at the same frequency and hence give a single transition frequency. Figure 15.2 shows the splitting of a $^4F_{3/2}$ in the magnetic field of a crystal.

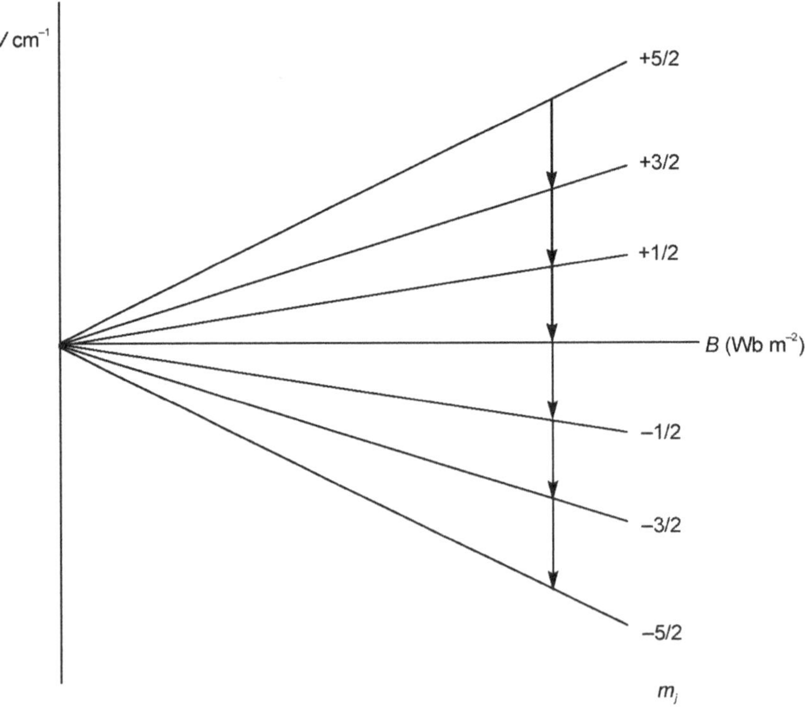

Figure 15.1   Magnetic components in a $^6S_{5/2}$

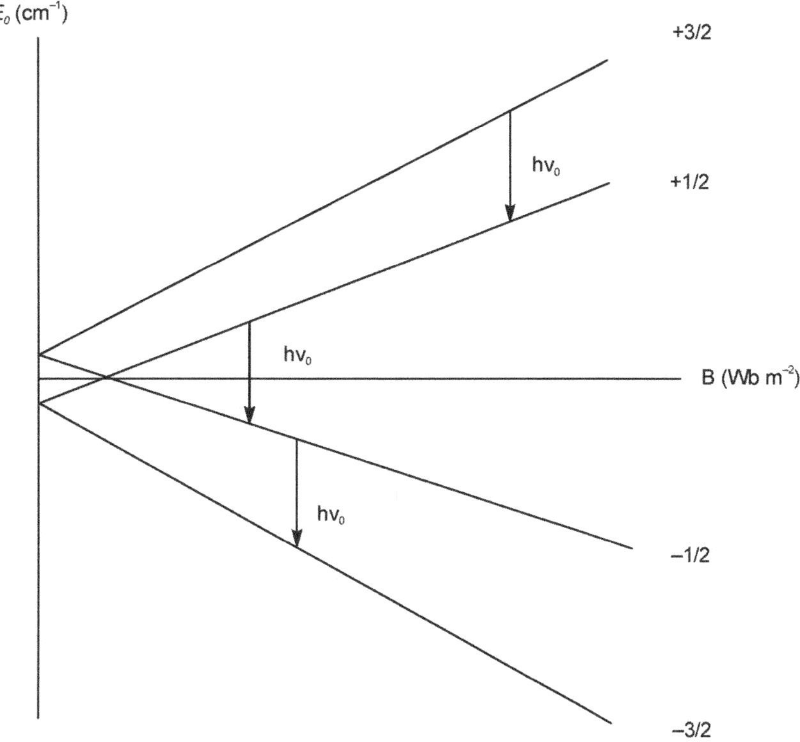

Figure 15.2   Magnetic components of a $^4F_{3/2}$ level with initial splitting in a crystal field

15.1.1 Principle of ESR

We have seen already in Chapter 7 that the magnetic moment associated with a spinning electron is obtained from equation 7.27.

$$\frac{\mu_s}{p_s} = -2\frac{e}{2m}$$

$$\mu_s = -2\frac{e}{2m}\cdot\frac{h}{2\pi}S = -2\frac{eh}{4\pi m}S$$

and in CGS unit $\mu_s = -2\dfrac{eh}{4\pi mc}S$

i.e., $\mu_s = -g\mu_B S$

where $g = 2$ in the case of spinning electron and $\mu_B = \dfrac{eh}{4\pi m}$

The magnetic moment associated with orbital motion is

$$\mu_L = -g\mu_B L$$

where, $g = 1$

The total magnetic moment is thus $\mu = \mu_L + \mu_s = -2\mu_B S - \mu_B L$.

If the electron has a magnetic moment originating from spin only we can calculate the interaction energy as

$$E = -\mu_s \cdot B = -\mu_s B \cos \theta$$

where θ is the angle between μ and B.

$$E = +g\mu_B B S \cos \theta$$
$$= g\mu_B B m_s$$

Since $S = \dfrac{1}{2}$ for an electron we get two energy levels corresponding to $m_s = +\dfrac{1}{2}$ and $m_s = -\dfrac{1}{2}$.

The energies of these two levels are

$$E_{+\frac{1}{2}} = g\mu_B B \cdot \frac{1}{2} = \frac{1}{2} g\mu_B B \quad \text{and}$$

$$E_{-\frac{1}{2}} = -\frac{1}{2} g\mu_B B$$

The splitting of unpaired electron in a magnetic field is shown in Figure 15.3.

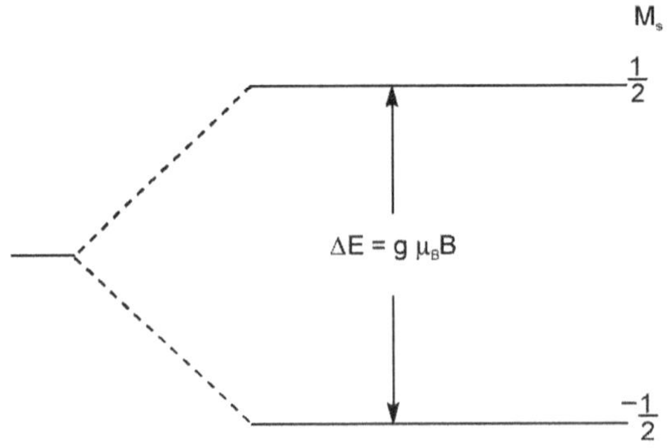

Figure 15.3   Splitting of an unpaired electron in a magnetic field

If a radiation of frequency satisfying the condition

$$h\nu = E_{\frac{1}{2}} - E_{-\frac{1}{2}} = g\mu_B B$$

falls on the sample, transition between these levels occur and we get an electron spin resonance spectrum. For a free electron $g = 2.0023$ and in a field of 0.34 Tesla the frequency of transition

$$\nu = \frac{2.0023 \times 9.274 \times 10^{-24}\, JT^{-1} \times 0.34T}{6.626 \times 10^{-34}\, J \cdot S}$$

where $\mu_B = 9.274 \times 10^{-24}$ JT^{-1}. The frequency thus obtained is 9528 MHz which lie in the microwave region. Normally microwave techniques are employed in ESR experiments.

The number and positions of the resonances depend on the type of the ion, symmetry of the crystal and the orientation of the external magnetic field. This is also sometimes called electron paramagnetic resonance (EPR). If the paramagnetic ion has a nuclear spin the levels will be further split.

In electron spin resonance spectroscopy the magnetic field can either be changed or the frequency can be changed keeping a constant magnetic field. In ESR experiments the second method is adopted usually. A block diagram of a balanced bridge ESR spectrometer is shown in Figure 15.4.

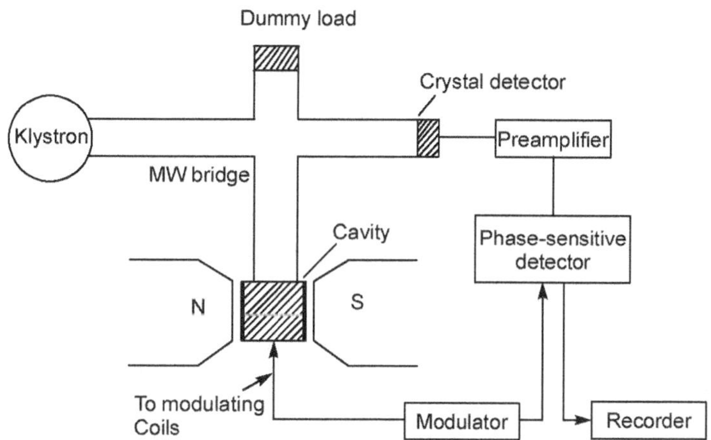

Figure 15.4   Block diagram of ESR spectrometer

Microwave radiation from a klystron source is passed through a microwave impedance bridge. The microwave cavity containing the sample is kept in between the poles of an electromagnet. A dummy load is kept in the third arm and a crystal detector in the fourth arm of the microwave bridge. The radiations that arrive at the fourth arm is detected by the crystal detector and fed to a preamplifier, phase sensitive detector and then to the recorder. By keeping the frequency coming from the klystron constant, the magnetic field is swept over a small range at the resonance condition varying the current in a pair of sweep coils mounted on the microwave cavity walls.

At a balanced position of the bridge, the microwave power flows only through two arms, the cavity and the dummy load. We will get power in the fourth arm only when the bridge is not balanced.

Thus if balancing of the bridge exists, there is no signal on the detector, however, when the sample in the microwave cavity absorbs the frequency, the balancing of the bridge is lost and power appears on fourth arm. This is amplified and detected by a lock-in amplifier and then recorded. Usually in ESR a derivative spectrum is obtained which has the advantage of measuring the frequency accurately.

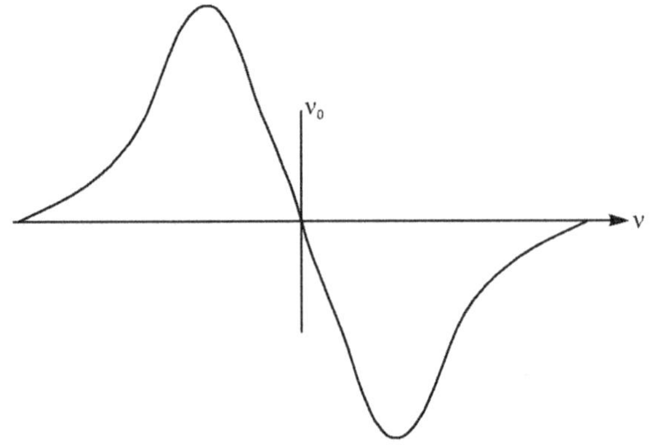

Figure 15.5 Derivative spectrum

15.2 NUCLEAR MAGNETIC RESONANCE SPECTROSCOPY

15.2.1 Background

Nuclear magnetic resonance spectroscopy, commonly referred to as NMR, has become an important technique for determining the structure of organic compounds. Magnetic resonance imaging in medical science is an outcome of NMR.

When the nuclei of certain atoms are placed in a static magnetic field and exposed to a second oscillating magnetic field nuclear magnetic resonance occurs. Some nuclei experience this phenomenon, and others do not, and is dependent upon whether they possess a nuclear spin or not.

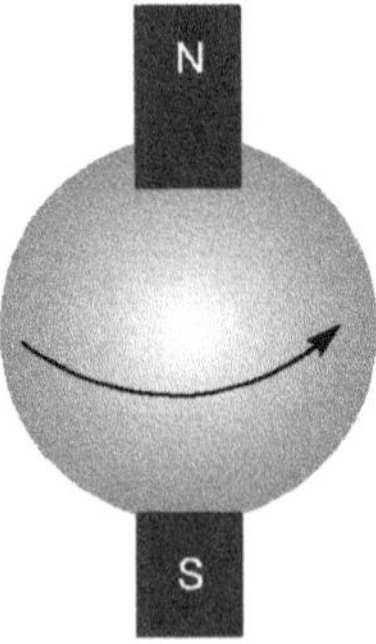

Figure 15.6 Spinning nucleus

We already know that the nuclei of many elements or their isotopes have a characteristic spin which we call it as I. These are intrinsic spin like the electron spin. Some nuclei have integral spins like $I = 1, 2, 3 \dots$ etc., some others have fractional spins like $I = 1/2, 3/2, 5/2 \dots$etc., and a few have no spin at all, $I = 0$ (e.g. ^{12}C, ^{16}O, ^{32}S, ...). However nuclei of particular interest and use to organic chemists and to NMR are ^{1}H, ^{13}C, ^{19}F and ^{31}P, all of which have $I = 1/2$. Since the analysis of this spin state is fairly straightforward, we will discuss here nuclear magnetic resonance in atoms having nuclear spin $I = 1/2$.

It should be remembered that the basic principles in ESR and NMR are very similar. However, while ESR is due to the interaction of external magnetic field with the electron spin, the NMR is due to the nuclear spin. NMR frequencies are lower by a factor of atleast 10^{-3} compared to ESR frequencies.

The following features lead to the NMR phenomenon:

We know that a spinning charge generates a magnetic field (*See* Figure 15.6). The resulting magnetic moment μ is proportional to spin. If we place the spinning nucleus in a magnetic field, the nuclear spin orients itself along the external magnetic field. The angular momentum vector I of the nucleus can take $2I+1$ orientations in space while spinning around the external magnetic field. These are quantized and the angular momentum is $M_I \hbar$ where M_I are magnetic quantum numbers which can take $2I+1$ values $I, I-1, I-2, \dots -I+1, -I$, i.e., M_I can take values from $-I$ to $+I$ by changing by one unit. Suppose we deal with an atom having a nuclear spin $I = 1/2$, there can be only two nuclear orientations in a magnetic field, viz., $+1/2$ and $-1/2$ and hence two states with $M_I = -1/2$ and $+1/2$.

Thus in the presence of a magnetic field say B_0, there exists two spin states $+1/2$ and $1/2$. $M_I = +1/2$ has lower energy than $M_I = -1/2$. The magnetic moment of the lower energy $+1/2$ state is aligned with the external field whereas the one with $-1/2$ spin state is opposed to the external field.

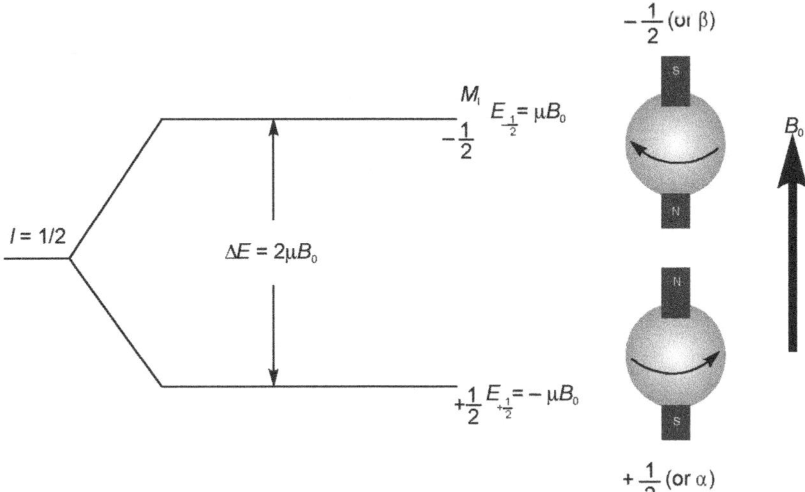

Figure 15.7   Spin energy states—spin state 1/2 orients along the field and –1/2 orients in the opposite direction of the field

Thus there are two spin states in a magnetic field different in energy and the difference depends on the applied magnetic field. This is shown in Figure 15.7. The figure illustrates that the two spin states have the same energy when the external field is zero, but diverge as the field increases. Now imagine that constant external field is applied. In this case we obtain two energy states corresponding to the two spin states $+1/2$ and $-1/2$. If we now apply radio frequency field of correct orientation and frequency, transition between these states can be induced. We know that the classical expression of the energy of the nuclear spin orientation in the external magnetic field is $-\mu_0 B_0 \cos\theta$. The quantum mechanical equivalent of this energy is $-\left(\dfrac{\mu}{I}\right) B_0 M_I$. Here B_0 is the strength of the magnetic field and M_I is the magnetic quantum number. The energy difference between the two levels is

$$-\left(\frac{\mu}{I}\right) B_0 \left[M_I' - M_I''\right] = -\left(\frac{\mu}{I}\right) B_0 \left[-\frac{1}{2} - \frac{1}{2}\right] = h\nu.$$

$$\text{Thus } \Delta E = \left(\frac{\mu}{I}\right) B_0 = g_N \mu_N B \text{ and } \nu = -\left(\frac{\mu}{Ih}\right) B_0.$$

Since we are considering $I = \dfrac{1}{2}$, the resonance condition reduces to

$$h\nu = 2\mu B_0$$

splitting of energy states with varying magnetic field is shown in Figure 15.8

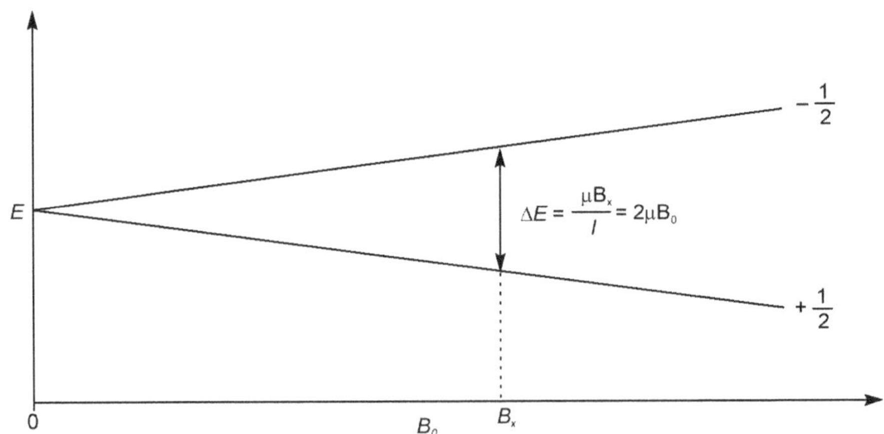

Figure 15.8   Nuclear spin splitting in a magnetic field

As stated above the energy splittings are very small in NMR spectroscopy and hence we require strong magnetic fields. Powerful magnets having fields of 1 to 20 T are used nowadays. Even with these high fields, the energy difference between the two spin states is less than 0.1 cal/mole. Note that infrared transitions involve 1 to 10 kcal/mole and electronic transitions are nearly 100 times greater.

The small energy difference (ΔE) is usually given in frequency in units of MHz (10^6 Hz), ranging from 20 to 900 MHz, depending on the magnetic field strength and the specific nucleus being studied. If we send through the sample radio frequency (rf) energy corresponding exactly to the spin state separation of a specific set of nuclei, it will cause excitation of those nuclei in the $+1/2$ state to the higher $-1/2$ spin state (*See* Figure 15.8). The nucleus of a hydrogen atom (the proton) has nuclear spin 1/2 and a magnetic moment $\mu = 2.7927$ and has been studied more than any other nucleus.

The energy difference between the two spin states at a given magnetic field strength will be proportional to their magnetic moments of the nuclei. Table 15.1 gives the magnetic moments of the four common nuclei, hydrogen, fluorine, phosphorous, carbon, nitrogen and chlorine in the unit of nuclear magnetons. 1 nuclear magneton = 5.05078×10^{-27} JT^{-1}.

Table 15.1 The magnetic moments of the nucleai

Atom	Nuclear spin	Nuclear moment (μ)	ν (MHz) $B_0 = 1$ Tesla	2.35T
^{1}H	1/2	2.7927	42.577	100.06
^{19}F	1/2	2.6273	40.055	94.13
^{31}P	1/2	1.1305	17.236	40.50
^{13}C	1/2	0.7022	10.705	25.15
^{2}H	1	0.8574	6.536	15.35
^{14}N	1	0.4036	3.076	7.23
^{35}Cl	3/2	0.8209	4.172	9.80
^{37}Cl	3/2	0.6833	3.472	8.16

For proton $\mu = 2.7927$ μ_N and hence $\nu = \dfrac{2 \times 2.7927 \times 5.05078 \times 10^{-27}\,\text{JT}^{-1}}{6.626 \times 10^{-34}\,\text{J.S}} B_0$

$$= \left(42.5772 \times 10^6\,\text{T}^{-1}\text{S}^{-1}\right) B_0$$

Figure 15.9 gives the spin state energy separations for each of the above nuclei in an external magnetic field with a field strength of 2.35 Tesla.

It is also possible to hold the radiofrequency constant and vary the strength of the magnetic field in order to get resonance.

A typical NMR spectrometer is shown in Figure 15.10. It is also sometimes referred to as the continuous wave (CW) spectrometer. A solution of the sample is kept 5 mm glass tube and is oriented between the poles of a powerful magnet. Radio frequency radiation of appropriate energy is sent into the sample from an antenna coil. A receiver coil is wound around the sample tube which receives the signal, and emission of absorbed rf energy is monitored by electronic devices and fed into a computer. An NMR spectrum is acquired by varying or sweeping the magnetic field over a small range while observing the rf signal from the sample. An equally effective technique is to vary the frequency of the rf radiation while holding the external field constant. That is in one case, one keeps the radiofrequency constant and change the strength of the magnetic field and in the second method one can keep the magnetic field constant and change the radiofrequency.

$$v = \frac{\mu B_0}{hI} = \frac{4.68\,\mu}{h}$$

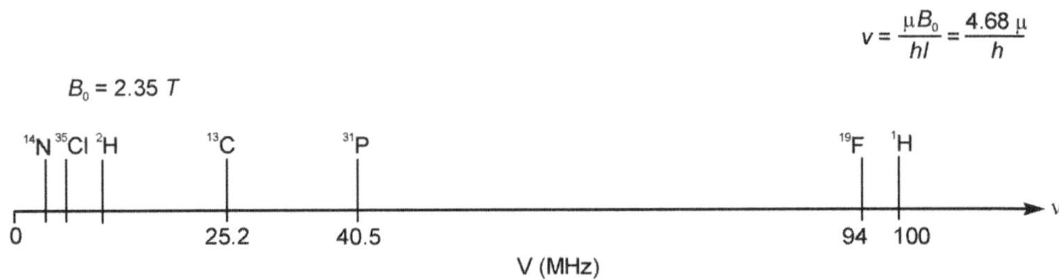

Figure 15.9   The spin-state energy separations

A crude form of the spectrometer can be depicted in terms of a Wheatstone's bridge. The sample is kept in a small glass tube which is placed in a strong magnetic field. The sample is kept inside a coil of inductance which functions as one arm of a radiofrequency balanced bridge as shown in the Figure 15.10. The radiofrequency is partly absorbed by the sample which is balanced by

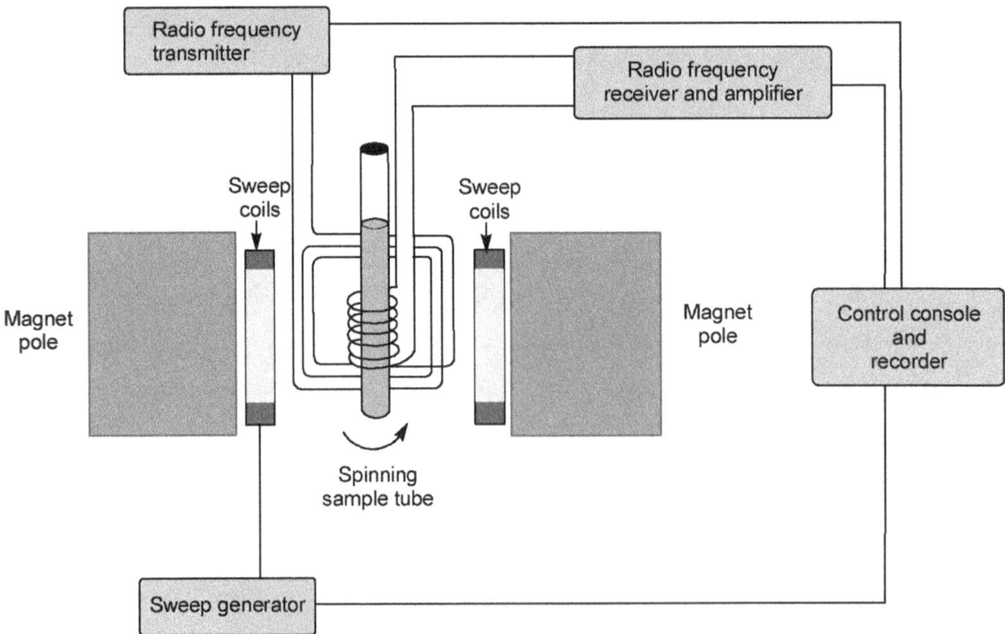

Figure 15.10   A crude form of the spectrometer

change in the resistance. The loss of energy creates an increase in the resistance thereby an unbalance in the bridge. A signal appears on the detector which can be amplified by suitable means. The magnetic field can be varied and a resonance appears when the radiofrequency signal exactly matches with the transition frequency. This is the resonance condition. Thus keeping the frequency constant, the magnetic field can be varied till it resonates or the magnetic field can be kept constant, the radiofrequency field can be changed till it reaches the resonance frequency at the field.

As an example, consider a sample of water in a 2.3487 T external magnetic field, irradiated by 100 MHz radiation. If the magnetic field is smoothly increased to 2.3488 T, the hydrogen nuclei of the water molecules will at some point absorb rf energy and a resonance signal will appear.

15.2.2 Chemical Shift

We have already seen that the frequency $v = -\left(\dfrac{\mu}{Ih}\right) B$. This would suggest that all atoms of an element in a molecule (having the same nuclear spin) will have the same resonance frequency in the same static magnetic field. But this is not always the case. An atom at a particular part of the molecule sometimes shows a different resonance frequency compared to the same type of atom at a different location in the molecule. The answer to this question lies with the electrons surrounding the proton (nucleus) in covalent compounds and ions. Since electrons are charged particles, they are also affected by the external magnetic field. They move in response to the external magnetic field (B_o) so as to generate a secondary field that opposes the much stronger applied field. This secondary field shields the nucleus from the applied field, and hence B_o must be increased in order to achieve resonance (absorption of rf energy). This magnetic field range is very small compared with the actual field strength (only about 0.0042%). It is customary to refer to small increments such as this in units of parts per million (ppm). For example the difference between 2.3487 T and 2.3488 T is therefore about 42 ppm. Instead of designating a range of NMR signals in terms of magnetic field differences (as above), it is more common to use a frequency scale, even though the spectrometer may operate by sweeping the magnetic field. Using the terminology basis as frequency and if the resonance occurs at 100 MHz for a free proton at a magnetic field of 2.34 and the proton signals extend over a 4,200 Hz frequency range we would find that at 2.34 T the change is 42 ppm (for a 100 MHz rf, 42 ppm is 4,200 Hz). Most organic compounds exhibit proton resonances that fall within a 12 ppm range.

Thus the resonance frequency depends on the surrounding atoms or the molecular environment. Though the change is small which is 1 in 10^6 for hydrogen or even smaller for other nuclei, they can be measured by modern instruments. The difference is called chemical shift, i.e., it occurs due to chemical environment. Thus $v = -\left(\dfrac{\mu}{Ih}\right) B (1+\delta)$ where δ is the chemical shielding factor. As mentioned above the change can be explained as due to the motion of the electrons in the surroundings which produces an internal magnetic field which could shift the magnetic levels.

In the case of ordinary spectroscopy like UV or visible, the transition frequencies are uniquely determined whereas in NMR, also in ESR, the resonance signals depend both on the applied magnetic field and the radio frequency and also on the surroundings. Since there can be variations in the magnetic fields due to instrumental problem and since no two magnets will have exactly the same field, the resonances may slightly vary when using different magnets. In order to overcome this difficulty one can use a reference sample along with the experimental sample. Thus a spectrum relative to a reference signal from a standard compound is added along with the sample spectrum. Such a reference standard should be chemically inactive, and easily removable from the sample after the measurement. Also, it should give a single sharp NMR signal that does not interfere with the resonances normally observed for organic compounds. Tetramethylsilane, $(CH_3)_4Si$, usually referred to as TMS, meets all these requirements, and has become the reference compound of choice for proton and carbon NMR.

REFERENCES

Aruldhas, G. (2009). *Molecular Structure and Spectroscopy*. PHI Learning Pvt. Ltd, New Delhi.

Eisberg, R.M. (1961). *Fundamentals of Modern Physics*. John Wiley and Sons, Inc. New York.

Foot, C.J. (2005). *Atomic Physics*. Oxford University Press.

Herberg, G. (1945). *Atomic Spectra and Atomic structure*, Dover, New York.

Slichter, C.P. (1963). *Principles of Magnetic Resonance*. Harper and Row, New York. pp. 1–22.

Straughan, B.P. and Walker, S. (1976). *Spectroscopy*, Chapman and Hall, London.

16

LASER COOLING

One of the greatest achievements in the last decade of 20th century was undoubtedly the one where scientists were able to cool a bunch of atoms to temperatures very close to the absolute zero. It has made a milestone in physics which led to producing the so called Bose–Einstein condensates (BEC). As the name suggests laser cooling is a technique that uses light to cool atoms to very low temperature. The name may sound like an oxymoron since the lasers are always thought to be associated with intense light and heat. Even in the medical field the heat produced from the lasers are used in medical surgery to burn away unwanted tissues. Hence it is a surprise to common man to know that lasers can be used for cooling atoms.

Starting with the work of Steven Chu and others[1] during the time around 1985, the use of lasers to achieve extremely low temperatures has advanced to the point that temperatures of 10^{-9} K have been reached. The advancement in tunable diode lasers and developments in high vacuum technology have made possible to cool atoms to very low temperatures. Interaction between a laser beam of appropriate wavelength (frequency) and atomic vapour is the main process in obtaining cooling of atoms. It is related to the transitions between quantum states of an atom in the presence of laser beams.

There are three properties which are important and involved in laser cooling process. They are

1. Doppler effect
2. Light quantum hypothesis by Planck
3. The atom–photon interaction—the resonance phenomena

16.1 DOPPLER EFFECT

The hardest part to laser cooling was to figure out how to avoid hitting the slow atoms with light at the same time hitting the fast ones to slow them down. The problem was resolved with the idea that the colour or the light is Doppler shifted by the atoms' motion. It says that if an atom is going towards the laser light, it sees the light shifted to a blue colour, and if it is going away from the laser, it sees the light as redder than it really is. And the amount of the shift depends on the speed.

So if the laser is just the right colour, the Doppler shift of a fast atom will make the light look the right colour for exciting it, and so photons will bounce off and slow it down. But if the atom is moving slowly, or in the wrong direction, the Doppler shift will be different, and the laser light will be the wrong colour to excite the electron. In that case, the laser light just goes right by the atom.

So if we have the laser just the right colour, it will slow down the fast atoms without pushing the slow ones backwards because of the Doppler shift. But this is just in one direction. The atoms in the box are bouncing around in all directions. How do we slow all of them down?

We need to take our laser and send the beam at the atom from all the different directions. Then if we adjust it to exactly the right colour all the atoms will get cold. Thus by choosing the right laser colour, one can get big bunches of very cold atoms.

The Doppler effect is more familiar to us in the case of sound waves. It was first proposed by the Austrian Physicist Christian Doppler. He pointed out that whenever there is a relative change of the source and the observer, there is a change in frequency. Thus it is the change in observed frequency when there is relative motion between the source and the observer. It is often experienced when a vehicle sounding a siren or horn approaches, passes and recedes from an observer. The fractional relative change in frequency is proportional to the ratio of the source velocity to the speed of waves in the medium, measured relative to the observer. We know that the speed of sound wave in air is 330 m/s (1200 km/hr) and the speed of light is 3×10^8 m/s (1 million km/hr). In the case of a train moving with a speed of 120 km/hr is about 10% of the speed of sound and is a negligible fraction of the speed of light. However in the case of atoms, they move with a speed of atleast 500 m/s at room temperatures and they are more sensitive to see an incoming radiation due to the process of resonance. The received frequency is always higher when the source of the wave is approaching the observer and is the same when it just passes and is lower when it moves away.

The relationship between observed frequency f and emitted frequency f_0 is given by:

$$f = \left(\frac{v + v_r}{v + v_s} \right) f_0 \qquad (16.1)$$

where

    v is the velocity of waves in the medium

    v_s is the velocity of the source relative to the medium and

    v_r is the velocity of the receiver relative to the medium.

From the expression it can be seen that the observed frequency increases when either the source is moving towards the observer or the observer is moving towards the source. The reverse is the case when the source or the observer moving away from the other. The frequency in this case is decreased. If the speed of the wave is much faster than the speed of the source and the observer the relationship between observed frequency f and emitted frequency f_0 is given by:

$$\text{Observed frequency } \quad f = \left(1 - \frac{v_{s,r}}{c}\right) f_o \tag{16.2a}$$

$$\text{Change in frequency } \quad \Delta f = -\frac{v_{s,r}}{c} f_o = -\frac{v_{s,r}}{\lambda_o} \tag{16.2 b}$$

where $v_{s,r}$ is the velocity of the source relative to the receiver: it is negative when the source is moving towards the receiver, positive when moving away, c is the speed of wave (e.g., 3×10^8 m/s for electromagnetic waves travelling in a vacuum) and λ_0 is the wavelength of the transmitted wave in the reference frame of the source.

This is the case with electromagnetic radiations. In this case also when the source and the observer are moving away or the source moves away from the observer we observe a lowering of frequency or in terms of wavelength, an increase in wavelength. The Doppler effect in electromagnetic waves is very important in astrophysics. The Doppler effect is recognizable in the fact that the absorption or emission lines obtained from distant stars of interstellar media are not always at the frequencies that are obtained from the spectrum of a stationary light source. Since blue light has a higher frequency than red light, the spectral lines of an approaching astronomical light source exhibit a blue shift and those of a receding astronomical light source exhibit a redshift. However, most of the astrophysical objects show redshifts. The redshift observed in many of the spectra from astrophysical objects and interstellar space reveal that they are receding from us. The redshift and the velocity of the distant objects have helped the scientists to extrapolate to the origin of the universe and to the Bigbang theory. Both the redshift and the blueshift are made use in laser cooling of atoms as we shall see in the later part of this chapter.

The second property of light relates to the idea put forward by Marx Planck introducing the quantum concept namely that the light waves composed of packets called the photons which contain quantum of energy $h\nu$. We have already seen that by considering that each photon carries an energy of $h\nu$, Albert Einstein could successfully explain the black body radiation. In his historical paper on the "quantum theory of radiation" of 1917 he also showed that each photon also carries a unit of momentum which is equal to h/λ where λ is the wavelength of light. This implies that when an atom is absorbed or emitted there is a change in momentum as per conservation of momenta and this change is equal to h/λ. If an absorption takes place it causes an increase in momentum of this amount and if an emission takes place the atom experiences a decrease in momentum. The light forces on atoms occur due to these momentum changes. Thus if an atom moving in the (+X) direction is hit by a laser beam in the opposite (–X) direction and if it is in resonance condition, an increase in momentum is experienced by the atom in the direction of the laser beam.

The third property of interest is that of resonance in atom–photon interactions. The vast majority of photons that come anywhere near a particular atom are almost completely unaffected by that atom. That means the atom is almost completely transparent to most frequencies (colours) of photons or incident electromagnetic radiation.

However a few photons having the right frequency tries to resonate with the atoms. This is a narrow band of frequency or one can say single frequency or a single colour. When one of those photons comes close to the atom, the atom typically absorbs that photon (absorption spectrum) for a brief period of time, then emits an identical photon (emission spectrum) in some random, unpredictable direction. (Other sorts of interactions between atoms and photons do exist, but are not relevant for the present discussion.)

Thus normally the electric field of the incident light wave tries to interact with the electrons in an atom, but the light transfers energy to the atom when the resonance condition is achieved. It implies that there is a transfer of energy only when the difference in energy of two states is exactly equal to the energy of the incident photon,. i.e., the frequency of the light exactly matches with the internal frequency of the atom. If we plot the photon absorption in terms of the frequency of light we obtain a curve as shown in Figure 16.1. The curve is known as "Lorentzian curve". The Lorentzian profile can be represented by a Lorentzian function given by

$$\phi(\nu) = \frac{1}{\pi} \frac{\Gamma}{\left(\nu - \nu_o\right)^2 + \Gamma^2} \tag{16.3}$$

where Γ is the width of the curve at half maximum intensity and Γ is usually of the order of 10 MHz in the optical region. Any resonance line can be represented by a Lorentzian curve.

Already in 1920s Albert Einstein (14 March, 1879–18 April, 1955)[2] and Indian Physicist Satyendra Nath Bose (1 January, 1894–4 February, 1974) had predicted such low temperature conditions, in which the atoms fall into the same quantum state and essentially behave as a single atom. It was at a time when quantum mechanics was still new. Einstein wondered if BEC's were a reality even though he himself had thought about it. However, with the experiments of Chu and co-workers and from the work of Cornell (b. 1961)[3], Carl Wieman (b. 1951)[4] and their colleagues at the University of Colorado during 1985, BEC has become a reality.

The phenomenon of Bose–Einstein condensation was predicted by S. N. Bose and A. Einstein in 1924, during the early days of quantum mechanics while studying the statistical properties of photons and massive particles with integer spin, which are now known as bosons. It was found that not only it is possible for two or more bosons to share the same quantum state, but that the bosons actually prefer to be in the same quantum state. It was predicted that at a finite temperature, almost all the particles of a bosonic system would occupy the ground state as soon as the quantum wave functions of the particles start to overlap. In a Bose–Einstein condensate millions of atoms can occupy a single quantum state, thus bringing the quantum world into the macroscopic regime.

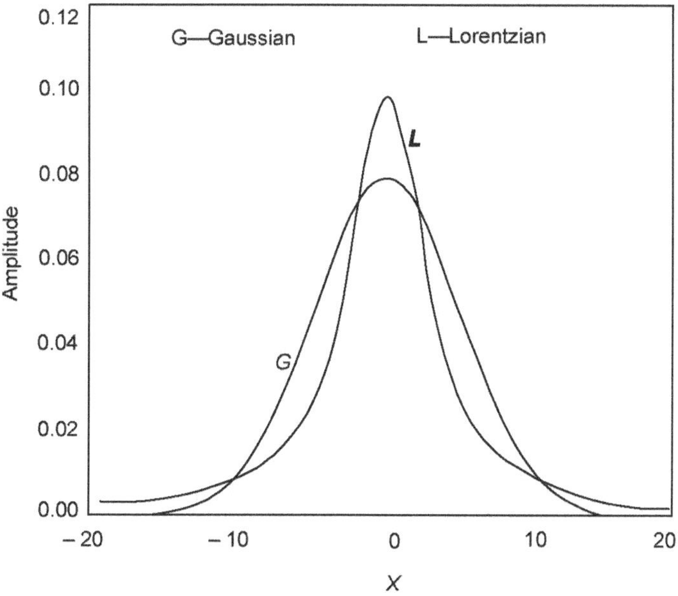

Figure 16.1 Typical Gaussian and Lorentzian profile

16.2 BOSE–EINSTEIN STATISTICS

Before going into the details of the Bose–Einstein condensation it will be helpful to understand Bose–Einstein statistics.

Two forms of quantum statistics had been proposed in order to explain statistical distribution of indistinguishable particles. They are Bose–Einstein statistics and Fermi–Dirac statistics. In order to explain the statistical distribution of particles of distinguishable nature Maxwell–Boltzmann distribution laws were derived.

The fundamental postulate in Bose–Einstein and Fermi–Dirac statistics is that similar particles cannot be distinguishable from one another. The ultimate justification of any postulate lies with its ability to reproduce the observed phenomena. The two statistics are dependent on the principle of symmetry and antisymmetry. In Bose–Einstein statistics only symmetric solutions are acceptable whereas in Fermi–Dirac statistics only antisymmetric solutions are accepted. It should be noted that the symmetric character of an assembly of similar systems is not altered by vicissitudes of either mutual interaction or interaction with the surroundings. This is also true in the case of antisymmetric character. It has been shown by Pauli, as a necessary consequence of relativistic spin theory together with group theoretical postulates that particles for which the wave function is always symmetrical must obey Bose–Einstein statistics and the particles for which the wave function is always antisymmetrical must obey Fermi–Dirac statistics.

Experience shows that photons, nuclei of atoms with even number of elementary particles like helium nucleus of mass 4 and mesons of spin 0 and 1 give rise to symmetric wave functions and hence obey Bose–Einstein statistics whereas elementary particles like electrons, protons and neutrons with half integral spins have antisymmetric wave functions. The former group is called bosons.

Observation of the Bose–Einstein condensates in the cooled atomic gases in the year 1985 has successfully proved the quantum statistical theory of non-interacting identical particles proposed by Albert Einstein and S.N. Bose during 1924–25.

Let us consider a system with n indistinguishable particles or elements. Consider that these particles are divided into quantum groups or levels such that there are n_1, n_2...n_i number of elements having energies ε_1, ε_2 ... ε_i respectively. If the degeneracy of the ith state is g_i, the total number of eigen states available for the n_i elements is given by the number of ways in which the elements can be distributed among the g_i functions. The complete eigen function in the Bose–Einstein case being symmetric there is no restriction as to the number of particles associating with any of the functions. This can be viewed in the manner in which there is no restriction of a number of indistinguishable particles are to be distributed in a box of g_i compartments, without restrictions to the number of particles in any of the compartments. Let us now consider a compartment where there are g_i compartments, or a box having g_i sections. The indistinguishable particles are to be distributed in these sections. The choice that which of the section will head the sequence is made in g_i ways. Once this is done the total number of permutations of n_i particles and the remaining $g_i - 1$ sections is $(n_i + g_i - 1)!$. The distributions that can be obtained from one another by mere permutations of the cells by themselves, or the particles themselves do not produce different states, the above number should be divided by $g_i! \, n_i!$. Thus the required number of ways of n_i particles can be distributed in the remaining $g_i - 1$ sections becomes

$$\frac{g_i(n_i + g_i - 1)!}{g_i! \, n_i!} = \frac{(n_i + g_i - 1)!}{n_i!(g_i - 1)!} \tag{16.4}$$

Since the particles we are considering are indistinguishable, the arrangements of the total number of particles into groups n_1, n_2, ...n_i... can be done only in one way. For each group the number of eigen states are given by an expression given above. Hence the total number of eigen states G for the whole system of elements corresponding to the specified groups can be written as

$$G = \frac{(n_1 + g_1 - 1)!}{n_1!(g_1 - 1)!} \cdot \frac{(n_2 + g_2 - 1)!}{n_2!(g_2 - 1)!} \cdots \frac{(n_i + g_i - 1)!}{n_i!(g_i - 1)!} \tag{16.5}$$

$$= \prod \frac{(n_i + g_i - 1)!}{n_i!(g_i - 1)!} \tag{16.6}$$

On the basis of the postulate on the equal apriori probability of eigen states, the probability W of the system is proportional to the total number of eigen states. Therefore

$$W \; = \; \text{constant} \; \times \prod \frac{(n_i + g_i - 1)!}{n_i!(g_i - 1)!} \tag{16.7}$$

If we take the logarithm on both sides

$$\ln \; W = \text{Constant} \; + \sum_i \ln(n_i + g_i - 1)! - \ln n_i! - \ln(g_i - 1)! \tag{16.8}$$

In order to have maximum probability $\delta \ln \; W = 0$ $\tag{16.9}$

The well known Stirling's approximation for large n and g is

$$\ln \; n! = n \ln \; n - n \tag{16.10}$$

and hence

$$\ln \; W = \text{Constant} +$$

$$\sum_i (n_i + g_i - 1)\ln(n_i + g_i - 1) - (n_i + g_i - 1) - n_i \ln n_i + n_i - (g_i - 1)\ln(g_i - 1) + (g_i - 1) \tag{16.11}$$

Since n_i and g_i are large we can approximate

$$\ln \; W = \text{Constant} + \sum_i (n_i + g_i)\ln(n_i + g_i) - (n_i + g_i) - n_i \ln n_i + n_i - (g_i \ln g_i) + g_i$$

$$= \text{Constant} + \sum_i (n_i + g_i)\ln(n_i + g_i) - n_i \ln n_i - g_i \ln g_i \tag{16.12}$$

The condition for maximum probability is given by

$$\delta \ln \; W = \sum_i \{\{\ln n_i - \ln(n_i + g_i)\}\partial n_i = 0 \tag{16.13}$$

$$\text{or} \; \sum \left(\ln \frac{n_i}{n_i + g_i} \right) \partial n_i = 0 \tag{16.14}$$

The other conditions are $\partial n = \sum \partial n_i = 0$ and $\tag{16.15}$

$$\partial \varepsilon = \sum \varepsilon_i \partial n_i = 0 \tag{16.16}$$

The equations 16.15 and 16.16 represent the two subsidiary conditions which must be satisfied simultaneously with the condition for maximum probability. Applying the Langrange method of undetermined multipliers by multiplying the first subsidiary condition by α and the second by β and adding the resultant expression to expression (16.14) we get

$$\sum_i \left(\ln \frac{n_i}{n_i + g_i} + \alpha + \beta \varepsilon_i \right) \partial n_i = 0 \tag{16.17}$$

Since the auxiliary condition that all the variations ∂n_i must add up to zero is already contained in expression (16.17) the variations in ∂n_i may now be treated as independent of each other. Therefore equation 16.17 will be satisfied only when each term in c is separately equal to zero. Hence

$$\ln \frac{n_i}{n_i + g_i} + \alpha + \beta \varepsilon_i = 0$$

or

$$\ln \frac{n_i + g_i}{n_i} = \alpha + \beta \varepsilon_i$$

or

$$1 + \frac{g_i}{n_i} = e^{\alpha + \beta \varepsilon_i}$$

i.e.,

$$n_i = \frac{g_i}{e^{\alpha + \beta \varepsilon_i} - 1} \tag{16.18}$$

The expression represents the most probable distribution of particles among various energy levels of a system obeying Bose–Einstein statistics.

We have already stated that Bose–Einstein statistics is applicable to photons. If it is so it should give correct derivation to Planck's equation for the distribution of energy in the case of a black body at a given temperature. Let us consider an ensemble of masses each having three degrees of freedom. We can consider that these consist of monoatomic molecules of a gas enclosed in a vessel of volume V. By applying Heisenberg's uncertainty relation it is possible to write the element of volume in the momentum space as

$$\sigma = \frac{h^3}{V} \tag{16.19}$$

The particles in this volume are indistinguishable. Hence it represents an eigen state. At any instant all particles having a momentum between p and $p + dp$ will lie in a small volume $4\pi p^2 dp$. Thus the total number of eigen states is given by

$$g(p)\, dp = \frac{4\pi p^2}{\dfrac{h^3}{V}} dp \tag{16.20}$$

For a photon $\lambda = \dfrac{h}{p}$ and hence $p = \dfrac{h}{\lambda} = \dfrac{h\nu}{c}$ (16.21)

$$dp = \frac{hd\nu}{c}$$ (16.22)

Thus we get $g(\nu)d\nu = 4\pi\, V\dfrac{\nu^2}{c^3}d\nu$ (16.23)

Because of the polarization in two independent directions the total number of eigen states are two times the expression (16.23).

Hence $g(\nu)d\nu = 8\pi\, V\dfrac{\nu^2}{c^3}d\nu$ (16.24)

This represents the total number of eigen states lying in the frequency range ν and $\nu+d\nu$. Applying Bose–Einstein distribution law we get

$$dn = \frac{g(\nu)d\nu}{e^{\alpha+\beta\varepsilon}-1} = 8\pi\, V\frac{\nu^2}{c^3}\frac{d\nu}{e^{\alpha+\beta\varepsilon}-1}$$ (16.25)

$$\frac{dn}{V} = 8\pi\,\frac{\nu^2}{c^3}\frac{d\nu}{e^{\alpha+\beta\varepsilon}-1}$$ (16.26)

where $\dfrac{dn}{V}$ represents the number of photons per unit volume. Hence energy per unit volume is $\dfrac{dn}{V}\times h\nu$. The energy density du in the frequency range ν and $\nu+d\nu$ is given by

$$du = 8\pi\frac{h\nu^3}{c^3}\frac{d\nu}{e^{\alpha+\beta\varepsilon}-1}$$ (16.27)

It can be shown that $\beta = 1/kT$. If we substitute for β and if $\alpha = 0$ we obtain Planck's radiation formula. The condition that $\alpha = 0$ implies that the auxiliary condition $\partial n = \sum \partial n_i = 0$ must be abandoned. And in the case of photon there is ample reason to believe so as every process of emission results in the creation of a photon and in every absorption, an absorption of photon occurs which may be converted into other forms of energy. And hence the condition $\partial n = \sum \partial n_i = 0$ is not applicable to photon and thus we obtain the above result given in equation (16.27).

16.2.1 Results of Three Statistics

The most probable distribution in the case of the three statistics can be expressed in the following forms:

$$\frac{g_i}{n_i} = e^{\alpha + \beta \varepsilon_i} \quad \text{Maxwell–Boltzmann} \tag{16.28}$$

$$\frac{g_i}{n_i} + 1 = e^{\alpha + \beta \varepsilon_i} \quad \text{Bose–Einstein} \tag{16.29}$$

$$\frac{g_i}{n_i} - 1 = e^{\alpha + \beta \varepsilon_i} \quad \text{Fermi–Dirac} \tag{16.30}$$

If $\dfrac{g_i}{n_i}$ is very large in comparison to unity we can write

$$\frac{g_i}{n_i} = \frac{g_i}{n_i} + 1 = \frac{g_i}{n_i} - 1 \tag{16.31}$$

That means for $\dfrac{g_i}{n_i}$ is very large all statistics give the same distribution.

16.3 BOSE–EINSTEIN CONDENSATION

According to Bose–Einstein statistics we have for the most probable distribution, the expression

$$n_i = \frac{g_i}{e^{\alpha + \beta \varepsilon_i} - 1}$$

$$n = \sum n_i = \sum_i \frac{g_i}{e^{\alpha + \beta \varepsilon_i} - 1} \tag{16.32}$$

For free particles of mass m

$$g(p) = \frac{\frac{4}{3}\pi p^3}{h^3 / V} \tag{16.33}$$

The total number of eigen states is given by

$$g(p)\, dp = \frac{4\pi p^2}{\dfrac{h^3}{V}}\, dp = \frac{4\pi V}{h^3}\, p^2 dp \tag{16.34}$$

$\beta = 1/kT$ and writing $\varepsilon = \dfrac{p^2}{2m}$ and integrating from $\varepsilon = 0$ to $\varepsilon = \infty$ we obtain

$$n = \sum n_i = \sum_i \frac{g_i}{e^{\alpha + \beta \varepsilon_i} - 1} = \frac{4\pi V}{h^3} \int_0^\infty \frac{p^2 \, dp}{e^{\alpha + p^2/2mkT} - 1} \tag{16.35}$$

Assuming $\dfrac{p^2}{2mkT} = z$, then $dz = \dfrac{p \, dp}{mkT}$

$$dp = \frac{mkT}{p} dz \tag{16.36}$$

$$n = \frac{V}{h^3} 4\pi 2mkTz \cdot \int_0^\infty \frac{mkT}{(2mkT)^{1/2} z^{1/2}} \cdot \frac{dz}{e^{\alpha + z} - 1} \tag{16.37}$$

Thus $n = \dfrac{V}{h^3}(2\pi mkT)^{3/2} \left[\dfrac{2}{\sqrt{\pi}} \displaystyle\int_0^\infty \dfrac{z^{1/2} dz}{e^{\alpha + z} - 1} \right] = \dfrac{V}{h^3}(2\pi mkT)^{3/2} F(\alpha)$ \hfill (16.38)

$$F(\alpha) = \left[\frac{2}{\sqrt{\pi}} \int_0^\infty \frac{z^{1/2} dz}{e^{\alpha + z} - 1} \right] = \left[\frac{2}{\sqrt{\pi}} \int_0^\infty \frac{z^{1/2} dz}{(1/A)e^z - 1} \right] \tag{16.39}$$

$$= A + A^2 \frac{1}{2^{3/2}} + A^3 \frac{1}{3^{3/2}} + \ldots \tag{16.40}$$

Comparing (16.38) and (16.40) $F(\alpha) = \dfrac{h^3}{(2\pi mkT)^{3/2}} \dfrac{n}{V} = A + A^2 \dfrac{1}{2^{3/2}} + A^3 \dfrac{1}{3^{3/2}} + \ldots$ \hfill (16.41)

If A is very much less than 1 ($A \leq 1$) then $A = F(\alpha)$ \hfill (16.42)

Hence in general we can write $A = F(\alpha) - F(\alpha)^2 \dfrac{1}{2^{3/2}} - F(\alpha)^3 \dfrac{1}{3^{3/2}} - \ldots$ \hfill (16.43)

Case I When $(A \leq 1)$, then $A = F(\alpha) = e^{-\alpha} = \dfrac{h^3}{(2\pi mkT)^{3/2}} \dfrac{n}{V}$ \hfill (16.44)

where $n_i = \dfrac{g_i}{\dfrac{1}{A} e^{\beta \varepsilon_i} - 1}$ and for higher energy values $\left(\dfrac{1}{A} \right) e^{\beta \varepsilon_i}$ becomes very large (remember A

is very much less than 1) and hence we can write $\dfrac{g_i}{n_i} = \dfrac{1}{A} e^{\beta \varepsilon_i} - 1 \approx \dfrac{1}{A} e^{\beta \varepsilon_i}$. \hfill (16.45)

Under these conditions the Bose–Einstein formula is the same as for classical Maxwell–Boltzmann distribution law.

Case II $A \sim 1$. In this case $\dfrac{1}{A} e^{\beta \varepsilon_i}$ is no longer large for the lower energy values and deviations occur from the classical distribution law. Hence in the expression $n_i = \dfrac{g_i}{\dfrac{1}{A} e^{\beta \varepsilon_i} - 1}$, the term unity cannot be neglected and its presence makes the number of molecules for a Bose–Einstein gas greater than the gas obeying classical distribution law. This means that the effect of Bose–Einstein statistics is to increase the number of particles in the lower energy states or to condense them into lower energies, which is now commonly called Bose–Einstein condensation. Another way of saying is that the gas is degenerate and A is called degeneracy parameter.

$$A = \frac{h^3}{(2\pi mkT)^{3/2}} \frac{n}{V} \tag{16.46}$$

The maximum value of A admissible is 1 and thus for large value of A $(A \sim 1)$ degeneracy may be expected which occur for low temperatures and small volume V, i.e., at high pressures. The value greater than 1 is not admissible for A, because $n_i = \dfrac{g_i}{\dfrac{1}{A} e^{\beta \varepsilon_i} - 1}$ for the lowest cell of zero energy (zero point energy neglected) becomes infinity when $A \to 1$. For $A > 1$, n_i becomes negative which is physically absurd. Hence a can never be negative. For low energies the limiting case for Bose–Einstein gas is reached when $A = 1$, that is when the temperature is low and pressure is high. The maximum value of $F(\alpha)$ is

$$F(0) = 1 + \frac{1}{2^{3/2}} + \frac{1}{3^{3/2}} + \ldots = 2.612 \tag{16.47}$$

The limiting case has been employed to explain many strange properties of the liquid helium II. The main workers in this field were Einstein, London and Tisza.

16.4 BOSE–EINSTEIN CONDENSATE

We have seen above that the phenomenon of Bose–Einstein condensation is based on the indistinguishability of the particles so also its wave nature. Wave nature of the particles is an outcome of de Broglie's principle. We can consider the atoms in a gas as quantum mechanical wave packets which have an extent of the order of de Broglie wavelength given by

$$\lambda_{dB} = \frac{h^2}{(2mk_B T)^{1/2}} \tag{16.48}$$

where λ_{dB} is the de Broglie wavelength, T is the absolute temperature of the gas, m is the mass of the atom and k_B is the Boltzman constant. λ_{dB} can also be thought of the uncertainty associated with the thermal momentum distribution. From expression it is clear that as the temperature is lower λ_{dB} is larger. When the atoms are cooled to a temperature at a level where λ_{dB} is comparable to inter-atomic separation, the atomic wave packets overlap from one atom to the other. In this case it should be remembered that the particles are also indistinguishable. At this temperature the bosonic particles undergo a phase transition as the wave packets overlap and the atoms occupy the same quantum mechanical state. The phenomenon is called Bose–Einstein condensation (BEC). In order to achieve BEC, the number density and temperature of the bosonic atoms should reach the values such that phase space density satisfy the condition

$$n\lambda_{dB}^3 > 2.61$$

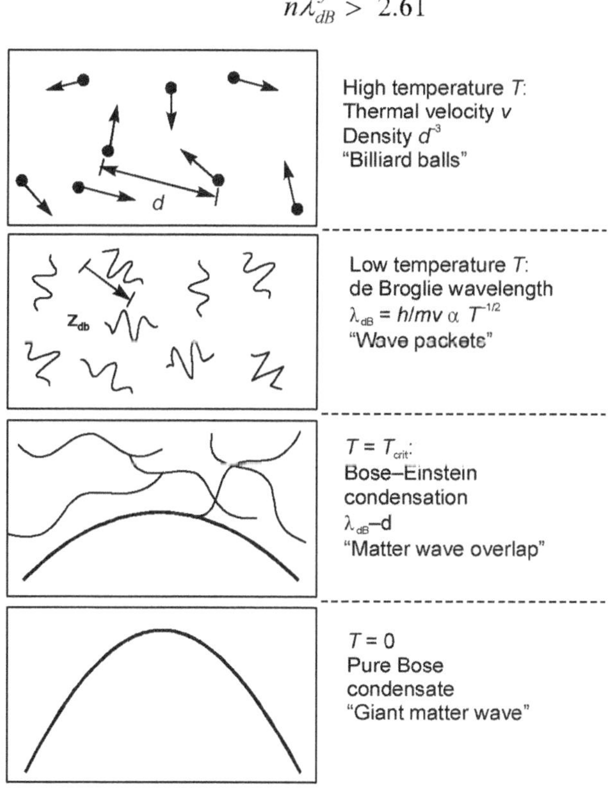

Figure 16.2   Bose–Einstein condensation

As temperature is lowered the de Broglie wavelength increases. When it becomes comparable to interatomic separation thermal cloud start converting into BEC. This is depicted in Figure 16.2.

When a beam of light falls on your body you feel hotter rather than colder. This is because light is absorbed and turns into heat. However we can feel colder if the light is made to bounce back.

This is the principle in laser cooling. When atoms are hit by light which is nothing but , collection of photons, they are made to bounce off with more energy than when it hits the atoms. It took a while for scientists to figure out how to do it.

We have the usual idea that we can heat up and vaporize objects with a laser. This is the principle behind laser ablation experiments. But the popular idea that one can heat up and vaporize objects with a laser is not exactly true when we look at individual atoms.

The phenomena of laser cooling was first proposed by Winelan and Dehmelt and simultaneously by Theodor W. Hänch and A.L. Schawlow already in 1975. Practical demonstration of laser cooling was achieved in 1976 by the work of Letokhov, Minogin and Pavlik. Three physicists shared Nobel Prize for laser-cooling and atom-trapping. Steven Chu, Claude Cohen–Tannoudji and W.D. Phillips were pioneers in this field and they were awarded the 1997 Nobel Prize for their pioneering work in laser cooling and trapping of atoms. Chu, at Stanford University (Stanford, CA) along with colleagues at Bell Laboratories (Holmdel, N.J.), developed a way to cool atoms using laser light in 1985. Using six pairwise orthogonal laser beams they created a small pod of about a million sodium atoms moving in a thick "optical molasses" at 240 µK—a technique called Doppler cooling. Later they constructed a magneto-optical trap that counteracted the force of gravity and securely captured atoms for study and experiment. The simple form of laser cooling is referred to as optical molasses because of its resemblance to the viscous drag of a body moving through molasses.

16.5 ATOMS FLOATING IN OPTICAL MOLASSES

At room temperature the atoms and molecules of which air consists move in different directions at a speed of about 4,000 km/hr. It is hard to study these atoms and molecules because they disappear all too quickly from the area being observed. However, by lowering the temperature one can reduce the speed, but the problem is that when gases are cooled down they normally condense into liquids first and then freeze into a solid form. In liquids and solids, study is made more difficult by the fact that single atoms and molecules get too close to one another. If, however, the process takes place in a vacuum the density can be kept low enough to avoid condensation and freezing. But even a temperature as low as –270°C involves speeds of about 400 km/hr. Only as one approaches absolute zero (–273°C) does the speed fall greatly. When the temperature is one-millionth of a degree from this point (termed 1 µK, microkelvin) free hydrogen atoms, for example, move at speeds of less than 1 km/hr (= 25 cm/s).

Steven Chu, Claude Cohen-Tannoudji, and William D. Phillips have developed methods of using laser light to cool gases to the µK temperature range and keeping the chilled atoms floating or captured in different kinds of "atom traps". The laser light functions as a thick liquid, dubbed optical molasses, in which the atoms are slowed down. Individual atoms can be studied there with very great accuracy and their inner structure can be determined. As more and more atoms are captured in the same volume a thin gas forms, and its properties can be studied in detail. The new methods of investigation that the Nobel Laureates have developed have contributed greatly to increasing our knowledge of the interplay between radiation and matter. In particular, these developments have opened the way to a deeper understanding of the quantum-physical behaviour of gases at low temperatures. The methods may lead to the design of more precise atomic clocks for use in, e.g., space navigation for accurate determination of position. A start has also been made on the design of atomic interferometers with which, e.g. very precise measurements of gravitational forces can be made, and atomic lasers, which may be used in the future to manufacture very small electronic components.

We know that the laser light consists of stream of photons. These photons are very light, so to speak. In laser cooling the atoms, which are moving with thermal velocity, are hit by bunch of photons. If we could compare the atoms to a bowling ball, we could consider the tiny photons as ping-pong balls. If one wants to stop or slow down a moving bowling ball, probably it is not possible to do it by a single ping-pong ball. However if you shoot it by a bunch of ping-pong balls it may be possible to push it around by bouncing ping-pong ball away. Thus it would be possible to push the atoms by a bunch of photons, the photons being bounced back or scattered.

When the photons are bouncing off the atom, the electron in the atom absorbs the photon and jumps up to a higher level if it encounters resonance condition (inelastic collision). Then it quickly jumps back down and spits the photon back out. Scientifically saying the photon excites the electron vis-à-vis the atom in to an upper excited state. Once the electron is in the excited state it spontaneously emits and returns to the ground state. One needs to have exactly the right colour of light vis-à-vis the right frequency to make an atom to do that. A big part of the difficulty in making laser cooling works was to get enough light of just the right colour vis-à-vis the right frequency. Now lasers are available in which one can adjust the colour very precisely to match the colour that the atom wants. That means the laser frequency should match the transition frequency of the transition. If the colour is wrong the photons just go right through the atoms (elastic collision).

Thus in order to avoid laser photons hitting the slow atoms instead of hitting the faster ones and to slow them down, one can easily use the technique of Doppler effect. We already know that an atom moving towards the laser light will see the light blue-shifted and if it is of the right frequency for absorption, the atom will be excited. The atoms moving away from the laser photons will be unaffected.

So if the laser light has just the right colour so that the fast atoms are slowed down and the Doppler shifted frequency for that speed, however, does not affect the slow ones thus it slows down only the fast atoms without pushing the slow ones backwards. But this is just in one direction. The atoms in the box are bouncing around in all directions. How we will slow all of them down? One needs to send the laser beam at the atom from all the different directions. Then if we adjust it to exactly the right colour all the atoms will get cold. One could get "optical molasses" in this way.

Cornell and Wieman did experiments with rubidium atoms in their little glass cell. By choosing the right laser colour, they could get big bunches of very cold atoms.

However, there are possibilities that the atoms can wander away from the laser beams and move away from the centre (where the laser beams crossed), however, they can be pushed back into the middle by more light hitting them from other laser beams. This is known as a laser trap.

16.5.1 Brief Explanation

This technique works by sending a laser light having a frequency slightly below an electronic transition in the atom. Because the light is detuned to the "red" (i.e., at lower frequency) of the transition, the atoms moving towards the laser beam will see the frequency uplifted by Doppler shift (blue shifted- to higher frequency) and becomes suitable for an absorption to the higher level. Thus the atoms will absorb more photons if they move towards the light source, due to the Doppler Effect. Thus if one applies light from two opposite directions, the atoms will always scatter more photons from the laser beam pointing opposite to their direction of motion. When the atom is excited it loses momentum. In each scattering event the atom loses a momentum equal to the momentum of the photon. Once the atom is in the excited state it emits spontaneously. As spontaneous emission is in random directions, the atom does not gain any momentum, not even the one which it lost during absorption. Thus the atom, which is making a transition from the excited state by spontaneous emission, will be kicked by the same amount of momentum but in a random direction. The net result of the absorption and emission process is to reduce the speed of the atom, provided its initial speed is larger than the recoil velocity from scattering a single photon. If the absorption and emission are repeated many times, the mean velocity, and therefore the kinetic energy of the atom will be reduced. We know that the temperature depends upon the kinetic energy and since the temperature of an ensemble of atoms is a measure of the random internal kinetic energy, this is equivalent to cooling the atoms.

It should be remembered that the vast majority of photons that come anywhere near a particular atom are almost completely unaffected by that atom as the frequency supplied may not be equal to the transition frequency. Thus the atom is almost completely transparent to most frequencies (colours) of photons.

When a few photons happen to "resonate" with the atom, in a few very narrow bands of frequencies (a single colour rather than a mixture like white light), the atom is excited to the excited state. The atom typically absorbs that photon when its frequency is close to resonance and then emits an identical photon (emission spectrum) in some random, unpredictable direction.

16.5.2 Limitations

1. The atoms perform a random walk in momentum space in steps which is equal to the photon momentum because of the spontaneous emission and photon absorption. The net effect is that it produces a heating effect which counteracts the cooling processes.

2. The optical transition used for cooling has a frequency width, with the likelihood that it may result in a scattering from the beam outside the correct beam and thereby limits the velocity discrimination and hence the cooling. The temperature is called Doppler temperature. The effect of the recoil temperature can be reduced by sub-Doppler cooling. Beyond the sub-Doppler cooling evaporative cooling is employed to get ultracold atoms.

3. The concentration of atoms must be minimal. If two atoms happen to collide and if one is in the excited state, then there is a possibility that the excited electron drops back to the lower state

with extra energy liberated which may heat the atoms. This works against the cooling process. Thus it creates a limit to the maximum concentration of atoms which can be cooled.

4. Only certain atoms have energy states vis-à-vis optical transitions amenable to laser beams. Moreover other finer interactions in atoms like the hyperfine interactions create hyperfine levels and hence the photon absorbed from the laser may not drop back to the original state. Thus it goes to a "dark state" removing it from the cooling process. This is one of the reasons that laser cooling is more difficult in molecules as it contains many ro-vibrational levels and their fine or hyperfine structures.

16.5.3 Quantitative Ideas

The laser cooling works on the principle that when an atom sees a laser beam it absorbs a photon if the frequency matches with the resonance frequency of the atom. When the atom absorbs a photon from the laser, it will be slowed down because of the fact the photon has a momentum $p = E/c = h/\lambda$. If we consider sodium atoms for example, and consider that they are freely moving in a vacuum chamber, at 300K, the rms velocity of a sodium atom from Maxwell distribution would be about 570 m/s. We have already seen in previous chapter that sodium has two prominent lines, viz., the sodium D-lines (589.0 and 589.6 nm, about 2.1 eV), and if a laser is tuned just below one of the sodium D-lines a sodium atom travelling towards the laser absorbs a laser photon as it is blue shifted and the frequency as observed by the atom becomes equal to the resonance frequency. By absorbing the laser photon the atom would have its momentum reduced by the amount of the momentum of the photon. When the momentum of the atom is reduced it is slowed down and with large number of such absorptions the sodium atoms gets retarded or slowed down. The velocity corresponding to absorption is given by

$$v_p = (2kT/M)^{1/2} \tag{16.49}$$

From the expression it is clear that when the speed is reduced the atoms get cooled. It can be calculated that one absorption would slow down a sodium atom by about 3 cm/s out of a speed of 570 m/s. It means that it requires that atleast 20,000 absorbing photons are necessary to reduce the sodium atom momentum to zero.

It is now useful to understand the scattering process in some detail. We have already seen earlier that each absorbed photon takes the atom from the ground state to the particular excited state by absorption followed by a momentum kick h/λ in the direction of the laser beam (opposite to the direction of the atomic beam). Before the atom can absorb another photon it has to be de-excited back to the ground state. Once the atom in the excited state we already know that the atom does not remain in the excited state and there is a tendency to transfer back to the original state. This is mainly accomplished by spontaneous emission though stimulated emission is also possible. In any case the photon emitted by the atom will have the energy $h\nu$ and it would look like that there a momentum kick by this process according to the expression 16.52 (*See* Figure 16.3). However, at low intensities the spontaneous emission is predominant. The spontaneous emission is a random process and the photon can be emitted in any direction. There are equal probabilities of spontaneous emission in any of the 4π directions in space. Hence on the average this process does not contribute to any momentum change in the atom. The stimulated emission if present will always occur along the axis of the cell in the direction of the laser beam and an absorption-stimulated emission cycle may not contribute any gain or loss in atom velocity. Thus on the average the emission process does not result in any momentum change in the atom. Thus we see that each absorption-emission process results in a net momentum transfer of $h\lambda$ to the atom in the direction of the exciting laser beam.

Let us consider for the time being a one-dimensional case where the atoms are restricted to move along the X-axis. Let us also consider that two identical laser beams (it is possible by splitting the same laser beam by a beam splitter) are sent in opposite direction along the X-axis. The frequency of the identical laser beam is chosen slightly below the resonance frequency of atoms. This is known as detuning and in this case it is red detuning as the wavelength chosen is towards the red side of the resonance wavelength. An atom moving the +X direction sees the laser beam in the –X direction as Doppler shifted to a higher frequency and the laser beam in the + direction (i.e., in the direction of atom beam) to a lower frequency. Therefore the photons from the –X direction beam scatters ending in resonance absorption and thereby the atoms towards the +X direction gets a momentum kick in the –X direction. It means that the atom in the +X direction is slowed down. The opposite happens for an atom moving in the –X direction which sees that the laser frequency in the +X beam as Doppler shifted thereby in the resonance range. By resonance absorption a momentum kick in the + direction is experienced by this atom proceeding in the –X direction. Thus the beam in the +X direction gives a momentum kick down to the atom moving in the –X direction and the beam in the –X direction gives a momentum kick down to atom moving in the +X direction. Thus it is just like having a frictional force which always opposes the motion of atoms. The net effect is that the velocity of atoms and thereby the kinetic energy is reduced finally bringing the atoms to "rest". Since the kinetic energy is a measure of temperature, the reduction in kinetic energy amounts to cooling. It should be remembered that the atoms comes to a complete rest only at absolute zero, however, in this case the "rest" means that mean velocity is zero.

The change in speed from the absorption of one photon can be calculated from

$$\Delta_{p/p} = p_{photon} / mv = \Delta v / v \tag{16.50}$$

$$\Delta v = p_{photon} / m \tag{16.51}$$

This would look like a lot of photons, but according to Chu, a laser can induce on the order of 10^7 absorptions per second so that an atom could be stopped in a matter of milliseconds.

It should be remembered that an absorption can also speed up an atom if it catches it from behind, i.e., if the laser beam is in the direction of the movement of the atoms. Hence it is necessary to have more absorption from head-on photons since our goal is to slow down the atoms. In practice this is accomplished by tuning the laser slightly below the resonance absorption of a stationary sodium atom in order to compensate for the Doppler shift. From the atom's perspective, the head-on photon is seen as Doppler shifted upward toward its resonant frequency and it therefore more strongly absorbed than a photon travelling in the opposite direction which is Doppler shifted away from the resonance. In the case of sodium atom at room temperature, the incoming photon would be Doppler shifted up in frequency by 0.97 GHz, and hence in order to get the head on photon to match the resonant frequency it would require that the laser be tuned below the resonant peak by that amount. Theodore Hansch and Arthur Schawlow at Stanford University proposed this method in 1975 and achieved by Chu at AT and T Bell Labs in 1985. Sodium atoms were cooled from a thermal beam at 500K to about 240 μK. The experimental technique is to direct laser beams from opposite directions upon the sample, linearly polarized at 90° with respect to each other. Six lasers could then provide a pair of beams along each coordinate axis. Chu then named it as "optical molasses" comparing the "viscous" effect of the laser beams in slowing down the atoms.

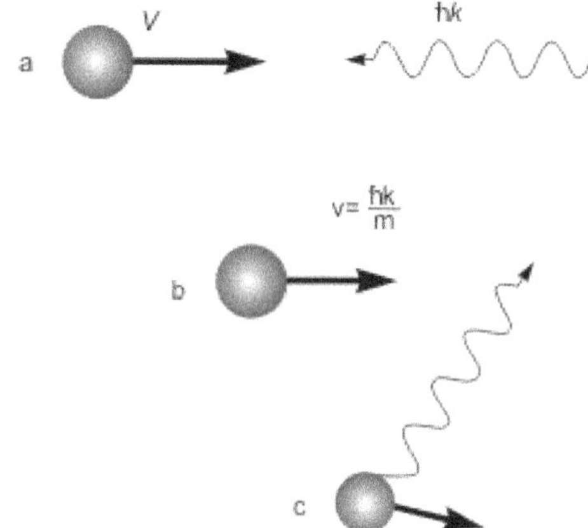

Figure 16.3   (a) An atom with velocity v encounters a photon with momentum $\hbar k = h / \lambda$ (b) after absorbing the photon, the atom is slowed by $\hbar k / m$ (c) after re-radiation in a random direction, on average the atom is slower than in (a)

We should now remember that a step by step process is necessary in tuning the laser frequency. Continuing to cool the atoms requires the tuning of the laser upward in frequency toward the atomic resonant frequency step by step because the Doppler shift will be smaller as the atoms are slowed down. This places a practical limit on the cooling which can be achieved, as the differential cooling rate is reduced and at a certain point the cooling mechanism is destroyed by heating due to the random absorption and re-emission of photons. This practical limit can be characterized by $2kT = E_{\text{resonant photon}}$. At low temperatures of approximately 240 μK this would correspond to photon energies around 4×10^{-8} eV. This is almost of the order in magnitude of the Zeeman splitting of energy levels of the atoms in the magnetic fields produced by the laser photons.

It was found that these splittings which limited the original laser cooling processes could be exploited to lower the ultimate temperatures below these limits. With the "optical molasses" and the polarization gradient in the region of opposing laser beams, temperatures as low as 35 μK for sodium and 3 μK for cesium were obtained.

Let us now consider an atom having two energy levels and also assume that the atom is at rest. We bombard this atom with a photon having a resonant energy (i.e., sufficient energy to excite it into the upper state). The atom absorbs the photon and also gets a kick in the direction of the original photon (*See* Figure 16.4). The atom acquires a momentum which is

$$mv = \frac{hv}{c} = \frac{\hbar\omega}{c} \tag{16.52}$$

The corresponding energy is given by $E = \dfrac{p^2}{2m} = \dfrac{\hbar^2\omega^2}{2mc^2}$ (16.53)

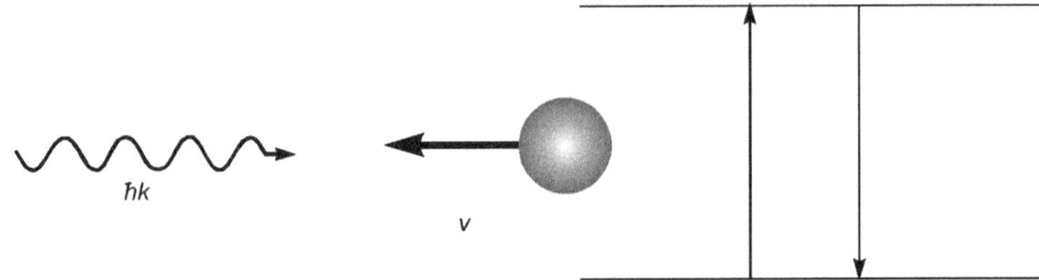

Figure 16.4   Momentum acquired by an atom

If we consider a sodium atom with mass number $A = 23$, then $E = \dfrac{(2eV)^2}{2 \times 23 \times 10^9 \text{eV}} = 10^{-10}$ eV.

If we convert this energy into degree Kelvin $T = 10^{-10}$ eV $\cdot 10^4 \dfrac{K}{eV} = 1\mu K$.

When the atom is excited to the upper state by absorption of the photon, it suddenly decays to the ground state at a rate $\gamma_0 = \dfrac{1}{\tau_o}\cdot$ Spontaneous emission process is in random direction and hence the atom recoils in a random direction.

Thus when the atom absorbs photons and then emitting photons in random directions, we get atoms moving in the direction of laser beam. In this case we may also "heat" atoms in the orthogonal direction as a result of the "random walk" kicks.

Let us think of an atomic beam in which all atoms move in the same direction with the same velocity v_b. We now send a laser beam head on with the atoms in the atomic beam. This is shown in Figure 16.5.

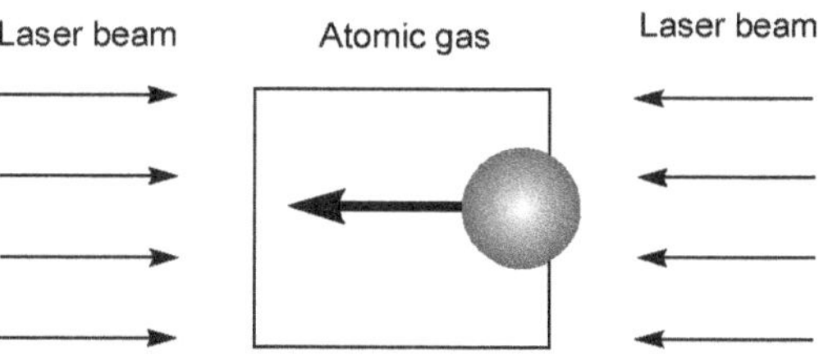

Figure 16.5   Red-tuned laser beam

The laser beam is having the right frequency to excite the atoms to their excited state. Once they reach the excited state, spontaneous emission occurs. We can tune the laser frequency to be in tune with the frequency of the atomic transition v_0. Remember we will have to compensate for the Doppler shift in the absorption frequency since the atoms are moving. We can calculate the right frequency needed from the Doppler shift

$$v_0 = v\left(1 + \frac{v_b}{c}\right) \tag{16.54}$$

From the above equation we note that the frequency of the laser must be slightly smaller than the atomic resonance frequency v_0 in order to compensate for the Doppler shift (*See* Figure 16.6a). It means that the laser has to be "red-tuned".

Thus the atoms are slowed down. We have already remarked that as the velocity of the "cold" atoms is less now, the Doppler shift is different now than earlier. Changing the laser frequency little by little can solve this problem. This is called chirping (increasing the laser frequency gradually). There are other methods as well, like "Zeeman slowing", applying varying magnetic fields. The changing Zeeman effect compensate for the slowing effect, keeping the atoms in resonance with the fixed laser frequency. The energy states change in this case but the transitions can be adjusted to tune to the fixed laser frequency by suitably adjusting the magnetic field. This is further discussed in the section "stopping of atoms in atomic beams".

Yet another quantitative picture of the process can be obtained by considering the absorption rates in the two counter propagating beams as a function of velocity of atoms. The force which is nothing but rate of change of momentum in this case is the product of the absorption rate and the momentum transferred per absorption–emission cycle. Figure 16.6 b shows a graphical picture. The force F+ from the beam travelling in the +X direction is positive and is centred at the negative velocity side as seen in the figure . In this case we have already seen that the atoms moving in the –X direction are maximum scattered (absorbed) as the frequency of the laser beam is below resonance and the resonance is obtained by the Doppler shift to higher frequency. Similarly the force F– from the laser beam travelling in the –X direction is negative and is centred in the positive velocity side. The net effect or the sum of these two forces is linear in velocity near the origin having a negative gradient. And such a force is a frictional force.

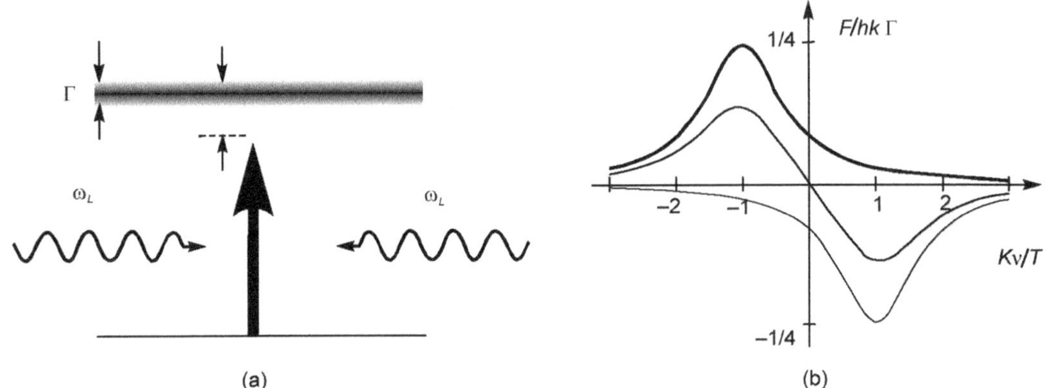

Figure 16.6   One-dimensional Doppler cooling

Figure 16.6 shows the case of one-dimensional Doppler cooling: (a) Two identical counter propagating laser beams are incident on the atom cloud. The frequency of the standing laser field ω_L is detuned by an amount Δ below the resonance with a transition to an excited state, which has a linewidth, Γ, equal to the inverse of the natural lifetime of the excited state; (b) Each of the counter-propagating beams exerts a force with a Lorentzian velocity dependence. For an intensity equal to the saturation intensity the maximum force corresponds to one photon momentum transferred every four natural lifetimes. The green curve shows the combined force from the two beams, displaying the viscous damping around $u = 0$.

The above descriptions are restricted in one-dimensional and the scheme can be extended to three dimensions by considering three orthogonal set of axes and in each axes counter propagating beams are sent. We can name the one dimensional laser beam-atomic beam configuration as one dimensional optical molasses. In three dimension the velocity of atoms propagating in any direction can be resolved into three orthogonal directions X, Y and Z. The argument in the one dimensional case can be applied in this case as well and the atoms can be brought to "rest" in all three dimensions.

If we consider a laser beam hitting an atom moving in the same direction it can happen that the absorption speeds up the atom. Conceptually if the laser beam catches the atom from behind the momentum imparted to the atom is in the same direction of its movement and hence increase its speed. So it is necessary to have more absorption from head-on photons in order to slow down the atoms.

16.6 VARIATION OF ATOMIC ABSORPTION IN AN ELECTROMAGNETIC FIELD

The interaction of electromagnetic radiation with an atom and the absorption of radiation by the atom depend on the frequency and intensity of the radiation field. We shall now consider the absorption of an electromagnetic radiation of an atom by assuming two energy levels having energies E_1 and E_2

having N_1 and N_2 atoms respectively in each state as shown in Figure 16.7. We shall also assume that the atoms are in a collision free environment and only transitions are due to radiative process, viz., absorption, spontaneous emission and stimulated emission.

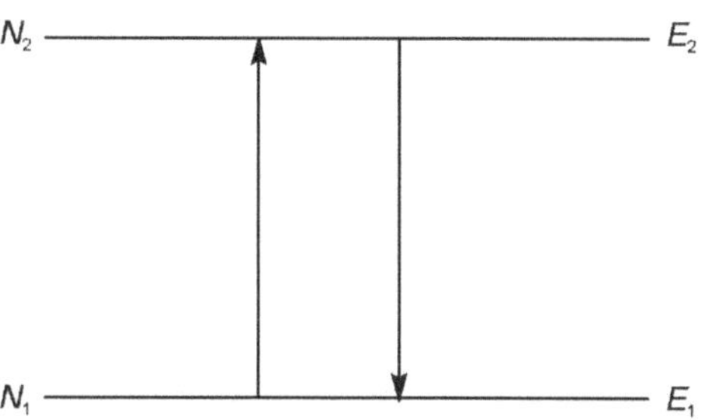

Figure 16.7   Variation of atomic absorption in an electromagnetic field

We know that $\quad E_2 - E_1 = h\nu = \dfrac{h}{2\pi}\omega = \hbar\omega \quad$ as $\quad 2\pi\nu = \omega$ $\hfill$ (16.55)

The rate of absorption from level 1, $\quad \dfrac{dN_1}{dt} = B_{12}N_1\rho(\omega)$ $\hfill$ (16.56)

where B_{12} is the Einstein's coefficient of absorption and $\rho(\omega)$ is the energy density of the radiation field.

The rate of spontaneous and stimulated emission from state 2 is written as

$$\frac{dN_2}{dt} = AN_2 + B_{21}N_2\rho(\omega) \tag{16.57}$$

In thermal equilibrium, the number of upward transitions is equal to number of downward transitions.

$$\text{i.e.,} \quad \frac{dN_1}{dt} = \frac{dN_2}{dt} \tag{16.58}$$

$$\text{i.e., } B_{12}N_1\rho(\omega) = B_{21}N_2\rho(\omega) + AN_2 \tag{16.59}$$

We have seen in Chapter 3 that $B_{12} = B_{21} = B$

$$BN_1\rho(\omega) \quad BN_2\rho(\omega) = AN_2 \tag{16.60}$$

Therefore $\quad B(N_1 - N_2)\,\rho\,(\omega) = AN_2$

The population difference $\Delta N = N_1 - N_2$ can be obtained as

$$\Delta N = \frac{N_1 + N_2}{1 + \dfrac{2B}{A}\rho(\omega)} \tag{16.61}$$

We consider that all atoms are either in the state E_2 or in the state E_1, that means all other states are empty, then $N_1 + N_2 = N$

If $\rho(\omega) = 0$ there is no absorption and we obtain from above that $\Delta N = N_1 + N_2 = N$ as there are no atoms in the E_2 state.

That means that all atoms are in the ground energy state with energy E_1. The intensity of radiation field $I(\omega) = \rho(\omega).c$ where c is the velocity of light.

$$\text{i.e., } \rho(\omega) = \frac{I(\omega)}{c} \tag{16.62}$$

$$\Delta N = \frac{N_1 + N_2}{1 + \dfrac{2B}{A}.\dfrac{I(\omega)}{c}} = \frac{N}{1 + \dfrac{I}{I_S}} \tag{16.63}$$

where we have assumed $I_S = \dfrac{A.c}{2B}$ and is called saturation intensity. If $I = I_s$ we can see from the above expression that $\Delta N = \dfrac{N}{2}$. It means at saturation intensity half of the atoms is in the excited state and it is the saturation limit.

The intensity profile of a spectral line is generally described by Lorentzian shape which is defined by the expression

$$L(\omega - \omega_0) = \frac{\Gamma^2}{\Gamma^2 + 4(\omega - \omega_0)^2} = \frac{1}{1 + \dfrac{4(\omega - \omega_0)^2}{\Gamma^2}} \tag{16.64}$$

where ω_0 is the centre of the line and Γ is the full line width at half maximum.

The absorption probability of a monochromatic radiation of frequency ω is given by

$$R_a = B\rho(\omega)\, L(\omega - \omega_0) \tag{16.65}$$

We have already seen that $\Delta N = \dfrac{N_1 + N_2}{1 + \dfrac{2B}{A}\rho(\omega)} = \dfrac{NA}{A + 2B\rho(\omega)}$

$$\Delta N \, A \; + \Delta N \; 2B\rho(\omega) = NA$$
$$\Delta N \; 2B\rho(\omega) = NA - \Delta NA \tag{16.66}$$

$$B\rho(\omega) = \frac{NA - \Delta NA}{2\Delta N} = \frac{A(N - \Delta N)}{2\Delta N} = \frac{A}{2}\left(\frac{N}{\Delta N} - 1\right) \tag{16.67}$$

We have seen that $\quad \Delta N == \dfrac{N}{1 + \dfrac{I}{I_S}}$

Hence $\quad \dfrac{N}{\Delta N} = 1 + \dfrac{I}{I_S} = \dfrac{I_S + I}{I_S} \tag{16.68}$

Thus we get $\quad B\rho(\omega) = \dfrac{A}{2}\left[\dfrac{I_S + I}{I_S} - 1\right] = \dfrac{A}{2}\dfrac{I}{I_S} \tag{16.69}$

The half line width of the spectral line Γ is related to the life time of the excited state by the relation $\Gamma_i = A_i = \dfrac{1}{\tau_i}$ where τ_i is the life time of the excited state and A_i is the Einstein's coefficient of spontaneous emission.

Therefore $\quad R_a = B\; \rho(\omega)\; L(\omega - \omega_0) = \dfrac{\dfrac{A}{2}\left(\dfrac{I}{I_S}\right)}{1 + 4\left[\dfrac{(\omega - \omega_0)}{A}\right]^2} = \dfrac{\dfrac{\Gamma}{2}\left(\dfrac{I}{I_S}\right)}{1 + 4\left[\dfrac{(\omega - \omega_0)}{A}\right]^2} \tag{16.70}$

We define $(\omega - \omega_0) = \delta$ which is called as detuning function.

Thus $\quad R_a = \dfrac{\dfrac{\Gamma}{2}\left(\dfrac{I}{I_S}\right)}{1 + 4\left[\dfrac{\delta}{A}\right]^2} \tag{16.71}$

The equation (16.71) epresents the absorption rate in terms of the intensity of the incident radiation and its frequency difference from the central frequency ω_0.

16.7 RATE OF ABSORPTION–SPONTANEOUS EMISSION

Atoms move around the cell with random velocity v depending upon the temperature. The electromagentic wave in the rest frame is represented by the expression

$$\vec{E} = \vec{E}_0 \exp\, i(\omega t - \vec{k} \cdot \vec{r}) \qquad (16.72)$$

where ω is the frequency of radiation. The frequency is changed from the rest frame to the frame of the moving atom moving with a velocity v such that frequency is $\omega' - \vec{k}\vec{v}$ where k is the wave vector along the propagation of the electromagnetic wave and its magnitude is equal to $\dfrac{2\pi}{\lambda}$.

As the atoms are in motion with random velocity, the moving atom can absorb a frequency only when the transition frequency ω_0 is equal to ω'.

Thus for absorption $\omega_0 = \omega - \vec{k} \cdot \vec{v}$

or $\qquad \omega = \omega_0 + \vec{k} \cdot \vec{v} \qquad (16.73)$

From this equation we find that the frequency of the absorbed radiation by the atom is greater that its original transition frequency ω_0. Thus if $\vec{k} \cdot \vec{v} > 0$, the direction of atomic motion has finite component in the direction of propagation of electromagnetic radiation. In this case the actual frequency ω of the atomic absorption is higher. That means that the transition is lower in wavelength and hence the transition is said to be blue shifted (an effect of Doppler shift). Contrary if the atomic velocity is having a nonzero component in the direction opposite to the direction of light wave, i.e., $\vec{k} \cdot \vec{v} < 0$ the absorption frequency has a lower value meaning thereby that the absorption is red shifted with respect to transition frequency ω_0, the frequency if the atom is at rest. These shifts are generally not observed in conventional spectroscopy as these shifts are well inside the line width. However in Doppler free spectroscopy, this becomes visible by a dip in the line profile. At ordinary room temperature the atoms have random velocities and in accordance with Maxwellian distribution of velocities the largest number of atoms having zero component are in the Z direction.

We consider again the Lorentzian profile with ω_0 replaced by $\omega_0 + \vec{k} \cdot \vec{v}$. In this case we define

$$\delta(v) = \omega - \omega_0 - \vec{k} \cdot \vec{v} = \delta - \vec{k} \cdot \vec{v}. \qquad (16.74)$$

Therefore

$$L[\delta(v)] = \frac{\Gamma^2}{\Gamma^2 + 4\delta(v)^2} = \frac{1}{1 + \dfrac{4\delta(v)^2}{\Gamma^2}} \qquad (16.75)$$

Thus the atomic absorption is inhomogenously broadened by the random motion of the atoms and the total line profile is the net effect of all Lorentzian profiles contributed by atoms with different velocities (*See* Figure 16.8). Three such Lorentzian profiles are shown in the Figure 16.7. $\omega = \omega_0$ corresponds to atoms with zero velocity, $\omega = \omega_0 + k \cdot v$ corresponds to atoms moving with a velocity in the direction of light propagation and $\omega = \omega_0 - k \cdot v$ to those moving away from the direction of light propagation.

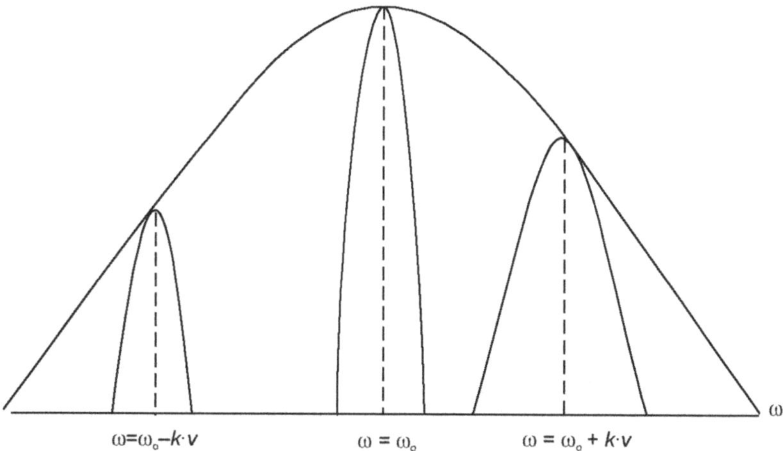

Figure 16.8 The Lorentizian profiles contributed by atoms

From equation (16.71) we have seen that the absorption rate for atoms in thermal equilibrium is

$$R_a = \frac{\dfrac{\Gamma}{2}\left(\dfrac{I}{I_S}\right)}{1+4\left[\dfrac{\delta}{A}\right]^2} \tag{16.76}$$

In the present case it becomes

$$R_a = \frac{\dfrac{\Gamma}{2}\left(\dfrac{I}{I_S}\right)}{1+4\left[\dfrac{\delta(v)}{\Gamma}\right]^2} \tag{16.77}$$

The rate equation (16.59) can now be written as

$$\frac{dN_1}{dt} = -\frac{dN_2}{dt} = -R_a N_1 + (A + R_a)N_2 = 0$$

Thus $\dfrac{N_2}{N_1} = \dfrac{R_a}{A+2R_a}$ $\tag{16.78}$

The probability of spontaneous emission from state of energy E_2 following an absorption from the state with energy E_1 is

$$R_{sp} = A\frac{N_2}{N_1} = A\frac{R_a}{A+2R_a} \tag{16.79}$$

We have seen already that $A_i = \Gamma_i = \dfrac{1}{\tau_i}$ $\hfill$ (16.80)

Hence $\quad R_{sp} = \dfrac{\Gamma R_a}{\Gamma + 2R_a} = \dfrac{\dfrac{\Gamma}{2}\left(\dfrac{I}{I_S}\right)\Gamma}{1+4\left(\dfrac{\delta(\nu)}{\Gamma}\right)^2}{\Bigg/}\left[\Gamma + 2\dfrac{\dfrac{\Gamma}{2}\left(\dfrac{I}{I_S}\right)}{1+4\left(\dfrac{\delta(\nu)}{\Gamma}\right)^2}\right] = \dfrac{\dfrac{\Gamma}{2}\left(\dfrac{I}{I_S}\right)\Gamma}{1+4\left(\dfrac{\delta(\nu)}{\Gamma}\right)^2\Gamma} \cdot \dfrac{1+4\left(\dfrac{\delta(\nu)}{\Gamma}\right)^2}{1+4\left(\dfrac{\delta(\nu)}{\Gamma}\right)^2\Gamma + 2\dfrac{\Gamma}{2}\left(\dfrac{I}{I_S}\right)}$ $\hfill$ (16.81)

$$= \dfrac{\dfrac{\Gamma}{2}\left(\dfrac{I}{I_S}\right)}{1+\left(\dfrac{I}{I_S}\right)+4\left[\dfrac{\delta(\nu)}{\Gamma}\right]^2} \hfill (16.82)$$

where R_{sp} determines the rate of decrease in the thermal velocity of atoms in the direction of propagation of the laser beam. If $\dfrac{I}{I_S}$ is very large, i.e., the intensity of the laser beam is large, the above expression takes the form

$$R_{sp} = \dfrac{\dfrac{\Gamma}{2}s_o}{1+s_o+4\left[\dfrac{\delta(\nu)}{\Gamma}\right]^2} = \dfrac{s_o\dfrac{\Gamma}{2}}{1+s_o\left[1+4\left[\dfrac{\delta(\nu)}{\Gamma(1+s_o)^{1/2}}\right]^2\right]} \hfill (16.83)$$

i.e., $\quad R_{sp} = \left[\dfrac{s_0}{1+s_0}\right]\left[\dfrac{\dfrac{\Gamma}{2}}{1+4\left[\dfrac{\delta(\nu)}{\Gamma_s}\right]^2}\right] \hfill (16.84)$

where $\dfrac{I}{I_S} = s_0$ is called the saturation parameter and $\Gamma_s = \Gamma(1+s_0)^{1/2}$ is called power broadened line width.

In absorption experiment when one scans the frequency spread of atomic transitions one could see that for large values of s_0 the line centre is saturated whereas it is not saturated at the wings (*See* Figure 16.9). The absorption at the wings continues to increase with increase in laser power, but at the centre it has already attained saturation. That means half of the atoms are in the excited state and hence there is no further increase in absorption at the centre.

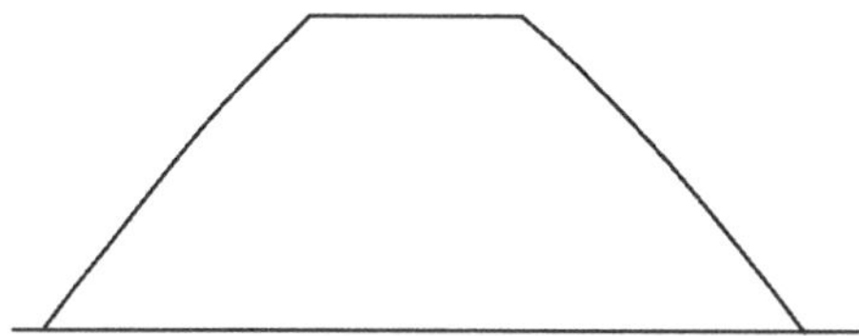

Figure 16.9   Shape of the saturated line

16.8   POLARIZATION GRADIENT OF COUNTER PROPAGATING LASER BEAMS

We have already seen in the introductory section of this chapter a qualitative explanation of the interaction of two counter propagating laser beams with the atoms in the Z direction, the direction of propagation of the laser beams. The interaction is complicated as the two laser beams form the two stationary wave patterns. The stationary wave pattern makes the polarization of light beams to change and hence a polarization gradient is found in the Z-axis. Atomic transitions depend on the polarization of light beams by which atoms interact. A transition induced by a left circular polarized light may not be induced by a right polarized light beam. Thus the atomic absorptions are different at different points along the Z-axis.

Case I   Beams of identical linear polarization

Let us consider the case of counter propagating laser beams of identical linear polarization with angular frequency ω_L. The light beam propagating in the +Z direction can be represented by the expression

$$\vec{E}_+ = E_0 \vec{\varepsilon} \cos\,(\omega_L t - kz) \tag{16.85}$$

where $k = \dfrac{2\pi}{\lambda}$ is the wave vector with its direction along the propagation of the electromagnetic wave. $\vec{\varepsilon}$ is the unit polarization vector perpendicular to the direction of wave propagation.

An identical linearly polarized light propagating in the –Z direction is represented by

$$\vec{E}_- = E_0\,\vec{\varepsilon} \cos\,(\omega_L t + kz) \tag{19.86}$$

The resultant electric field due to the two light beams is then

$$\vec{E} = \vec{E}_+ + \vec{E}_- = 2\,E_0\,\vec{\varepsilon}[\cos\,(\omega_L t - kz) + \cos(\omega_L t + kz)]$$

$$= 2\,E_0\,\vec{\varepsilon} \cos\omega_L t\,\cos kz = 2\,E_0\,\vec{\varepsilon} \cos\,\omega_L t\,\cos\frac{2\pi}{\lambda} z \tag{19.87}$$

Thus we find that the resultant $\vec{E}$ changes with a periodicity of $\dfrac{\lambda}{2}$ along Z direction though the resultant electric vector is oriented along $\vec{\varepsilon}$.

Case 2   Counter propagating linearly polarized light beams with their $\vec{\varepsilon}$ vectors at right angles.

Let us consider two light beams propagating in $+Z$ and the other in $-Z$ direction. The unit polarization vector $\vec{\varepsilon}$ in the first case is assumed to be in X-direction and in the second case it is assumed to be in the Y-direction.

The resultant electric field is then

$$\vec{E} = \vec{E}_+ + \vec{E}_- = E_0\, \vec{x}\cos(\omega_L t - kz) + E_0\, \vec{y}\cos\,(\omega_L t + kz) \tag{16.88}$$

$$= E_0\left[(\vec{x}+\vec{y})\cos\,\omega_L t \cos kz + (\vec{x}-\vec{y})\sin\,\omega_L t\,\sin\,kz\right] \tag{16.89}$$

If $z = 0$, the equation (16.89) gives

$$\vec{E} = \sqrt{2}E_0(\vec{x}+\vec{y})\cos\omega_L t \tag{16.90}$$

This represents a linearly polarized light having $\sqrt{2}E_0$ as the magnitude of the electric field and its electric field oriented at $\dfrac{\pi}{4}$ to the x-axis. This is a π-polarized light.

For $z = \dfrac{\lambda}{8}$, $kz = \dfrac{2\pi}{\lambda}\cdot\dfrac{\lambda}{8} = \dfrac{\pi}{4}$ and the equation (16.88) gives

$$\vec{E} = E_0\left[\vec{x}\,\cos\left(\omega_L t - \dfrac{\pi}{4}\right) + \vec{y}\cos\left(\omega_L t + \dfrac{\pi}{4}\right)\right] \tag{16.91}$$

In this case the electric vector of magnitude E_0 rotates about the z-direction in the positive sense and thus represents a circularly polarized light. It is called a left circular polarized light and is denoted by σ^+ polarization.

For $z = \dfrac{\lambda}{4}$, $kz = \dfrac{2\pi}{\lambda}\cdot\dfrac{\lambda}{4} = \dfrac{\pi}{2}$ and $\vec{E} = \sqrt{2}E_0(\vec{x}-\vec{y})\sin\omega_L t$ $\tag{16.92}$

In this case the magnitude of the electric vector is $\sqrt{2}E_0$ and is oriented $-\dfrac{\pi}{4}$ with respect to the X-axis. This is also called π-polarized light.

For $z = \dfrac{3\lambda}{8}$, $kz = \dfrac{2\pi}{\lambda}\cdot\dfrac{3\lambda}{8} = \dfrac{3\pi}{4}$ and the equation (16.88) becomes

$$\vec{E} = E_0 \left[\vec{x} \cos\left(\omega_L t - \frac{\pi}{4} \right) + \vec{y} \cos\left(\omega_L t - \frac{\pi}{4} \right) \right] \tag{16.93}$$

In this case the electric field of magnitude E_0 rotates in the negative sense. It is called a right circular polarized light and is denoted by σ^- polarization.

For $z = \dfrac{\lambda}{2}$, $kz = \dfrac{2\pi}{\lambda} \cdot \dfrac{\lambda}{2} = \pi$, the equation (16.88) becomes

$$\vec{E} = \sqrt{2} E_0 (\vec{x} + \vec{y}) \cos \omega_L t$$

the polarization in this case is the same as in the case of $z = 0$

Case 3   Counter propagating Light beams with circular polarization in opposite sense.

We shall now consider a left circular polarization (LCP) σ^+ beam propagating in the $+Z$ direction and right circular polarization beam σ^- propagating in the $-Z$ direction. The resultant electric field in this case is

$$\begin{aligned}
\vec{E} = \vec{E}_+ + \vec{E}_- &= E_0\, \vec{x} \cos(\omega_L t - kz) - E_0\, \vec{y} \sin\,(\omega_L t - kz) \\
&\quad + E_0\, \vec{x}\,[\cos(\omega_L t + kz) + E_0\, \vec{y} \sin(\omega_L t + kz)] \\
&= 2E_0\, \vec{x} \cos\, \omega_L t\,[\vec{x} \cos\,kz + \vec{y} \sin\,kz]
\end{aligned} \tag{16.94}$$

For $z = 0$, $kz = 0$ the equation represents a linearly polarized light and the magnitude of the electric field is $2E_0$ and the orientation is along the X-axis.

In the case where $z = kz = \dfrac{2\pi}{\lambda} \cdot \dfrac{\lambda}{8} = \dfrac{\pi}{4}$ and the equation (16.94) gives

$$\vec{E} = \sqrt{2} E_0 (\vec{x} + \vec{y}) \cos \omega_L t \tag{16.95}$$

Again it represents a linearly polarized light and the magnitude of the electric vector is $\sqrt{2} E_0$ at angle $\dfrac{\pi}{4}$ with the x-axis.

For $z = \dfrac{\lambda}{4}$, $kz = \dfrac{2\pi}{\lambda} \cdot \dfrac{\lambda}{4} = \dfrac{\pi}{2}$ we obtain a linearly polarized light and the magnitude of the electric vector is $\sqrt{2} E_0$ and is oriented along the Y-axis.

For $z = \dfrac{3\lambda}{8}$, $kz = \dfrac{2\pi}{\lambda} \cdot \dfrac{3\lambda}{8} = \dfrac{3\pi}{4}$ and the equation (16.94) becomes

$$\vec{E} = \sqrt{2} E_0 (\vec{y} - \vec{x}) \cos \omega_L t$$ and is polarized at an angle $-\dfrac{\pi}{4}$ with the electric vector of magnitude $\sqrt{2} E_0$.

For $z = \dfrac{\lambda}{2}$, $kz = \dfrac{2\pi}{\lambda} \cdot \dfrac{\lambda}{2} = \pi$, the electric vector is polarized along the X-axis as in the case of $z = 0$.

It has been found that the polarization change has a spatial periodicity of $\dfrac{\lambda}{2}$ along Z-axis and has a strong polarization gradient. This is shown in Figure 16.10.

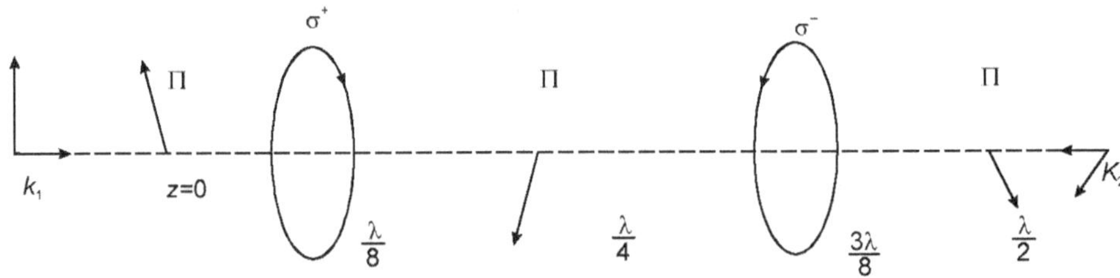

Figure 16.10   The polarized electric vector

16.9 OPTICAL PUMPING

We shall now consider two states of the atom having the total angular momentum including spin $F' = 3/2$ as excited state and $F'' = 1/2$ as the lower state. In this case $\Delta F = 1$ and hence the transition between these states is allowed by dipole selection rules. In a magnetic field the $F' = 3/2$ will split into $M_F' = 3/2,\ 1/2,\ -1/2,\ -3/2$ magnetic sublevels. Or we can say that the $F' = 3/2$ state is four-fold degenerate in field free case. Similarly the lower state with $F'' = 1/2$ is two-fold degenerate. In this case $M_F'' = 1/2$ and $-1/2$.

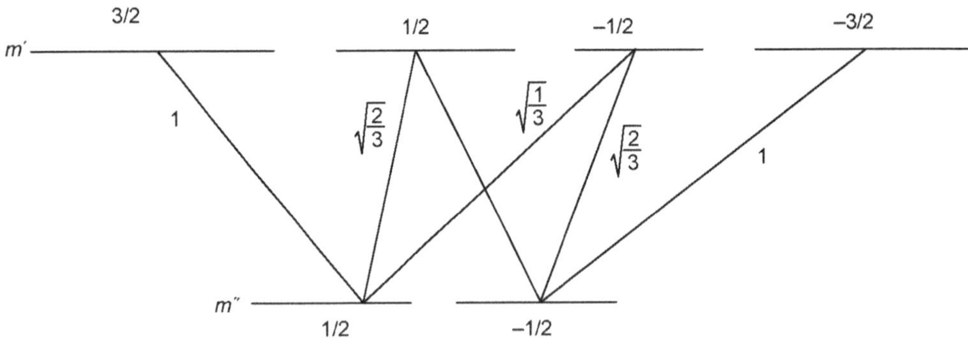

Figure 16.11   The transition between two states of the atom

The transitions between the magnetic levels depend on the nature of polarizibility of the incident light though they satisfy the Bohr's conditions.

For left circular polarization (LCP) light the selection rule is

$$m' - m'' = +1 \; (\sigma^+ \text{ transitions})$$

and for RCP $m' - m'' = +1 \; (\sigma^- \text{ transitions})$

for linearly polarized light $m' - m'' = 0, \pm 1 \; (\sigma^+ \text{ transitions})$

The spontaneous emission can take place according to the selection rule $m' - m'' = 0, \pm 1$.

The transition probabilities of transitions are generally calculated with the aid of Clebsch–Gordon coefficients. There are $3j$, $6j$ and $9j$ symbols associated with Clebsch–Gordon coefficients and they can be easily calculated by the aid of computers. These are indicated by the numbers on the inclined lines in Figure 16.11. It can be shown that the absorption due to a left circular polarized light (LCP) and the subsequent spontaneous emission from the excited magnetic sublevels can deplete the number of atoms in the $m'' = -1/2$ and transfer them into $m'' = +1/2$ by selective absorption-spontaneous emission processes. For example as per the selection rules mentioned above $m' = 1/2 \leftarrow m'' = -1/2$ and $m' = 3/2 \leftarrow m'' = 1/2$ are only possible with absorption due to an LCP light. Once they are in the excited state they return to the ground state magnetic sublevels by spontaneous emission with the selections rules $m' - m'' = 0, \pm 1$. Thus the atoms in the $m' = 3/2$ state can return to $m'' = 1/2$ by spontaneous emission, so also the atoms in the excited state $m' = 1/2$ state c to $m'' = -1/2$ and $m' = 1/2$ state c to $m'' = 1/2$ by spontaneous emission. Thus it is evident that the absorption by an LCP light transfer more atoms to the magnetic sublevel $m'' = \frac{1}{2}$ through $m' = 3/2$ and $m' = 1/2$ thereby reducing the number of atoms in the $m'' = -1/2$ magnetic sublevel. After several absorption-spontaneous emission processes almost all atoms in the $m'' = -1/2$ level can be transferred to $m'' = 1/2$. This is called optical pumping. This was first demonstrated by A.Kastler for which he received the Nobel Prize in 1966.

If we use an RCP light the optical pumping would deplete the $m'' = 1/2$ magnetic sublevel and populate $m'' = -1/2$ magnetic sublevel. If we use a linearly polarized light there will be no selective absorption and hence no population transfer to other levels.

16.10 THEORETICAL DESCRIPTION OF DOPPLER COOLING AND OPTICAL MOLASSES

Let us now have a quantitative description of two counter propagating identical laser beams along the Z-axis. The transition frequency is ω_0 and let the two beams have the same intensity and the detuning $\delta = \omega_l - \omega_0$. The detuning experienced by the atom moving in the $+Z$ direction is

$$\delta^+(v) = \delta - kv \tag{16.96}$$

and for the atom moving in the $-Z$ direction, the detuning experience is

$$\delta^-(v) = \delta + kv \tag{16.97}$$

During absorption–emission process the scattering rates for the two beams will be different. The scattering rate for the laser beam travelling in the +Z direction (*See* equation 16.82).

$$R_{sp}^+ = \frac{\frac{\Gamma}{2}\left(\frac{I}{I_S}\right)}{1+\left(\frac{I}{I_S}\right)+4\left[\frac{\delta^+(v)}{\Gamma}\right]^2} \tag{16.98}$$

and the scattering rate for the laser beam travelling in the –Z direction

$$R_{sp}^- = \frac{\frac{\Gamma}{2}\left(\frac{I}{I_S}\right)}{1+\left(\frac{I}{I_S}\right)+4\left[\frac{\delta^-(v)}{\Gamma}\right]^2} \tag{16.99}$$

For a red-detuned laser δ is negative and hence R_{sp}^+ is smaller than R_{sp}^-.

The net acceleration for the atom in the +Z direction in an absorption–spontaneous emission cycle

$$a = v_R(R_{sp}^+ - R_{sp}^-) \tag{16.100}$$

where v_R is the recoil velocity. Since R_{sp}^+ is smaller than R_{sp}^- the net acceleration 'a' is negative for an atom moving in the +Z direction and is decelerated.

In a similar way the net acceleration for the atom in the –Z direction in an absorption–spontaneous emission cycle

$$a = v_R(R_{sp}^- - R_{sp}^+) \tag{16.101}$$

and as $\delta^-(v)$ has reversed magnitude than the atom moving in the +Z direction and hence R_{sp}^+ is larger than R_{sp}^- and hence these atoms will also be decelerated.

From equations 16.96 and 16.101, R_{sp}^+ will be minimum and R_{sp}^- maximum for atoms moving in the +Z direction if the red-detuned laser beams satisfy the condition

$$|\delta| = kv \tag{16.102}$$

Under the above condition R_{sp}^+ is maximum and R_{sp}^- minimum for atoms moving in the –Z direction and thus the deceleration atoms moving in the Z-direction are maximum. It has been found experimentally that the maximum deceleration of atoms and hence the maximum cooling is achieved for atoms whose mean thermal velocities are in the range for few hundred to few thousand cm/s.

If $\dfrac{kv}{\delta} < 1$, we can neglect higher terms in the equations 16.98.

$$R^{+}_{sp} = \frac{\dfrac{\Gamma}{2}\left(\dfrac{I}{I_S}\right)}{1+\left(\dfrac{I}{I_S}\right)+\dfrac{4\delta^{2}}{\Gamma^{2}}\left(1-\dfrac{2kv}{\delta}\right)}$$

$$R^{-}_{sp} = \frac{\dfrac{\Gamma}{2}\left(\dfrac{I}{I_S}\right)}{1+\left(\dfrac{I}{I_S}\right)+\dfrac{4\delta^{2}}{\Gamma^{2}}\left(1+\dfrac{2kv}{\delta}\right)} \tag{16.103}$$

From 16.101 and 16.103 we get

$$a = v_R(R^{+}_{sp} - R^{-}_{sp}) = v_R\left[\frac{\dfrac{\Gamma}{2}\left(\dfrac{I}{I_S}\right)}{1+\left(\dfrac{I}{I_S}\right)+\dfrac{4\delta^{2}}{\Gamma^{2}}\left(1-\dfrac{2kv}{\delta}\right)} - \frac{\dfrac{\Gamma}{2}\left(\dfrac{I}{I_S}\right)}{1+\left(\dfrac{I}{I_S}\right)+\dfrac{4\delta^{2}}{\Gamma^{2}}\left(1+\dfrac{2kv}{\delta}\right)}\right]$$

$$a = \frac{8v_R\left(\dfrac{I}{I_s}\right)\dfrac{\delta}{\Gamma}kv}{\left(1+\dfrac{I}{I_s}+\dfrac{4\delta^{2}}{\Gamma^{2}}\right)^{2}} \tag{16.104}$$

$$v_R = \frac{hv}{Mc} = \frac{h}{2\pi}\cdot\frac{2\pi}{\lambda}\cdot\frac{1}{M} = \frac{\hbar k}{M}$$

and hence we get

$$a = \frac{8\hbar k^{2}}{M}\frac{\left(\dfrac{I}{I_s}\right)\dfrac{\delta}{\Gamma}v}{\left(1+\dfrac{I}{I_s}+\dfrac{4\delta^{2}}{\Gamma^{2}}\right)^{2}} = \frac{\alpha}{M}v \tag{16.105}$$

where,

$$\alpha = 8\hbar k^2 \frac{\left(\dfrac{I}{I_s}\right)}{\left(1 + \dfrac{I}{I_s} + \dfrac{4\delta^2}{\Gamma^2}\right)^2} \cdot \frac{\delta}{\Gamma}$$

δ is negative for red-detuned laser beams and hence we can write the above expression as

$$a = -\frac{\alpha}{M} \mathbf{v} \tag{16.106}$$

where,

$$\alpha = 8\hbar k^2 \frac{\left(\dfrac{I}{I_s}\right)}{\left(1 + \dfrac{I}{I_s} + \dfrac{4\delta^2}{\Gamma^2}\right)^2} \cdot \frac{|\delta|}{\Gamma}$$

Thus the acceleration acting on the atoms are in opposite to their direction of propogation. This is valid for atoms moving in $+Z$ and $-Z$ directions.

The optical force $$F = Ma = M - \frac{\alpha}{M} \mathbf{v} = -\alpha \mathbf{v} \tag{16.107}$$

Thus two oppositely propagating identical laser beams can slow down the thermal motion of atoms and the damping force is proportional to the velocity of the atoms. It is as if two counter propagating laser beams form a medium for high viscosity where the thermal motion of the atoms is considerably damped. This set-up is therefore called optical molasses.

Atomic clouds subjected to pair of identical laser beams in three orthogonal directions will be slowed down in all directions. In such an arrangement atoms are cooled and confined in a small volume determined by the spatial overlap of the laser beams.

16.11 STOPPING OF ATOMS IN AN ATOMIC BEAM

We have already discussed in the section that the atoms can be stopped in a laser beam by means of three pairs of identical laser beams in three orthogonal directions. Generally the velocities of atoms in an atom beam of the order of 10^5 cm/sec and hence they are to be first reduced before subjecting the atoms to interactions of optical molasses for further cooling.

We have already seen that an atom moving in $+Z$ direction will experience a detuning for the two beams by

$$\delta^+(v) = \delta - kv$$

and for atoms in the $-Z$ direction.

$$\delta^-(v) = \delta + kv$$

We assume that the atomic beam is propagating in the +Z direction and if $\delta^-(v) = 0$, it is sufficient to slow down the beams in the –Z direction effectively. There are three different methods to make $\delta(v) = 0$ as the velocity of the atom is decreased.

1. The laser frequency ω_L of the red-tuned laser beam should be increased as the velocity of the atoms decreases.

2. The laser frequency ω_L can be fixed and the atomic transition ω_0 can be decreased by the application of an external magnetic field as the atomic velocity decreases. The transitions are then between the Zeeman levels of the atom.

3. Yet another method is to make the light beam slightly divergent so that the changing Doppler shift due to the decrease in velocity of the atomic beams is compensated by light propagating at increasing angle with respect to the atomic beam.

The second method of Zeeman trapping (magnetic trapping) of atoms is easier to adopt. Let us take the example of cooling of sodium atoms by considering the Sodium D-lines. The energy level diagram is shown in Figure 16.12. Sodium atom ^{23}Na has a nuclear $\dfrac{3}{2}$ spin.

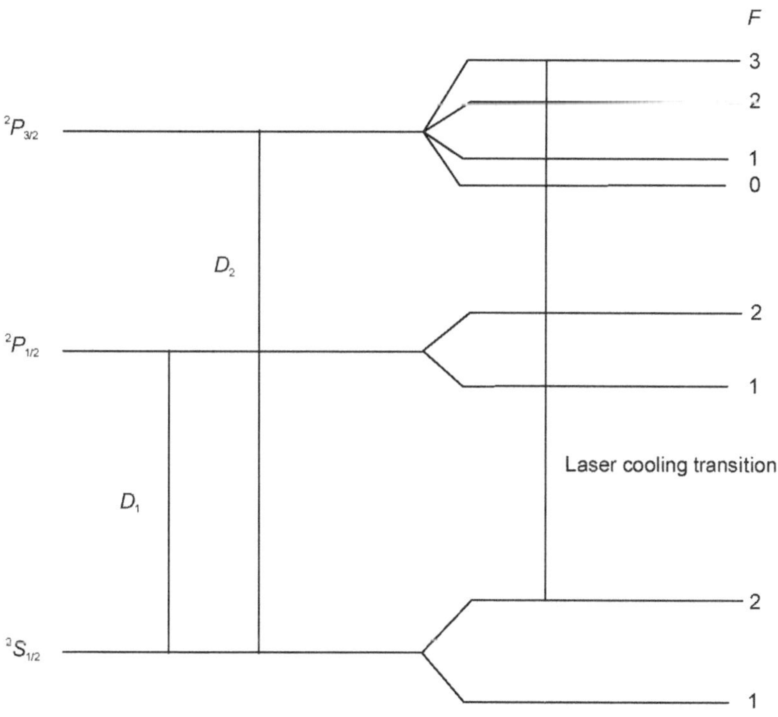

Figure 16.12 The energy level diagram of ^{23}Na

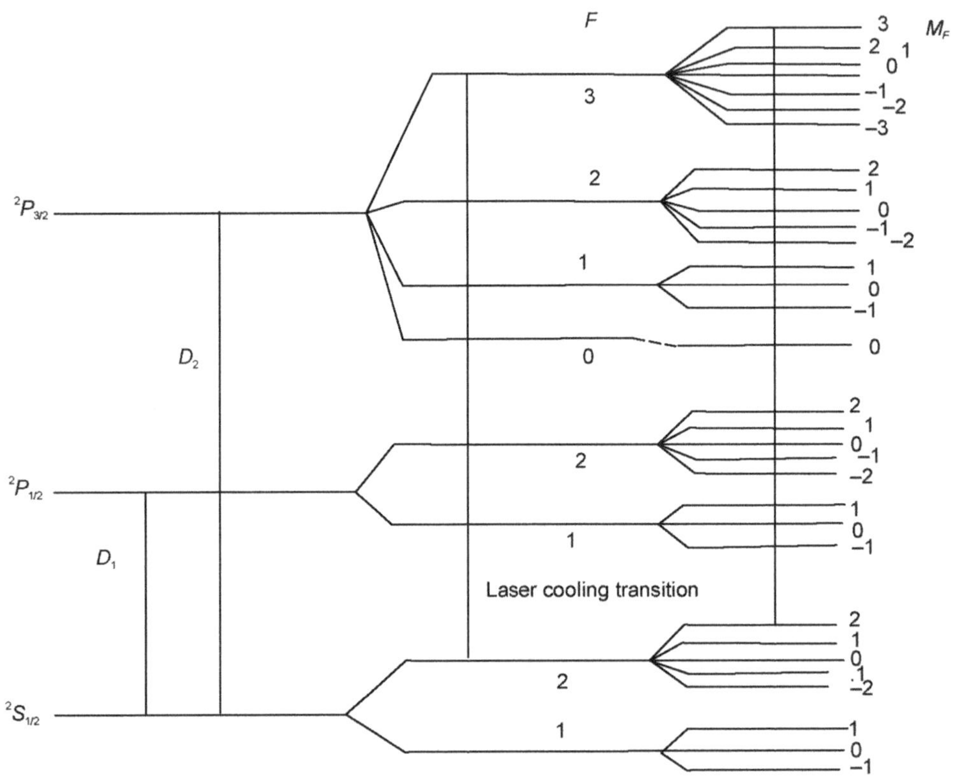

Figure 16.13   Laser cooling transition of ^{23}Na

We have already seen in Chapter 6 that the famous Sodium line corresponds to the transitions $^2P_{\frac{1}{2}} - {}^2S_{\frac{1}{2}}$ and $^2P_{\frac{3}{2}} - {}^2S_{\frac{1}{2}}$. In Figure 16.13 it is shown that each spin components split into hyperfine components due to the interaction of the nuclear spin and the resultant levels are described by F quantum numbers. And in a magnetic field each of the hyperfine components exhibits Zeeman patterns and each level splits into different M_F components $M_F = F,\ F{-}1,\ \ldots\ {-}F$. The allowed transitions are those with $\Delta M_F = 0,\ \pm 1$. In the laser cooling experiment the Zeeman component of the D_2 hyperfine structure corresponding to the transition $F' = 3,\ M_F' = 3$ to $F'' = 2,\ M_F'' = 2$ were selected. The incident laser light is left circular polarized so that it induces the transition $M_F' = 3$ to $M_F'' = 2$ The frequency of $F' = 3,\ M_F' = 3$ to $F'' = 2,\ M_F'' = 2$ transition, say ω_0, is increased with the increasing magnetic field. If we keep the detuning parameter as zero, ω_0 must decrease as the velocity of the atoms decrease by laser cooling. Still increasing laser cooling means decrease in velocity and the absorption will take place at a lower frequency ω_0. The lowering of frequency is again achieved by lowering the magnetic field. This can be easily achieved by having a higher magnetic field at the entrance of the atomic beam and slowly decreasing the magnetic field towards the exit end (*See* Figure 16.14).

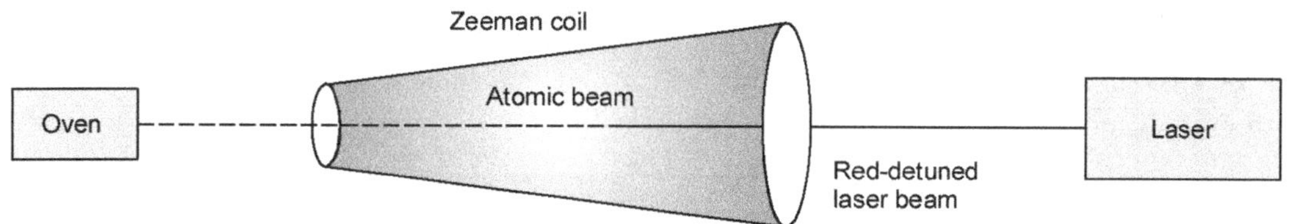

Figure 16.14   Stopping of atoms in an atomic beam

16.12 TRAPPING THE ATOMS

In the preceeding sections we have seen two distinct methods of laser cooling. The Doppler cooling is obtained by absorption–spontaneous emission cycles. The other process is the sub-Doppler cooling. The atoms slowed by sub-Doppler cooling method have very low thermal velocities, but they continue to move as they are not confined in a limited region. A magnetic trap uses a magnetic gradient in order to trap neutral particles with a magnetic moment. The magnetic trap as a way of trapping very cold atoms was first proposed by David Pritchard.

Neutral atoms in their ground state do not have electric dipole moment because of their spherical symmetry but they possess magnetic dipole moment.

Thus many atoms have a magnetic moment, and their energy shifts in a magnetic field. We have already seen in Chapter 9 that if an atom is placed in a strong magnetic field, its magnetic moment will be aligned along the magnetic field. If a number of atoms are placed in the same field, they will be distributed over the various allowed values of magnetic quantum numbers for that atom.

If we now superimpose a magnetic field gradient on the uniform magnetic field already applied, those atoms whose magnetic moments are aligned with the field will have lower energies in a higher field. Like a ball rolling down a hill, these atoms will tend to occupy locations with higher field (i.e., lower energy). They are called "high-field-seeking" atoms.

On the other hand, those atoms with magnetic moments aligned opposite to the field will have higher energies in a higher field, and will tend to occupy locations with lower field, and so are called "low-field-seeking" atoms.

It is generally not possible to produce a local maximum of the magnetic field magnitude in free space. However, it is possible to produce a local minimum. This minimum can trap atoms which are low-field seeking if they do not have enough kinetic energy to escape the minimum. Typically, magnetic traps have relatively shallow field minima. They can trap atoms whose kinetic energies correspond to temperatures of a fraction of a Kelvin.

If an atom is placed in an inhomogeneous magnetic field it is subjected to a force given by

$$F = -\text{grad}\,(\mu.B) \tag{16.108}$$

where μ is the magnetic moment of the atom and B is the spatially variable magnetic field.

A quadrupole magnetic trap is one of the simplest systems of trapping the atoms.

The most simple magnetic trap is formed by the magnetic quadrupole field that is created in the centre of two coils which carry currents in opposite directions (*See* Figure 16.15).

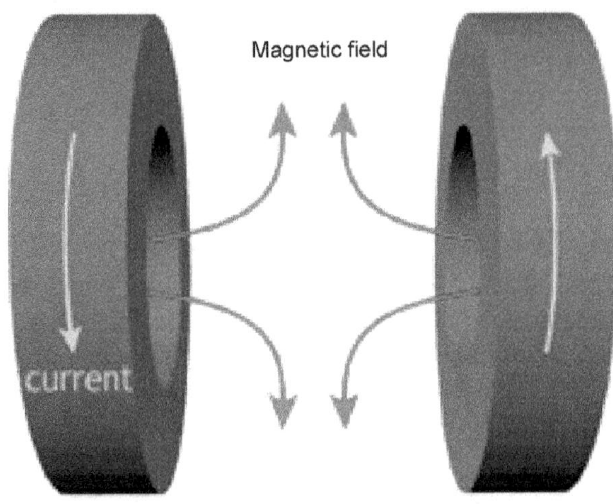

Figure 16.15　A quadrupole magnetic trap

Such a system was first conceived and constructed by Wolfgang Paul in 1950's. Wolfgang Paul shared Nobel Prize in 1989 with H.G. Dehmelt and N.F. Ramsey for this work. If we take the Z-axis along the centre of the coils and the XY plane parallel to the planes of the coils, then the inhomogeneous magnetic field is given by

$$B(x, y, z) = B_0 \ (\rho^2 + 4 \ z^2)^{1/2} \tag{16.109}$$

where $\rho^2 = x^2 + y^2$

It can be seen from the above expression that the magnetic field is zero at the origin and the field gradient is constant along any line passing through the origin but with different values along different directions. For the trapping of the atoms it is important that their magnetic moments are so oriented that they are repelled from regions of large magnetic field towards the centre of the trap. In order to achieve this the atoms are transferred to suitable magnetic levels by optical pumping. The potential experienced by the atoms near the centre of the magnetic trap in a particular magnetic sublevel is

$$U = \mu \ B_0 (\rho^2 + 4 \ z^2)^{1/2} \tag{16.110}$$

Circular orbits of radius ρ in the $z = 0$ plane is found when the magnetic force is equal to the centrifugal force $\dfrac{MV^2}{\rho}$.

After the atoms have been laser cooled in the magneto-optical trap, they are transferred into a magnetic trap. Here they are levitated in a magnetic field without any contact to the surrounding vacuum chamber, just as a spinning magnetic top toy.

The magnetic field grows linearly with the distance from the centre, where the magnetic field is zero. The magnetic moment of a trapped atom is oriented such that it experiences a force towards the minimum of the magnetic field. Due to Lamor precession, the magnetic moment of the atom maintains its orientation with respect to the local field. The major shortcoming of the quadrupole trap is that cold atoms which are close to the trap centre are removed from the trap due to non-adiabatic spin flips. This loss process has to be circumvented in order to achieve Bose–Einstein condensation. In magnetic traps of the Ioffe-type geometry the minimum of the magnetic field is non-zero.

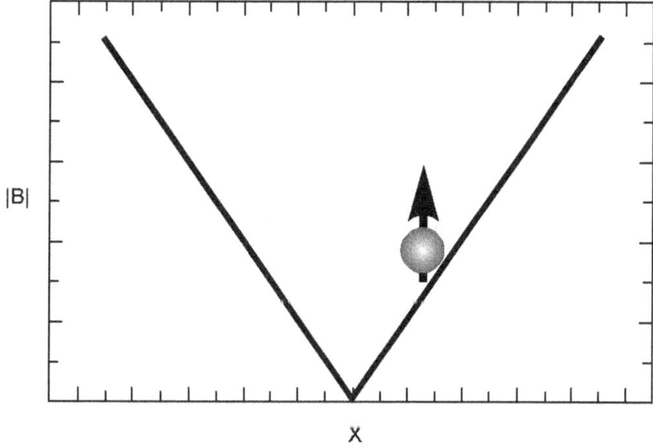

Figure 16.16   A conical potential magneto-optical trap

Magneto-optical trap   A magneto-optical trap (abbreviated MOT) is a device that cools down non-charged atoms to temperatures near absolute zero and traps them at a certain place using magnetic fields and circularly polarized laser light. Charged particles can be trapped in a Penning trap or a Paul trap using a combination of electric and magnetic fields, but these traps do not work for neutral atoms. In a magneto-optical trap, neutral atoms (non-charged) can be cooled down and stored by using the optical force of laser light. A MOT requires the atom to have a laser cooling transition in order to work. A typical MOT for sodium atoms can trap and cool the atoms down to 300 μK or 0.0003 degrees above absolute zero. Trapping of atoms in the conical potential formed by magnetic field in a MOT is shown in Figure 16.16.

16.13 EVAPORATIVE COOLING

The final stage of the cooling is obtained by evaporative cooling.

Evaporative cooling is a well known technique we come across every day, for example in cooling a cup of hot coffee. When you try to cool a cup of coffee you blow away the steam, thereby removing the most energetic particles from your coffee. The coffee then comes back to thermal equilibrium at a lower temperature. By repeating the process again and again we will be able to cool our coffee to room temperature. Now, could we cool our coffee any further ? Yes, it is possible provided our coffee wouldn't be in thermal contact with the outside world, just as the atoms in our vacuum chamber. In that case we always have the possibility of removing the hottest atoms, thereby cooling the coffee to very low temperatures.

This is exactly what we are doing our atoms in the magnetic trap. In order to remove atoms, we drive radio-frequency transitions between magnetically trapped and untrapped states of the atoms. By setting the radio-frequency to the right value we are able to limit the depth of our magnetic trapping potential very precisely, so that only the hottest atoms are allowed to escape the trap.

A pictorial representation is shown in Figure 16.17.

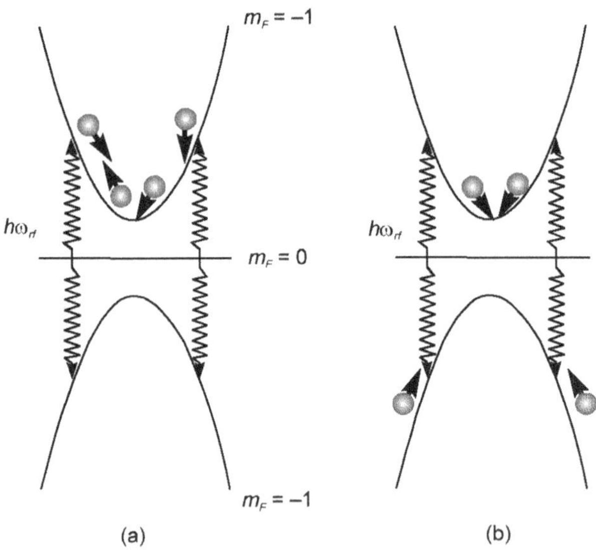

Figure 16.17   Thermalization through elastic collisions

With evaporative cooling we try to remove the hottest atoms from a velocity distribution at a temperature T_1. After thermalization through elastic collisions, thermal equilibrium is restored with a temperature $T_2 < T_1$ (*See* Figure 16.18). If we repeat this process over and over again we are able to cool the atoms and also build up phase-space density. This technique is so successful that we can reach Bose–Einstein condensation at temperatures around 1 µK with it !

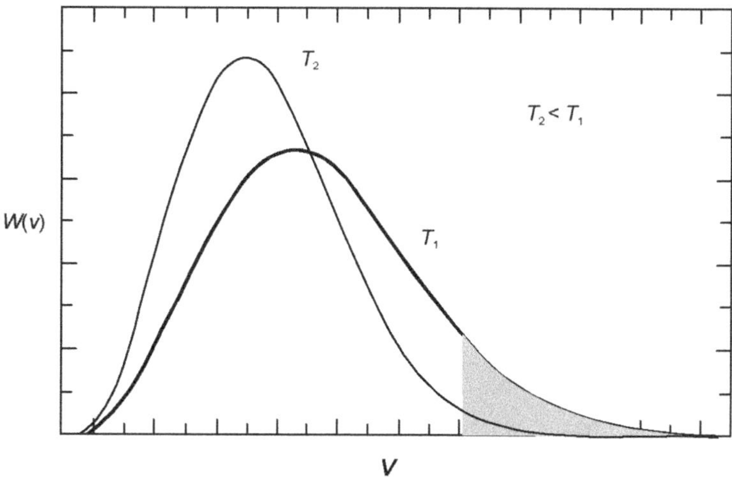

Figure 16.18   The graph representing restoration of thermal equlibrium

Thus in order to achieve BE condensate in a low density atomic vapour we have to follow systematic and sensitive experimental stages.

1. The atoms are cooled first by Doppler cooling.

2. Atoms from the velocity tail of the Maxwellian velocity distribution are captured by MOT thereby the atoms are cooled down to a temperature of 10^{-3}K.

3. The optical molasses are then cooled to 10^{-6} K by polarization gradient cooling.

4. The final stage is evaporative cooling. When the magnetic field in MOT is turned on the cold atoms are confined purely to magnetic trap. The magnetic field creates a conical potential. The hotter atoms are on the top of the cone and the colder atoms are at the bottom. The height of the cone can be suddenly changed by changing the strength of the magnetic field. The hotter atoms escape by evaporation in the case of a coffee cup.

16.14 OPTICAL TWEEZERS

One of the most important applications of the laser manipulation of atoms is in the development of a technique called "Optical tweezers". They are being extensively used for many novel experiments in biological sciences. They are used for the micro manipulation at cellular and sub-cellular levels of research.

Optical tweezers is a technique by which we use light to manipulate microscopic objects as small as a single atom. The radiation pressure gained for a laser beam is able to trap small particles.

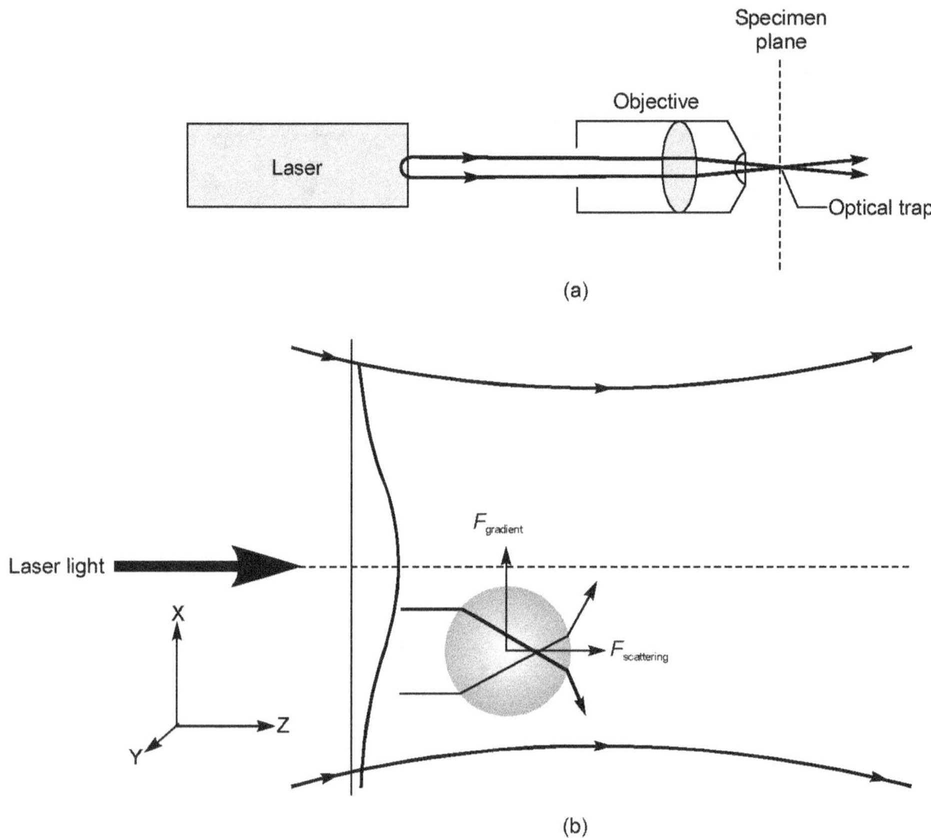

Figure 16.19 Optical tweezers

The basic principle of optical tweezers is given in Figure 16.19 a and b. A laser beam is focused through the objective of a microscope to a spot in the specimen plane. This spot creates an "optical trap" which is able to hold a small particle at its centre. The forces felt by this particle consist of the light scattering and gradient forces due to the interaction of the particle with the light.

The basic principle behind optical tweezers is the momentum transfer associated with bending light. We already know that light always has a momentum that is proportional to its energy and is in the direction of its propagation. Any change in the direction of light, by reflection or refraction, will result in a change of the momentum of the light. If an object bends the light, changing its momentum, conservation of momentum requires that the object must undergo an equal and opposite momentum change. This gives rise to a force acting on the object.

Usually the incoming light from a laser beam can be considered to have a "Gaussian intensity profile". It means that the light at the centre of the beam is brighter than the light at the edges. When the light falls on a bead it interacts with it, and the light rays are bent according the laws of reflection and refraction. Examples of two rays are shown in Figure 16.18b. The sum of the forces from all such rays can be split into two components: One is the scattering force, $F_{scattering}$ which points in the direction of the incident light (z-axis in Figure 16.18b), and the other is the gradient force, $F_{gradient}$, which arises from the gradient of the Gaussian intensity profile and directs in x–y plane pointing towards the centre of the beam (dotted line). The gradient force is a restoring force that pulls the bead into the centre. If the contribution to $F_{scattering}$ of the refracted rays is larger than that of the reflected rays then a restoring force is also created along the z-axis, and a stable trap will exist. The image of the bead can be projected onto a quadrant photodiode to measure nm-scale displacements.

Optical tweezers have been used to trap dielectric spheres, viruses, bacteria, living cells, organelles, small metal particles, and even strands of DNA.

ENDNOTES

1. Ashkin, A., Dziedzic, J.M., Bjorkholm, J.E. and Chu, S. (1986). "Observation of a single-beam gradient force optical trap for dielectric particles."

2. Einstein received Nobel Prize in Physics in 1921 " for his services to theoretical physics, especially for his discovery of the law of the photoelectric effect.

3. Eric, A. Cornell and Carl, E. Wieman shared Nobel Prize in Physics in 2001 along with Wolfgang Ketterle.

4. Anderson, M.H., Ensher, J.R., Matthews, M.R., Wieman, C.E. and Cornell, E.A. (1995). "Observation of Bose-Einstein Condensation in a Dilute Atomic Vapour." *Science* 269, 198.

REFERENCES

Anderson, M.H., Ensher, J.R., Mathews, M.R., Wieman, C.E. and Cornel, E.A. (1995). "Observation of Bose-Einstein condensation in a dilute atomic Vapour." *Science*. 269, 198–201.

Ashkin, A. (1978). "Trapping of atoms by resonance radiation pressure." *Phys. Rev. Lett.* 40, 729.

Ashkin, A. (2000). "History of optical trapping and manipulation of small-neutral particle, atoms, and molecules." IEEE Journal of Selected Topics in Quantum Electronics 6(6):841–856.

Ashkin, A., Dziedzic, J.M., Bjorkholm, J.E. and Chu, S. (1986). "Observation of a single-beam gradient force optical trap for dielectric particles." *Opt. Lett.* 11 (5): 288–290.

Block, S.M. (1992). "Making light work with optical tweezers." *Nature*. 360(6403):493–5.

Chu, S. *et al.* (1985). "Three-dimensional viscous confinement and cooling of atoms by resonance radiation pressure." *Phys. Rev. Lett.* 55, 48.

Chu, Steven. (1991). "Laser manipulation of atoms and particles." *Science.* 253, 861–866.

Cohen-Tannoudji, C.N. and Phillips, W.D. (1990). "New Mechanisms for Laser Cooling." *Physics Today.*

David, E. Pritchard. (1983). "Cooling neutral atoms in a magnetic trap for precision spectroscopy. " *Phys. Rev. Lett.* 51, 1336.

Foot, C.J. (2004). *Atomic Physics.* Oxford University Press, Oxford, U.K.

Gittes, F. and Schmidt, C.F. (1998). "Interference model for back-focal-plane displacement detection in optical tweezers." *Optics Letters.* 23(1):7–9.

Hänsch, T.W. and Schawlow, A.L. (1975). "Cooling of gases by laser radiation." *Opt. Commun.* 13, 68 .

Horikoshi, M. and Nakagawa, K. (2006). *"Atom chip based fast production of Bose–Einstein condensate".* *Applied Physics.* B 82 (3): 363–366.

Ishii, Y., Ishijima, A. and Yanagida, T. (2001)." Single molecule nanomanipulation of biomolecules." *Trends Biotechnol.* 19(6):211–6.

Itano, Wayne M., Bergquist, J.C. and Wineland, D.J. (1987). "Laser spectroscopy of trapped atomic ions." *Science.* 7. 237: 612–617.

Kuo, S.C. (2001). "Using optics to measure biological forces and mechanics." *Traffic.* 2(11): 757–63.

Lang, M.J., Asbury, C.L., Shaevitz, J.W. and Block, S.M. (2002). "An automated two-dimensional optical force clamp for single molecule studies." *Biophys. J.* 83(1):491–501.

Mark Wilson, "Experiment reveal a Bose-Einstein condensate of Photons." *Physics Today,* 64, Feb.2011, 10–11

Metcalf, Harold, J. and Straten, Peter van der. (1999). *Laser Cooling and Trapping.* Springer-Verlag, New York, Inc.

Monroe, C., Swann, W., Robinson, H. and Wieman, C. (1990). "Very cold trapped atoms in a vapor cell." *Physical Review Letters.* 65 (13): 1571.

Neuman, K.C., Chadd, E.H., Liou, G.F., Bergman, K. and Block, S.M. (1999). "Characterization of photodamage to Escherichia coli in optical traps." *Biophys. J.* 77(5):2856–63.

Paul, D. Lett, Richard, N. Watts, Christoph, I. Westbrook, William, D. Phillips, Phillip, L. Gould, and Harold, J. Metcalf. "Observation of atoms laser cooled below the doppler limit." *Phys. Rev. Lett.* 61, 169.

Peters, I.M., de Grooth, B.G., Schins J.M., Figdor, C.G. and Greve, J. (1998). "Three dimensional single-particle tracking with nanometer resolution." *Review of Scientific Instruments.* 69(7): 2762–6.

Phillips, D. William, Gould, L. Phillip. and Lett, D. Paul. (1988). "Cooling, stopping, and trapping Atoms." *In Science*.

Pralle, A., Prummer, M., Florin, E.L., Stelzer, E.H. and Horber, J.K. (1999). "Three-dimensional high-resolution particle tracking for optical tweezers by forward scattered light." *Microsc. Res. Tech.* 44(5):378–86.

Simmons, R.M., Finer, J.T., Chu, S. and Spudich, J.A. (1996). "Quantitative measurements of force and displacement using an optical trap." *Biophys. J.* 70(4):1813–22.

Smith, S.P., Bhalotra, S.R., Brody, A.L., Brown, B.L., Boyda, E.K. and Prentiss, M. (1998). "Inexpensive optical tweezers for undergraduate laboratories." *Am. J. Phys.* 67(1):26–35.

Visscher, K., Gross, S.P. and Block, S.M. (1996). "Construction of multiple-beam optical traps with nanometer-resolution position sensing." IEEE Journal of Selected Topics in Quantum Electronics 2(4):1066–1076.

Visscher, K. and Block, S.M. (1998). "Versatile optical traps with feedback control." *Methods Enzymol.* 298:460–89.

William, D. Phillips and Harold Metcalf. "Laser deceleration of an atomic beam." *Phys. Rev. Lett.* 48, 596.

Wineland, D.J. and Dehmelt, H. (1975). "Laser cooling and double resonance spectroscopy of stored ions." *Bull. Am. Phys. Soc.* 20, 637.

Wineland, D.J., Drullinger, R.E. and Walls, F.L. "Radiation-pressure cooling of bound resonant absorbers." *Phys. Rev. Lett.* 40, 1639.

17

THE ATOM LASER

17.1 DISCOVERY OF AN ATOM LASER

No one could have foreseen the potential when a small experiment was performed way back in 1934 on the inversion doublets of ammonia which developed after 20 years into the most outstanding experiment of the 20th century, viz., the **maser** (microwave amplification by stimulated emission of radiation). Maser is the forerunner of **laser** (light amplification by stimulated emission of radiation).

Similarly, no one could have realized all of the laser's applications when the first working model was built in 1960, but it went on to revolutionize telecommunications, medicine, information storage, and many other areas of science and technology. Now the scientists at MIT have created a rudimentary version of an "atom laser," based on matter waves, a device that does for matter what ordinary lasers do for light. The MIT researchers have verified the key property of a laser with the atom-laser beam, viz., coherence. Over and above improving measurements with atoms, the atom laser may lead to major innovations in nanotechnology, the manipulation of matter at the atomic level.

We have already seen how the researchers used laser cooling to produce Bose–Einstein condensates. As the name suggests laser cooling is a technique that uses light to cool atoms to very low temperature. The invention of laser and developments in high vacuum technology have made possible to cool atoms to very low temperatures. Interaction between a laser beam of appropriate wavelength (frequency) and atomic vapour has made possible to obtain cooling of atoms. Albert Einstein (14 March1879 –18 April, 1955)[1] and Indian Physicist Satyendra Nath Bose (1 January 1894 – 4 February 1974) had predicted such a low temperature condition already in 1920s, in which the atoms fall into the same quantum state and essentially behave as a single atom. The use of lasers to achieve extremely low temperatures has already started in about 1985 with the work of Steven Chu and others[2], and scientists were able to achieve extremely low temperatures of the order of 10^{-9} K. It refers to the transitions between quantum states of an atom in the presence of laser beams.

Thus after cooling atoms to a temperature of a few hundreds of nano-Kelvin and trapping them for several seconds, an exotic state of matter known as a "Bose–Einstein Condensate" was created in 1995 by Cornell (b. 1961), Carl Wieman (b. 1951)[3,4] and their colleagues at the University of Colorado.

We have already seen in the previous chapter that many advancements have been made like magnetic trap, magneto-optical trap so on to obtain Bose–Einstein condensate in a dense cloud of atoms to a temperature of nearly absolute zero, where all the atoms are collected in the quantum mechanical ground state and described by a single macroscopic wavefunction. Scientists led by Wolfgang Ketterle (b. 1957)[5] at the Massachusetts Institute of Technology (MIT) have shaped these novel materials into pulses of atoms that have the hall mark of laser beam.

A Bose–Einstein condensate is a group of a few million atoms that merge to make a single matter-wave about a millimetre or so across. In 1995, Ketterle created BECs in his lab by cooling a gas made of sodium atoms to a few hundred billionths of a degree above absolute zero—more than a million times cooler than interstellar space! At such low temperatures the atoms became more like waves than particles. Held together by laser beams and magnetic traps, the atoms overlapped and formed a single "giant" (by atomic standards) matter wave.

Comparing the different parts of the conventional laser, the magnetic trap acts as a resonator in atomic lasers and the condensate of atoms confined to it, is used as the gain medium. The output of "atom laser beam" is obtained by creating a leak in the trap by an external magnetic field to take a fraction of the condensate to come out. We already know that the essential characteristic of a laser, whether it is conventional or atomic, is its coherence. Using the sodium atom condensate, the researchers at MIT have observed the sharp interference fringes between the two Bose–Einstein condensates taken from the same trap on overlapping and hence showed that a Bose–Einstein condensate is coherent.[6]

Working independently in 1995, Eric Cornell (National Institute of Standards & Technology) and Carl Wieman (University of Colorado) also created BECs; theirs were made of super-cold rubidium atoms. Cornell and Wieman shared the 2001 Nobel Prize with Ketterle "for the achievement of Bose–Einstein condensation in dilute gases of alkali atoms, and for early fundamental studies of the properties of the condensates."

17.2   ORDINARY LASERS

An ordinary laser produces an extremely special form of light vastly different from that which emerges from a light bulb. Unlike all other light sources, the laser creates photons (particles of light) that are in exactly the same quantum state. Laser light is coherent, meaning that its wavefront varies predictably in time and space. Light from a light bulb, in contrast, is incoherent: later wavefronts have no predictable relationship to earlier ones. In most conventional laser designs, light builds up inside the laser by reflecting many times from mirrors at either end of the laser tube. Extracting the light is possible because one of the mirrors is only partially silvered and lets some of the light escape to produce an output beam. The beam travels in a single, well-defined direction, unlike sunlight or lamplight, which shines in all directions. This combination of unique properties makes a laser beam more intense than an equivalent stream of light emanating for example from the sun.

Coherence can be understood by comparing the uniform rhythmic march of a military troop or the uniform movements of the hands in canoeing or in a boat race.

17.2.1   What is an Atom Laser?

An atom laser is analogous to an optical laser, but it emits matter waves instead of electromagnetic waves. The output of an atomic laser is a coherent matter wave, a beam of atoms which can be focused to a pinpoint or can be collimated to travel large distances without spreading. Because of its coherency, the atom laser beams can interfere with each other. Compared to an ordinary beam of atoms, the beam of an atom laser is extremely bright. One can describe laser-like atoms as atoms "marching in lockstep". Although there is no rigorous definition for a laser, all lasers whether it is optical or atomic laser, they are characterized by its brightness and coherence as the essential features.

17.2.2   Historical Roots

"To see something which nobody else has seen before is thrilling and deeply satisfying. Those are the moments when you want to be a scientist," were the words of Wolfgang Ketterle, one of the first scientists who created the new kind of matter called Bose–Einstein condensates.

17.3   EVENTS LEADING UP TO THE ATOM LASER

The atom laser is based on the quantum-mechanical wave nature of particles. Louis Victor de Broglie (1892–1987), during his Ph.D. thesis in 1923, predicted that atoms—and all matter in general—can also act as waves that spread out in space and combine with other waves to produce interference patterns and exhibit other wavelike phenomena. He introduced the famous formula which states that the wavelength of a particle is inversely proportional to the product of its mass and its speed. (The wavelength equals Planck's constant divided by the mass and the speed of the particle.)

The de Broglie's equation and the wavelike nature of matter have been confirmed in countless physics experiments that have followed. In 1917, Albert Einstein discovered theoretically the stimulated emission of light which is the basic mechanism generating laser light. In what was then unrelated work, in 1924, Einstein and Satyendra Nath Bose predicted a novel form of matter that forms at very low temperatures and is now known as Bose–Einstein condensate.

Charles Townes of Columbia University and Arthur Schawlow, then at Bell Laboratories, build the first maser, a laser for microwave light way back in 1950's. The maser was the precursor to the optical laser. Theodore Maiman of Hughes Aircraft Corporation created the first working optical laser in 1960. At the time of its invention, most physicists could not foresee any practical applications for the laser. Yet, optical lasers have now profoundly changed telecommunications, data storage, and numerous areas of medicine and surgery.

Eric Cornell, Carl Wieman, and their co-workers at the National Institute of Standards and Technology and the University of Colorado [4,7] created the first Bose–Einstein condensate in a dilute gas of atoms in July, 1995. Thereafter, Randall Hulet and his co-workers at Rice University, and Wolfgang Ketterle and his colleagues at MIT produced Bose–Einstein condensates of their own. Ketterle and his colleagues created a technique for extracting a beam of atoms from the Bose–Einstein condensate verifying that their beam has properties directly analogous to an optical laser beam.

17.4 THE MAIN INGREDIENT FOR AN ATOM LASER

Since 1995, scientists have been making the main ingredient for an atom laser: the most important is a Bose–Einstein condensate, a collection of gas atoms with temperatures just billionths of a degree above absolute zero, colder than anything observed before. The basic result of Bose–Einstein theory was that when a sufficiently densely packed group of particles becomes very cold they would, under certain conditions, collectively enter a single quantum state and act as a single, coherent wave, central requirements for a laserlike beam of atoms. Although previous experiments provided clear evidence for the formation of a Bose–Einstein condensate, they could not directly verify that the condensate formed a single coherent wave. In addition, they were unable to extract a beam of atoms from the Bose–Einstein condensate, which exists as a fragile clump of atoms trapped by magnetic fields. In the 27 January 1997 issue of Physical Review Letters, Wolfgang Ketterle and his colleagues at MIT announced that they have created an "output coupler" which allows them to pluck a controlled fraction of atoms from a BEC of sodium atoms to produce a beam that falls in the direction of gravity. Meanwhile, in the 31 January Issue of Science, the same group showed that the beam has coherence properties analogous to those of a laser light beam. Demonstrating coherence was "the most important step," says Ketterle, "to show that a Bose condensate with an output coupler acts as an atom laser."

17.4.1 The Parts of an Atom Laser

A laser requires a cavity (resonator), an active medium, and an output coupler. In the MIT atom laser, the "resonator" is a magnetic trap in which the atoms are confined by "magnetic mirrors". The active medium is a thermal cloud of ultracold atoms, and the output coupler is an rf pulse which controls the "reflectivity" of the magnetic mirrors.

17.4.2 The Gain Process in an Atom Laser

The analogy to spontaneous emission in the optical laser is elastic scattering of atoms (collisions similar to those between billiard balls). In a laser, stimulated emission of photons causes the radiation field to build up in a single mode. In an atom laser, the presence of a Bose–Einstein condensate (atoms that occupy a "single mode" of the system, the lowest energy state) causes stimulated scattering by atoms into that mode. More precisely, the presence of a condensate with N atoms enhances the probability that an atom will be scattered into the condensate by N + 1.

In a normal gas, atoms scatter among the many modes of the system. But when the critical temperature for Bose–Einstein condensation is reached, they scatter predominantly into the lowest energy state of the system, a single one of the myriad of possible quantum states. This abrupt process is closely analogous to the threshold for operating a laser, when the laser suddenly switches on as the supply of radiating atoms is increased.

In an atom laser, the "excitation" of the "active medium" is done by evaporative cooling—the evaporation process creates a cloud which is not in thermal equilibrium and relaxes towards colder temperatures. This results in growth of the condensate. After equilibration, the net "gain" of the atom laser is zero, i.e., the condensate fraction remains constant until further cooling is applied.

Unlike optical lasers, which sometimes radiate in several modes (i.e., at several nearby frequencies) the matter wave laser always operates in a single mode.

17.4.3 The Output of an Atom Laser

The output of an optical laser is a collimated beam of light. For an atom laser, it is a beam of atoms. Either laser can be continuous or pulsed but so far, the atom laser has only been realised in the pulsed mode. Both light and atoms propagate according to a wave equation. Light is governed by Maxwell's equations, and matter is described by the Schroedinger equation. The diffraction limit in optics corresponds to the Heisenberg uncertainty limit for atoms. In an ideal case, the atom laser emits a Heisenberg uncertainty limited beam.

17.5 DIFFERENCES BETWEEN ATOM AND OPTICAL LASERS

Wolfgang Ketterle has explained some of the differences between a laser producing beams of atoms and a laser producing beams of photons. He says that "Photons can be created, but not atoms,". "The number of atoms in an atom laser is not amplified. What is amplified is the number of atoms in the lowest-energy quantum state, while the number of atoms in other states decreases." In addition, he says, "Atoms interact with each other—this creates additional spreading of the atom-laser beam. Unlike light, an atom-laser beam cannot travel far through air." Furthermore, Ketterle says, "an atom-laser beam will fall like a beam of ordinary atoms.

Thus

- ✧ Photons can be created, but not atoms. The number of atoms in an atom laser is not amplified. What is amplified is the number of atoms in the ground state, while the number of atoms in other states decreases.

✧ Atoms interact with each other—that creates additional spreading of the output beam. Unlike light, a matter wave cannot travel far through air.

✧ Atoms are massive particles. They are therefore accelerated by gravity. A matter wave beam will fall like a beam of ordinary atoms.

✧ A Bose condensate occupies the lowest mode (ground state) of the system, whereas lasers usually operate on very high modes of the laser resonator.

✧ A Bose condensed system is in thermal equilibrium and characterized by extremely low temperature. In contrast, the optical laser operates in a non-equilibrium situation which can be characterized by a negative temperature (which means "hotter" than infinite temperature!). There is never any population inversion in evaporative cooling or Bose condensation.

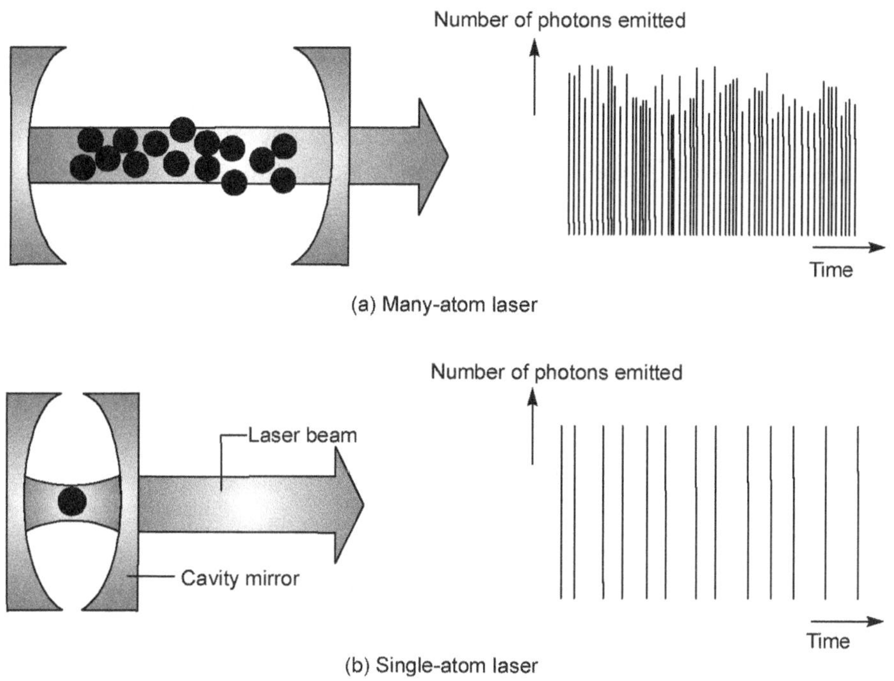

Figure 17.1   Photon to order

In a conventional, many-atom laser, stimulated emission of photons from atoms inside a mirrored cavity gererates an intense beam of radiation. If the make-up of the beam is analysed over time, the flux of photons is seen to fluctuate considerably (Figure 17.1a). In the single-atom laser demonstrated by McKeever, *et al.*, the photons are "anti-bunched", spread out over time. But the number of photons emitted over a given time interval is more predictable than for the many-atom laser (Figure 17.1 b). A conventional many-atom laser and a single-atom laser is shown in Figure 17.1.

17.6 POTENTIAL APPLICATIONS OF AN ATOM LASER

17.6.1 Potential Applications

Although an atom laser has now been demonstrated, major improvements are necessary before it can be used for applications, especially in terms of increased output "power" and reduced overall complexity. Laser-like atoms exist only in an ultrahigh vacuum environment, and so it is unlikely that the atom laser will ever improve supermarket scanners or CD players! However, there are many applications in fundamental research and industry where atomic beams are used, e.g., atomic clocks, atom optics, precision measurements of fundamental constants, tests of fundamental symmetries, atomic beam deposition for chip production (atom lithography), and, more generally, nanotechnology. The atom laser may have an impact on all of these applications. Today, if you have a demanding job for light, you use a laser. In the future, if there is a demanding job for an atomic beam, you may be able to use an atom laser.

Atom lasers can be used for atom holography. Similar to conventional holography atom holography uses the diffraction of atoms. The de Broglie wavelength of the atoms is much smaller than the wavelength of light, and hence atom laser can create much higher resolution holographic images. Atom holography might be used to project complex integrated-circuit patterns, just a few nanometres in scale, onto semiconductors. Another kind of application, is atom interferometry. In an atom interferometer an atomic wave packet is coherently split into two wave packets that follow different paths before recombining. Atom interferometers, which can be more sensitive than optical interferometers, could be used to test quantum theory, and have so high precision that they may even be able to detect changes in space-time. This is because the de Broglie wavelength of the atoms is smaller than the wavelength of light, the atoms have mass, and because the internal structure of the atom can also be exploited.

17.6.2 Some Anticipated Applications

Just as the laser has greatly improved optical experiments, the atom laser promises to increase the precision of many atomic beam experiments. For instance, it may be possible to improve dramatically the already impressive precision of atomic clocks. It may enable more powerful tabletop tests aiming to determine the relationships between the fundamental forces in nature. It is likely to increase the precision in measurements of fundamental physical constants. In particular, it promises to improve the technique known as atom interferometry in which an atom from a conventional beam is split into wavelets and recombined to form interference patterns which provide precise information about the atom. Using atoms from atom-laser beams will greatly improve the technique, which is already rivalling conventional optical interferometers for measuring earth's rotation and testing relativity.

17.7 THE STEPS FROM A BOSE CONDENSATE TO AN ATOM LASER

An important intermediate step towards the atom laser was the realization of Bose–Einstein condensation (BEC), which was achieved in 1995 by a group at Boulder and Ketterle's group at

MIT. (In 1996, two more groups, a group at Rice and a second group at Boulder, observed BEC). The Bose condensate has frequently been compared to photons in a laser beam, but what was missing was a controlled way of extracting a beam of atoms and a method for determining whether the Bose condensed atoms are coherent as the photons in a laser beam. Both these steps have now been taken by the MIT team, thus realising the atom laser.

17.7.1 Realisation of an Output Coupler for a Bose Condensate

In the 27 January 1997 issue of Physical Review Letters, Ketterle and his colleagues describe their "output coupler," which allows them to extract a controlled fraction of atoms from a Bose–Einstein condensate of sodium atoms. An output coupler is one of the essential elements of a laser[8]. It allows the controlled extraction of atoms from the Bose condensate, i.e., the generation of a (quasi-) continuous beam or multiple pulses. Before the MIT group realized an output coupler, the entire condensate was either trapped or freely expanding.

The MIT group achieved the controlled extraction of atoms in the following way: Magnetically trapped atoms can be regarded as atoms bouncing back and forth between magnetic mirrors. The magnetic mirror is 100% reflective for atoms with their magnetic moment anti-parallel to the magnetic field, and fully transmissive for the opposite orientation. The MIT group tilted the magnetic moment of the atoms by a variable angle, thus adjusting the reflectivity of the magnetic mirror. This was done by using short pulses of an oscillating magnetic field.

When the MIT group realised the output coupler in July 1996, they had all the elements for an atom laser together. However, a crucial feature of a laser had yet to be demonstrated: the coherence of the condensed atoms. This was achieved in November 1996 through the observation of high-contrast interference between two Bose condensates.

The researchers extract atoms by applying radio frequency (rf) radiation to the BEC, which is held in a magnetic trap.

17.7.2 Demonstration of Coherence of a Bose Condensate

It should be noted that laser light has two important features: brightness and coherence. Brightness does not necessarily mean high absolute power, but the concentration of power into the direction of propagation and in a small frequency interval (monochromatic light). This is the reason why a laser pointer is brighter. The second important feature is coherence, i.e., all the photons in a laser form one macroscopic wave (they "oscillate synchronously").

In the case of atoms, a Bose condensate is very cold and coherent. Coldness corresponds to brightness in the optical case, because a very low temperature restricts the quantum states which are accessible to the atoms to the lowest states of the system (Brightness in the optical case also means restricting the photons to a few modes of the laser resonator). It is the low energy of the condensate which was studied in previous experiments and used to identify the Bose condensate. However, although coldness and coherence are related, there has been some controversy about how coherent the atomic Bose condensate would be. It has been argued that the atoms first become very cold, but then it would take much longer (may be forever) for the coherence to build up. Furthermore, collisions among the atoms and with background gas were predicted to destroy the coherence. The MIT results resolve these issues. They prove that a Bose condensate is coherent, and that a coherent beam of atoms can be extracted from it.

The proof of the coherence was obtained by observing a high-contrast interference pattern when two Bose condensates overlapped. The MIT researchers could directly photograph this pattern which had a period of 15 micrometre, a gigantic length for matter waves. (Room temperature atoms have a matter wavelength of 0.04 nm, 400,000 times smaller). The interfering condensates were propagating with an energy of 0.5 nanokelvin—the coldest temperature ever reported. However, temperature has lost its meaning in this regime, it is only used as a measure for the residual (non-thermal) energy of the atoms.

When matter waves interfere destructively, it is as if one atom plus one atom give zero atoms! Of course, the matter is not destroyed, and the atoms appear elsewhere. Nevertheless, the interference of streams of atoms from separate sources is a dramatic phenomenon.

17.8 CREATING A BEAM OF ATOMS

By applying a radio frequency radiation to BEC which is held in a magnetic trap it is possible to extract atoms. Each of the atoms acts as a tiny magnet in the presence of the trap's magnetic fields. The atom possesses the property known as the "spin"; the value of spin describes how it will respond to an external magnetic field. Initially, the atoms all have the same spin value, corresponding to a state in which they are all pushed towards the centre of the trap. But the rf radiation, which contains magnetic fields of its own, will make some of the atoms "flip" their spin. This reverses the magnetic forces on them, and these atoms are expelled from the trap, forming a beam that falls in the direction of gravity.

17.8.1 Other Techniques used to Realise the Atom Laser

A variety of schemes to realise an atom laser have been discussed during the last several years. The MIT group chose a particularly simple way. They cooled an atomic gas to extremely low temperatures until it spontaneously formed a Bose–Einstein condensate with "laser-like" properties, and then extracted these atoms into output pulses.

The MIT work was based on powerful cooling techniques which were used to reduce the temperature of a sodium gas by a factor of a billion, from the temperature of an oven to around one microkelvin. These cooling techniques are laser cooling [the key techniques were invented at NIST (W. Phillips), Bell Labs/Stanford (S. Chu), MIT (D. Pritchard)] and evaporative cooling [developed at MIT (T. Greytak, D. Kleppner)]. Many other groups in the atomic physics and condensed matter communities have contributed to these efforts (e.g., Amsterdam, Boulder, Cornell, Harvard, Paris).

Between 1992 and 1995, Ketterle's group pioneered ways to combine laser cooling and evaporative cooling. The combined cooling was the key to the observation of Bose–Einstein condensation in Boulder in June and at MIT in September of 1995.

In laser cooling, the atoms are bombarded with laser light. The details of the laser cooling was discussed in earlier sections.

The atoms have not only to be cooled, but also very well insulated from the room-temperature environment. This is accomplished by purely magnetic confinement inside an ultrahigh vacuum chamber.

The first pulsed atom laser was demonstrated at MIT by Professor Wolfgang Ketterle *et al.*, and co-workers in November 1996. Ketterle used an isotope of rubidium and used an oscillating magnetic field as their output coupling technique, letting gravity pull off partial pieces looking much like a dripping tap.

From the creation of the first atom laser there has been a surge in the recreation of atom lasers along with different techniques for output coupling and in general research. The current developmental stage of the atom laser is analogous to that of the optical laser during its discovery in the 1960s. To that effect the equipment and techniques are in their earliest developmental phases and still strictly in the domain of research laboratories. The radio frequency output coupler for a Bose-Einstein condensate is given in Figure 17.2.

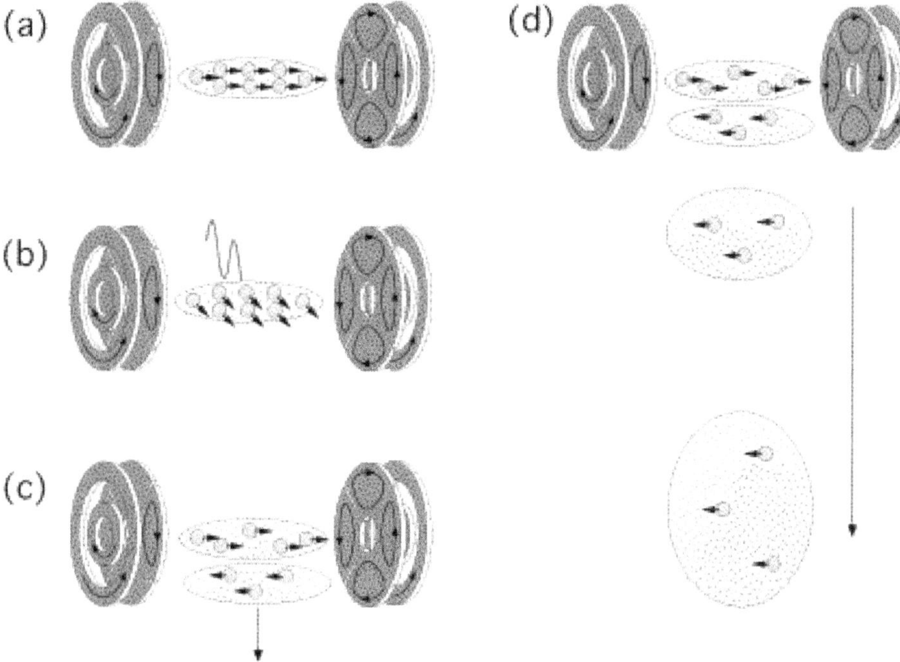

Figure 17.2   RF output coupler for a Bose condensate

Figure 17.2a shows a Bose condensate trapped in a magnetic trap. All the atoms have their (electron) spin up, i.e., parallel to the magnetic field. A short pulse of rf radiation tilts the spins of the atoms (Figure 17.2b). Quantum-mechanically, a tilted spin is a superposition of spin up and down (Figure 17.2c). Since the spin-down component experiences a repulsive magnetic force, the cloud is split into a trapped cloud and an out-coupled cloud. Several output pulses can be extracted, which spread out and are accelerated by gravity (Figure 17.2d).

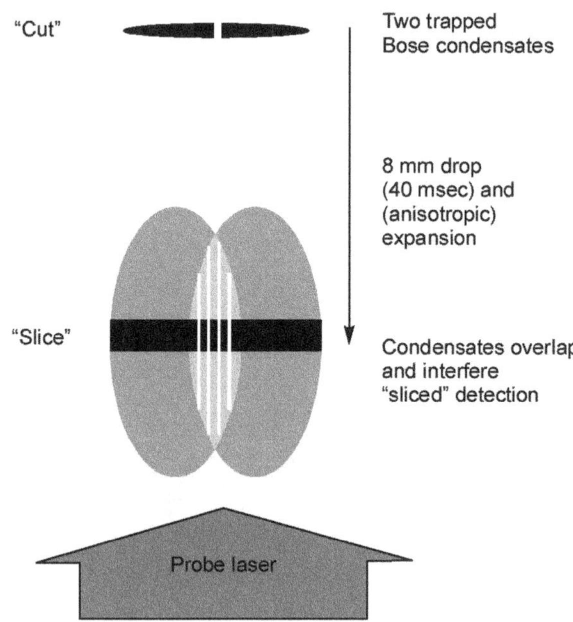

Figure 17.3 Interference of two condensates

Figure 17.3 shows the set-up for observing the coherence of a Bose condensate. This is done by creating two independent Bose condensates in a special trap which uses magnetic and optical forces and has two separated "pockets". The two condensates are separated by a laser beam which "cuts" the cloud into two pieces. When the trap is switched off, the condensates fall down, spread out and eventually overlaps. In the overlap region, a high-contrast interference pattern was observed, a clear proof for the coherence of the Bose condensates. A "sliced" observation was chosen to avoid blurring of the interference pattern by integrating along the direction of propagation of the probe laser beam.

The interference experiment of Figure 17.4 was also done with two condensates coupled out from a trap holding two independent condensates. The observation of an interference pattern proved that the rf output coupler preserved the coherence.

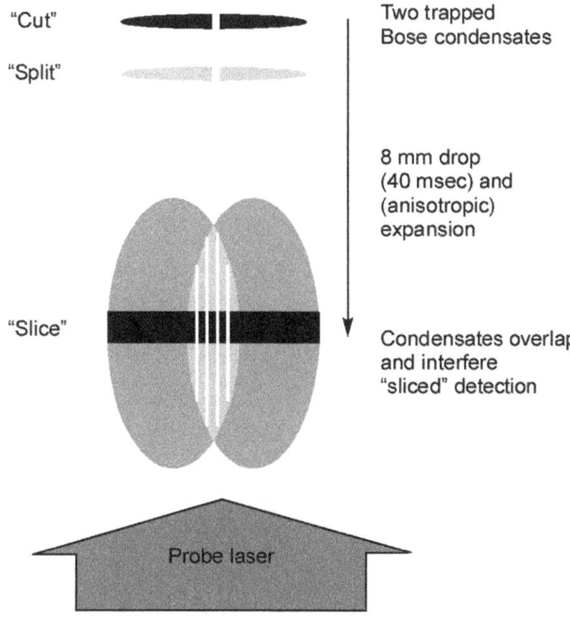

Figure 17.4   Interference of two out-coupled condensates

17.9 FUTURE DESIGNS OF ATOM LASERS

Work on the atom laser has just begun. Ketterle and his colleagues are already thinking of improvements to their device. For instance, the atom laser currently produces a beam that falls in the direction of gravity. Future steps are to make the atom beam travel in other directions by combining the output coupler with "atom mirrors" that use optical or magnetic forces to direct the atom. In addition, the MIT atom beam diffracts, or spreads out, somewhat as it emerges from the trap; upcoming designs may reduce such diffraction effects. Finally, the current design produces only bursts of atoms; future designs may produce continuous beams.

17.10 NANOTECHNOLOGY

Finally, the atom laser holds exciting possibilities for nanotechnology, the manipulation of matter at the atomic level. With an atom laser beam, it would be possible to deposit atoms onto surfaces with unprecedented precision, potentially allowing scientists to create more sophisticated nanostructures than ever before. Ketterle points out, however, that these first atom lasers will only be able to make nanostructures at a very slow rate. According to Ketterle, the fluxes of atoms emerging from an atom laser are currently too small to lead to a practical nanofabrication scheme in which nanostructures could be mass produced. And he notes that the atom laser must operate in extreme vacuum conditions, unlike ordinary lasers whose light can be used in all types of environments. Nonetheless, the atom laser has the potential to become a tool with unexpected and widespread consequences.

ENDNOTES

1. Einstein received the 1921 Nobel Prize in Physics "for his services to Theoretical Physics, and especially for his discovery of the law of the photoelectric effect.

2. Ashkin, A., Dziedzic, J.M., Bjorkholm, J.E. and Chu, S. (1986). "Observation of a single-beam gradient force optical trap for dielectric particles." *Opt. Lett.* 11 (5) 288–290.

3. Eric, A.Cornell and Carl, E.Wieman shared Nobel Prize in Physics in 2001 along with Wolfgang Ketteerle.

4. Anderson, M.H., Ensher, J.R., Matthews, M.R., Wieman, C.E. and Cornell, E.A. (1995). "Observation of Bose–Einstein condensation in a dilute atomic vapor." *Science* 269, 198.

5. Nobel Prize 2001.

6. *Science*, Jan. 1997).

7. Ensher, J.R., Jin, D.S., Matthews, M.R., Wieman, C.E. and Cornell, E.A. (1996). "Bose–Einstein condensation in a dilute gas: measurement of energy and ground-state occupation." *Phys. Rev. Lett.* 77,4984.

8. Phys. Rev. Lett., January 27, 1997.

REFERENCES

Cornell, E.A., Ensher, J.R. and Wieman, C.E. (1999). "Experiments in dilute atomic Bose–Einstein condensation in atomic gases." Proceedings of the International School of Physics "Enrico Fermi" Course CXL (Inguscio, M., Stringari, S. and Wieman, C.E. (eds.). *Italian Physical Society*, pp.15–66.

Ensher, J.R., Jin, D.S., Matthews, M.R., Wieman, C.E. and Cornell, E.A. (1996). Bose-Einstein condensation in a dilute gas: measurement of energy and ground-state occupation. *Phys. Rev. Lett.* 77.

Jin, D.S., Ensher, J.R., Matthews, M.R., Wieman, C.E. and Cornell, E.A. (1996). "Quantitative studies of Bose-Einstein condensation in a dilute atomic vapor." Proc. of XXI Intl. Conf. on Low Temp. *Phys., J. Czech. Phys.* 46 S6.

Jin, D.S., Ensher, J.R., Matthews, M.R., Wieman, C.E. and Cornell, E.A. (1996) "Collective excitations of a Bose-Einstein condensate in a dilute gas." *Phys. Rev. Lett.* 77, 420.

Myatt, C.J., Burt, E.A., Ghrist, R.W, Cornell, E.A. and Wieman, C.E. (1997). "Production of two overlapping Bose-Einstein condensates by sympathetic cooling." *Phys. Rev. Lett.* 78, 586.

Newbury, N.R., Myatt, C.J. and Wieman, C.E. (1995). "S-wave elastic collisions between cold ground-state ^{87}Rb atoms." *PRA* 51, R2680.

JOURNAL ARTICLES

Anderson, M.H., Ensher, J.R., Matthews, M.R., Wieman, C.E. and Cornell, E.A. (1995). "Observation of Bose-Einstein condensation in a dilute atomic vapor." *Science.* 269, 198.

Andrews, M.R., Townsend, C.G., Miesner, H.J., Durfee, D.S. Kurn, D.M. and Ketterle, W. (1997). "Observation of interference between two Bose condensates." *Science.*

Andrews, M.R., Mewes, M.O., van Druten, N.J., Durfee, D.S., Kurn, D.M. and Ketterle, W. (1996). "Science, direct, nondestructive observation of a Bose condensate." *Science.* 273, 84.

Howard Carmichael and Luis A. Orozco. (2003). "Single atom cases orderly light." *Nature.* Vol 425, 246.

Mewes, M.O., Andrews, M.R., Kurn, D.M., Durfee, D.S., Townsend, C.G. and Ketterle, W. (1997). "An output coupler for Bose-condensed atoms." *Physical Review Letters.* 78, 582.

INDEX

V

W

X

Z

www.ingramcontent.com/pod-product-compliance
Lightning Source LLC
LaVergne TN
LVHW080420200726
843507LV00004B/669